AF829264

Algorithms for Intelligent Systems

Series Editors

Jagdish Chand Bansal, Department of Mathematics, South Asian University, New Delhi, Delhi, India

Kusum Deep, Department of Mathematics, Indian Institute of Technology Roorkee, Roorkee, Uttarakhand, India

Atulya K. Nagar, School of Mathematics, Computer Science and Engineering, Liverpool Hope University, Liverpool, UK

This book series publishes research on the analysis and development of algorithms for intelligent systems with their applications to various real world problems. It covers research related to autonomous agents, multi-agent systems, behavioral modeling, reinforcement learning, game theory, mechanism design, machine learning, meta-heuristic search, optimization, planning and scheduling, artificial neural networks, evolutionary computation, swarm intelligence and other algorithms for intelligent systems.

The book series includes recent advancements, modification and applications of the artificial neural networks, evolutionary computation, swarm intelligence, artificial immune systems, fuzzy system, autonomous and multi agent systems, machine learning and other intelligent systems related areas. The material will be beneficial for the graduate students, post-graduate students as well as the researchers who want a broader view of advances in algorithms for intelligent systems. The contents will also be useful to the researchers from other fields who have no knowledge of the power of intelligent systems, e.g. the researchers in the field of bioinformatics, biochemists, mechanical and chemical engineers, economists, musicians and medical practitioners.

The series publishes monographs, edited volumes, advanced textbooks and selected proceedings.

More information about this series at http://www.springer.com/series/16171

Sunil Dutt Purohit · Dharm Singh Jat ·
Ramesh Chandra Poonia ·
Sandeep Kumar · Saroj Hiranwal
Editors

Proceedings of International Conference on Communication and Computational Technologies

ICCCT-2019

Springer

Editors
Sunil Dutt Purohit
Rajasthan Technical University
Kota, Rajasthan, India

Ramesh Chandra Poonia
Department of ICT and Natural Sciences
Norwegian University of Science
and Technology
Alesund, Norway

Saroj Hiranwal
Rajasthan Institute of Engineering
and Technology
Jaipur, Rajasthan, India

Dharm Singh Jat
Department of Computer Science
University of Science and Technology,
Namibia
Windhoek, Namibia

Sandeep Kumar
Department of Computer Science
Amity University
Jaipur, Rajasthan, India

ISSN 2524-7565 ISSN 2524-7573 (electronic)
Algorithms for Intelligent Systems
ISBN 978-981-15-5076-8 ISBN 978-981-15-5077-5 (eBook)
https://doi.org/10.1007/978-981-15-5077-5

© Springer Nature Singapore Pte Ltd. 2021
This work is subject to copyright. All rights are reserved by the Publisher, whether the whole or part of the material is concerned, specifically the rights of translation, reprinting, reuse of illustrations, recitation, broadcasting, reproduction on microfilms or in any other physical way, and transmission or information storage and retrieval, electronic adaptation, computer software, or by similar or dissimilar methodology now known or hereafter developed.
The use of general descriptive names, registered names, trademarks, service marks, etc. in this publication does not imply, even in the absence of a specific statement, that such names are exempt from the relevant protective laws and regulations and therefore free for general use.
The publisher, the authors and the editors are safe to assume that the advice and information in this book are believed to be true and accurate at the date of publication. Neither the publisher nor the authors or the editors give a warranty, expressed or implied, with respect to the material contained herein or for any errors or omissions that may have been made. The publisher remains neutral with regard to jurisdictional claims in published maps and institutional affiliations.

This Springer imprint is published by the registered company Springer Nature Singapore Pte Ltd.
The registered company address is: 152 Beach Road, #21-01/04 Gateway East, Singapore 189721, Singapore

Preface

Nature continuously provides inspiration for ways to solve the real-life problems. Nature has progressed over thousands of years and now aids the development of innovative models, techniques and tools for many disciplines of scientific aspiration. Humankind has been continuously trying to comprehend and adapt nature in order to develop innovative techniques and tools. The area of nature-inspired computing is a combination of computing science with knowledge from different streams such as mathematics, biology, chemistry, physics and engineering. Basically, the main source of inspiration is nature, and thus, they are called nature-inspired. More than hundred algorithms inspired by nature are available today. These algorithms are able to solve the complex optimization problems very easily and with high efficiency. They always provide optimal or near-to-optimal solutions. Smart and intelligent computing is not complete without nature-inspired algorithms. Therefore, this conference has the main objective of circulating the innovations and advancements of smart and intelligent computing.

The International Conference on Communication and Computational Technologies-2019 (ICCCT-2019) was held at Rajasthan Institute of Engineering and Technology, Jaipur, India, from August 30 to 31, 2019. This proceedings consists of 59 research papers from ICCCT-2019. We are thankful to Soft Computing Research Society (SCRS) for providing technical sponsorship for this event. We are also very thankful to Springer for supporting ICCCT-2019.

We hope that this conference proceedings will prove to be useful.

Jaipur, India

Dr. Sunil Dutt Purohit
Prof. Dharm Singh Jat
Dr. Ramesh Chandra Poonia
Dr. Sandeep Kumar
Dr. Saroj Hiranwal

About This Book

This book features selected papers presented at the International Conference on Communication and Computational Technologies-2019 (ICCCT-2019), held at Rajasthan Institute of Engineering and Technology, Jaipur, India, from August 30 to 31, 2019. Covering recent advances in communication and computing intelligence, the book focuses on the recent developments in intelligent computing, such as intelligent computational technologies, intelligent communication and intelligent computing paradigms with applications in every aspect of life.

Contents

About the Editors

Sunil Dutt Purohit obtained his M.Sc. (Gold Medalist) and Ph.D. degree from the Faculty of Science at Jai Narayan Vyas University, Jodhpur. Currently, he is an Associate Professor of Mathematics at the Department of HEAS, RTU, Kota. His research interests include special functions, mathematical physics, geometric function theory and computational mathematics. He has published more than 120 research papers in respected international journals, and serves as a reviewer for Mathematical Reviews, USA (AMS), and Zentralblatt MATH, Berlin.

Dharm Singh Jat is a Professor at the Department of Computer Science, Namibia University of Science and Technology. His research interests include multimedia communications, wireless technologies, mobile communication systems, roof computing and video transmission over wired–wireless networks, software-defined networks, network security, Internet of Things and ICTs applications. He is the author of four books and editor of 16 books, and has authored/co-authored over 135 research publications in respected peer-reviewed journals and conference proceedings.

Ramesh Chandra Poonia is currently an Associate Professor at Amity University Rajasthan, Jaipur; and a Postdoctoral Fellow at CPS Lab, Department of ICT and Natural Sciences, NTNU, Alesund, Norway. He has published more than 100 research papers in various journals and conference proceedings, and is an active member of the IEEE, CSI and ACM.

Sandeep Kumar is currently an Assistant Professor of Computer Science & Engineering at Amity University Rajasthan, Jaipur, India. Holding a Ph.D. in Computer Science & Engineering, he has published more than 50 research papers in various journals/conference proceedings, authored two books and six book chapters, and edited two books and two conference proceedings. He has been Guest Editor for special issues of IJIIDS, IJARGE, IJESD, RPCS and IJGUC.

Saroj Hiranwal is currently a Principal at Rajasthan Institute of Engineering and Technology, Jaipur. Holding a Doctor of Philosophy in CSE, her research interests include high-performance scientific computing, real-time systems and cloud computing. Her work has been published in various peer-reviewed journals.

Chapter 1
Ability Study of Proximity Measure for Big Data Mining Context on Clustering

Kamlesh Kumar Pandey and Diwakar Shukla

1 Introduction

Rapidly growing digital and communication technology produces a huge amount of structured, unstructured, and semi-structured data in the real-time environment. According to literature of [1], these technologies produced five exabytes of data every two days. The uses of this technology change the behavior and meaning of traditional data to big data respect to Volume, Variety, and Velocity with data management technology. Jin et al. [2] gave two categories for big data sources. The first source of big data is the physical world through sensors, scientific experiments such as biological analysis, neural analysis, astronomical analysis, business analysis, and so on. The second source of big data is the human and social world through social networks, internet of things, cloud computing, the transportation sector, health sector, and finance and economics sectors, and so on [2].

In big data mining, various applications such as social network analysis, customer segmentation, scientific data analysis, bioinformatics, and target marketing used to cluster analysis. Data mining is one of the tools for big data analysis and clustering is one of the techniques for data mining. Clustering is used for finding similar relations or patterns or groups the data together based on their characteristics, similarities, or features for unlabeled objects and provides high homogeneity within a cluster and reduces the data volume through the making of groups [3, 4]. In the big data perspective, the clustering algorithm must support Volume, Variety, and Velocity with parallel and distributed computation. [5]. For improving traditional clustering algorithm in the context of big data, clustering needs to improve various fields such as

K. K. Pandey (✉) · D. Shukla
Department of Computer Science & Applications, Dr. Hari Singh Gour Vishwavidyalaya, Sagar, MP, India
e-mail: kamleshamk@gmail.com

D. Shukla
e-mail: diwakarshukla@rediffmail.com

© Springer Nature Singapore Pte Ltd. 2021

S. D. Purohit et al. (eds.), *Proceedings of International Conference on Communication and Computational Technologies*, Algorithms for Intelligent Systems,
https://doi.org/10.1007/978-981-15-5077-5_1

scalability, handling outline, parallel and distributed computation, under the big data dimensions, and so on. The cluster construction is one of the cluster research fields for improving clustering efficiency and algorithm accuracy. Proximity measure such as dissimilarity and similarity is the basis of constructing the clustering algorithm. Dissimilarity is preferred for dealing with quantitative data features and similarity is preferred for dealing with qualitative data features [6].

This paper has organized into five sections for archiving the paper objectively. The objective of this paper is identifying a proximity measure for big data mining and how to create a cluster with help of proximity measure. Section 2 defines the background of big data and big data mining. Section 3 presents various proximity measures under the Minkowski, $L(1)$, $L(2)$, Inner product, Shannon's entropy, Combination, Intersection, and Fidelity family. Section 4 analyzed all discussed proximity measures for big data mining under volume, variety, and velocity. Section 5 describes how to create a cluster with the help of a proximity measure under the partition, hierarchical, density, grid, model, fuzzy, and Graph-based cluster taxonomy.

2 Background

2.1 Big Data

Big Data researchers tried to develop capturing, processing, analyzing, and visualizing techniques for large amounts of structured, unstructured, and semi-structured data with the different sources within a limited processing time through mathematics, statistics, optimization methods, data mining, and machine learning and so on. These big data analysis technologies and tools are grouped into three groups as a batch, streaming, and interactive [7].

The literature of [8] described the big data terms according to Laney (2001) and Gartner (2012). The big data term is introduced by Laney (2001) through three V's as Volume, Variety, and Velocity. After that Gartner (2012) summarizes these three V's as "high volume, velocity, and Variety information assets cost-effective demand, innovative forms of information processing for enhanced insight and decision making" [8]. High-volume represents the size of the data with respect to scale above to terabytes or petabytes. High-variety represents the heterogeneous data source as well as a data type respect to structured, unstructured, and semi-structured data with multiple sources. High velocity represents the speed of the data generation and analysis respect to technologies related to data captured and processed. Various reputed organizations gave another V's for big data due to the accuracy of the analyzed results. IBM gave fourth V's as Veracity, SAS gave fifth V's as Variability, and Oracle given sixth V's as Value. The Veracity represents the unreliability, untrustworthiness, and uncertainty sources of data for measuring accuracy. Variability represents the data whose meaning is continuously is continuously changing dues to time and source. Value represents the attribute identification for data mining or analysis. The last V's

Fig. 1 7V's characteristics for big data

of big data is Visualization, which represents the mine and analysis' results according to user expectation [9–11]. Figure 1 summarized all V's of big data

2.2 Big Data Mining

The big data architecture provides four logical layers as big data sources, data massaging and store, analysis, and consumption for effective big data mining [8]. Big data mining or analysis task has to meet the biggest challenges due to scalable, volume, variety, and velocity. Efficient mining algorithms deal with scalability, large volume capacity, with heterogeneous data and source, and noise handling in less time. These requirements need to change traditional data management techniques as well as mining algorithms with respect to three dimensions of big data and motivate design a new data mining technique [12].

Big data mining is the technique of discovering relevant information, knowledge, hidden patterns, and complex relations from high-volume, high-variety, and high-velocity dataset with scalability through parallel and distributed approach execution. For the big data storages perspective, NoSQL databases and MapReduce model most popular for big data mining because of this database and model deal with high-volume, high-variety, and high-velocity with parallel and distributed computing [13–15].

3 Proximity Measure Taxonomy

Clustering is one of the applications of proximity measure. All clustering algorithm depends upon the proximity measure for cluster creation and validation. Various authors study proximity measures in various fields such as pattern recognition, classification, retrieval related problems [16], text classification and clustering [17], uncertain data classification [18] and clustering [19], network intrusion anomaly detection [20], analyzing inconsistent information [21], heavy aggregation operators [22], combining density [23] and spatial concentration [24], author profiling [25], location fingerprinting [26], document content comparison [27], time-series data management [28], biocryptic authentication in cloud databases [29], help to the derivation of new distance measure [30], and so on.

Proximity Measure is categorized into two groups as similarity and dissimilarity. Similarity measure in terms of how alike one data point to another point and given a value between 0 and 1, where closed value 1 indicated to more similarity and closed value 0 indicated to less similarity. Dissimilarity measures in terms of how unlike one data point to another point and given a value between 0 and ∞, where greater value shows more dissimilarity and less value shows less dissimilarity. Various dissimilarity and similarity functions are available under statistics. Proximity measure is defined by Zero distance or Reflexivity, Non-negativity, Symmetry, and Triangle inequality property. If any proximity measure covers all properties they know the dissimilarity or distance function and dissimilarity or distance metric otherwise known as similarity or divergence function. Sometimes the proximity measure is not covered only fourth property as triangular inequality, but satisfying all the other properties as reflexivity, non-negativity, and symmetry, then these types of proximity measure are known as semi-metric (distance or similarity function) [31–34].

In present time, various proximity measures are available for cluster construction and these proximity measure categories under the Minkowski, $L(1)$, $L(2)$, Inner product, Shannon's entropy, Combination, Intersection, and Fidelity family [16, 25, 35]. In this section, the paper describes various distances and similarity functions under these families [16–30, 32–34]. This paper used is dis( ) function for defining proximity measure requirements and this function takes two distinct data points A and B and returns their distance value.

3.1 *Minkowski Family*

This family defines the base of proximity measure and considers as the derive Manhattan distance, Euclidean distance, Minkowski distance, and Chebyshev distance on the bases of the p-value.

$$\mathrm{dis}(A, B) = \left(\sum_{i=1}^{n} |A_i - B_i|^P\right)^{\frac{1}{P}} \tag{1}$$

If a *p*-value fixed is 1, is known as Manhattan distance. Manhattan distance measures the two data points along the axes at optimal angles and given absolute value with a hyper-rectangular shape of the cluster. Manhattan distance formulation shows as Eq. 2 based upon the Eq. 1 and it also defines $L1$ distance family.

$$\mathrm{dis}_{\mathrm{manhattan}}(A, B) = \left(\sum_{i=1}^{n} |A_i - B_i|^P\right) \tag{2}$$

If a *p*-value fixed is 2, is known as Euclidean distance. Euclidean distances measure the two data points on the basis of straight-line in the coordinate axes. This distance measure is given a compact or isolated cluster with higher accuracy. Euclidean distance formulation shows as Eq. 3 based upon the Eq. 1 and it also defines $L2$ distance family.

$$\mathrm{dis}_{\mathrm{euclidean}}(A, B) = \sqrt[2]{\sum_{i=1}^{n} |A_i - B_i|^2} \tag{3}$$

If a *p*-value is normalized between 2 to ∞, then this distance measure known as Minkowski distance. This distance is given isolated or compacted clusters with good efficient distance. Minkowski distance formulation shows as Eq. 4 based upon Eq. 1.

$$\mathrm{dis}_{\mathrm{minkowski}}\,(A, B) = \sqrt[p]{\sum_{i=1}^{n} |A_i - B_i|^p} \tag{4}$$

If a *p*-value fixed is ∞ (maximum), it's known as Chebyshev distance. This distance measure is used for when two data points are greatest of their absolute magnitude along with data dimension. Chebyshev distance formulation shows as Eq. 3 based upon Eq. 1.

$$\mathrm{dis}_{\mathrm{chebyshev}}(A, B) = \sqrt[\infty]{\sum_{i=1}^{n} |A_i - B_i|^{\infty}} = \max_{i=1}^{n} |A_i - B_i|^P \tag{5}$$

3.2 L(1) Family

This family member based on Manhattan distance and known as variants of the $L(1)$. These members are given the solution regarding the normalization and figure out of small and large distances. The variously proposed solution is available for these difficulties and these solutions are shown as Eqs. 6–12 on the bases of Eq. 2.

$$\mathrm{dis}_{\mathrm{sorensen}}(A, B) = \frac{\left(\sum_{i=1}^{n} |A_i - B_i|\right)}{\left(\sum_{i=1}^{n} |A_i + B_i|\right)} \tag{6}$$

$$\mathrm{dis}_{\mathrm{soergel}}(A, B) = \frac{\left(\sum_{i=1}^{n} |A_i - B_i|\right)}{\left(\sum_{i=1}^{n} \max(A_i B_i)\right)} \tag{7}$$

$$\mathrm{dis}_{\mathrm{kulczynski}}(A, B) = \frac{\left(\sum_{i=1}^{n} |A_i - B_i|\right)}{\left(\sum_{i=1}^{n} \min(A_i B_i)\right)} \tag{8}$$

$$\mathrm{dis}_{\mathrm{motyka}}(A, B) = \frac{\left(\sum_{i=1}^{n} \max(A_i B_i)\right)}{\left(\sum_{i=1}^{n} |A_i + B_i|\right)} \tag{9}$$

$$\mathrm{dis}_{\mathrm{canberra}}(A, B) = \sum_{i=1}^{n} \frac{|A_i - B_i|}{|A_i + B_i|} \tag{10}$$

$$\mathrm{dis}_{\mathrm{lorentzian}}(A, B) = \sum_{\mathrm{i}=1}^{\mathrm{n}} \ln(1 + |A_i - B_i| \tag{11}$$

$$\mathrm{dis}_{\mathrm{wavehedge}}(A, B) = \sum_{i=1}^{n} \frac{|A_i - B_i|}{\max(A_i B_i)} \tag{12}$$

3.3 L(2) or χ^2 Family

This family member based on Euclidian distance and known as variants of the $L(2)$. These members are given the distance value after the normalization. Some proposed $L(2)$ family member is Matusita, Clark, Divergence, Squared Euclidean, squared χ^2, Pearson χ^2 Neyman χ^2 shown as Eqs. 13–19, respectively, on the basis of Eq. 3.

$$\mathrm{dis}_{\mathrm{matusita}}(A, B) = \sqrt{\sum_{i=1}^{n} \left(\sqrt{A_i} - \sqrt{B_i}\right)^2} \tag{13}$$

$$\mathrm{dis}_{\mathrm{clark}}(A, B) = \sum_{i=1}^{n} \left(\frac{|A_i - B_i|}{(A_i + B_i)}\right)^2 \tag{14}$$

$$dis_{\text{divergence}}(A, B) = 2\sum_{i=1}^{n} \frac{(A_i - B_i)^2}{(A_i + B_i)^2} \tag{15}$$

$$\text{dis}_{\text{squared\_euclidean}}(A, B) = \left(\sum_{i=1}^{n}(A_i - B_i)^2\right) \tag{16}$$

$$\text{dis}_{\text{squared\_chi}}(A, B) = \sum_{i=1}^{n} \frac{(A_i - B_i)^2}{(A_i + B_i)} \tag{17}$$

$$\text{dis}_{\text{pearson\_chi}}(A, B) = \sum_{i=1}^{n} \frac{(A_i - B_i)^2}{B_i} \tag{18}$$

$$\text{dis}_{\text{neyman\_chi}}(A, B) = \sum_{i=1}^{n} \frac{(A_i - B_i)^2}{A_i} \tag{19}$$

3.4 *Inner Product Family*

The inner product distance measure gives the distance on the basis of multiplication. Their formulation shows as Eq. 20 and normalized distances shown as Eqs. 21–24, respectively, on the basis of Eq. 20. These all distance measure is known as the variance of the Inner product.

$$\text{dis}_{\text{inner\_product}}(A, B) = \sum_{i=1}^{n} A_i B_i \tag{20}$$

$$\text{dis}_{\text{cosine}}(A, B) = \frac{\sum_{i=1}^{n} A_i B_i}{\sqrt{\sum_{i=1}^{n} A_i^2}\sqrt{\sum_{i=1}^{n} B_i^2}} \tag{21}$$

$$\text{dis}_{\text{Jaccard}}(A, B) = 1 - \frac{\sum_{i=1}^{n} A_i B_i}{\sum_{i=1}^{n} A_i^2 + \sum_{i=1}^{n} B_i^2 - \sum_{i=1}^{n} A_i B_i} \tag{22}$$

$$\text{dis}_{\text{dice}}(A, B) = 1 - \frac{2\sum_{i=1}^{n} A_i B_i}{\sum_{i=1}^{n} A_i^2 + \sum_{i=1}^{n} B_i^2} \tag{23}$$

$$\text{dis}_{\text{Harmonic\_mean}}(A, B) = 2\sum_{i=1}^{n} \frac{A_i B_i}{A_i + B_i} \tag{24}$$

3.5 Shannon's Entropy Family

This distance family is based on probabilistic uncertainty or entropy. Here all data point A_i must be non-negative and their sum is always equal to 1 and the logarithm base is fixed at 2. Formulation of Shannon's distance measure is shown as Eq. 25 and it is also known as Kullback–Leibler divergence. The normalization of Eq. 25 shown as Eqs. 26–30, respectively, on the basis of Eq. 25. These all distance measure is known as the variance of Shannon's entropy.

$$\text{dis}_{\text{kullback\_leibler}}(A, B) = \sum_{i=1}^{n} A_i \ln\left(\frac{A_i}{B_i}\right) \tag{25}$$

$$\text{dis}_{\text{jeffreys}}(A, B) = \sum_{i=1}^{n} A_i - B_i \ln\left(\frac{A_i}{B_i}\right) \tag{26}$$

$$\text{dis}_{\text{k\_divergence}}(A, B) = \sum_{i=1}^{n} A_i \ln\left(\frac{2A_i}{A_i - B_i}\right) \tag{27}$$

$$\text{dis}_{\text{topsoe}}(A, B) = \sum_{i=1}^{n}\left(A_i \ln\left(\frac{2A_i}{A_i - B_i}\right) + B_i \ln\left(\frac{2A_i}{A_i - B_i}\right)\right) \tag{28}$$

$$\text{dis}_{\text{jensen\_shannon}}(A, B) = \frac{1}{2}\left(\sum_{i=1}^{n} A_i \ln\left(\frac{2A_i}{A_i - B_i}\right) + \sum_{i=1}^{n} B_i \ln\left(\frac{2A_i}{A_i - B_i}\right)\right) \tag{29}$$

$$\text{dis}_{\text{jensen\_difference}}(A, B) = \sum_{i=1}^{n}\left(\frac{A_i \ln A_i + B_i \ln B_i}{2} - \frac{A_i - B_i}{2} \ln \frac{A_i - B_i}{2}\right) \tag{30}$$

3.6 Combination Family

This family member based on the combination of two or more distance measures. Taneja, and Kumar Johnson distance measures come under these family and their formulation is shown as Eqs. 31 and 32.

$$\text{dis}_{\text{taneja}}(A, B) = \sum_{i=1}^{n} \frac{A_i + B_i}{2} \ln\left(\frac{A_i + B_i}{\sqrt[2]{A_i B_i}}\right) \tag{31}$$

$$\text{dis}_{\text{kumar\_johnson}}(A, B) = \sum_{i=1}^{n} \frac{\left(A_i^2 - B_i^2\right)^2}{2(A_i B_i)^{\frac{3}{2}}} \tag{32}$$

3.7 Intersection Family

This family member based on the intersection between two data points. The formulation of the intersection distance measure is shown as Eq. 33 and their normalized formulation is shown as Eqs. 34–36.

$$\mathrm{dis}_{\mathrm{intersection}}(A, B) = \sum_{i=1}^{n} \min(A_i B_i) \tag{33}$$

$$\mathrm{dis}_{\mathrm{wavehedges}}(A, B) = \sum_{i=1}^{n} \frac{|A_i - B_i|}{\max(A_i B_i)} \tag{34}$$

$$\mathrm{dis}_{\mathrm{ruzicka}}(A, B) = \frac{\sum_{i=1}^{n} \mathrm{mix}(A_i B_i)}{\sum_{i=1}^{n} \max(A_i B_i)} \tag{35}$$

$$\mathrm{dis}_{\mathrm{tanimoto}}(A, B) = \frac{\sum_{i=1}^{n} \max(A_i B_i) - \sum_{i=1}^{n} \mathrm{mix}(A_i B_i)}{\sum_{i=1}^{n} \max(A_i B_i)} \tag{36}$$

3.8 Fidelity Family or Squared-Chord Family

This family member based on geometric means. Fidelity distance measure formulation shows as Eq. 37 and their variance is shown as Eqs. 38 and 41, respectively.

$$\mathrm{dis}_{\mathrm{fidelity}}(A, B) = \sum_{i=1}^{n} \sqrt{A_i B_i} \tag{37}$$

$$\mathrm{dis}_{\mathrm{bhattacharyya}}(A, B) = -\ln \sum_{i=1}^{n} \sqrt{A_i B_i} \tag{38}$$

$$\mathrm{dis}_{\mathrm{hellinger}}(A, B) = \sqrt[2]{1 - \sum_{i=1}^{n} \sqrt{A_i B_i}} \tag{39}$$

$$\mathrm{dis}_{\mathrm{matusita}}(A, B) = \sqrt{2 - 2\sum_{i=1}^{n} \sqrt{A_i B_i}} \tag{40}$$

$$\mathrm{dis}_{\mathrm{squared\_chord}}(A, B) = \sum_{i=1}^{n} \left(\sqrt{A_i} - \sqrt{B_i}\right)^2 \tag{41}$$

4 Analysis of Proximity Measure for Big Data Mining

In the above section of this paper are studied 41 proximity or distance measures (Eqs. 1–41) under the eight groups. After that this section, paper recognized which proximity measure use as similarity, dissimilarity, and semi-metric functions in Table 1. Table 1 given the choice of stable proximity measure for clustering under the big data dimensions such as volume, variety, and velocity. In the big data mining perspective, Fahad et al. [36] and Pandove et al. [6] describe clustering algorithm criteria on the bases of the Volume, Velocity, and Variety dimensions of big data. Volume-related criteria such as cluster is must be dealt huge scale size, high dimensional and noisy of the dataset, Variety related criteria such as cluster is must be recognized as dataset categorization and clusters Shape, and Velocity related criteria define the complexity, scalability, and performance of the clustering algorithm during the execution of the real dataset. This paper takes Volume as high scale dataset, Variety as data types, and Velocity defines time complexity for identification of big data enable distance measures [6, 36].

Clustering deal continuous (numerical) and categorical (nominal and binary) data that are defined as a Variety of clustering. If distance value is small that is defined data point has a high relation and they are close to each other otherwise they have a large relation [33]. In general, the similarity function is used for categorical data because it's given to distance value is large. When data is used as a continuous form and fulfill to all distance measure properties that it is used as a dissimilarity function. Sometimes data are organized in the mixing of categorical and continuous data, then archive the suitable distance measures through dichotomize. Dichotomize process complete by two phases. First apply to similarity function for categorical data and apply to rescale on all the data, after that replaces the same data on the bases of their ranks and used to dissimilarity function for continuous data [32–34].

According to Table 1 Eqs. 2–5, 7, 10–14, 34, 38–41 is more stable and universal for cluster creation under the big data mining and other equation used for comparing two clusters on the basis of similarity.

5 Proximity Measures Enable Clustering

This section of the paper describes the various clustering taxonomy such as Partition, Hierarchical, Density, Grid, Model, Fuzzy, and Graph-based on their working process, behaviors, and cluster nature. The basis of cluster creation is given in Sects. 5.1–5.7. References [4–8, 32–36] and Table 2 describe which clustering algorithm used to which type of proximity measures.

Table 1 Ability analysis of proximity measure for big data mining [16–30, 32–34].

Distance measure	Type of function	Volume (high data set)	Variety (type of data)	Velocity (time complexity)
Minkowski family				
Equation 2	Distance	Large	Continuous	0(n)
Equation 3	Distance	Large	Continuous	0(n)
Equation 4	Distance	Large	Continuous	0(n)
Equation 5	Distance	Large	Continuous	0(n)
L(1) family				
Equation 6	Semi-metric	Large	Categorical	0(2n)
Equation 7	Distance	Large	Continuous	0(n)
Equation 8	Semi-metric	Medium	Categorical	0(n)
Equation 9	similarity	Medium	Categorical	0(n)
Equation 10	Distance	Large	Continuous	0(n)
Equation 11	Distance	Large	Continuous	0(n)
Equation 12	Distance	Large	Continuous	0(n)
L(2) or χ^2 *family*				
Equation 13	Distance	Large	Continuous	0(n)
Equation 14	Distance	Large	Continuous	0(n)
Equation 15	Semi-metric	Large	Categorical	0(n)
Equation 16	Semi-metric	Large	Categorical	0(n)
Equation 17	Semi-metric	Large	Categorical	0(2n)
Equation 18	Semi-metric	Large	Categorical	0(2n)
Equation 19	Semi-metric	Large	Categorical	0(2n)
Inner product family				
Equation 20	Similarity	Medium	Categorical	0(3n)
Equation 21	Similarity	Medium	Categorical	0(3n)
Equation 22	Semi-metric	Large	Categorical	0(2n)
Equation 23	Semi-metric	Large	Categorical	0(2n)
Equation 24	Similarity	Large	Categorical	0(3n)
Shannon's entropy family				
Equation 25	Similarity	Medium	Categorical	0(n)
Equation 26	Semi-metric	Large	Categorical	0(n)
Equation 27	Similarity	Medium	Categorical	0(n)
Equation 28	Semi-metric	Large	Categorical	0(n)
Equation 29	Semi-metric	Large	Categorical	0(n)
Equation 30	Semi-metric	Large	Categorical	0(n)
Combination family				

(continued)

Table 1 (continued)

Distance measure	Type of function	Volume (high data set)	Variety (type of data)	Velocity (time complexity)
Equation 31	Semi-metric	Large	Categorical	0(2*n*)
Equation 32	Semi-metric	Large	Categorical	0(2*n*)
Intersection family				
Equation 33	Similarity	Medium	Categorical	0(2*n*)
Equation 34	Distance	Large	Continuous	0(2*n*)
Equation 35	Similarity	Medium	Categorical	0(2*n*)
Equation 36	Semi-metric	Large	Categorical	0(2*n*)
Fidelity family or Squared-chord family				
Equation 37	Similarity	Medium	Categorical	0(3*n*)
Equation 38	Distance	Medium	Continuous	0(3*n*)
Equation 39	Distance	Large	Continuous	0(3*n*)
Equation 40	Distance	Large	Continuous	0(3*n*)
Equation 41	Distance	Large	Continuous	0(3*n*)

Table 2 Clustering Taxonomy with their Distance measure [4–8, 32–36]

Clustering algorithm	Distance measure
Partitioning-based method	Dissimilarity
Hierarchical-based clustering	Similarity
Density-based clustering	Similarity
Model-based Clustering	Dissimilarity
Grid-based clustering	Dissimilarity
Fuzzy-based clustering	Similarity
Graph-based clustering	Similarity

5.1 Partitioning-Based Clustering

The Partitioning-based method constructs the clusters on the bases of center in the choice of k number of clusters. This clustering method used proximity [37] measures to find out the center of the cluster creation. Some typical algorithms of this kind of clustering algorithm are K-Mean, K-Medoids, K-parameter PAM, CLARA, and CLARANS.

5.2 *Hierarchical-Based Clustering*

This clustering method constructs the clusters by recursively splitting the dataset by using Agglomerative and merging the dataset by using the Divisive method. Agglomerative and Divisive methods obtained by the desired cluster through similarity function and normalized them according to need such as convert its distance function or replace similarity to dissimilarity function. Some typical algorithms of this kind of clustering algorithm are BIRCH, CURE, ROCK, Chameleon, ECHIDNA, WARDS, and SNN.

5.3 *Density-Based Clustering*

This clustering method scans the spatial databases and used probability distribution based distance measure. Here distance measurements used for creating the core, border, and noise point of the density cluster. Some typical algorithms of this kind of clustering algorithm are DBSCAN, OPTICS, Mean-Shift, DENCLUE, and GDBSCAN.

5.4 *Model-Based Clustering*

This clustering method optimizes the dataset into the mathematical model-based on the mixing of probability distributions based distance measure, and distance measure is also measuring the parameter of the selected model. Some typical algorithms of this kind of clustering algorithm are COBWEB, SLINK, SOM, ART, and EM.

5.5 *Grid-Based Clustering*

This clustering algorithm splits the data space into a grid structure, calculate the cell density and structure with the help of distance measure. Some typical algorithms of this kind of clustering algorithm are STING, MAFIA, CLIQUE, ENCLUS, Wave Cluster, ORCLUS, OptiGrid, PROCLUS, and STIRR.

5.6 *Fuzzy-Based Clustering*

This clustering algorithm used to membership function for constructs the cluster. These membership functions same as similarity measure which gives the value between 0 and 1. Some typical algorithms of this kind of clustering algorithm are FCM (Fuzzy C-Means), FCS, and MM.

5.7 Graph-Based Clustering

This clustering algorithm first constructs the graph and after that applies the clustering algorithm to partition the graph through a similarity function. Spectral clustering is one of the variances of Graph-based clustering. Some typical algorithms of this kind of clustering algorithm are CLICK and MST.

6 Conclusions

This paper reviewed the core idea of big data, big data mining, clustering, proximity measures, and their uses under the clustering taxonomy in the respect of big data characteristics, and identified which proximity measures are suitable for the big data mining, and which clustering taxonomy used to which type of proximity measures. This work is very helpful for choosing good proximity measure for cluster construction under the big data mining. The first section of this paper presents how to change data to big data and what is the role of proximity measures for clustering. The second section gives the background details about big data dimensions, big data mining, and big data storage for the mining purpose. The third section reviewed various proximity measures categories under the Minkowski, $L(1)$, $L(2)$, Inner product, Shannon's entropy, Combination, Intersection, and Fidelity family and describes which proximity measures are known as similarity, dissimilarity, and semi-metric function. The fourth section is analyzed all studied proximity measures for big data mining characteristics as Volume (dataset size) Variety (dataset type, and Velocity (time complexity) and find Eqs. 2–5, 7, 10–14, 34, 38–41 are more scalable for big data mining. The fifth section describes clustering taxonomy with the uses of proximity measures for cluster creation.

References

1. Rouhani S, Rotbei S, Hamidi H (2017) What do we know about the big data researches? A systematic review from 2011 to 2017. J Decis Syst 26(4):368–393. https://doi.org/10.1080/12460125.2018.1437654
2. Jin X, Wah BW, Cheng X, Wang Y (2015) Significance and challenges of big data research. Big Data Res 2(2):59–64. https://doi.org/10.1016/j.bdr.2015.01.006
3. Chen M, Mao S, Liu Y (2014) Big Data: a survey. Mobile Netw Appl 19(2):171–209. https://doi.org/10.1007/s11036-013-0489-0

4. Chen W, Oliverio J, Kim JH, Shen J (2018) The modeling and simulation of data clustering algorithms in data mining with big data. J Ind Integr Manage 4:1850017. https://doi.org/10.1142/s2424862218500173
5. Zhao X, Liang J, Dang C (2019) A stratified sampling based clustering algorithm for large-scale data. Knowl-Based Syst 163:416–428. https://doi.org/10.1016/j.knosys.2018.09.007
6. Pandove D, Goel S (2015) A comprehensive study on clustering approaches for big data mining. In: Proceedings of IEEE 2nd international conference on electronics and communication systems. IEEE Xplore Digital Library, pp 1333–1338. https://doi.org/10.1109/ecs.2015.7124801
7. Chen CP, Zhang C (2014) Data-intensive applications, challenges, techniques and technologies: a survey on Big Data. Inf Sci 275:314–347. https://doi.org/10.1016/j.ins.2014.01.015
8. Amado A, Cortez P, Rita P, Moro S (2018) Research trends on Big Data in Marketing: a text mining and topic modeling based literature analysis. European Res Manage Bus Econ 24(1):1–7. https://doi.org/10.1016/j.iedeen.2017.06.002
9. Lee I (2017) Big data: dimensions, evolution, impacts, and challenges. Bus Horiz 60(3):293–303. https://doi.org/10.1016/j.bushor.2017.01.004
10. Gandomi A, Haider M (2015) Beyond the hype: big data concepts, methods, and analytics. Int J Inf Manage 35(2):137–144. https://doi.org/10.1016/j.ijinfomgt.2014.10.007
11. Sivarajah U, Kamal MM, Irani Z, Weerakkody V (2017) Critical analysis of Big Data challenges and analytical methods. J Bus Res 70:263–286. https://doi.org/10.1016/j.jbusres.2016.08.001
12. Bendechache M, Tari A, Kechadi M (2018) Parallel and distributed clustering framework for big spatial data mining. Int J Parallel Emergent Distrib Syst 34:1–19. https://doi.org/10.1080/17445760.2018.1446210
13. Chen M, Mao S, Liu Y (2014) Big data a survey. Mob Netw Appl 19(2):171–209. https://doi.org/10.1007/s11036-013-0489-0
14. Gole S, Tidke B (2015) A survey of Big Data in social media using data mining techniques. Proc IEEE ICACCS. https://doi.org/10.1109/ICACCS.2015.7324059
15. Elgendy N, Elragal A (2014) Big data analytics a literature review paper. In: LNAI, vol 8557, pp 214–227. https://doi.org/10.1007/978-3-319-08976-8_16
16. Cha S (2007) Comprehensive survey on distance/similarity measures between probability density functions. Int J Math Models Methods Appl Sci 4(1):300–307. https://doi.org/10.1109/icpr.2000.906010
17. Lin Y, Jiang J, Lee S (2014) A similarity measure for text classification and clustering. IEEE Trans Knowl Data Eng 26(7):1575–1590. https://doi.org/10.1109/tkde.2013.19
18. Tavakkol B, Jeong MK, Albin SL (2017) Object-to-group probabilistic distance measure for uncertain data classification. Neurocomputing 230:143–151. https://doi.org/10.1016/j.neucom.2016.12.007
19. Liu H, Zhang X, Zhang X, Cui Y (2017) Self-adapted mixture distance measure for clustering uncertain data. Knowl-Based Syst 126:33–47. https://doi.org/10.1016/j.knosys.2017.04.002
20. Weller-Fahy DJ, Borghetti BJ, Sodemann AA (2015) A survey of distance and similarity measures used within network intrusion anomaly detection. IEEE Commun Surv Tutor 17(1):70–91. https://doi.org/10.1109/comst.2014.2336610
21. Grant J, Hunter A (2017) Analysing inconsistent information using distance-based measures. Int J Approx Reason 89:3–26. https://doi.org/10.1016/j.ijar.2016.04.004
22. Merigó JM, Casanovas M, Zeng S (2014) Distance measures with heavy aggregation operators. Appl Math Model 38(13):3142–3153. https://doi.org/10.1016/j.apm.2013.11.036
23. Ikonomakis EK, Spyrou GM, Vrahatis MN (2019) Content driven clustering algorithm combining density and distance functions. Pattern Recogn 87:190–202. https://doi.org/10.1016/j.patcog.2018.10.007
24. Marcon E, Puech F (2017) A typology of distance-based measures of spatial concentration. Reg Sci Urban Econ 62:56–67. https://doi.org/10.1016/j.regsciurbeco.2016.10.004
25. Kocher M, Savoy J (2017) Distance measures in author profiling. Inf Process Manage 53(5):1103–1119. https://doi.org/10.1016/j.ipm.2017.04.004

26. Moghtadaiee V, Dempster AG (2015) Determining the best vector distance measure for use in location fingerprinting. Pervas Mobile Comput 23:59–79. https://doi.org/10.1016/j.pmcj.2014.11.002
27. Chim H, Deng X (2008) Efficient phrase-based document similarity for clustering. IEEE Trans Knowl Data Eng 20(9):1217–1229. https://doi.org/10.1109/tkde.2008.50
28. Wang X, Yu F, Pedrycz W (2016) An area-based shape distance measure of time series. Appl Soft Comput 48:650–659. https://doi.org/10.1016/j.asoc.2016.06.033
29. Ramya R, Sasikala T (2018) A comparative analysis of similarity distance measure functions for biocryptic authentication in cloud databases. Cluster Comput. https://doi.org/10.1007/s10586-017-1568-y
30. Abudalfa SI, Mikki M (2013) K-means algorithm with a novel distance measure. Turkish J Electr Eng Comput Sci 21:1665–1684. https://doi.org/10.3906/elk-1010-869
31. Nadler M, Smith EP (1993) Pattern recognition engineering. Wiley, New York. ISBN-13 978-0471622932
32. Gan G, Ma C, Wu J (2007) Data clustering: theory, algorithms, and applications. SIAM, Society for Industrial and Applied Mathematics, Philadelphia, PA
33. Everitt BS (2011) Cluster analysis, 5th edn. Wiley series in probability and statistics. Wiley, Southern Gate, Chichester, West Sussex, UK. ISBN 978-0-470-74991-3
34. Aggarwal CC, Reddy C (2014) Data clustering algorithms and applications. CRC Press, Taylor & Francis Group. ISBN 978-1-4665-5822-9
35. Manning CD, Raghavan P, Schütze H (2008) Introduction to information retrieval. Cambridge University Press, Cambridge
36. Fahad A, Alshatri N, Tari Z, Alamri A, Khalil I, Zomaya AY, Bouras A (2014) A survey of clustering algorithms for big data: taxonomy and empirical analysis. IEEE Trans Emerg Top Comput 2(3):267–279. https://doi.org/10.1109/tetc.2014.2330519
37. Gupta, S et al (2019) Tier application in multi-cloud databases to improve security and service availability. In: Handbook of research on cloud computing and big data applications in IoT. IGI Global, pp 82–93

Chapter 2
Not Yet Decided

Anil Kumar and Rajeev Kumar

1 Introduction

The objective of recommender system is to provide meaningful recommendation to a user from collection of data which can be film, music, book, news, video, jokes, social tags, restaurants, Twitter follower, people recommendation in facebook or in general product sell on various e-commerce Web site like flipkart, snapdeal, amazon, eBay, etc., that user like. Recommendation for books on Amazon, movies on Netflix [1], music on Last.fm, and Pandora Radio are real-world examples of recommendation system. The implementation of recommender system depends on the field and particular features of the dataset available. As the growth of social networking Web sites and Web 2.0, huge amounts of information or data are submitted daily on social networking Web sites and the Web. Similarly emerging technology is coming every week, month, or yearly. To obtain valuable information from the World Wide Web (www) according to user taste, a Web site can be customized.

Recommender systems are software tool which recommends product, movie, music, book, etc., to users. These systems employ users' taste and a suitable algorithm to get relevant or preferred products, movie, music, book, etc. Recommender systems handle information overload problems by filtering products that likely may similar to users' taste. These systems help users to efficiently overcome the problem by filtering [2] irrelevant information when users looking for desired information. Recommender systems apply filtering algorithms to offer recommendations to users.

A. Kumar (✉)
Department of Computer Science & Engineering, Government Engineering College, Bharatpur, Rajasthan, India
e-mail: anil_kpawar@yahoo.com

R. Kumar
Department of Information Technology, Hindustan College of Science & Technology, Mathura, UP, India
e-mail: rajeev.sahani@gmail.com

© Springer Nature Singapore Pte Ltd. 2021
S. D. Purohit et al. (eds.), *Proceedings of International Conference on Communication and Computational Technologies*, Algorithms for Intelligent Systems,
https://doi.org/10.1007/978-981-15-5077-5_2

These algorithms are divided into collaborative-based filtering, content-filtering, and hybrid approach [2].

Collaborative filtering (CF) refers to an algorithm that recommends products, movie, music, etc., to users depending on ratings of other users with similar taste in products, movie, music, etc., collectively. It collects users rating for product or movie and then calculates similarity [3] to find correlation among users, i.e., how some users are similar to another group of users. Those who are highly correlated to users are recommended on top and a similar approach is applied for other and recommended in decreasing order (Fig. 1).

Content-based filtering approach recommends products, movie, music, etc., to the user based on user taste in past (user's profile of taste) (Fig. 2).

It compares user profile of taste with product, movie, and music dataset and finds best-matched products and recommend them in decreasing [4] order. Similarity calculation can be done by various methods like cosine similarity, log likelihood, etc.

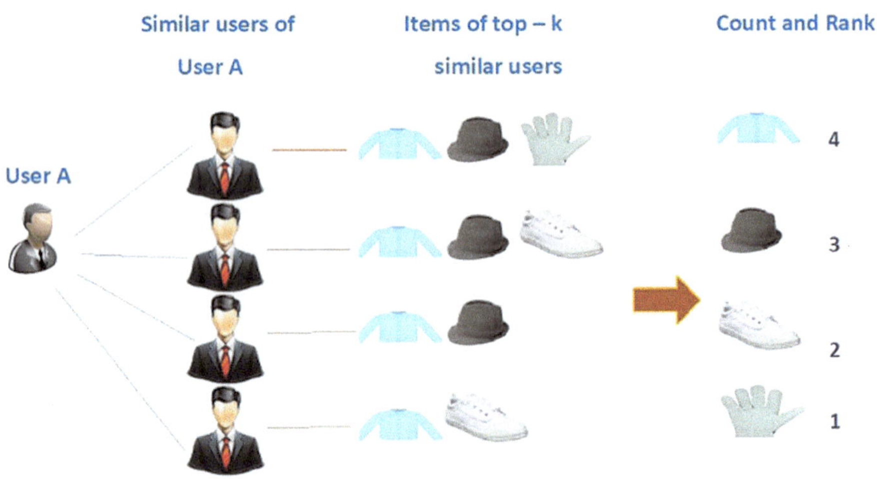

Fig. 1 Collaborative filtering

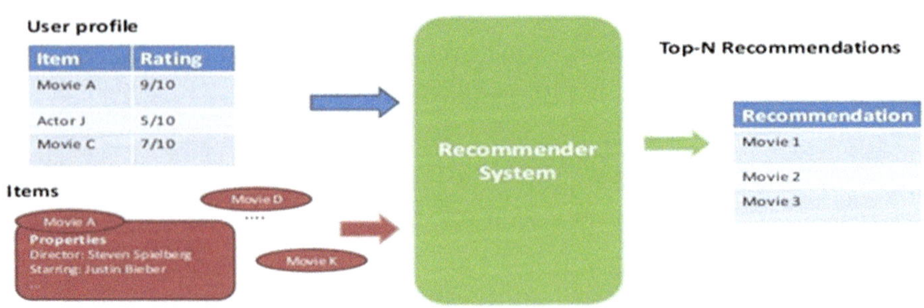

Fig. 2 Content-based filtering

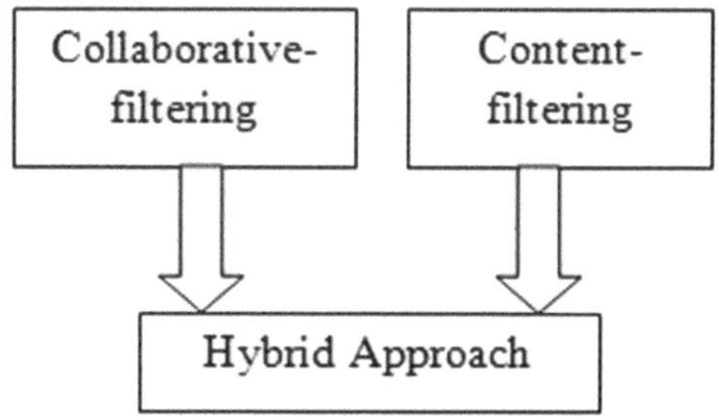

Fig. 3 Hybrid approach

Direct feedback from a user can be provided to give higher or lower scores on the significance of definite feature typically in the form of a like or dislike.

Hybrid approaches combine both content-filtering and collaborative-filtering techniques to create separate lists of recommendations and then combine their results to make a final list of recommendations. Hybrid approach [5] could be more effective in some cases. It can be implemented in numerous ways: by producing content-filtering and collaborative-filtering recommendation separately and then merging them; by attaching content-filtering potential to a collaborative-filtering approach or vice versa; or by combining the techniques into one model (Fig. 3).

The paper is ordered in the following manner. The next section talks about related work. Section 3 discusses the proposed work. Section 4 discusses experimental evaluation, and Sect. 5 discusses the results and discussion. In the last part, conclusion and future work are discussed.

2 Related Work

Goossen et al. [3] proposed CF-IDF-based approach to news personalization and tested in Athena framework which is the extension to the Hermes news personalization framework. Athena uses a user profile to store concepts or terms found in news items searched by the user. Chen et al. [1] proposed an agent-based customized recommendation system based on private dynamic user profile. The system gathers and extracts the private data of the user at the client-side, finds out, saves it and updates private dynamic user profile (DUP) at the client-side. The system obtains relevant data from the content server according to the dynamic user profile. It can be used as personalized advertising system in the RSS reader application. Pazzani and Billsus [2] present content-based recommendation systems where a profile is built and updated automatically in reaction to feedback on the interest of products that have been provided to the user. Van Meteren and Van Someren [4] present the user model based on the content-based approach which is very dynamic and is trained from positive feedback only. The relevance feedback technique is used for learning such a user model which is both efficient and dynamic.

Cantador et al. [6] present recommendation models that utilize user and item profiles described in forms of a stored list of social tags. Vector space and Okapi BM25 can be implemented in IR models. Hannon et al. [5] focus on the identifying relationship between users in social Web, like Twitter. This is done with the help of both content-based and collaborative-filtering approach. We try to attach the real-time social Web user profile and recommendation. Diaby and Viennet [7] present taxonomy-based recommendation systems that recommend relevant jobs to LinkedIn and Facebook users and it is being developed by San Francisco-based software organization called Work4. Wu et al. [8] present probabilistic retrieval model which interpret TF-IDF term scores to make relevance decisions. It replicates the local relevance for every location of a document and merges all of these local relevance results to make document result.

Zhang et al. [9] present a job recommendation based on collaborative-filtering-based approach. It uses both user-based and item-based collaborative approach for better online job hunting. Schafer et al. [10] present a collaborative-filtering approach for the adoptive Web that adopt their content to users, such as deciding which news articles a Web site should be recommended to a user. Malinowski et al. [11] present an approach which consider both interest of user and interest of recruiter foe good match between personnel and jobs. Ramanthan et al. made [12] a framework which incorporates trust and distrust information in the model and it provides better accuracy than rating-based model. This may be suitable in certain circumstances only. Sarwar et al. [13] present a method of model-based approach of collaborative filtering which decreases the dimension of matrix by dividing into parts. This method is very scalable and robust. Sarwar et al. [14] proposed a model for evaluation of the recommendation system. Evaluation can be done by various methods like mean square error (MSE) and root mean square error (RMSE) that are best suit for collaborative-filtering approach while content-based approach is evaluated using precision and recall.

3 Related Work

3.1 Luhn's Ideas

Luhn's ideas state that the frequency of word occurrence in a document determines the word importance. It further states that the relative position within a sentence of words containing known values of importance [15] provides a valuable measurement for deciding the importance of sentences. The important feature of a sentence will, therefore, be depending on a combination of these two measurements.

Suppose f be the frequency of occurrence of word and r be the rank order, i.e., the order of their frequency of occurrence. The curve has been plotted between f and r which make hyperbolic curve given in Fig. 4. This hyperbolic curve is similar to Zipf's laws [16] which state that the product of the frequency of words and the rank order is approximately constant.

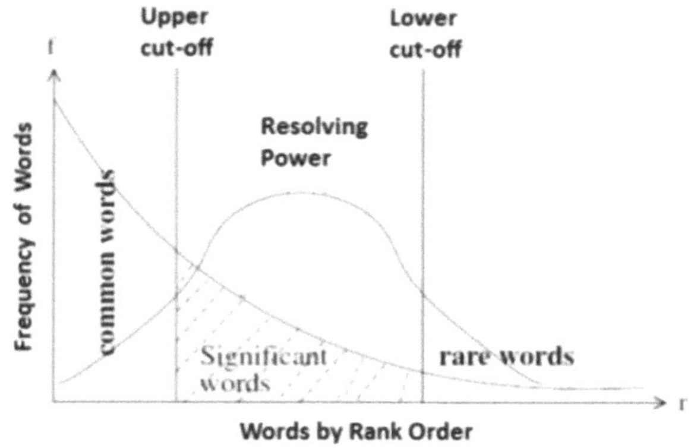

Fig. 4 Luhn's ideas

3.2 *Term Frequency-Inverse Document Frequency (tf-idf)*

The score of tf-idf is used in information retrieval (IR) and data mining. The score of tf-idf is a numerical computation utilized to evaluate how important a word is to a document in any dataset or corpus. The importance enhances proportionally to the number of times a word comes in the document but is a balance [17] by the frequency of the word in the corpus. The variant of the tf-idf scoring system is used by search engines as an essential tool in scoring and ranking a document's relevance specified a user query.

The tf-idf score is made of two terms, i.e., term frequency and inverse document frequency. Term frequency calculates the normalized term frequency, i.e., the number of times word appeasers in a Web site or document divided by the number of words in a Web site [18] or document. The inverse document frequency (IDF) calculates the logarithm of the number of the document or Web site in the World Wide Web (WWW) divided by the number of Web sites or document where the specific term appears.

- **Term frequency (tf)** measures how often a term appears in a Web site or document. It is possible that term appears much more times in long document or long Web site compared to shorter document or Web sites. This situation occurs due to a different length of document or Web site. Therefore, the term frequency (tf) is divided by the document or Web site length (aka. the total number of terms in the Web site or document) to make it normalized.

$$\text{tf} = \frac{\text{number of times a term appears in website or document}}{\text{total number of term in website document}} \quad (1)$$

- **Inverse document frequency (idf) m**easures the importance of a term. Term frequency (tf) gives equal importance to every term, though it is identified that stop word, such as "i", "big", "am", "m", "a", "we", "are", "it", "of", "this", "and", "is", "to", "at", "in", "was", "with", "doing", "It", "not", "our", may occur many times but have little importance. Thus, we have to minimize the effect of frequent terms while increasing the rare ones. Therefore, idf can be computed as

$$\text{idf} = \log \frac{\text{total number of documents websites}}{\text{number of documents or website which term occurs}} \tag{2}$$

by default base of log is 10.

3.3 Numerical Example

Suppose a Web site having 200 words where the word lion occurs 9 times. The term frequency (tf) for lion is then (9/200) = 0.045. Suppose there are 3 billion Web sites at a moment and the word lion occurs 30 thousand of these. Then, the inverse document frequency (i.e., idf) is computed as log 3,000,000,000/30,000 = 5. Therefore, score of lion is a product of tf and idf, i.e., 0.045 * 5 = 0.225.

4 Experiment Evaluation

The proposed models have been evaluated by using actual dataset by using two different evaluation matrices, i.e., precision and recall [15]. The dataset and evaluation metric are explained in this part. Then, results are presented in the next parts with the help of tables and curve. Finally, conclusions and future work are given.

4.1 Dataset

There is no publicly available dataset on the Web for the project done by B.Tech students. Therefore, we have collected various project synopsis from various institution library. Also, we have collected synopsis from [19] some well-known Web site like IBM which conducts the great mind challenge (TGMC) [18] content, India, every year and provides the synopsis for the entire project. We also collected synopsis from IEEE project, Nxtlogic software solution, Nevon projects software and embedded kits.

4.2 Evaluation Metrics

In information retrieval, when we put the query in search engine then result is retrieved which may be classified into two broad categories, i.e., relevant and non-relevant.

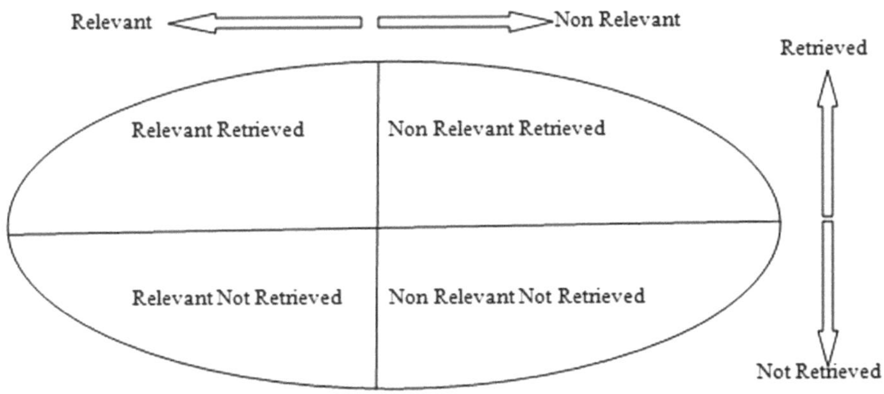

Fig. 5 Precisions and recall

Similarly, content-based recommender systems are also evaluated with the help of similar procedure. We have used precision and recall to evaluate our proposed model (Fig. 5).

Precision It is fractions of retrieved document that are relevant to the query. It is computed with the help of total retrieved document from which relevant documents are identified.

$$\text{precision} = \frac{\text{total relevant document retrieved}}{\text{total retrieved documents}} \tag{3}$$

Recall It is a fraction of relevant documents that are relevant to the query. It considers all relevant documents that are retrieved or not retrieved.

$$\text{recall} = \frac{\text{total relevant document retrieved}}{\text{total relevant documents}} \tag{4}$$

5 Result and Discussion

The accuracy of our proposed model in terms of precision and recall is promising. Precision and recall [20] are calculated on a different dataset of different types of project. We have collected dataset for [21] computer science (CS), electronic & communication (EC), mechanical engineering (ME), and civil engineering (CE). Results obtained on these datasets are promising (Tables 1 and 2).

Results show that precision and recall value on less amount of project is better but when size is increasing then still it provides better precision and recall value. Recall value is estimated manually on the basis of individual personality and interest.

Table 1 Precision and recall depend on number of project

	Number of projects					
	20	50	100	200	500	1000
Precision	0.31	0.28	0.26	0.25	0.24	0.24
Recall	23.00	0.18	0.17	0.15	0.15	0.16

Table 2 Precision and recall depend on types of project

	Types of project (1000 projects)			
	CS	EC	ME	CE
Precision	0.24	0.26	0.23	0.21
Recall	0.15	0.16	0.13	0.14

6 Conclusion and Future Work

Our proposed model is initially implemented to recommend projects based on user profile. After a certain course of time [21], we can include various other fields like arts, commerce, medical sciences, humanities, etc. We will also try to generate the more user-specific profile on the basis of their hobbies and will recommend [22] more courses and training program which are conducted worldwide. The value of precision and recall can be improved using user feedback. Feedback information can be used to learn the student interest and train recommendation engine. In Future, we can also extend this model for job recommendations.

References

1. Chen T et al (2007) Content recommendation system based on private dynamic user profile. In: 2007 International conference on machine learning and cybernetics, vol 4. IEEE
2. Pazzani MJ., Billsus D (2007) Content-based recommendation systems. In: The adaptive web. Springer, Berlin, Heidelberg, pp 325–341
3. Goossen F et al (2011) News personalization using the CF-IDF semantic recommender. In: Proceedings of the international conference on web intelligence, mining and semantics. ACM
4. Van Meteren R, Van Someren M (2000) Using content-based filtering for recommendation. In: Proceedings of the machine learning in the new information age: MLnet/ECML2000 workshop
5. Hannon J, Bennett M, Smyth B (2010) Recommending twitter users to follow using content and collaborative filtering approaches. In: Proceedings of the fourth ACM conference on recommender systems. ACM
6. Cantador I, Bellogín A, Vallet D (2010) Content-based recommendation in social tagging systems. In: Proceedings of the fourth ACM conference on recommender systems. ACM
7. Diaby M, Viennet E (2014) Taxonomy-based job recommender systems on Facebook and LinkedIn profiles. In: 2014 IEEE eighth international conference on research challenges in information science (RCIS). IEEE
8. Wu HC et al (2008) Interpreting tf-idf term scores as making relevance decisions. ACM Trans Inf Syst (TOIS) 26(3):13

9. Zhang Y, Yang C, Niu Z (2014) A research of job recommendation system based on collaborative filtering. In: 2014 Seventh international symposium on computational intelligence and design (ISCID), vol 1. IEEE
10. Schafer JB et al (2007) Collaborative filtering recommender systems. In: The adaptive web. Springer, Berlin, Heidelberg, pp 291–324
11. Malinowski J et al (2006) Matching people and jobs: a bilateral recommendation approach. In: Proceedings of the 39th annual Hawaii international conference on system sciences, 2006. HICSS'06, vol. 6. IEEE (2006)
12. Guha R et al (2004) Propagation of trust and distrust. In: Proceedings of the 13th international conference on World Wide Web. ACM
13. Sarwar B et al (2000) Application of dimensionality reduction in recommender system-a case study. No. TR-00-043. Department of Computer Science, Minnesota University Minneapolis
14. Sarwar B et al (2000) Analysis of recommendation algorithms for e-commerce. In: Proceedings of the 2nd ACM conference on electronic commerce. ACM
15. Gemulla R et al (2011) Large-scale matrix factorization with distributed stochastic gradient descent. In: Proceedings of the 17th ACM SIGKDD international conference on knowledge discovery and data mining. ACM
16. Gabaix X (1999) Zipf's law and the growth of cities. Am Econ Rev 89(2):129–132
17. Baltrunas L, Ludwig B, Ricci F (2011) Matrix factorization techniques for context aware recommendation. In: Proceedings of the fifth ACM conference on recommender systems. ACM
18. Ricci F, Rokach L, Shapira B (2011) Introduction to recommender systems handbook. Springer, US
19. Xu Bin et al (2012) An exploration of improving collaborative recommender systems via user-item subgroups. In: Proceedings of the 21st international conference on World Wide Web. ACM
20. Rendle S et al (2011) Fast context-aware recommendations with factorization machines. In: Proceedings of the 34th international ACM SIGIR conference on research and development in information retrieval. ACM
21. Paparrizos I, Barla Cambazoglu B, Gionis A (2011) Machine learned job recommendation. In: Proceedings of the fifth ACM conference on recommender systems. ACM
22. Lee DH, Brusilovsky P (2009) Reinforcing recommendation using implicit negative feedback. In: User modeling, adaptation, and personalization. Springer, Berlin, Heidelberg, pp 422–427

Chapter 3
An Approach to Implement MapReduce and Aggregation Pipeline Utilizing NoSql Technologies

Monika and Vishal Shrivastava

1 Introduction

The NoSQL databases give a medium to capacity and recovery of huge information that is displayed in non-tabular structure rather than unthinkable relations utilized in social databases. The information structures utilized in NoSQL databases are of different kinds instead of tabular structure utilized by social databases the executive framework, e.g., record, chart, key-value and so forth. These databases are schema-less, which makes them marvelously incredible in execution.

NoSQL databases are superior to social databases on the grounds that these are progressively adaptable and give exceptional execution [1]. There are primarily three characterizations of information organized, semi-organized and unstructured information. Social databases can just deal with organized information; it cannot deal with other two sorts of information. Social databases need data of the information before really putting away it in type of patterns, which fits inadequately with nimble advancement approaches in light of the fact that there will be a need to change the blueprint each time at whatever point there is need of new highlights which log jams the procedure if the database is substantial. This report makes an endeavor to break down the execution time of questions (of removing or inserting information) into the two record document-oriented databases MongoDB.

For organizations like Google, Facebook, Amazon and LinkedIn which collects information in terabytes (1e + 12) each and every day. It is inadequate to deal with this huge information with conventional databases (RDBMS). In the greater part of the cases, the information caught by these organizations are not organized which again restricts the utilization of social databases.

Monika · V. Shrivastava (✉)
Department of Computer Science, Arya College of Engineering and IT, Jaipur, Rajasthan, India
e-mail: Vishal500371@yahoo.co.in

Monika
e-mail: Monikabishnoi3@gmail.com

© Springer Nature Singapore Pte Ltd. 2021

S. D. Purohit et al. (eds.), *Proceedings of International Conference on Communication and Computational Technologies*, Algorithms for Intelligent Systems,
https://doi.org/10.1007/978-981-15-5077-5_3

This issue is explained by NoSQL databases by giving flat scaling which implies an expanding number of servers (hubs) in the group, which decreases the obligations of every server by giving extra endpoints to the customer associations. The thought behind level scaling is to expand the sharing of burden among the hubs results in an expansion in dick capacity [1].

1.1 Document Database

Document databases are a standout among the most utilized and prevalent NoSQL frameworks, where each record is thought of as a "document." Document is utilized to store a gathering of information that convert a type of client intelligible data to the standard organizations, for example, JSON, XML, BSON and so forth. These databases are a subclass of key-value databases [2]. To keep up region of information report is the best option in light of the fact that these are autonomous units, which improve execution on the grounds that the related information is perused adjoining off a disk. It additionally makes the circulation of information over various servers simple. The capacity of unstructured information is simple in light of the fact that the record contains just those keys and qualities which application rationale requires. It likewise gives incredible adaptability by not knowing data pattern ahead of time [2]. We are working at document databases MongoDB on different techniques.

1.1.1 MongoDB

MongoDB is cross-stage NoSQL document database. MongoDB is written in C, C++, JavaScript. It was first created by the product organization 10 gen (presently MongoDBInc) and moved to open-source network in 2009. MongoDB is generally utilized by numerous organizations, for example, Forbes, Bosch, MetLife and so on. Figure 1 demonstrates the MongoDB information model which depicts how the semi-organized information is put away in records as different fields. A collection is a gathering of documents though the database is a gathering of collection. This improves the comprehension of the databases [3].

MongoDB stores the information in archives like JSON. These fields can incorporate arrays, binary data and sub-documents. The choice of fields in an application can shift as per the prerequisite. This element enables engineers to change the information model much of the time as their application requires.

Documents can be gotten to through rich drivers accessible in practically all famous programming dialects, for example, Java, PHP and so on. MongoDB evacuates the need of independent ORM layer which implies designers need not to deal with mapping of articles from database to application which makes them progressively profitable.

MongoDB permits auto-sharding for even scaling of the database. To give high accessibility crosswise over server farms, it gives replication. Replication implies

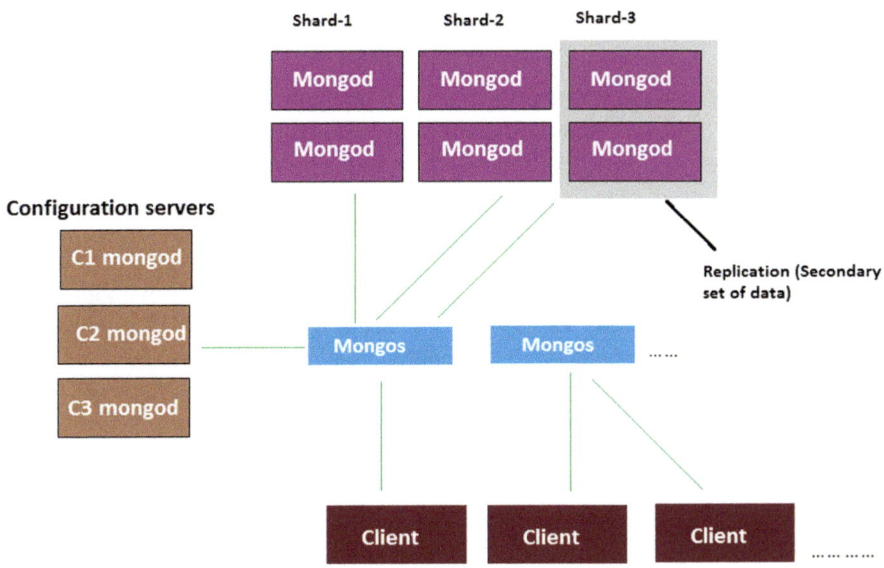

Fig. 1 Working of MongoDB

putting away a duplicate of information (second set) in servers. On the off chance that whenever the essential arrangement of information goes down, the auxiliary set consequently takes over as the essential arrangement of information. MongoDB additionally gives an in-memory instrument to accelerate the task by broad utilization of random access memory (RAM). The working of MongoDB further can be clarified by the accompanying figure.

Figure 2 appeared above demonstrates the average auto-sharding utilized by MongoDB to guarantee high accessibility of information. It likewise demonstrates a replication procedure which is utilized for putting away same information duplicate in at least one servers.

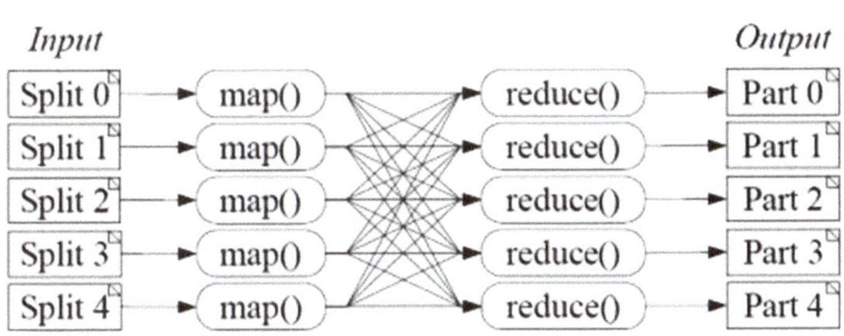

Fig. 2 MapReduce architecture

1.2 JSON

JavaScript object notations (JSON) is an open standard organization which gives an approach to transmit information objects comprising of property estimation combines in type of intelligible content. It is utilized as an option in contrast to extensible markup language (XML). It is a method for sending and accepting information between Web applications and servers. JSON was first determined by Douglas Crockford. The information is put away in records with .json augmentation [4].

These destinations are controlled by AJAX. These locales work without deferring page rendering process. This permits exchanging up the substance of specific component inside out designs without the need of invigorating the page.

As of late, the expansion in prevalence of Internet-based life, numerous sites depend on the substance given by Facebook, Twitter, Flickr and others. These sites give RSS channel, which is not anything but difficult to work with AJAX. JSON tackles the cross-space issue in JSON design isolate column group by utilizing wavy props. It is anything but difficult to peruse and get it.

1.3 MapReduce

MapReduce is a developing pattern of utilizations that handle big data. Big data breaking down is a difficult issue today. Big data applications, the MapReduce system has as of late pulled in a great deal of consideration. Google's MapReduce or its open-source counterparts Hadoop is a useful asset for structure such applications. In this instructional exercise, we will present the MapReduce structure. MapReduce is a system which utilizes the record ordering with mapping, arranging, rearranging and reducing by two capacities that is Map work and Reduce work. The MapReduce work is done in two fundamental activities: Map and Reduce. The Map work read sets of information and performs calculations on it. At that point, these came about middle of the road (key, value) sets are additionally passed to Reduce work. Reduce work bunches every one of the qualities for every exceptional key created by Map work.

To comprehend the working, the model MapReduce work is appeared. In this precedent, the Map work peruses the information of the record and parses out the words. In guide venture for each word, (key, value) pair is produced, for example, (word, 1). Here, the word is a key and value 1 demonstrates various event of a word is one in the archive [5].

2 Literature Review

Qi Lv and Wei Xie had a novel execution approach for an occasion driven document log analyzer are introduced, and execution examination of a query, sweep and collection tasks over MongoDB, HBase and My SQl is broke down. In their test results demonstrate that HBase performs best adjusted in all tasks, while MongoDB gives under 10 ms query speed in certain activities which is most reasonable for ongoing applications.

Hamoud Alshammari and Hassan Bajwa compose a paper and in this paper, they present a review of a created MapReduce calculation and its reenactment utilizing HBase as a NoSQL database. They portrayed and actualize the proposed enhanced Hadoop architecture with certain progressions on the first plan to meet various cases.

Chirag Patal and Mosin Hasan compose a paper about the Benchmark of MapReduce Framework in MongoDB. The fundamental angles behind this paper are to display the design of "Benchmark of MapReduce Framework in MongoDB," for example, "MongoMRBench" and "MongoGen" are a framework that produces semantic enormous data that are bolstered by MongoDB which is document-based database.

Jeffrey Dean and Sanjay Ghemawat presenting about MapReduce utilized on large clusters. Their execution of MapReduce keeps running on an enormous cluster of production machines and is very versatile: A run of the mill MapReduce calculation forms numerous terabytes of data on a large number of machines. Developers and the framework simple to utilize: Many MapReduce projects have been actualized and upwards of one thousand MapReduce employments are executed on Google's clusters each day.

3 Proposed Approach

MongoDB are NoSQL databases which are utilized when information is gigantic. In our proposed research work, we study the aggregation operation then compare the time of operation for two MongoDb functions, namely MapReduce and aggregation pipeline in python on the same dataset. Here, JSON document is utilized to store a lot of information. We analyze MapReduce and match aggregate pipeline performance for document database. We will utilize diverse collections (i.e., 5k, 10k, 50k and 100k records) in it. Then, we add parallelizing function with the aggregation work over subsets of the collection, using multiple threads with aggregate pipeline and called as modified aggregate pipeline. A function is defined with two parameters x and y, x is used for the limit query and y is used for skip query. The result of this aggregation operation was documented which contained two fields _id that is customer ID and sum which is the sum of recurrence of the given customer ID where the condition is that the customer ID should be from the United Kingdom. The system used had eight threads so the total documents per collection were divided into these eight threads and then joined. Each thread does the operation thus using the maximum potential of the system and doing it efficiently.

I demonstrated this modification with aggregate pipeline gives a better result in the MongoDB database in an unstructured format. This process performs the process of grouping metadata parameters present in the file, group and all the files into the parameters that belong to each file. The result of the matching process is identification, file info, status file, output tool, statistic and metadata group in categories document.

We will utilize the Windows 8 stage to play out the tasks on MongoDB. These databases ought to give the fast and high throughput when contrasted with social databases.

4 Results and Discussion

Evaluations were conducted on "E-commerce Data" dataset that focused on the document database. The data set contains Invoice No, Stock Code, Description, Quantity, Invoice Date, Unit Price, Customer ID and Country. Then, the process for setting up MongoDb was done (Windows 10 system). MongoDb was downloaded from its official Web site. Then, a database "mydb" was created and collections were created namely "fivek", "tenk", "twentyk" and "fiftyk. First the csv file dataset was opened then it was opened in a file iterator. Then, in a list, the different column names of Table 1 were stored. Using the iterator, each row of the table was saved as a document. The first 5000 rows were saved in the "fivek" collection, first 10,000 rows were saved in the "tenk" collection, first 20,000 rows were saved in the "twentyk" collection and first 50,000 rows were saved in the "fiftyk" collection.

First, we perform the aggregation function on the dataset. This aggregation operation took 0.223 s to happen for "fivek" collection. This aggregation operation took 0.412 s to happen for "tenk" collection. This aggregation operation took 0.842 s to happen for "twentyk" collection. This aggregation operation took 2.013 s to happen for "fiftyk" collection.

Next, we perform the MapReduce operation on the documents. The MapReduce operation took 1.416 s to happen for "fivek" collection. The MapReduce operation took 2.553 s to happen for "tenk" collection [6]. The MapReduce operation took

Table 1 Comparison between MapReduce and aggregation pipeline

Data records	MapReduce (s)	Aggregation pipeline (s)	Aggregation pipeline with parallelization (s)
5k	1.416	0.223	0.396
10k	2.553	0.412	0.387
20k	5.240	0.842	0.384
50k	12.102	2.013	0.401

5.240 s to happen for "twentyk" collection. The MapReduce operation took 12.102 s to happen for "fiftyk" collection. Thus, we can say that the aggregation pipeline function is much faster as compared to the MapReduce function and produces equally accurate results.

Then, we try to optimize or improve the performance of aggregation pipeline operation by using parallelization by using a multithreading approach. The aggregation pipeline with parallelization took 0.396 s to happen for "fivek" collection. The aggregation pipeline with parallelization took 0.387 s to happen for "tenk" collection. The aggregation pipeline with parallelization took 0.384 s to happen for "twentyk" collection. The aggregation pipeline with parallelization took 0.401 s to happen for "fiftyk" collection. Thus, the optimized aggregation performs the best specially for large datasets.

5 Conclusion

In this paper, we describe a task where we conducted a study to compare the time of operation for two Mongo Db functions, namely MapReduce and aggregation pipeline in python on the same E-commerce data dataset. That dataset is downloaded from Kaggle which was taken from UCI machine learning repository this data is a set of actual transactions between the year 2010 and 2011. The dataset contains Invoice No, Stock Code, Description, Quantity, Invoice Date, Unit Price, Customer ID and Country. In our examination process, we will perform curd operation on dataset by applying MapReduce, aggregate pipeline and modified aggregate pipeline method.

The modified aggregate pipeline method has better performance in processing data in the larger block size compared to both match aggregate pipeline and MapReduce. Due to the execution time can be reduced even further, by parallelizing, the aggregation work over subsets of the collection, using multiple threads. I demonstrated this for a single non-sharded database.

6 Future Scope

The present and just as future extent of NoSQL innovations are splendid. As there are numerous chances and incredible difficulties which should be defeated in big data. Later on, we can play out a correlation of these two document-based databases with the diverse sorts of records, for example, CSV, XML and JSON. We can likewise contemplate other report situated databases, for example, Redis, HBase, Postgresql and so forth. NoSQL advancements are likewise brilliant if there should arise an occurrence of pattern matching algorithms on account of big data.

References

1. Kumar N, Saxena S (2015) A preference-based resources allocation in cloud computing systems. In: 3rd International conference on recent trends in computing 2015. Procedia Computer Science, vol 57, pp 104–111
2. Xu Q, Arumugam RV, Yong KL, Wen Y, Ong YS, Xi W (2015) Adaptive and scalable load balancing for metadata server cluster in cloud-scale file system. Front Comput Sci 9(6):904–918
3. Teng F (2012) Management Des Donnees Et Ordinnnancement Des Taches Sur Architectures Distributes, Dissertation, Ecole Cenrale Paris Et Manufactures, Centrale Paris
4. Yu Z (2012) Research of conversion method of entity object and JSON data. In: The 2nd international conference on computer application and system modeling. Guizhou University for Nationalities, Guiyang 550025, China
5. Dean J, Ghemawat S (2008) MapReduce: simplified data processing on large clusters. ACM, Commune, pp 107–113
6. Gupta S et al (2019) Tier application in multi-cloud databases to improve security and service availability. In: Handbook of research on cloud computing and big data applications in IoT. IGI Global, pp 82–93

Chapter 4
Energy Hole Alleviation in Wireless Sensor Networks Through Node Behavior Modeling Using Firefly Algorithm

Sachin Sharma, Saroj Hiranwal, and Kapil Dev Sharma

1 Introduction

A wireless sensor network is a wireless network with (usually) low transmission range sensors and low bandwidth links. The most general form of WSN consists of a sink node and a number of spatially distributed sensing nodes connected in the form of clusters, which, in turn, are connected to the sink node. These networks find large applications in scientific, industrial, and consumer electronics domains. The nodes of the network can be fire sensors, radiation sensors, wildlife habitat monitoring sensors, etc., to name a few. The physical and MAC layer specifications for WSN are specified in IEEE 802.15.4 also known as low-rate wireless personal area networks (LR-WPAN) [1].

In Fig. 1, the sensor nodes are shown as a hollow circle and the grouped in the form of clusters. The cluster head is also a sensor node and shown as a solid circle. The cluster head is a full function device (FFD), and the sensing nodes may be reduced function device (RFD). The cluster head transmits the data to the sink node shown as a computer system in Fig. 1. Table 1 gives the typical application domains [2] and the measured physical quantities by the nodes in a WSN.

Internet of things (IoT) [3] is considered as the future of the Internet. It is a wireless sensor network which is connected to an IP network to provide entirely new kind of data as compared to what we are accessing today. It can extend the current Internet deep into the physical world in which we are living today. Such an IP network requires a new type of query mechanisms and user-level interfacing to handle the requests from the users and provide the responses that are usable for the end user. It is the Internet in which a number of sensor nets can communicate to the core IP network through a set of gateways as shown in Fig. 2

S. Sharma (✉) · S. Hiranwal · K. D. Sharma
Rajasthan Institute of Engineering and Technology, Bhankrota Ajmer Road, Jaipur 302026, Rajasthan, India
e-mail: rj14.sachin@gmail.com

© Springer Nature Singapore Pte Ltd. 2021

S. D. Purohit et al. (eds.), *Proceedings of International Conference on Communication and Computational Technologies*, Algorithms for Intelligent Systems,
https://doi.org/10.1007/978-981-15-5077-5_4

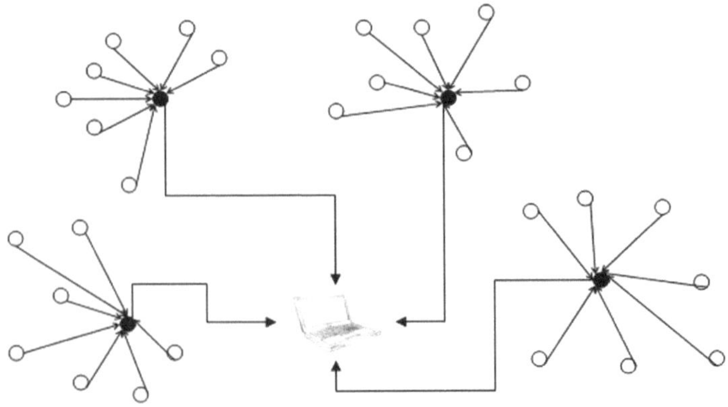

Fig. 1 Generic illustration of wireless sensor network

Table 1 Applications of wireless sensor networks

WSN deployment domain	Measured quantity
Medical diagnosis	Body temperature sensor, heart beat monitoring, blood pressure sensing and much more types of sensors in ICU's
Fire alarms, thermal power plants	Temperature and heat
Nuclear power stations	Radiation level
Departmental stores, malls	Pressure sensors on escalators, light sensors on automatic doors, light sensors on automatic flush in urinal pots, humidity sensors for cooling equipment's regulation
City automation	Light sensors for automatic on and off of road lights
Smart roads and highways	Automatic sensing of smartcards for toll charges deduction on highways. Sensing of fatigue in driver's eyes during long journeys
Home automation	Theft alarms, Internet-enabled household devices
Wildlife	Animal fingerprints monitoring, animal movement tracking
Military	Deployments in critical regions for continuous monitoring as per the need
Agriculture	Plant growth monitoring, humidity level monitoring
Pollution monitoring	The concentration (PPM) of the gases which are harmful to human population
Landslide detection	A slight movement in the soil levels and other changes that occur during a landslide
Future	WSN provides the foundation for Internet of things (IoT) [3]

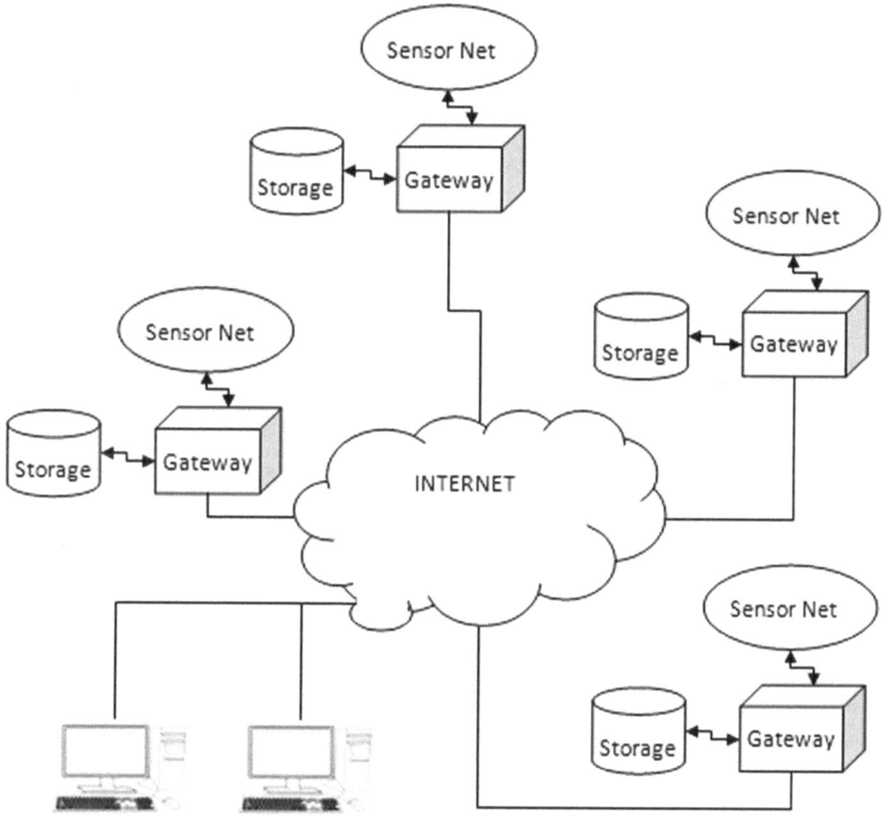

Fig. 2 Generic architecture of IoT network

The most common deployment of the WSN consists of spatially distributed wireless nodes with sink at the center of the area. The network can be visualized as nodes placed over concentric circles with the sink node at the middle of the system. Also, the transmission range of each node is nearly equal to the difference in the radius of the concentric circles. Thus, a number of the nodes over a ring chooses a cluster head in the next inner ring as depicted in Fig. 3.

The network model depicted in Fig. 3 is the one which is most commonly used in literature and considered for modeling in this paper. The network lifetime is defined as the time since the deployment of the network to the time when the very first node of the network runs out of its battery. The following are the key aspects controlling network lifetime.

1. Read operations per second by the node.
2. Transmission range.
3. The period of operation of a cluster, i.e., setup phase time plus steady-state phase time.

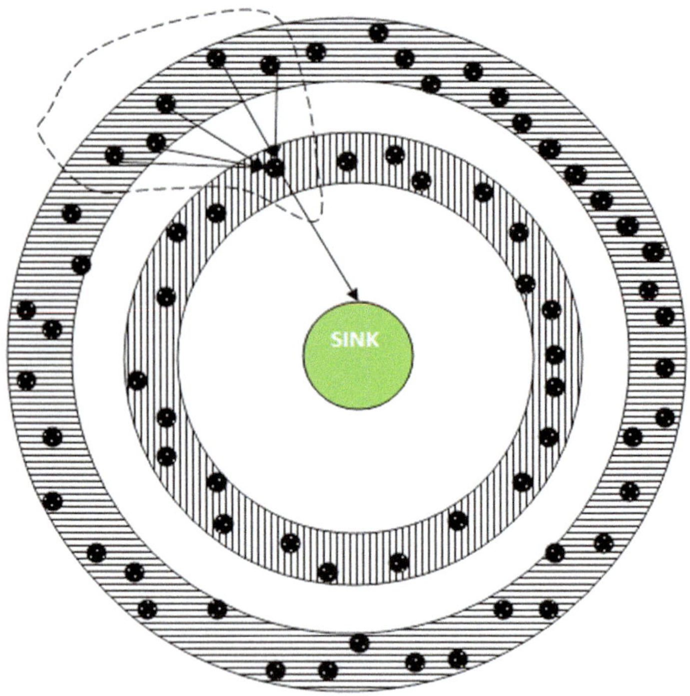

Fig. 3 Network illustration of most common deployments of WSN

4. Energy expenditure incurred in cluster setup phase.
5. Cluster size.
6. Data aggregation technique.

Techniques are proposed by several authors in literature considering complex parameters such as the size of the data packet used in communication. In this research, only those parameters are considered that can be altered through the APIs provided by the most common sensor node manufacturers. Changing one or more of the above parameters has a significant effect on network lifetime. However, the range of permissible values of the parameters above depends upon the type of deployment; viz., humidity sensor at smart home or radiation sensor at a nuclear reactor. Providing an optimal policy in the form of an equation, including all the parameters specified above is an NP-hard problem, that is also, of no practical use given its complexity and the associated requirement of computational hardware.

A sensor node can sense and transmit the data in asynchronous or synchronous way. Asynchronous transmission refers to the system in which the node transmits only when some critical event occurs. Synchronous transmission means that the node transmits data at regular time interval. Most of the commercially available sensors, e.g., Xbee humidity and temperature sensor, costing about $1, are synchronous and comes with the specifications of the number of reading operations per second. The reading rate can be configured through Zigbee API. Synchronous communication is widely used for WSN as it allows for cheap commodity sensors, thereby maintaining low costs and embedding entire logic of data processing and decision making at the sink node. The number of read operations per second may vary from once every millisecond, in case of radiation sensors, to once every 5 min for humidity sensors. Correspondingly, the battery life from 6 months to 2 years can be achieved. This parameter has a significant impact on the real-time characteristics of the network.

For any node, there are primarily two heads of energy consumption. The first corresponds to energy expenditure in transmitter circuit for data processing. The second corresponds to energy expenditure in amplification circuitry for data transmission. The longer the transmission range, the higher is the energy expenditure, and it grows with the square of the transmission range. Also, small transmission range requires multi-hop communication from the sensor node to the sink node. It requires additional routing logic and intermediate processing. Furthermore, reliable forwarding is necessary to prevent the network from intrusion. Multi-hop communication with intermediate data processing (aggregation) also introduces latency in the system. Therefore, an optimal transmission range is necessary for long battery life.

Clustering and data aggregation are techniques to reduce the overall energy consumption in the network. Setting up of a cluster requires energy expenditure which corresponds to network management overhead. Frequent cluster formation increases this overhead. On the other hand, a long steady-state phase of a cluster leads to more energy dissipation by cluster head and results in uneven battery consumption. Optimal clustering formation frequency must be computed as a function of energy expenditure incurred in cluster setup phase, which, in turn, is dependent on cluster size. It also depends on the number of reading operations per second by the node. Once the cluster head receives the data, the CH may process it to reduce its size. An analysis of clustering frequency in WSN is given by Sharma et al. in [4].

2 Review of Techniques Proposed for Extending Network Lifetime

Extension of the network lifetime in WSN is dependent on two primary factors:

1. Minimal energy consumption in individual nodes of the network.
2. Evenly energy consumption in all the nodes of the network.

To achieve the first objective, a number of techniques have been proposed in the literature. The IEEE 802.15.4 specifications for MAC sublayer come with the feature of battery life extension option, at the cost of CSMA-CA back-off exponent. Another important aspect related to the energy consumption in individual nodes is the optimal transmission range. Large transmission range needs a more powerful antenna and more battery power. Low transmission range needs low battery power but results in multi-hop communication in case of sizable networks. Despite the multi-hop routing expenses, the overall battery consumption in data transmission in far less in case of multi-hop communication as compared with single-hop long transmission range communication. Additionally, data aggregation techniques can be incorporated at intermediate routing nodes to aggregate data from multiple nodes, to reduce redundant data and to improve network throughput. However, multi-hop communication requires additional processing logic and more functional nodes to support routing mechanism. Multi-hop communication also increases the latency in the network. The importance of multi-hop communication, clustering and data aggregation can be figured out from a simple computation described as follows.

Consider the following values [5] for the transmission and processing circuitry energy consumption:

Energy required per bit in the processing circuit: 50nJ/bit
Energy required per bit in the transmission circuit: 10pJ/bit/m^2
For transmission of 1 kb for a distance of 1 km, the transmission circuit consumes $10 \times 10^{-12} * 1024 * 1000 * 1000 = 0.01024$ J.
For processing of 1 kb, the processing circuit requires $50 * 10^{-9} * 1024 = 5.12 \times 10^{-5}$. Thus, the ratio of energy consumed in transmission and processing is:

$$\frac{1024 \times 10^{-5}}{1024 \times 50 \times 10^{-9}} = 200$$

Thus, 200 instructions can be executed for the same energy consumption as for the transmission of one instruction over the distance of 1 km. Moreover, it increases with the square of the distance. Thus, it is promising to use clustering techniques in WSN to allow for prolonged battery life. Table 2 gives the comparison of single-hop and multi-hop communication in the context of WSN consisting of identical nodes.

To ensure evenly battery consumption in all the nodes of the network, the following measures are required:

1. Clustering on a regular basis so that all the nodes get a fair chance to become the cluster head.
2. Energy conservation policies for nodes which lies near to sink, as they need to work as routers for all the data packets from farthest corners of the network.

The former measure is essential because the node which is chosen as cluster head has the responsibility to receive data from all the cluster members and to transmit the data to the next-hop node toward the sink. It costs additional battery power as compared to other nodes of the cluster. This leads to rapid depletion of battery power as compared to the other nodes of the network.

The latter measure is the problem identified in this research paper. Xbee humidity sensors networks are considered for modeling as these sensors represent a large class of WSN deployments. The nodes which are located near the sink node deplete the battery power at a much higher rate as compared to the other nodes of the network. Much of the recent researches in WSN have been centered on the objective of alleviation of energy hole problem near the sink. Energy conservation techniques

Table 2 Comparison between single-hop and multi-hop communication

S. No.	Parameter	Single-hop communication: Large transmission range	Multi-hop communication: Small transmission range
1	Energy required in processing circuitry	Equal for all nodes	Equal for all nodes
2	Energy required in amplifier circuitry	More joules per node	Less joules per node
3	#Nodes	LESS: Less number of nodes required for a single packet transmission	MORE: More number of nodes required for a single packet transmission
4	Cost (per node)	MORE: Large transmission range radios LESS: Less complex as no routing logic is required Overall cost is insignificant due to cheap electronic components	LESS: Small transmission range radios MORE: More complex as routing logic is required Overall cost is insignificant due to cheap electronic components
5	Scalability	LESS: New nodes are seldom added to the network as these are of high cost	MORE: New nodes are frequently added to the network as these are of low cost
6	Network reliability	MORE: This is due to fewer chances of node failure due to the less number of nodes LESS: Individual sensor reading is important as the number of nodes is less MORE: In case of reliable and high-quality hardware LESS: In case of harsh and hostile conditions	LESS: This is due to more chances of node failure due to the usually dense network MORE: Individual sensor reading is less important as the number of nodes is large MORE: In case of cheap and low-quality hardware MORE: In case of harsh and hostile conditions
7	Uneven power consumption	LESS: Less routing is required, thus battery consumption is relatively even as compared to the multi-hop communication	MORE: Nodes near to the sink node depletes more power as these nodes need to work as routers to forward all the data packets from a large number of boundary nodes of the network toward the sink node
8	Clustering	Not important	Critically required
9	Data aggregation	Not important	Highly recommended
10	Latency	LOW: Direct data transfer to the sink node	HIGH: Due to multi-hop data communication Latency can be reduced using progressive data aggregation and processing

Table 3 Review of techniques for evenly battery consumption for prolonged network lifetime

S. No.	Algorithm/Techniques	Year	Features	Parameters enhanced	Parameters compromised
1	WSNEHA Xue et al. [6]	2014	Consideration of the network in ring topology	Balanced energy consumption (only for first radius of nodes), network lifetime	Overall energy consumption, latency
2	Energy-efficient geographic routing robust against location errors Popescu et al. [7]	2014	Geographic routing, based on location errors, energy efficient	Overall energy consumption in the network, latency	Evenly battery consumption, network lifetime
3	BECHA, Jan et al. [8]	2017	Energy-aware routing, energy level computation	Evenly battery consumption, network lifetime	Overall battery consumption, latency
4	EA-BECHA, Jan et al. [8]	2017	Energy-aware routing, residual battery level computation, choosing optimal forwarder one-hop neighbor	Evenly battery consumption, network lifetime	Overall battery consumption, latency

are critically required to be implemented to conserve the energy in the nodes at a one-hop distance from the sink. The proposed solution takes into account the number of read operations and the node behavior as a function of remaining battery life of the node, to enable the node to work longer.

Table 3 lists the techniques proposed by several authors for evenly battery consumption in the nodes.

3 Proposed Work

3.1 Network Model and Assumptions

The research work presented by several authors suggests controlling various parameters that cannot be customized or practically infeasible once the network is deployed for particular sensing task. For example, once the transmission range for a node is set for a certain distance, it will require a fixed amount of energy for packet transmission to any node lying at a certain distance from it within its transmission range. However, smaller distance gives better PSNR. Nevertheless, the energy requirement of the transmitter circuitry remains the same. In the same way, once the network is deployed, the data size of the sensor readings cannot be customized (without firmware reconfiguration) in cheap commodity sensors that are widely used in WSNs. This is indicated in the specification manual of the commercially amiable sensing nodes. For example, the Xbee devices available in the market for sensing temperature comes with specifications

One read per 30 s/1.5 years.
One read per minute/2.5 years.
One read per hour (or less frequent)/6 years.

We present the following mathematical model along with the underlying assumptions that hold for a large class of WSN deployments using commodity sensors, in which latency up to a specific value is permissible, i.e., excluding the networks with hard real-time requirements.

1. The network consists of N nodes arranged in the form of concentric circles each with radius r, $2r$, $3r\ldots$ and so on up to nr. The sink station is located in the center. The sink is externally powered, and therefore, energy dissipation is not an issue for sink node.
2. The nodes are preconfigured for certain read operations per second. This feature is implemented in firmware and it can be configured through application programming interface provided by Zigbee. However, the data packet size at the physical layer and MAC layer are IEEE 802.15.4 specified and will remain the same throughput the operation of the node.
3. The model assumes that the nodes are homogenously distributed across a geographical area with the sink at the central location of the area. This model holds for the applications in wildlife monitoring, agriculture applications and others in which a large number of sensors are homogenously distributed over a large area.
4. The number of nodes lying approximately on any ring is in proportion to the circumference of the circle. Therefore, there are a large number of nodes that lie on the outermost ring and a relatively less number of nodes lie on the inner circles.
5. The model assumes that all the nodes in the network are identical and have some fixed transmission range. Thus, energy dissipation in the successful data transmission between two nodes in the network over a certain distance is independent of the distance, provided the distance is within the transmission range. However, smaller distances achieve high SNR as compared to large distances.
6. The model assumes that the nodes which appear close to the sink node need to forward all the data packets of all the nodes lying from the second innermost circle to the outermost ring. As there is less number of nodes that are in close

proximity of the sink node, these nodes run out of battery power more rapidly as compared to the other nodes of the network.

The alleviation of energy holes in the closest circle and the extension in the battery life of the nodes which have the most significant overhead to function as the router is the primary goal of this research.

3.2 Model Analysis

Let there be a total of N nodes in the network which are distributed in proportion to the length of the circumference of the concentric circles. Thus, each of the circles has node density in proportion to the size of the ring.

The total length of the circumference of n concentric circles is

$$T = 2\pi r + 2\pi(2r) + 2\pi(3r) + \cdots + 2\pi(nr)$$
$$T = 2\pi r(1 + 2 + 3 + \cdots + n)$$
$$T = 2\pi r\frac{n(n+1)}{2}$$

The number of nodes per unit length is:

$$N_{\text{avg}} = \frac{2N}{2\pi rn(n+1)}$$

The nodes at the outermost circle always perform sensing. The nodes of the inner circle perform sensing as well as routing of the data packets toward the sink nodes of the WSN. The largest pressure is on the innermost nodes which transfer data packets from all the nodes to the sink node.

Let N_1, N_2,... N_n refer to the nodes on the circumference of innermost to the outermost nodes, respectively, then

$$N_1 = 2\pi r N_{\text{avg}} = \frac{2N}{n(n+1)}$$
$$N_2 = 2\pi(2r)N_{\text{avg}} = \frac{4N}{n(n+1)}$$

Likewise

$$N_n = 2\pi(nr)N_{\text{avg}} = \frac{2nN}{n(n+1)}$$

It is easy to verify that N_1, N_2, ..., N_n add up to N.

Let

$$E_{\text{node}} = kPE_{\text{elec}} + kPE_{\text{amp}}r^2$$

It is the energy expenditure in each of the nodes of the network every second on account of sensing and transmission operation. Here, k is the number of read operations per second and P is the number of bits in the data packet that is formed as a consequence of the read operation.

For all the nodes in the hypothetical circles of the network, the energy consumption can be tabulated straightforwardly. Let

E_n = Energy consumption in all the nodes on the outermost circle of WSN

$$E_n = N_n * E_{\text{node}}$$

E_{n-1} = Energy consumption in all the nodes on the second outermost circle of WSN

$$E_{n-1} = N_{n-1} * E_{\text{node}} + N_n * E_{\text{node}} = (N_n + N_{n-1}) * E_{\text{node}}$$

The second term of RHS in the above expression is due to the fact that the nodes in the inner circle work as router for all the nodes in the outer circle. In the above expression, it is assumed that the energy consumption in the read and transmission of the sensed data is the same as the energy consumption in the receive-and-transmit operation of the node which works as router.

Similarly,

$$E_{n-2} = (N_n + N_{n-1} + N_{n-2}) * E_{\text{node}}$$

Thus, the energy consumption in the kth circle from the outermost ring is

$$E_{n-k} = (N_n + N_{n-1} + N_{n-2} + \cdots + N_{n-k}) * E_{\text{node}}$$

The above expression clearly indicates that with the growth in the density of the network, in multi-hop communication, the nodes which are near to the sink node rapidly depletes the battery power.

Thus,

$$E_{n-k} = \left[\frac{2nN}{n(n+1)} + \frac{2(n-1)N}{n(n+1)} + \cdots + \frac{2(n-1)N}{n(n+1)}\right] * E_{\text{node}}$$

$$E_{n-k} = \frac{2N}{n(n+1)}[n + (n-1) + \cdots + (n-k)] * E_{\text{node}}$$

$$E_{n-k} = \frac{2N}{n(n+1)}\left[\frac{n(n+1)}{2} - \frac{(n-k-1)(n-k)}{2}\right] * E_{\text{node}}$$

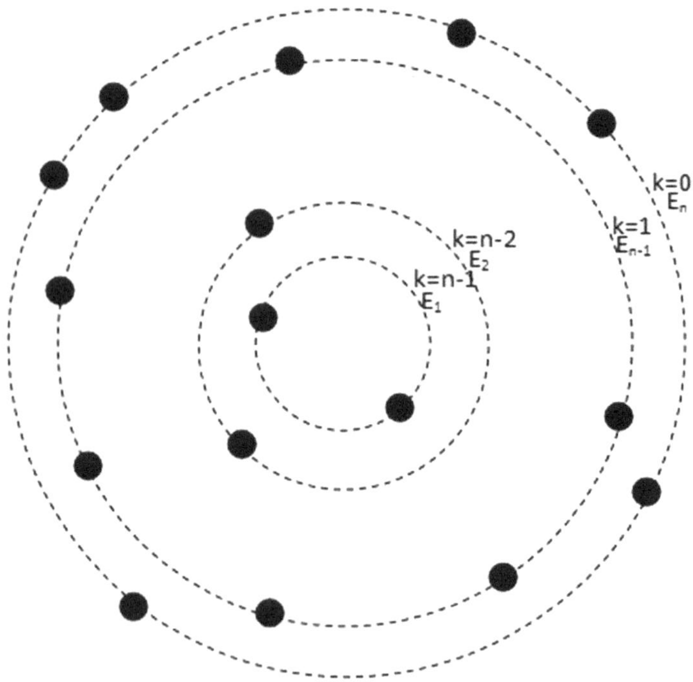

Fig. 4 Physical interpretation of the values of k and n

where $k \in I^+$ and $k \in [0, n-1]$.

The indicative values of k and n are shown in Fig. 4.

Here, $k = 1$ gives the cumulative energy dissipation for all the nodes of second outermost circle and $k = n - 1$ for the innermost nodes. It is important to note that the above expression gives the energy dissipation for nodes on the kth outermost ring.

The overall energy consumption for all the nodes in the innermost circle, corresponding to the one read operation by every node of the entire network is given by $k = n - 1$ in the above expression. This gives

$$E_1 = N * E_{\text{node}}$$

Comparing it with the energy dissipation rate of the nodes at the outermost periphery of the network, the following ratio can be obtained:

$$\eta = \frac{E_1}{E_n} = \frac{N}{N_n} = \frac{n+1}{2}$$

One important thing to note is that the expression is independent of N. This result is reasonable as the comparative energy consumption in all the innermost and outermost nodes is dependent only on the number of hops. For $n = 10$ circles (hops), the multi-hop communication without any network lifetime extension technique gives $\eta = 5.5$. In this configuration, the overall energy consumption in innermost nodes is about 5.5 times faster as compared to the outermost nodes.

It is evident that small transmission range results in more number of hops in hop-based communication, leading to more energy consumption in the nodes closer to the sink node. Large transmission range needs potent amplifiers and radio leading to large battery consumption per read operation. Also, large transmission range results in more energy depletion in the nodes which are situated in the boundary region. This is because the amplifier circuit needs energy in proportion to square of the transmission range. Also, for nodes with amplifiers having transmission range *R*, the transmitter and the receiver should be kept at a distance smaller than *R* so that the signals can be receipted under permissible SNR. If the distance is increased, it results in a decrease in SNR and there may be a need for retransmission. Therefore, an optimal transmission range and a dynamic routing strategies are essential for evenly battery consumption in all the nodes of the network so as prolong network lifetime.

3.3 Firefly Algorithm

Nature inspired algorithms; particularly, those based on swarm intelligence have found extensive applications in several computing applications in recent years. It appears that metaheuristics like firefly algorithm give much useful results as compared to traditional static configuration-based networks. Using these algorithms, one can achieve a global optimum solution as compared to a local optimum solution based on static or dynamic configurations based on traditional techniques. We will first outline the firefly algorithm and then focus on its application in the extension of network lifetime in WSN consisting of multi-hop communication configuration. Furthermore, we also explore the balance of exploration and exploitation of FA in given application.

Firefly algorithm was first developed by Xin-She Yang [3] in late 2007 at Cambridge University. It is based on the flashing pattern of fireflies. FA uses the following assumptions:

1. Fireflies are unisex so that one firefly will be attracted to other fireflies regardless of their sex.
2. The attractiveness is proportional to the brightness, and they both decrease as their distance increases. Thus, for any two flashing fireflies, the less bright one will move toward the brighter one. If there is no brighter one than a particular firefly, it will move randomly.
3. The brightness of a firefly is determined by the landscape of the objective function.

As a firefly's attractiveness is proportional to the light intensity seen by adjacent fireflies, we can now define the variation of attractiveness β with the distance r by

$$\beta = \beta_0 e^{-\gamma r^2}$$

where β_0 is the attractiveness at $r = 0$.

The movement of a firefly i is attracted to another more attractive (brighter) firefly j is determined by

$$x_i^{t+1} = x_i^t + \beta_0.e^{-\gamma r^2}\left(x_i^t - x_j^t\right) + \alpha_t \in_i^t$$

In the above equation, the second term is due to attraction and the third term is due to randomness. In the third term, the parameter α is the randomization parameter and $\in_i^t$ is a vector of random numbers drawn from a Gaussian distribution or uniform distribution at time t. If $\beta_0 = 0$, it becomes a simple random walk. On the other hand, if $\gamma = 0$, it reduces to a variant of particle swarm optimization. A simple way to model the randomness parameter is to use

$$\alpha_t = \alpha_0 \delta^t, 0 < \delta < 1$$

where t is the iteration counter, α_0 is the initial randomness scaling factor and δ is essentially a cooling factor. For most applications, we can use $\delta = 0.95$ to 0.97. The parameter β controls the attractiveness, and parametric studies suggest that $\beta_0 = 1$ can be used for most applications.

3.4 Proposed Work

To compensate for the uneven battery consumption in all the nodes of the network, a balanced energy consumption algorithm based on energy saving modes using firefly algorithm is proposed. The network is assumed to be homogenous and as per the assumptions given in Sect. 3.2. It is important to note that once a node becomes the personal area network coordinator (PANC), it starts depleting its battery much more rapidly as compared to the other nodes of the cluster. A node is chosen as PANC for a specific duration, called the steady-state phase. After this period, the setting up of new cluster starts again, and a new node is chosen as PANC.

In the proposed research, the residual energy threshold levels at which the node changes its behavior are obtained using FA. The scaling in this application is made over the threshold levels of residual battery life. The parameters α_0 and γ of the standard FA depends on the scaling variable L.

The proposed algorithm has the following specifications:

1. Let there be L fireflies generated using some distribution over [0, B] where 0 and B correspond to minimum and maximum battery level of the least powered node in each circle. Each firefly corresponds to a vector having n elements, obtain using the initial distribution. Let it be denoted by
   $L = [\rho_1, \rho_2, \ldots, \rho_L]$.

2. For each such vector, we can specify the following node behavior:
   (a) The number of reading operations per second.
   (b) Transmission range.

   For each of the node in any circle. Both (a) and (b) can also be defined as distribution over $[R_1, R_2]$ and $[T_1, T_2]$ where R_1 and R_2 are the minimum and maximum allowable read operations per second and T_1 and T_2 specifies minimum and maximum achievable transmission range.
3. Compute the objective function as follows:

$$\text{O.F.} = \rho/R + \rho/T$$

4. The objective function needs to be maximized in view to attain the maximum network lifetime.
5. There exists two threshold levels T_1 and T_2 ($T_1 > T_2$) for each node of the network.
6. For nodes belonging to a specific circular periphery, any of the nodes in the inner circle can be chosen as the PANC. The choice of the PANC is based on the residual battery life of the node. The node with the residual battery greater than T_1 is eligible to become PAN coordinator. Once the battery level of a node goes below T_1, the node will work either as sensor node or a relay node. This role is decided at the time of PAN setup. It is done using the PAN addressing mechanism. A 16-bit short address is assigned to each of the devices by the PANC at the time of establishment of the PAN. Every alternate address is marked as a relay node or a sensor. This addressing is dynamic, and hence, each of the nodes has the role of either sensor or relay node for a steady-state phase of the PAN coordinator and then has equal probability for entering into the other role in the next steady-state phase. Thus, for each of the cluster, fifty percent of the nodes become relay nodes and fifty percent of the nodes become sensing nodes after reaching the first threshold critical battery level.
7. When the battery level of the node goes below T_2, the number of read operations that the node has been configured to perform per second reaches the minimum permissible value, thus saving the battery life.

Practically, using low-cost hardware, the transmission range also fluctuates with the battery power. Most of the commercially available wireless motes come with the following two customization options:

1. The number of reading operations per second.
2. Transmission range.

In this research work, the same has been taken into account and analysis has been made on prolonged battery lifetime. The proposed algorithm is given below:

Algorithm: BECA-ESM

1. Initialize the variables E_1, E_2...., E_N representing the residual battery power of the nodes.
2. Maintain a list of possible neighbors at each of the nodes. The neighbors of a node are the nodes which are at one-hop distance from the given node. The sensing nodes are labeled provisionally as nodes or routers in the list.
   Initially, neighbor list $L_i = \{a_1, a_2.....a_k\}$
   Labeled neighbor list $= \{a_{1M}, a_{2M}.....a_{kM}\}$
   Here, M refers to the mark which indicates the role that the sensor node plays once the residual battery life reaches the critical level.
3. Each node, accepting the data packets from the other nodes, to be transited to the sink or the cluster head, aggregates data from all the nodes and the data sensed by its own and transmits the same to the next-hop node toward the sink node.
4. For all the devices operating in a PAN in any topology, each of the devices has an extended 64-bit address. This is unique address and embedded into the firmware of the device at the time of device manufacturing. Also, a short 16-bit address is also allocated to the device by the PAN coordinator. The short 16-bit address is allocated to each of the devices by adding 1 to the PAN coordinator address. Primarily, the addresses are marked as follows:

$$M = \begin{cases} S: The\, node\, will\, work\, as\, sensor\, if \\ 16\, bit\, dynamic\, address\, is\, even \\ R: The\, node\, will\, work\, only\, as\, a\, relay \\ if\, the\, 16\, bit\, dynamic\, address\, is\, odd \end{cases}$$

5. It is important to note that 16-bit dynamic address is assigned to the nodes at the time of initialization of the PAN. Thus, when the subsequent cluster head election takes place, the nodes receive new 16-bit addresses, which is the network address of the node.
6. For all the nodes having residual battery $E_k < T_1$, the role of the node will be exclusive as a sensor or as the relay node for the entire duration of the cluster operation.
7. For all the nodes having residual battery $E_k < T_2$, the number of reading operation of the node is reduced to the minimum specified threshold dependent on the application under consideration.

The network lifetime is the time when the first node of the network runs out of the battery life. It is computed as a function of the residual energy thresholds T_1 and T_2 and permissible minimum number of read operations per second for the sensing node.

4 Simulation Results

We consider the Xbee humidity sensor parameter specification for the simulation. The Xbee humidity sensor claims to have a battery life of 1.5 years with one read operation per 30 s [9]. Also, the application layer data packet size is 114 byte as in TinyOS and the energy consumption in the transmitter and the amplifier circuitry is 50 X10-9 J per bit 10X10-12 J per bit per meter, respectively. This computation suggests the initial energy level of each mote to be 4349 J. Also, the indoor/urban range is 40 m. Thus, the network deployment is considered in the form of concentric circles each having radius which is an integer multiple of 40. We assume that the nodes are uniformly placed over the circumference of the concentric circles. In this setting, the less number of nodes lies near the sink node and more is the number of nodes toward the outer circumferences. Figure 5 shows the density of nodes as a function of circle count and number of nodes.

We assume 1000 nodes network, arranged in 10 concentric circles, with identical linear density. The count of the nodes on each of the circle is depicted in Fig. 6.

In Fig. 6, the horizontal axis shows the circle ID and the vertical axis shows the node count. It is easy to verify that the sum of all the nodes over all the 10 circles is equal to one.

The energy overhead in each read operation by all the nodes of the network, as a function of circle ID, is shown in Fig. 7.

The circle with id = 1 refers to the circular region closest to the sink, and hence, has the responsibility to route all the packets from all the outer layers to the sink

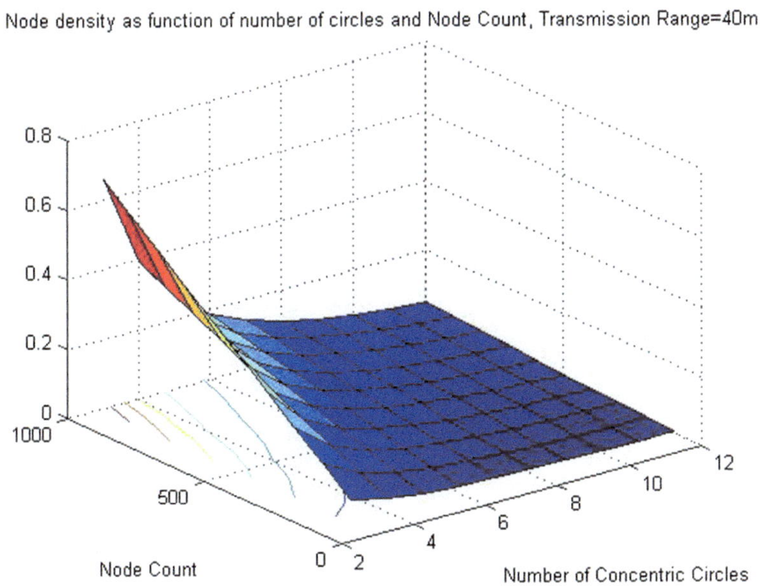

Fig. 5 Node density as function of node count and the number of concentric circles

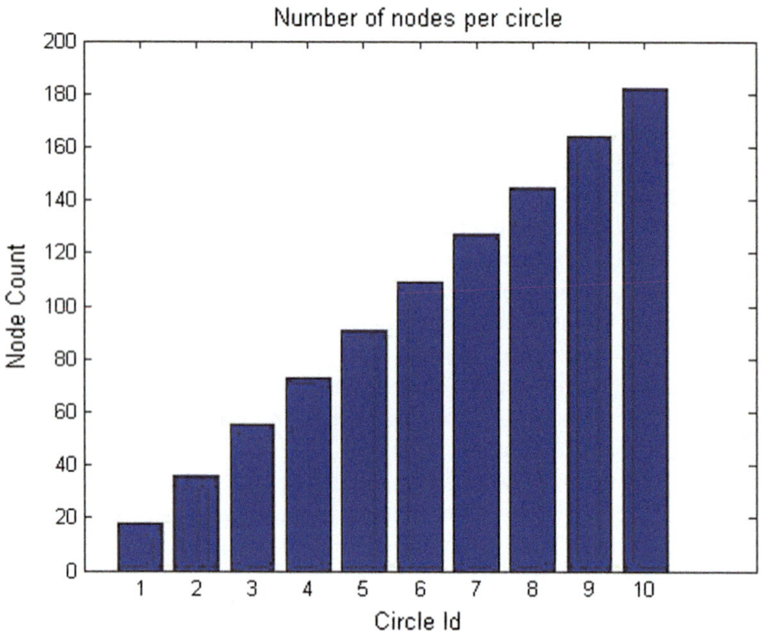

Fig. 6 Number of nodes distributed over the concentric circles in the network

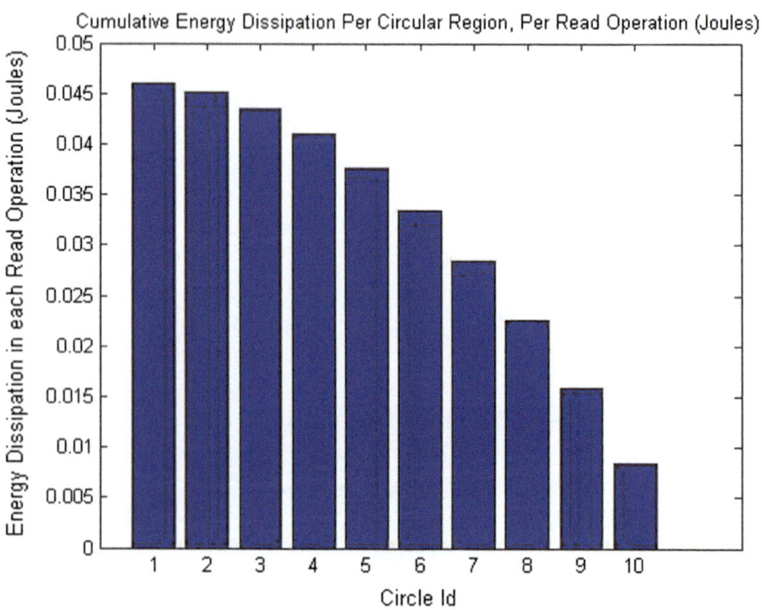

Fig. 7 Overall energy dissipation in each circular region per read operation

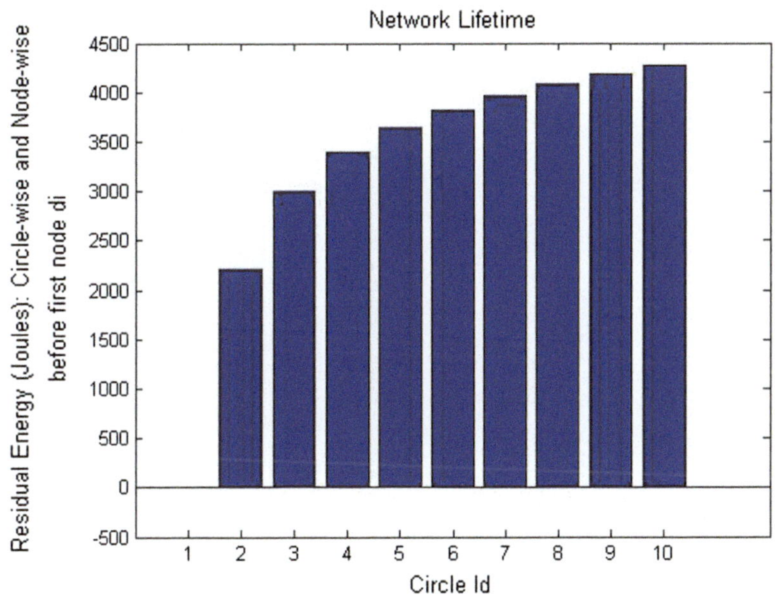

Fig. 8 Residual energy (cumulative) in all the nodes (circle-wise) after the first node runs out of the battery

node. Figure 7 indicates that largest energy expenditure is made in the layer closest to the sink node and least is the overhead on the nodes which are farthest from the sink node.

The residual energy of the nodes, after the very first node in the closest circle runs out of its battery life, without any energy-saving scheme is shown in Fig. 8.

It is clearly evident that in the absence of any energy-saving policy in the network, the nodes in the proximity of the sink node runs out of the battery very early. As indicated by Fig. 8, the nodes in the outermost circular region have almost full battery power, whereas the nodes near to the sink run entirely out of the battery. The above simulation gives a total of 1,703,086 read operations before the first node dies.

Figure 9 gives the residual energy status per node, when the first node dies, under the balanced energy consumption algorithm based on energy-saving modes. The node behavior is modeled as a function of remaining battery life. It specifies the role of the node as router or sensor with the number of reading operations decreased to a minimal extent.

The total number of reading operations achieved is 43,428,686. This clearly outperforms the number of read operations that are possible without any behavior modeling.

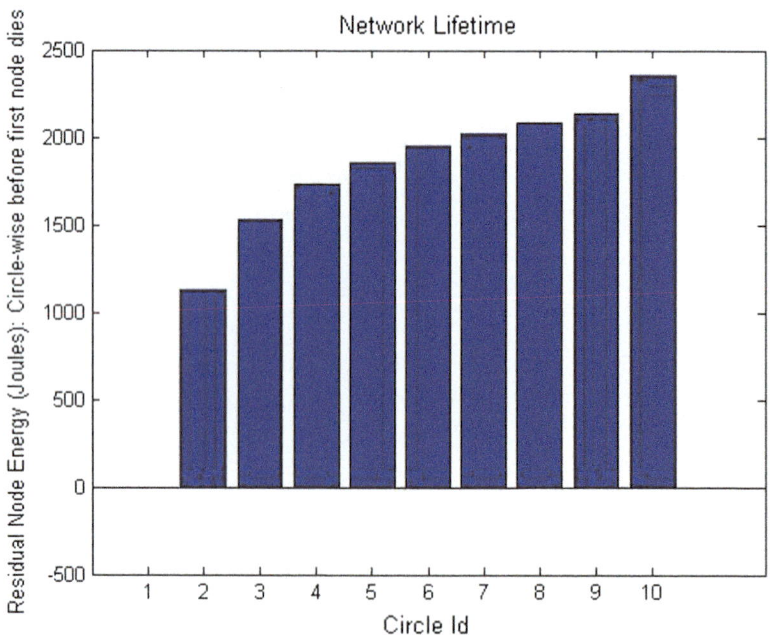

Fig. 9 Residual battery level of each node circle-wise, after the first node runs out of the battery

5 Conclusions

The present work focuses on balanced energy consumption considering clustering and energy consumption trade-off using the first-order radio energy dissipation model and provides analytical equation for the same. Based on this analysis, one can determine the optimal network configuration for a particular deployment of the WSN. This research follows the practical implementation approach that only the routing characteristic and the number of the read operation can be controlled through firmware programming at the time of network deployment. Some of the theoretical research in WSN is based on even controlling the data packet size as a function of remaining battery life [5] in the node. However, it is far from being feasible to implement this in the hardware provide the nodes run a minimum feature OS like Tiny OS.

As a future scope of the work, the specifics of the underlying hardware technology is also to be taken into account as the power dissipation varies significantly with the type of technology used for the realization of the sensor mote. However, the results presented in this analysis are of qualitative importance for every kind of realization for balanced energy consumption and hole alleviation.

References

1. Kothari S, Krishnan TSS, Ebenezer J, Satya Murty SAV (2015) Development of network debugging tools for IEEE 802.15.4 networks. In: 2015 Twelfth international conference on wireless and optical communications networks (WOCN), Bangalore, India, pp 1–5
2. Gavrilovska L, Krco S, Milutinovic V, Stojmenovic I, Trobec R (2011) Application and multidisciplinary aspects of wireless sensor networks: concepts, integration, and case studies. Springer, London, UK
3. Xia1 F, Yang LT, Wang L, Vinel A (2012) Internet of Things. Int J Commun Syst 25:1101–1102
4. Sharma S, Dr. Roy KC, Dr. Goyal D (2017) Balanced energy consumption model using clustering and energy consumption trade-off in wireless sensor networks. Int J Creative Res Thoughts (IJCRT) 5(4):58-67. ISSN 2320-2882
5. Heinzelman WR, Chandrakasan A, Balakrishnan H (2000) Energy-efficient communication protocol for wireless microsensor networks. In: IEEE proceedings of the Hawaii international conference on system sciences, January 2000, pp 1–10
6. Xue Y, Chang X, Zhong S, Zhuang Y (2014) An efficient energy hole alleviating algorithm for wireless sensor networks. IEEE Trans Consum Electron 60(3):347–355
7. Popescu AM, Salman N, Kemp AH (2014) Energy efficient geographic routing robust against location errors.. IEEE Sens J 14(6):1944–1951
8. Jan N et al (2017) A balanced energy consuming and hole alleviating algorithm for wireless sensor networks In: 2017 31st International conference on advanced information networking and applications workshops (WAINA). IEEE
9. Web URL. https://www.digi.com/products/xbee-rf-solutions/boxed-rf-modems-adapters/xbee-sensors#specifications

Chapter 5
Analysis of Modified LEACH Protocol in Wireless Sensor Network

Anurag Sharma, Jyoti Khandelwal, and Mukesh Kumar Choudhary

1 Introduction

Like living organisms, a ramification of present day devices and system's is predicated at the sensory information from the real international round it [1, 2]. These sensory records is provided by way of wireless sensor networks (WSN), which consists of several tiny sensor nodes to screen physical or environmental conditions, inclusive of temperature, vibration, stress, sound or motion, after which collectively ship those data to a vital computing system, referred to as the base station or sink [3]. Different routing protocols govern the motion of these records. Comprehensively, the directing conventions might be ordered as level-based absolutely steering, progressive-based thoroughly directing, and area-based steering. Low energy adaptive clustering hierarchy(LEACH) is a hierarchical-based totally routing protocol which makes use of random rotation of the nodes required to be the cluster heads to lightly distribute strength intake within the community [4]. The modern hobby in wireless sensor networks has introduced about the emergence of many utility orientated protocols of which LEACH is the most aspiring and considerably used protocol [5]. LEACH may be described as a combination of a cluster-based absolutely structure and multi-hop routing. The time period cluster-based totally absolutely can be defined through the fact that sensors using the LEACH protocol features are primarily based on cluster heads and cluster contributors [6]. Multi-hop routing is used for inter-cluster

A. Sharma · M. K. Choudhary (✉)
Rajasthan Institute of Engineering & Technology, Jaipur, India
e-mail: mukeshdesire@gmail.com

A. Sharma
e-mail: annurag.mudgal@gmail.com

J. Khandelwal
University of Engineering & Management, Jaipur, India
e-mail: jyoti.khandelwal19@gmail.com

© Springer Nature Singapore Pte Ltd. 2021

S. D. Purohit et al. (eds.), *Proceedings of International Conference on Communication and Computational Technologies*, Algorithms for Intelligent Systems,
https://doi.org/10.1007/978-981-15-5077-5_5

communiqué with cluster heads and base stations. Simulation effects shown in that multi-hop routing consumes less electricity at the same time as in assessment to direct transmission. We have stated that wireless sensors revel in records, mixture them after which deliver facts to the bottom station from a remote region the use of the radio transmission scheme as communiqué medium [7]. Data it is collected via the sensors and is dispatched to the bottom station. During this technique quite some elaborate issues arise, together with information collision and the information aggregation [8]. LEACH is well-proper to lessen the facts aggregation a trouble using a local information fusion which performs a compression of the quantity of facts this is gathered by the cluster head earlier than it sends it to the bottom station [9, 10]. All sensors form a self-prepared network with the aid of sharing the role of a cluster head as a minimum once. Cluster head is responsible for sending the statistics that is collected by the sensors to the bottom station [11]. It tries to stability the strength dissipation within the network and enhances the network's life time through enhancing the existence time of the sensors. The operations which are finished in the LEACH protocol are divided into two levels, the setup phase and steady phase [12, 13].

2 Problem Statement

On the off chance that it is distinguished that the circulation of hubs is not appropriate, a group may likewise have numerous hubs even as others may likewise have just a few. The bunch head (CH) of a huge assortment of hubs may furthermore in addition kick the bucket speedier. This marvel influences the proficiency of the system. Each drain and filter protection neglects to tackle this issue.

3 Proposed Methodology

In the proposed methodology, we prescribe another arrangement of guidelines that utilizes the idea of filter protection. The most new strategy is to distribute hubs to the nearest CH quality if the immense type of hubs effectively assigned to that CH is not exactly a particular sum X. On the off chance that the broad assortment is additional than x, a hub is allotted to the second one-closest CH, etc. X is determined through isolating the entire assortment of hubs with the guide of utilizing the wide assortment of CH. The changed-filter set of guidelines incorporates degrees: the setup segment and the regular nation segment. Inside the setup stage, the base station knows about the power notoriety and area of the majority of the hubs. The base station figures the normal intensity of the system. Every one of the hubs with quality more than the normal power is resolved on as CHs. The base station at that point assigns hubs to the closest CH handiest if the assortment of hubs effectively dispensed to the CH is significantly less than x. On the off chance that it is more than x, the base station at that point computes the second one-closest CH. In the event that the amount of hubs of the second CH is similarly more prominent than the amount x, the base station uncovers the 0.33 closest CH and the gadget keeps. When the hubs are assigned to the CH, the most reduced station sends to every hub the identifier in their CH. The equivalent quality channel on the CHs additionally will develope the execution and lifetime of the network. The standard nation segment is much the same as the consistent state fragment of the LEACH and LEACH-coverage conventions. The hubs convey insights to the CH. The CH totals the insights and advances it to the base station. After some predefined time, the resulting round begins and the setup and ordinary kingdom levels begin once again.

4 Results

4.1 *Simulation Parameters*

In this situation, we guess that a sensor organize spread into field of 100 X 100 m2. Vitality that is sufficient to transmit at far closures of field of 100 X 100 m^2 must be brought down multiple times for intra-bunch transmission. If a node elected as a cluster head, base station informs it to use high power amplification and in next round, when that node becomes a cluster member, routing protocol switches it to low-level power amplification simulations are conducted using MATLAB and to get precise plots, interval is taken.

Table 1 Simulation setup

Number of nodes	10, 20, 30, 40, 50
Topology	Dynamic
Simulation time	1000 s
MAC layer	802.11
Range	100 m
Simulation area	Simulation area^2
Routing protocol	AODV
Channel type	Wireless channel
Traffic model	Constant bit rate (CBR)
Packet size	512 Bytes
Interval	1 s

4.2 Performance Parameters

In order to evaluate the performance of our approach, we have used the following quality of services parameters:

1. **Throughput**: The throughput is the number of bytes transmitted or received per second. The throughput is denoted by T (Table 1),

$$\text{Throughput} = \text{received\_node}/\text{simulation\_time} \tag{1}$$

2. **Packet Delivery Ratio**: It is the ratio of the total number of data packets received by the destination nodes and the total number of data packets generated by the source nodes.

$$\text{PDR} = (\text{number\_of\_received\_packets}/\text{number\_of\_sent\_packets}) * 100 \tag{2}$$

3. **Average End-to-End Delay**: It can be defined as the time elapsed between the moment of sending of a bit by the source node and the moment of its reception by the destination node.

$$E\text{-}2\text{-}E\_\text{delay} = \text{received\_time-sent\_time} \tag{3}$$

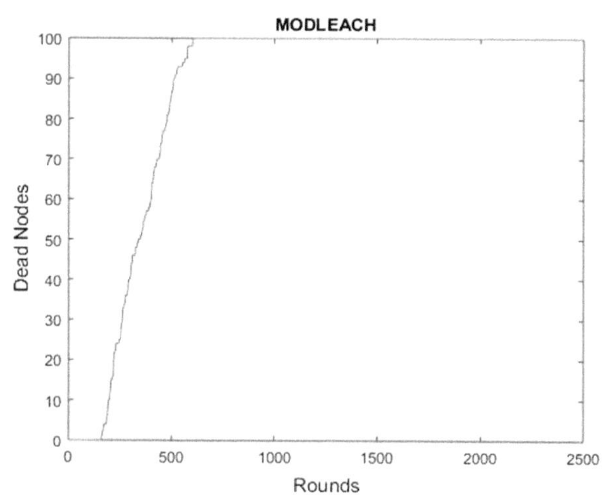

Fig. 1 Dead nodes in modified LEACH

4.3 Simulation Results and Analysis

1. **Dead Nodes in Modified LEACH**
   Figure 1 shows number of dead nodes is represented with the number of rounds. The performance curve of modified LEACH shows that the number of dead nodes are increasing as the number of rounds is increasing. In the beginning of simulation of modified LEACH there are no dead nodes because health's of the nodes are good due to power reserve in the beginning. As the number of iteration increases the energy of the power sources keep decreasing and the nodes getting dead per iteration. Toward the center of the absolute number adjusts the execution bends go astray; however, the execution of modified LEACH is great at limiting the quantity of dead hubs.
2. **Cluster Heads in Modified LEACH**
   In modified LEACH every node is related to some of the cluster because an individual node cannot transfer packer from source to destination if they are having very long distance. To resolve this problem in LEACH small no of clusters are created, which means group of nodes. Every cluster is having group of nodes to accomplish the data transmission in effective manner with optimum power utilization. Figure 2 shows that the no of cluster heads decreases in the proportion of number of rounds. The performance curve of modified LEACH shows that the no of cluster heads are increasing as the number of rounds is increasing. In the beginning, if simulations of modified LEACH there are sufficient no of cluster heads are present due to sufficient power reserve in the beginning (Fig. 3).
3. **Alive Nodes per round in Modified LEACH**
   Figure 1 shows number of alive nodes is represented with the number of rounds. The performance curve of modified LEACH shows that the no of dead nodes are increasing as the number of rounds is increasing. In the beginning of simulation

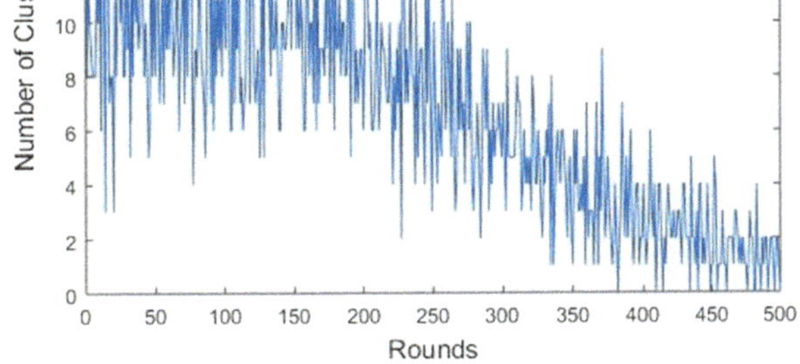

Fig. 2 Cluster heads in modified LEACH

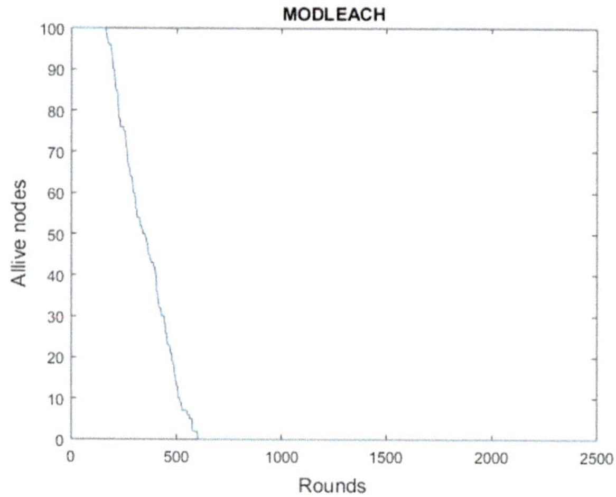

Fig. 3 Alive nodes per round in modified LEACH

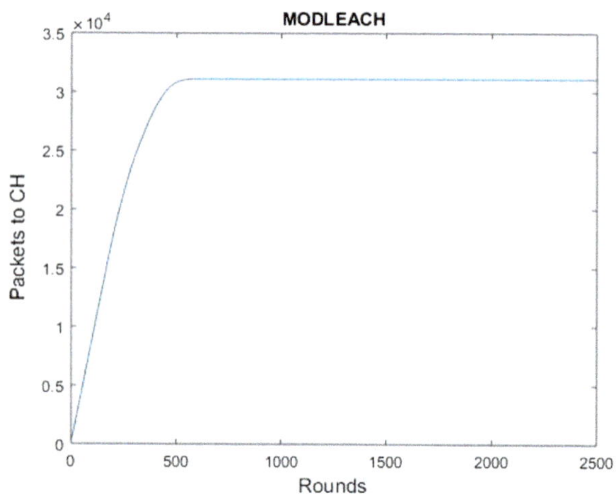

Fig. 4 Packets in CH in modified LEACH

of modified LEACH, there are 100 alive nodes because health of the nodes is good due to power reserve in the beginning. As the quantity of cycle builds the vitality of the power sources continue diminishing and the hubs getting dead per emphasis

4. **Packets in CH in Modified LEACH**
   The quantity of dead hubs to the quantity of rounds demonstrating that the quantity of dead hubs is littler for MM-LEACH than for LEACH. At cycle 700, all LEACH hubs are dead, yet a portion of the hubs stay live at cycle 2500 in MM-LEACH. The lifetime of the system is expanded in MM-LEACH. As appeared in Fig. 4, on the grounds that the lifetime of system is expanded by the nearness of the staying live hubs, CH determination is additionally executed until cycle 2000 in MM-LEACH, as is plainly plotted in Fig. 5.

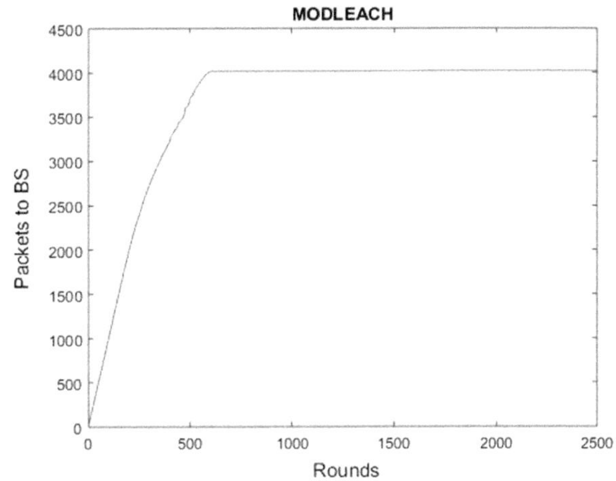

Fig. 5 Packets in BS in modified LEACH

5. **Packets in BS in Modified LEACH**
   Figure 5 unmistakably demonstrates the total number of parcels transmitted to the BS as a component of the quantity of rounds. The execution bend of modified LEACH coordinates that of LEACH for little quantities of rounds. Toward the center of the absolute number adjusts, the execution bends go astray, yet the execution of modified LEACH is great at limiting the quantity of dead hubs.

5 Conclusion

From our side, we presumed that if the enhancements for cluster head method and multi-hop procedure are incorporated with existing systems, the sensor system would have been considerably more vitality proficient. They can decrease the vitality utilization consequently increment the existence time. In this paper, we have overviewed different LEACH-based conventions and talked about how they lessen vitality utilization in WSNs and increment system's lifetime. In this task work, we give a concise exchange on rise of bunch-based directing in remote sensor systems. We too propose protocol, another variation of LEACH that can additionally be used in other bunching steering conventions for better proficiency. Proposed scheme will in general limit organize vitality utilization by acquainting strategy with raise arrange lifetime and throughput. In this plan, partitioning the system into four zone, one zone legitimately imparted to base station and staying three zones conveyed to base stations by means of entryway hub. In the trial, it is demonstrated that the proposed convention performs superior to LEACH. The outcome demonstrates that approximates twice improvement in LEACH and about 40% improvement appeared at the L-LEACH. In the L-LEACH convention, organize life time is around 1600 round, and in my proposed convention, arrange life time is around 2500 round. One of the principle challenges in the structure of directing conventions for WSNs is the way to effectively utilize the accessible vitality, since vitality assets are restricted. The primary point while plan a steering convention is to make the hubs lifetime to the extent that this would be possible. From our side, we presumed that if the enhancement for cluster head system, and multi-hop strategy is incorporated with existing strategies, the sensor system would have been significantly more vitality productive. They can decrease the vitality utilization accordingly increment the existence time. In this paper, we have studied different LEACH-based conventions and examined how they decrease vitality utilization in WSNs and increment system's lifetime. In this task work, we give a short exchange on rise of bunch-based steering in remote sensor systems. We likewise propose protocol, another variation of LEACH that can additionally be used in other bunching steering conventions for better proficiency.

6 Future Work

In the future work of the modified LEACH, we can implementation of proposed convention on heterogeneous remote sensor systems. Next progress can be conceivable by considering sink versatility and to guarantee fruitful conveyance of information. Plan of a superior directing convention on the off chance that when CH kicks the bucket before sending the information to the BS.

References

1. Dipakkumar JS, Srivastava AK, Vithlani SK (2016) Simulation based study of gray hole attack in MANET. In: 2016 3rd International conference on computing for sustainable global development (INDIACom), New Delhi, pp 3529–3534
2. Ahmed M, Hussain MA (2014) Performance of an IDS in an adhoc network under black hole and gray hole attacks. In: International conference on electronics, communication and instrumentation (ICECI), Kolkata, pp 1–4. https://doi.org/10.1109/ICECI.2014.6767377
3. Alheeti KMA, Gruebler A, McDonald-Maier KD (2015) On the detection of grey hole and rushing attacks in self-driving vehicular networks. In: 7th Computer science and electronic engineering conference (CEEC), Colchester, pp 231–236
4. Dharini N, Balakrishnan R, Renold AP (2015) Distributed detection of flooding and gray hole attacks in Wireless Sensor Network. In: 2015 International conference on smart technologies and management for computing, communication, controls, energy and materials (ICSTM), Chennai, pp 178–184
5. Dhende S, Musale S, Shirbahadurkar S, Najan A (2017) SAODV: black hole and gray hole attack detection protocol in MANETs. In: 2017 International conference on wireless communications, signal processing and networking (WiSPNET), Chennai, pp 2391–2394
6. Dongare SP, Mangrulkar RS (2015) Implementing energy efficient technique for defense against Gray-Hole and Black-Hole attacks in wireless sensor networks. In: 2015 International conference on advances in computer engineering and applications, Ghaziabad, pp 167–173
7. Soundaravalli D (2017) Identify intrigue Blackhole and Greyhole attacks in prorogation sophisticated networks. In: 2017 International conference on technical advancements in computers and communications (ICTACC), Melmaurvathur, pp 160–162

8. Geetanjali Gupta J (2017) Improved approach of co-operative gray hole attack prevention monitored by meta heuristic on MANET. In: 2017 4th International conference on signal processing, computing and control (ISPCC), Solan, pp 356–361
9. Kaur G (2016) A preventive approach to mitigate the effects of gray hole attack using genetic algorithm. In: 2016 International conference on advances in computing, communication, & automation (ICACCA) (Spring), Dehradun, pp 1–8
10. Khattak H, Nizamuddin Khurshid F, ul Amin N (2013) Preventing black and gray hole attacks in AODV using optimal path routing and hash. In: 10th IEEE international conference on networking, sensing and control (ICNSC), Evry, pp 645–648
11. Kumar S, Doohan NV (2016) A modified approach for recognition and eradication of extenuation of gray-hole attack in MANET using AODV routing protocol. In: Symposium on colossal data analysis and networking (CDAN), Indore, pp 1–5
12. Lupia A, De Rango F (2015) Energy consumption evaluation of SAODV with trust management scheme under gray-hole attacks. Wireless telecommunications symposium (WTS), New York, NY, pp 1–8
13. Pal S, Sikdar B, Chow JH (2018) An online mechanism for detection of gray-hole attacks on PMU data. IEEE Trans Smart Grid 9(4):2498–2507 July

Chapter 6
Recent Soft Computing Strategies for Induction Motor Analysis and Control: A Survey

Fani Bhushan Sharma and Shashi Raj Kapoor

1 Induction Motor Executions Using Recent Soft Computing Strategies

Various operations are carried out on IMs using conventional and non-conventional strategies. The significant work in executing IMs is reviewed in this section. The development of any society is based on the availability of electrical energy [3]. The electrical energy is mainly utilized by electrical motors [9, 27]. Monitoring of energy conservation of motors is quite a significant study. IMs are robust in design and require less maintenance, small area, and are easily controllable [19].

Due to its significance in the development of the society, the analysis and control of IMs become quite important. The analysis of IM is carried out using parameter estimation. The iterative methods of parameter estimation are presented by various researchers [15, 19, 25, 32]. The swarm intelligence (SI) motivated algorithms are iterative in nature. As reported in literature, the SI-based algorithms are performing well to solve real-world optimization problems. So, here also the SI-based algorithms are applied to solve parameter estimation of IM as well namely, ABC [1], GABC 1.5 [1], IABC [1], MABC [1], ModABC [1], BSFABC [1], EABC [1], CH-EABC [1]. The outcomes presented by these SI-based algorithms are quite promising and these outcomes established the applicability of SI-based algorithms in the analysis of IMs.

Further, the control of IM is carried out using speed control and vector control. The IMs speed control allows it to be used for an individual operation. The control of IMs extensive dynamic environment quite tedious. The nonlinearities perturbation in IM parameters due to rotor resistance alters the executing situations [10]. The IM on application of field-oriented control (FOC) or vector control gets decoupled torque and flux characteristics which leads to individualistic control of the torque and flux

F. B. Sharma (✉) · S. R. Kapoor
Rajasthan Technical University, Kota, India
e-mail: fbsharma.kota@gmail.com

© Springer Nature Singapore Pte Ltd. 2021

S. D. Purohit et al. (eds.), *Proceedings of International Conference on Communication and Computational Technologies*, Algorithms for Intelligent Systems,
https://doi.org/10.1007/978-981-15-5077-5_6

for a separately excited DC motor. Though the FOC strategies are fascinating but they lack with a basic problem (Their proneness to motor parametric variations such as the rotor time constant and an incorrect flux measurement or estimation at low speeds [34]).

A mathematical model is necessary to conventionally control the system. The performance of the system is not satisfactory, but it alters from the standard performance system parametric variations or environment disturbances [6]. The evaluation of mathematical modeling is quite tedious. To overcome this, identification strategies like system identification is applied to get exact mathematical model of the system. This increases the complexity and cost to implement the conventional control strategies. The implementation time in case of conventional control is also high. This ultimately reduces the reliability of the control system [6].

In expectation to increase the reliability, fuzzy-based techniques have been applied by researchers. Further, the negative impedance instability problem is introduced while performing efficient torque control of IM drives in combination with resonant DC-link input filters. The solution of above problem is proposed with application of the concept of nonlinear system stabilizing controller with plant [23]. The various variation schemes of amplitude and frequency of the basic supply voltage facilitates several significant control methods for IMs, namely scalar control, vector control or field-oriented control, direct torque and flux control, sliding mode control, and the adaptive control to operate at the steady state. An improved $V \div f$ control strategy for high voltage IMs and its stability was developed in 1998 [4].

Some other non-conventional techniques have also been practiced by various researchers [7, 8, 17, 22, 31]. The above methods were having limitations of unusual settling times and proper selection of rule base. In the year 2013, M. M. Eissa et al. applied particle swarm optimization (PSO) algorithm for speed control of IMs [13]. The simulation result is shown in Table represents a lesser settling and rise time by ABC and logistic ABC as compared to Mamdani control strategy [18] and PSO speed control algorithm [13].

The first implementations of vector control technology were essentially laboratory curiosities since the computing power did not exist in an embedded form to make this technique viable. But as fortune would have it, another technology was being developed that would change the face of motor control for ever the microprocessors. It was not long before several manufacturers began offering microprocessor-based vector control solutions over a wide range of power options. Soon the term vector control became synonymous with high performance and high tech motors. Later, a less ambiguous and more technically precise term was coined for it as field-oriented control (FOC).

Considering the enhanced performance and flexibility possible with FOC, it may very well turn out to be the most cost-effective solution for your application. This is why many low cost and high volume applications (such as appliances) are abandoning trapezoidal control altogether and embracing FOC faster than ever before. Various researchers have worked on the FOC as available in the literature. The control of IM was carried out for standard AC motor using microprocessors in year 1980 [14]. A strategy for improving reliability of field-oriented controlled IM was proposed in year 1993 [20]. Further, digital FOC for dual three-phase IM was proposed in year 2003 [5]. In year 2004, backstepping wavelet neural network control for indirect FOC of IM was presented [33]. Fuzzy self-tuning speed control of an indirect field-oriented control induction motor was presented in year 2008 [21]. In year 2007, GAPSO-based vector control of indirect three-phase induction motor was presented [16]. FOC has been implemented on stator-flux-oriented vector control for brushless doubly fed induction generator [28]. Various techniques for energy-efficient control of three-phase IM were discussed in year 2009 [26]. Genetic algorithm-based fuzzy speed controllers for indirect FOC of IM were proposed [11]. Optimized solutions for an optimization technique based on minority charge carrier inspired algorithm applied to selective harmonic elimination in IMs [30]. Artificial bee colony-based design of optimal online self-tuning PID controller was proposed in year 2014 [12]. In year 2015, IM control was supplied by wind turbine via imperialist competitive algorithm [2]. The intelligent modeling of three-phase induction machine was presented in the same year [24]. The PIDA controller design was proposed by spider monkey optimization algorithm in year 2016 [29]. As per the reviewed literature, the conventional strategies applied so far were doing well in the area of analysis and control of IMs. But with the advent of soft computing strategies, like artificial neural network, fuzzy logic control, and support vector machine, it is found that these strategies are performing far better in terms of accuracy of the results. Further, the applicability of recent soft computing strategies like SI-based ABC and PSO algorithms it is demonstrated that SI-based algorithms performed exceptionally well in this area.

2 Conclusion

This paper presents a review of recent soft computing strategies applied for the analysis and control of induction motors (IMs). As the IMs are playing a significant role in the area of societal development, it is necessary to understand and enlighten the analyzing and controlling requirements and strategies of IMs in recent years. Here, in this paper, the significant cutting edge strategies are reviewed and it is found that though the IMs were executed well by many conventional analysis and control methods; yet, in recent years, the non-conventional strategies have been incorporated to enhance the accuracy in operations. The soft computing strategies like neural network, fuzzy logic, and support vector machine performed very well to control the speed of IMs by optimizing the controller parameters. They also performed exceptionally well in the estimation of IM parameters. Further, it is found that the swarm intelligence (SI)-based algorithms were also applied in search of more better performances and it can be concluded that among many SI-based algorithms artificial bee colony algorithm (ABC) and particle swarm optimization algorithm (PSO) are outperforming than all other available strategies.

References

1. Abro AG, Saleh JM (2014) Multiple-global-best guided artificial bee colony algorithm for induction motor parameter estimation. Turkish J Electr Eng Comput Sci 22(3):620–636
2. Ali ES (2015) Speed control of induction motor supplied by wind turbine via imperialist competitive algorithm. Energy 89:593–600
3. Bazmi AA, Zahedi G (2011) Sustainable energy systems: role of optimization modeling techniques in power generation and supply–A review. Renew Sustain Energy Rev 15(8):3480–3500
4. Ben-Brahim L (1998) Improvement of the stability of the v/f controlled induction motor drive systems. In: IECON'98. Proceedings of the 24th annual conference of the IEEE industrial electronics society (Cat. No. 98CH36200), vol 2. IEEE, pp 859–864
5. Bojoi R, Lazzari M, Profumo F, Tenconi A (2003) Digital field-oriented control for dual three-phase induction motor drives. IEEE Trans Ind Appl 39(3):752–760
6. Bose BK (2009) Power electronics and motor drives recent progress and perspective. IEEE Trans Ind Electron 56(2):581–588
7. Bose BK, Patel NR, Rajashekara K (1997) A neuro-fuzzy-based on-line efficiency optimization control of a stator flux-oriented direct vector-controlled induction motor drive. IEEE Trans Ind Electron 44(2):270–273
8. Chen T-C, Hsu J (1994) A fuzzy sliding mode controller for induction motor position control. In: 20th International conference on industrial electronics, control and instrumentation, IECON'94, vol 1. IEEE, pp 44–49
9. Dandil E et al (2013) Artificial immunity-based induction motor bearing fault diagnosis. Turkish J Electr Eng Comput Sci 21(1):1–25
10. Denai MA, Attia SA et al (2002) Fuzzy and neural control of an induction motor. Appl Math Comput Sci 12(2):221–234
11. Douiri MR, Cherkaoui M, Essadki A (2012) Genetic algorithms based fuzzy speed controllers for indirect field oriented control of induction motor drive. Int J Circ Syst Sig Process 6(1):21–28
12. Ebrahim EA (2014) Artificial bee colony-based design of optimal on-line self-tuning PID-controller fed AC drives. Int J Eng Res 3(12):807–811
13. Eissa MM, Virk GS, AbdelGhany AM, Ghith ES (2013) Optimum induction motor speed control technique using particle swarm optimization. Int J Energy Eng 3(2):65–73
14. Gabriel R, Leonhard W, Nordby CJ (1980) Field-oriented control of a standard ac motor using microprocessors. IEEE Trans Ind Appl 2:186–192
15. Gupta RA, Wadhwani AK, Kapoor SR (2011) Early estimation of faults in induction motors using symbolic dynamic-based analysis of stator current samples. IEEE Trans Energy Convers 26(1):102–114
16. Kim DH (2007) GA-PSO based vector control of indirect three phase induction motor. Appl Soft Comput 7(2):601–611
17. Kim D-K, Kim H-S, Kim J-M, Won C-Y, Kim S-C (1996) Induction motor servo system using variable structure control with fuzzy sliding surface. In: Proceedings of the 1996 IEEE IECON 22nd international conference on industrial electronics, control, and instrumentation, vol 2. IEEE, pp 977–982
18. Ashok Kusagur, SF Kodad, and B V Sankar Ram. Ai based design of a fuzzy logic scheme for speed control of induction motors using svpwm technique. *Int. Jr. Comp. Sci. & Network Security*, 9(1):74–80, 2009
19. Lindenmeyer D, Dommel HW, Moshref A, Kundur P (2001) An induction motor parameter estimation method. Int J Electr Power Energy Syst 23(4):251–262
20. Liu T-H, Fu J-R, Lipo TA (1993) A strategy for improving reliability of field-oriented controlled induction motor drives. IEEE Trans Ind Appl 29(5):910–918
21. Masiala M, Vafakhah B, Salmon J, Knight AM (2008) Fuzzy self-tuning speed control of an indirect field-oriented control induction motor drive. IEEE Trans Ind Appl 44(6):1732–1740

22. Mohamed HAE, Lau EL, Yang SS Moghavvemi M (2008) Fuzzy-SMC-PI flux and speed control for induction motors. In: 2008 IEEE conference on robotics, automation and mechatronics. IEEE, pp 325–330
23. Mosskull H, Galic J, Wahlberg B (2007) Stabilization of induction motor drives with poorly damped input filters. IEEE Trans Ind Electron 54(5):2724–2734
24. Noei AR, Kholerdi HA (2015) Practical intelligent modeling of three-phase induction machine. In 2015 7th Conference on information and knowledge technology (IKT). IEEE, pp 1–6
25. Pedra J, Corcoles F (2004) Estimation of induction motor double-cage model parameters from manufacturer data. IEEE Trans Energy Convers 19(2):310–317
26. Raj CT, Srivastava SP, Agarwal P (2009) Energy efficient control of three-phase induction motor-a review. Int J Comput Electr Eng 1(1):61
27. Sakthivel VP, Bhuvaneswari R, Subramanian S (2011) An accurate and economical approach for induction motor field efficiency estimation using bacterial foraging algorithm. Measurement 44(4):674–684
28. Shao S, Abdi E, Barati F, McMahon R (2009) Stator-flux-oriented vector control for brushless doubly fed induction generator. IEEE Trans Ind Electron 56(10):4220–4228
29. Sharma A, Sharma H, Bhargava A, Sharma N (2016) Optimal design of pida controller for induction motor using spider monkey optimization algorithm. Int J Metaheur 5(3–4):278–290
30. Sreedhar M, Upadhyay N-M, Mishra S (2012) Optimized solutions for an optimization technique based on minority charge carrier inspired algorithm applied to selective harmonic elimination in induction motor drive. In: 2012 1st International conference on recent advances in information technology (RAIT). IEEE, pp 788–793
31. Takagi T, Sugeno M (1985) Fuzzy identification of systems and its applications to modeling and control. IEEE Trans Syst Man Cybern 1:116–132
32. Toliyat H, Levi E, Raina M et al (2003) A review of rfo induction motor parameter estimation techniques. IEEE Trans Energy Convers 18(2):271–283
33. Wai R-J, Chang H-H (2004) Backstepping wavelet neural network control for indirect field-oriented induction motor drive. IEEE Trans Neural Netw 15(2):367–382
34. Yi W, Kaiqi Z (2005) Field-oriented vector control of induction motor for electric vehicles. In: 31st Annual conference of IEEE industrial electronics society, 2005. IECON 2005, IEEE, p 5

Chapter 7
Protection of Wind Farm Integrated Double Circuit Transmission Line Using Walsh–Hadamard Transform-Based Artificial Intelligence Technique

Gaurav Kapoor

1 Introduction

The faults on wind farm integrated double circuit transmission lines (WFIDCTLs) have to be identified quickly so as to renovate the faulted phase, restore the electricity supply, and decrease the outage time as soon as possible. In the latest years, lots of studies have been dedicated to the problem of fault detection, classification, and location estimation in the DCTLs [1–7]. A brief literature review of various recently reported methods is introduced, henceforth. In [1], authors presented the fault detection and location method for the protection of three terminals DCTL. In [2], the WT has been introduced for the protection of DCTL. HHT has been applied in [8] for the WFISCCTL. In [3, 9], authors employed WT for the protection of SCCDCTL and TPTL, respectively. WT and MLP have been used in [4] for locating faults in DCTLs. In [5, 6], respectively, MM and DPA have been used for the DCTLs. In [7], a fault position estimation technique has been illustrated for DCTL compensated with capacitors. In [10] and [11], a technique based on WT has been suggested for six-phase transmission line.

In this work, the Walsh–Hadamard transform (WHT) is utilized and it is executed for the detection of faults in the WFIDCTL. Such type of work has not been depicted so far to the best of the information of the author. The results exemplify that the WHT competently detects the faults, and the uniformity of the WHT is not sensitive to the deviation in fault factors.

This article is structured as: Sect. 2 reports the specifications of the WFIDCTL. Section 3 describes the flowchart for the WHT. Section 4 presents the outcomes of the investigations which are carried out in this work. Section 5 concludes the article.

G. Kapoor (✉)
Department of Electrical Engineering, Modi Institute of Technology, Kota, India
e-mail: gaurav.kapoor019@gmail.com

© Springer Nature Singapore Pte Ltd. 2021

S. D. Purohit et al. (eds.), *Proceedings of International Conference on Communication and Computational Technologies*, Algorithms for Intelligent Systems,
https://doi.org/10.1007/978-981-15-5077-5_7

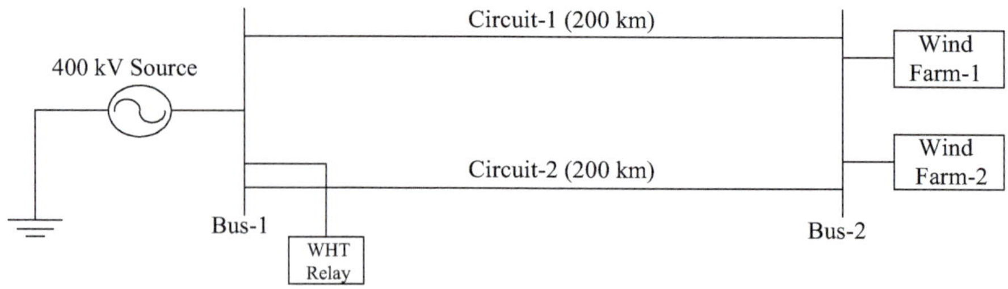

Fig. 1 Schematic of WFIDCTL

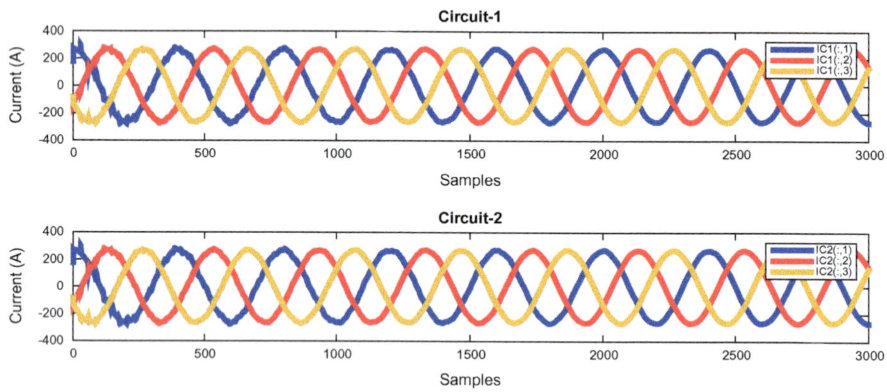

Fig. 2 Circuit-1 and circuit-2 currents for no-fault

2 The Specifications of WFIDCTL

Figure 1 depicts the schematic of WFIDCTL. The WFIDCTL has a rating of 400 kV, 50 Hz. The WFIDCTL has a total length of 200 km. The WFIDCTL is alienated into two zones of length 100 km each. The current measurement blocks are connected at bus-1 for measuring the currents of both circuits of the WFIDCTL. The simulation model of a 400 kV WFIDCTL is designed using MATLAB.

Figure 2 illustrates the currents of circuit-1 and circuit-2 for no-fault. The WHT coefficients of circuit-1 currents for no-fault are shown in Fig. 3. Figure 4 exemplifies the WHT coefficients of circuit-2 currents for no-fault. Table 1 reports the results of WHT for no-fault.

3 Walsh–Hadamard Transform (WHT)

Figure 5 illustrates the process for the WHT. The steps for the same are shown beneath.

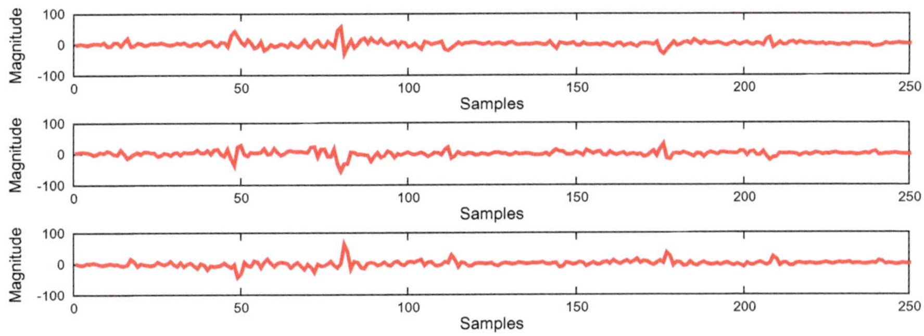

Fig. 3 WHT coefficients of circuit-1 currents for no-fault

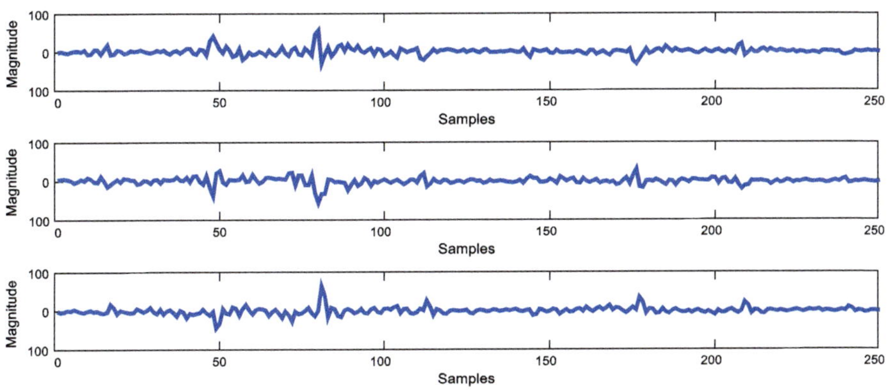

Fig. 4 WHT coefficients of circuit-2 currents for no-fault

Table 1 Results of WHT for no-fault

WHT coefficients					
Phase-A1	Phase-B1	Phase-C1	Phase-A2	Phase-B2	Phase-C2
56.5056	30.2316	63.4948	56.5056	30.2316	63.4948

Step 1 Simulate the WFIDCTL for creating faults and produce the post-fault currents for both circuits.

Step 2 Use WHT to examine the post-fault currents of both circuits for characteristics retrieval and determine the range of WHT coefficients.

Step 3 The phase will be proclaimed as the faulted phase if its WHT coefficient has the larger amplitude as compared to the healthy phase.

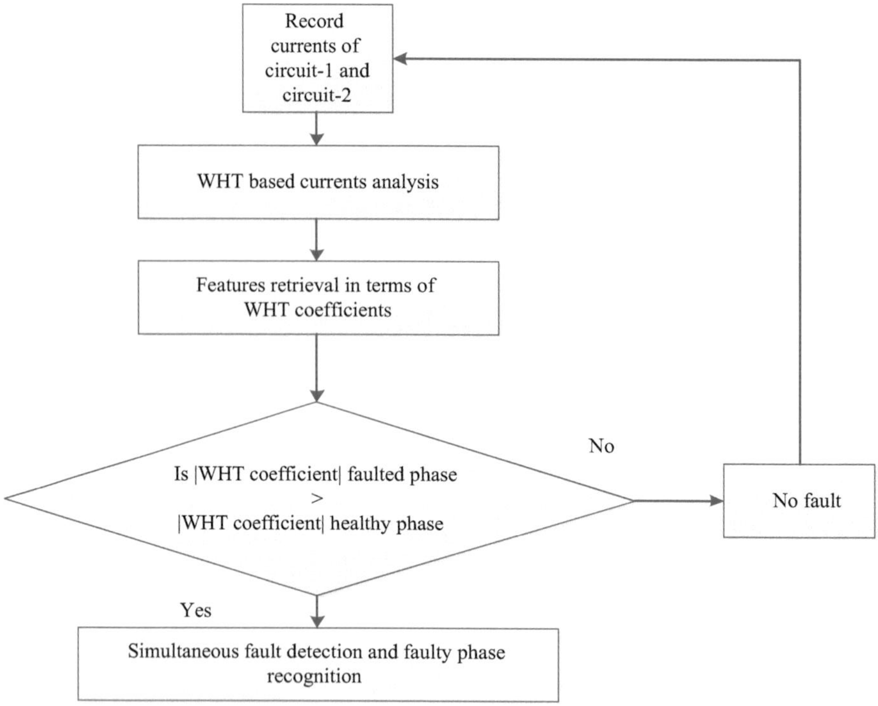

Fig. 5 Schematic for WHT

4 Performance Valuation

The simulation studies have been carried out for the near-in relay faults, far-end relay faults, converting faults and cross-country faults with the objective of verifying the feasibility of the WHT. In the separate subcategories, the outcomes of the work are investigated.

4.1 The Efficacy of WHT for the Near-in Relay Faults

The efficiency of the WHT is investigated for the near-in relay faults on the WFIDCTL. Figure 6 exemplifies the circuit-1 and circuit-2 currents for A1G near-in relay fault at 5 km at 0.04 s among $R_F = 0.5\ \Omega$ and $R_G = 0.25\ \Omega$. Figure 7 illustrates the WHT coefficients for circuit-1 currents. Figure 8 exemplifies the WHT coefficients for circuit-2 currents. The fault factors for the other fault cases are: $T = 0.04$ s, $R_F = 0.5\ \Omega$, and $R_G = 0.25\ \Omega$. Table 2 details the results of the WHT for the five different near-in relay faults. It is confirmed from Table 2 that the WHT has the ability to detect the near-in relay faults precisely.

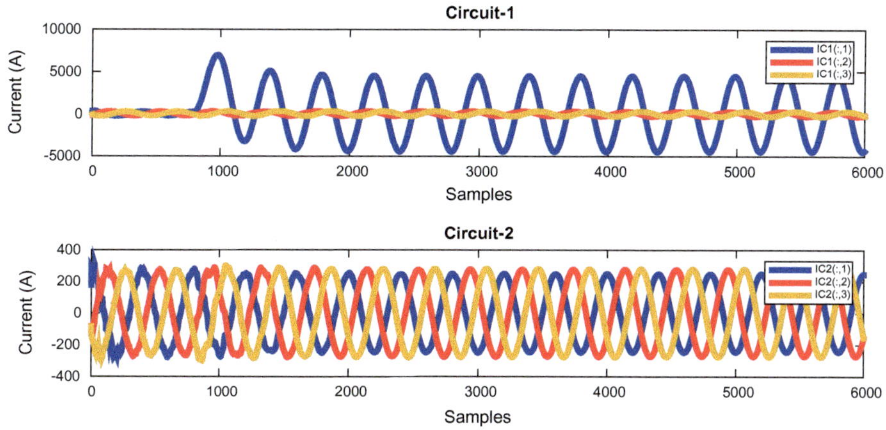

Fig. 6 Circuit-1 and circuit-2 currents for A1G near-in relay fault at 5 km at 0.04 s

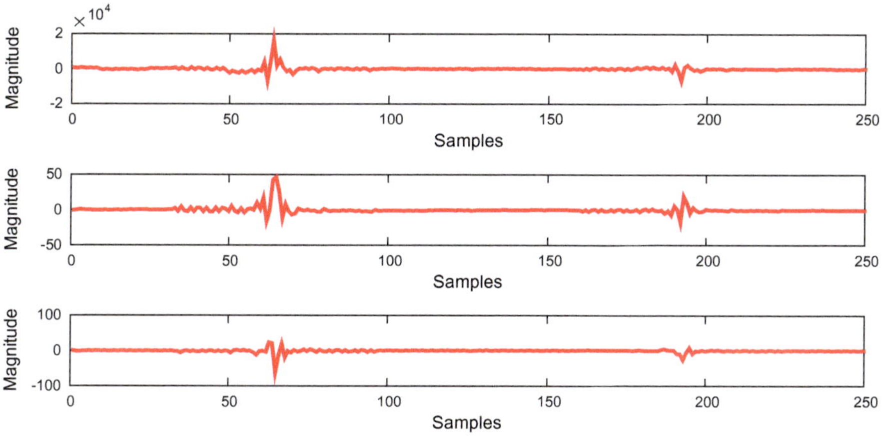

Fig. 7 WHT coefficients of circuit-1 currents for A1G near-in relay fault at 5 km

4.2 *The Efficacy of WHT for the Far-End Relay Faults*

The WHT has been explored for the far-end relay faults. Figure 9 illustrates the circuit-1 and circuit-2 currents for A1B1C1G far-end relay fault at 190 km at 0.0715 s with $R_F = 1.15\ \Omega$ and $R_G = 1.45\ \Omega$. The fault factors chosen for the further fault cases are: $T = 0.0715$ s, $R_F = 1.15\ \Omega$, and $R_G = 1.45\ \Omega$. Table 3 presents the results for the five cases of the far-end relay faults. It is inspected from Table 3 that the effectiveness of the WHT remains unaffected for the far-end relay faults (Figs. 10 and 11).

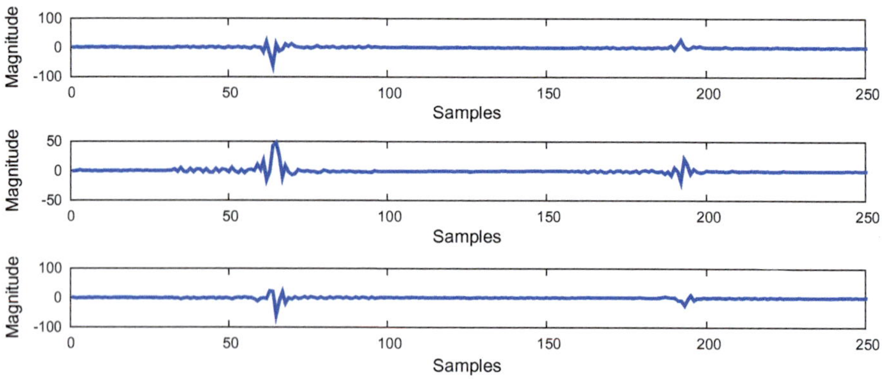

Fig. 8 WHT coefficients of circuit-2 currents for A1G near-in relay fault at 5 km

Table 2 Results of WHT for different near-in relay faults

Fault type	WHT coefficients					
	Phase-A1	Phase-B1	Phase-C1	Phase-A2	Phase-B2	Phase-C2
A1G (5 km)	1.6249×10^4	46.4565	23.3345	26.3957	47.6273	23.1216
A2B2C2G (10 km)	43.4310	53.2309	16.8560	8.5820×10^3	1.0589×10^4	1.2794×10^4
A1B1C1A2B2G (15 km)	3.8803×10^3	2.7833×10^3	3.1713×10^3	3.6965×10^3	2.8831×10^3	15.4958
A1A2B2G (20 km)	2.8998×10^3	18.4540	25.5934	4.3789×10^3	3.1753×10^3	29.1000
A1B1A2B2G (25 km)	6.3086×10^3	3.8245×10^3	28.8602	6.3086×10^3	3.8245×10^3	28.8602

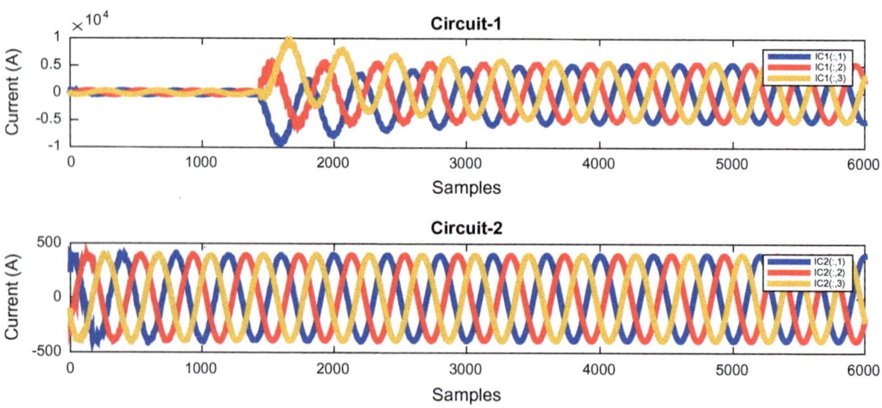

Fig. 9 Circuit-1 and circuit-2 currents for A1B1C1G far-end relay fault at 190 km

Table 3 Results of WHT for different far-end relay faults

Fault type	WHT coefficients					
	Phase-A1	Phase-B1	Phase-C1	Phase-A2	Phase-B2	Phase-C2
B2G (200 km)	70.1070	52.0569	31.2872	69.2892	371.9621	29.8233
A1B1C1G (190 km)	546.2430	463.1088	475.5963	73.9558	70.9462	55.5056
A1B1A2C2G (185 km)	937.0186	876.7034	92.4960	705.8787	70.4377	1.7063×10^3
A1B1C1A2B2G (180 km)	820.7797	518.7788	1.1211×10^3	590.1356	630.9077	50.3303
A1C2G (175 km)	661.8164	70.8136	63.4771	56.9060	67.9571	671.8474

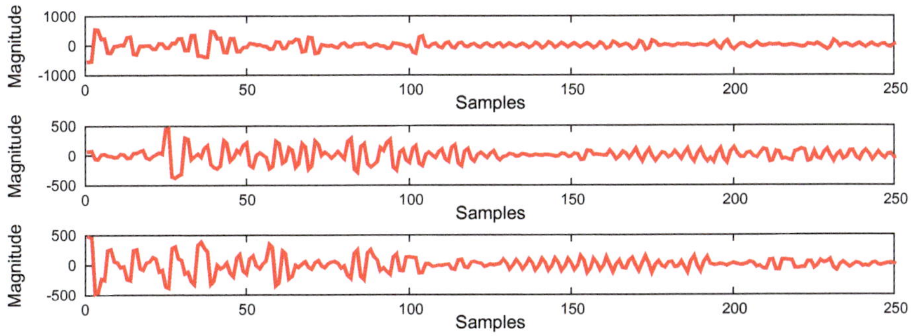

Fig. 10 WHT coefficients of circuit-1 currents for A1B1C1G far-end relay fault at 190 km

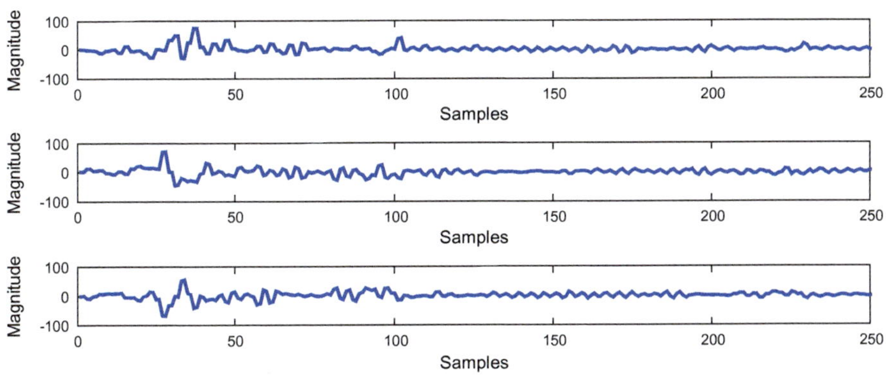

Fig. 11 WHT coefficients of circuit-2 currents for A1B1C1G far-end relay fault at 190 km

4.3 The Efficacy of WHT for the Converting Faults

The WHT has been investigated for the converting faults. Figure 12 exemplifies the circuit-1 and circuit-2 currents of the WFIDCTL when initially the A1G fault at 120 km at 0.0545 s is converted into the A1B1C1G fault at 120 km at 0.2055 s with $R_F = 2.5\ \Omega$ and $R_G = 3.5\ \Omega$. Figures 13 and 14, respectively, exemplify the WHT coefficients of currents of circuit-1 and circuit-2 when the A1G fault is converted into the A1B1C1G fault. The fault factors preferred for the additional fault cases are: $F_L = 120$ km, $R_F = 2.5\ \Omega$, and $R_G = 3.5\ \Omega$. Table 4 reports the results for different converting faults. It can be viewed that the WHT is insensitive to the converting faults.

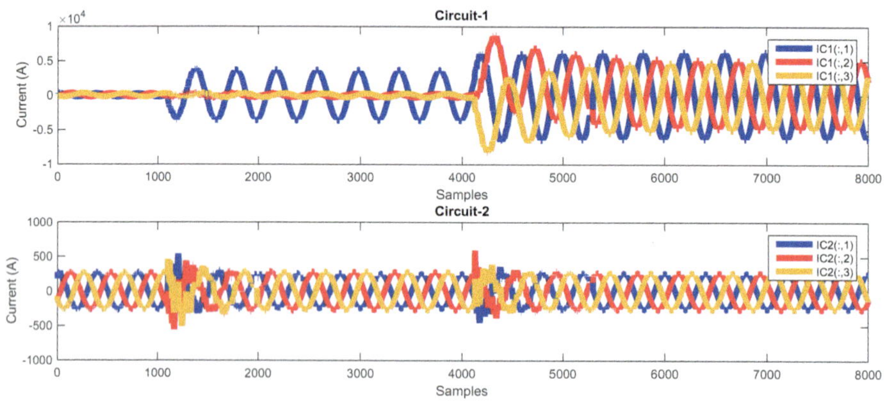

Fig. 12 Circuit-1 and circuit-2 currents when A1G fault at 120 km at 0.0545 s is converted into A1B1C1G fault at 120 km at 0.2055 s with $R_F = 2.5\ \Omega$ and $R_G = 3.5\ \Omega$

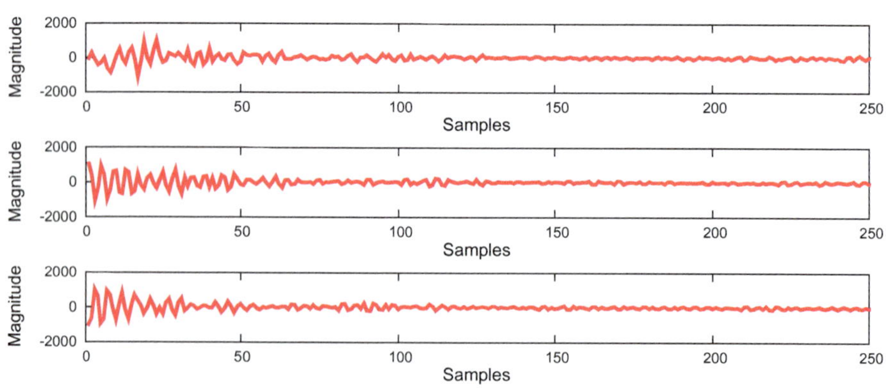

Fig. 13 WHT coefficients of circuit-1 currents when A1G fault at 0.0545 s at 120 km is converted into A1B1C1G fault at 120 km 0.2055 s with $R_F = 2.5\ \Omega$ and $R_G = 3.5\ \Omega$

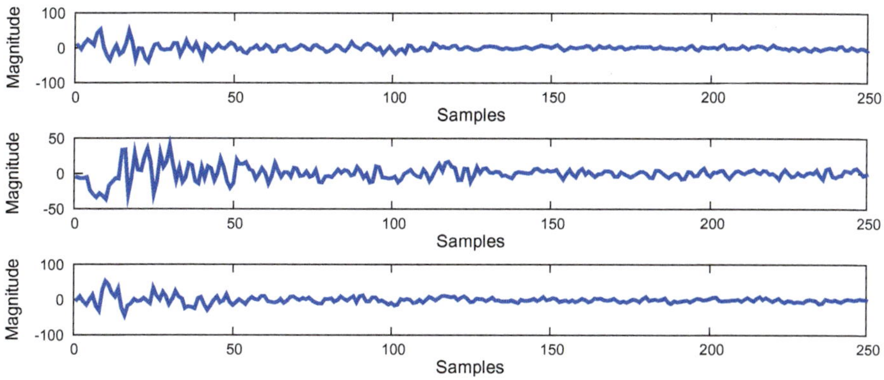

Fig. 14 WHT coefficients of circuit-2 currents when A1G fault at 0.0545 s at 120 km is converted into A1B1C1G fault at 120 km at 0.2055 s with $R_F = 2.5\ \Omega$ and $R_G = 3.5\ \Omega$

Table 4 Results of WHT for different converting faults

Fault type with time (s)	Converted fault type with time (s)	WHT coefficients					
		Phase-A1	Phase-B1	Phase-C1	Phase-A2	Phase-B2	Phase-C2
A1G (0.0545)	A1B1C1G (0.2055)	999.7395	1.0349×10^3	1.0048×10^3	51.8491	39.4384	52.7331
B2G (0.062)	A2B2G (0.1555)	51.8447	40.1817	40.8375	642.0853	995.2054	50.4008
B1C1G (0.078)	A2B2C2G (0.15)	42.8533	1.4002×10^3	1.3333×10^3	798.6775	539.1558	482.3032
A1B1C1G (0.0625)	A2B2G (0.1558)	976.7502	681.2808	625.7038	1.1547×10^3	1.0909×10^3	32.4719
A1G (0.075)	A1B1C1A2B2G (0.157)	940.1964	1.1480×10^3	787.7190	792.9871	818.6838	48.3282

4.4 *The Efficacy of WHT for the Cross-Country Faults*

The WHT is tested for different cases of cross-country faults. Figure 15 depicts the currents of circuit-1 and circuit-2 when the WFIDCTL is simulated for the cross-country A1G fault at 70 km and A1B1C1G fault at 130 km at 0.1 s with $R_F = 0.001\ \Omega$ and $R_G = 0.001\ \Omega$. The WHT coefficients of the circuit-1 and circuit-2 currents are illustrated in Figs. 16 and 17, respectively [12, 13]. Table 5 presents the results for the cross-country faults. It is examined from Table 5 that the WHT performs well for the detection of cross-country faults.

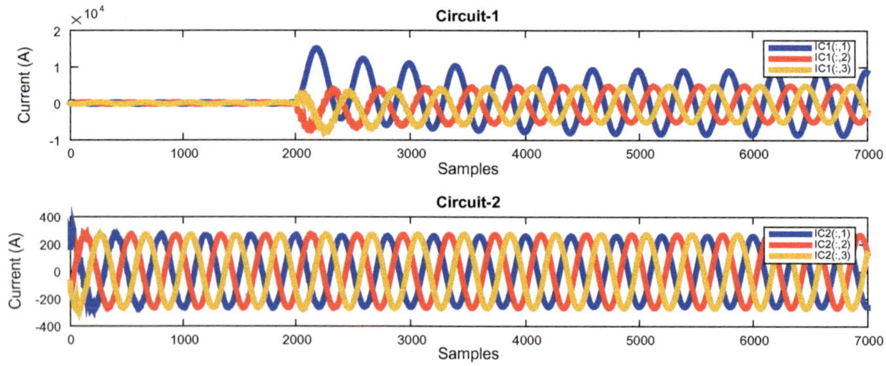

Fig. 15 Circuit-1 and circuit-2 currents for cross-country fault A1G at 70 km and A1B1C1G at 130 km at 0.1 s with $R_F = 0.001\ \Omega$ and $R_G = 0.001\ \Omega$

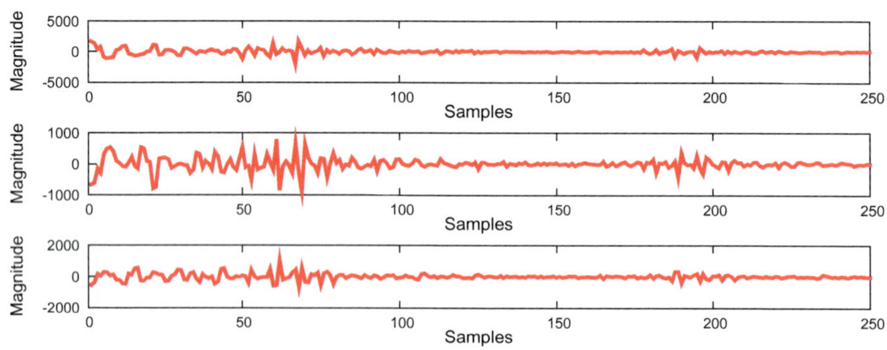

Fig. 16 WHT coefficients of circuit-1 currents for cross-country fault A1G at 70 km and A1B1C1G at 130 km at 0.1 s with $R_F = 0.001\ \Omega$ and $R_G = 0.001\ \Omega$

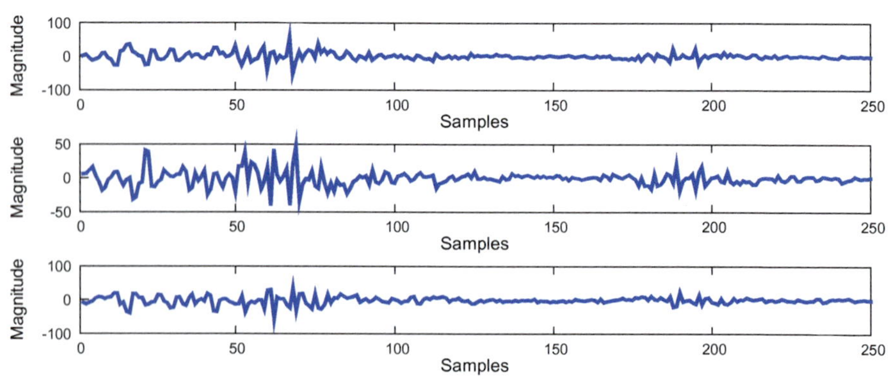

Fig. 17 WHT coefficients of circuit-2 currents for cross-country fault A1G at 70 km and A1B1C1G at 130 km at 0.1 s with $R_F = 0.001\ \Omega$ and $R_G = 0.001\ \Omega$

Table 5 Results of WHT for different cross-country faults

Fault-1 with location (km)	Fault-2 with location (km)	WHT coefficients					
		Phase-A1	Phase-B1	Phase-C1	Phase-A2	Phase-B2	Phase-C2
A1G (70 km)	A1B1C1G (130 km)	1.7353×10^3	782.6782	1.0484×10^3	58.5117	48.2713	36.0582
B1C1G (150 km)	A1B1G (50 km)	1.2867×10^3	918.4795	707.6427	35.7157	41.6298	38.6792
A1G (80 km)	B1G (120 km)	2.2456×10^3	675.5009	56.5531	54.6013	45.0850	52.2121
B2C2G (125 km)	A2G (75 km)	41.3742	27.6446	27.1770	406.1254	730.0407	596.0344
A2G (95 km)	C2G (105 km)	57.5504	55.8867	53.1563	2.0237×10^3	59.9666	1.1462 × 10^3

5 Conclusion

The Walsh–Hadamard transform (WHT) is seemed to be very efficient under varied fault categories for the WFIDCTL. The WHT coefficients of fault currents for both the circuits are assessed. The fault factors of the WFIDCTL are varied and it is discovered that the variation in fault factors do not influence the fidelity of the WHT. The outcomes substantiate that the WHT has the competence to protect the WFIDCTL beside different fault categories.

References

1. Gaur VK, Bhalja B (2017) New fault detection and localisation technique for double-circuit three-terminal transmission line. IET Gener Trans Distrib 12(8):1687–1696
2. Kapoor G (2019) A protection technique for series capacitor compensated 400 kV double circuit transmission line based on wavelet transform including inter-circuit and cross-country faults. Int J Eng Sci Technol 11(2):1–20
3. Gautam N, Ali S, Kapoor G (2018) Detection of fault in series capacitor compensated double circuit transmission line using wavelet transform. In: Proceedings of the IEEE international conference on computing, power and communication technologies (GUCON). IEEE, Greater Noida, India, pp 769–773
4. Hosseini K, Tayyebi A, Ahmadian MB (2017) Double circuit transmission line short circuit fault location using wavelet transform and MLP. In: Proceedings of the IEEE Iranian conference on electrical engineering. IEEE, Tehran, Iran, pp 1336–1342
5. Kapoor G (2018) Mathematical morphology based fault detector for protection of double circuit transmission line. ICTACT J Microelectron 4(2):589–600
6. Gawande PN, Dambhare SS (2016) Enhancing security of distance relays during power swing unblocking function for double circuit transmission lines: a differential power approach. IET Trans Distrib 1–6 (2016)

7. Kang N, Chen J, Liao Y (2015) A fault-location algorithm for series-compensated double-circuit transmission lines using the distributed parameter line model. IEEE Trans Power Delivery 30(1):360–367
8. Sharma N, Ali S, Kapoor G (2018) Fault detection in wind farm integrated series capacitor compensated transmission line using Hilbert Huang transform. In: Proceedings of the IEEE international conference on computing, power and communication technologies (GUCON). IEEE, Greater Noida, India, pp 774–778
9. Kapoor G (2018) Wavelet transform based detection and classification of multi-location three phase to ground faults in twelve phase transmission line. Majlesi J Mechatron Syst 7(4):47–60
10. Kapoor G (2018) Six-phase transmission line boundary protection using wavelet transform. In: Proceedings of the 8th IEEE India international conference on power electronics (IICPE). IEEE, Jaipur, India
11. Kapoor G (2018) Fault detection of phase to phase fault in series capacitor compensated six-phase transmission line using wavelet transform. Jordan J Electr Eng 4(3):151–164
12. Panwar D, Tomar P, Singh V (2018) Hybridization of Cuckoo-ACO algorithm for test case prioritization. J Stat and Manag Syst 21(4):539–546
13. Kumar A et al (2019) An enhanced quantum key distribution protocol for security authentication. J Discr Math Sc Cryptog 22(4):499–507

Chapter 8
Analysis of Ensemble Classifiers with Feature Selection for an Effective Intrusion Detection Model

Atreya Dyaram, V. Manojkumar, J. Sairabanu, and I. Sumaiya Thaseen

1 Introduction

There is a substantial increase in the use of Internet with increase in data rates and number of network devices. The current era of connected world has mandatorily moved all our sensitive data like financial data, personal data and data pertaining to even national security to devices connected to the network. All entities, persons, institutions and nations have a digital footprint and sensitive data in the Internet. The presence of the sensitive data encourages many malicious users and opens the door to different types of attacks. Network attacks are being carried out regularly and many new attacks are discovered by security analysts. Monitoring data traffic using conventional packet inspection methods become increasingly difficult and time consuming. Hence, it would be very beneficial to determine efficient and effective ways to reduce the number of attacks and to increase the safety of host computers and the network.

Intrusion detection system is an application or a device which monitors a system or a network of systems for malicious activities or any kind of policy violations. Intrusion detection is of two types: network intrusion detection system and host intrusion detection system [1]. Machine learning (ML) is one of the popular tools useful for prediction [2]. By definition, machine learning is a study of computer algorithms that learn by experience, much like humans do. Machine learning is especially useful and accurate when there are substantial amounts of relevant data to work on which provides a rich set of examples to learn from.

A. Dyaram · V. Manojkumar · J. Sairabanu
School of Computer Science and Engineering, Vellore Institute of Technology, Vellore, TamilNadu, India

I. Sumaiya Thaseen (✉)
School of Information Technology and Engineering, Vellore Institute of Technology, Vellore, TamilNadu, India
e-mail: sumaiyathaseen@gmail.com

© Springer Nature Singapore Pte Ltd. 2021

S. D. Purohit et al. (eds.), *Proceedings of International Conference on Communication and Computational Technologies*, Algorithms for Intelligent Systems,
https://doi.org/10.1007/978-981-15-5077-5_8

The remainder of this paper is divided as Sect. 2 discusses related work. The dataset is described in Sect. 3. The proposed model is discussed in Sect. 4. Section 5 describes the proposed model. The results are illustrated in Sect. 6. The conclusion is specified in Sect. 7.

2 Related Work

In this section, various supervised and ensemble algorithms are reviewed in the field of intrusion detection systems. In [3], the authors Pavel Laskov, Patrick D¨ussel, Christin Schäfer and Konrad Rieck held experiments to evaluate both supervised and unsupervised algorithms. Their experiments demonstrate that supervised algorithms outperform unsupervised algorithms if the test data contains unknown attack types. The authors suggested using semi-supervised approach for better results. In [4], *M. Bahrololum, E. Salahi and M. Khaleghi* introduced the concept of feature reduction to remove redundancy and irrelevant features. Their work suggests that decision trees outperform the other methods, particle swarm optimization and flexible neural tree.

Liu Hui and CAO Yonghui highlighted the importance of introduction of ML into the domain of intrusion detection as one of the major concerns to improve the performance of intrusion detection system [5]. They have introduced the basic structure of intrusion detection system and also analyzed four methods based on ML including SVM which is a supervised algorithm. They suggested using various kernel functions for SVM such as linear, polynomial and Gaussian as the kernel plays a very important role. In [6], Divyatmika and Manasa Sreekesh proposed a novel approach, two-tier architecture to build a network-based intrusion detection system using machine learning. The authors suggested that K-nearest neighbor (KNN) is statistically stronger than the other classifiers used in their work with an accuracy of 99.75%. Mohammad Almseidin, Maen Alzubi, Szilveszter Kovacs and Mouhammd Alkasassbeh performed several experiments and different Machine Learning classifiers using the KDD intrusion dataset [7]. Out of the supervised algorithms used, their results show that Decision Trees had the best results. Soubhik Das and Manisha J. Nene choose Machine Learning algorithms and 'human-like' methods over normal methods and techniques as they are too tedious to handle and sometimes don't even render fruitful results [8]. They described each algorithm and briefly explained when and how to use the algorithms. Marzia Zaman and Chung Horng Lung performed intrusion detection using seven algorithms. They have evaluated the algorithms on Kyoto 2006 dataset and used few supervised, unsupervised and ensemble learners. KNN gives excellent results in comparison with all the seven supervised algorithms used. In addition, it is also visible from the results that supervised algorithms produce better results in comparison with unsupervised algorithms. They also deployed to use information entropy as a feature selection followed by machine learning classifiers to show improved performance [9]. Unal Cavugoslu in [10] proposed a layered and hybrid intrusion detection system using different machine learning and feature selection techniques. It has been shown that the proposed system has better accuracy rate and low false positive rates in all attack types. The proposed model uses supervised algorithms along with other algorithms.

Hence, from the literature, it is evident that feature selection along with machine learning approaches yields better results in the detection of accuracy for intrusion detection system. This is the motivation of our proposed model.

3 Dataset

The KDD Cup 1999 dataset is a benchmark for intrusion detection techniques. But KDD dataset suffers from a few major flaws.

1. The attack rate within the dataset is unnatural.
2. The attack distribution is highly unbalanced.
3. Analysis of the KDD dataset has shown that there are 78–75% of the records which are duplicated [11]. This will cause learning algorithms to be biased toward frequently occurring results.
4. The dataset was created in 1999. Thus, the dataset is old and outdated.

Another dataset under consideration is the CIDDs dataset. CIDDs dataset is a labeled flow-based dataset for anomaly-based network intrusion detection system. The data was captured using OpenStack. However, it was not the chosen dataset because, out of the 14 features, many were discarded as they affected the learning environment negatively. Hence, very few features for the algorithm were remaining to learn the model. The chosen dataset is Kyoto 2006–2016. This dataset has 24 features, which is near the ideal number of features required in a dataset model for learning algorithm. Additionally, the dataset ranges from 2006–2016 which is the most recent dataset within a ten-year span. The Kyoto dataset has 24 features, 14 of those features are similar to the KDD dataset.

4 Proposed Model

Figure 1 shows the architecture of our proposed model and the subsection below provides the detailed description.

4.1 Feature Selection

In this phase, 24 supposedly relevant features for the dataset are chosen. However, it was observed that often some features in a dataset are not good for intrusion detection and just increase the data size. The following three features were not useful in intrusion detection.

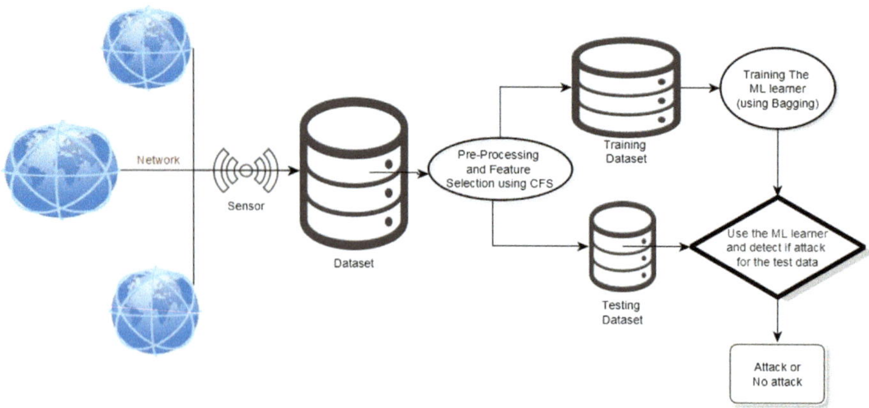

Fig. 1 Simple mechanism in a machine learning model

1. Source IP Address: This indicates the IP address of the sender. There are innumerable addresses in this world. So many different Source IP addresses can only play a little role in predicting the labels.
2. Destination IP Address: Similarly, this attribute will not contribute anything for our algorithm.
3. Start Time: This is a string denoting time. These strings need to be converted to integers for analysis. Conversion of these strings is likely to be error prone as it will lead to multiple values for the attribute.

Many of the features will be related to other attributes in some or the other way, though not visible to the user. But in such cases, it will lead to redundancy by increasing the computation time and complicating the algorithm, thus leading to an inefficient machine learning model. To eliminate redundant datasets, there are many feature selection methods like information gain (IG), chi-squared test, recursive elimination, ridge regression and CFS. In the proposed model, CFS is deployed on each of the algorithm and the performance of the model is analyzed with and without the application of CFS.

4.2 *Pre-processing*

Non-integer features present a unique difficulty that the compiler cannot execute the code for strings and mixed data types [12]. So, in order to support non-integer features, labels are used to encode the non-integer features.

The following features are to be encoded into labels:

1. Service
2. Flag
3. Duration.

Inconsistencies were found in other features like IDS detection, malware detection and Ashula detection which have values other than mentioned in the description of the dataset. For example: IDS detection can have only two possible values 0 or 1 corresponding to IDS alert being triggered and not being triggered, respectively. But for some instances, the data was not from the domain like 20008(1). This value has a 1 inside the parenthesis but this kind of output is not expected as per the given description of the dataset [13]. For such instances, we have replaced the feature values with 1. In addition, some of the values were junk string values with a '1' inside parenthesis. Each of these values has been considered as a 1 for the analysis of the paper.

Now in the label, there are three possible categories:

1 Normal session
−1 Attacked session
−2 Unknown attack.

As there are neglible instances with label −2, the −2 s are also replaced with −1 s. This algorithm, therefore, predicts if the session was a normal one or an attack one. The chosen dataset is then split into training and testing dataset in the ratio of 60:40.

4.3 Choosing the Classifier/Learner

Supervised algorithms became an obvious choice because of the better results as compared to unsupervised algorithms. Hence, this paper uses supervised algorithms. Also, ensemble methods have shown good results. The model is trained with three different classifiers, namely KNN, AdaBoost and bagging. The bagging classifier performs superior in comparison with KNN and AdaBoost when integrated with CFS. Hence, the proposed model is developed using CFS + Bagging. However, the analysis section below specifies all the classifier results with and without CFS.

How the Models Were Evaluated

The result of each sample in the dataset is a binary value denoted by TRUE or FALSE. The model evaluates the given sample based on the given features and predicts a TRUE or FALSE.

The models were evaluated based on the following performance measures

- Accuracy
  Accuracy is the ratio of correct predictions to the total number of samples. It is indicative of the overall correctness of the learning algorithm.
- Precision
  Precision is the ratio of correctly predicted TRUE observations to the total predicted TRUE observations.

- Recall
  Recall is the ratio of correctly predicted positive observations to all observations in actual class.

5 Results

The three machine learning algorithms: KNN, AdaBoost and bagging have been evaluated. Results have been analyzed with each of these three algorithms with and without CFS. AdaBoost and bagging utilize all the processors in the working system and it is not possible to modify or change the number of cores. But for KNN, we can vary the number of cores to be used. Thus, for KNN, the difference between computation time for parallel processing and simple processing has also been recorded.

5.1 *Analysis of Classifiers with and Without CFS*

Table 1 characterizes the results of implementing the KNN learning algorithm (without CFS) on nine samples from the Kyoto dataset. Each sample is data for a given day used from a year. The average values of accuracy, precision and recall, respectively, are 98.4%, 92.57% and 87.81%. The average time taken is 1 min and 48 s when using all cores while using only one core takes 2 min 10 s on average. In the

Table 1 KNN without CFS

Year	Size(in MB)	Accuracy (in %age)	Precision (in %age)	Recall (in %age)	Error metrics (in %age)	Time using njobs = −1	Time without using njobs
2006	18.5	98.8	90.9	93.55	0.46	1 m 53 s	2 m 15 s
2007	11.4	99.04	98.24	98.5	0.38	38 s	56 s
2008	9	98.12	98.27	98.34	7.5	28 s	40 s
2009	16.5	98.8	98.72	98.78	4.5	1 m 12 s	1 m 39 s
2011	20	99.18	96.79	94.60	3.27	1 m 37 s	2 m
2012	14	98.85	98.11	95.62	4.6	1 m	1 m 11 s
2013	27	99.68	78.55	63.4	1.2	2 m 50 s	3 m 13 s
2014	33	94.95	80.93	83.34	20.1	3 m 04 s	3 40 s
2015	29	98.59	92.67	64.23	5.6	3 m 14 s	3 m 19 s
Avg. values	19.82	98.4	92.57	87.81	5.29	1 m 48 s	2 m 10 s

Table 2 KNN with CFS

Year	Size (in MB)	Accuracy(in %)	Precision (in %)	Recall (in %)	Mean absolute error (in %)	Time with njobs = −1	Time without njobs
2006	18.5	99.3	96.46	94.05	0.14	2 m 05 s	2 m 22 s
2007	11.4	99	98.41	98.5	2	46 s	52 s
2008	9	96.44	97.68	95.88	8	36 s	42 s
2009	16.5	98.8	98.64	98.67	2.4	1 m 08 s	1 m 30 s
2011	20	98.8	93.84	97.61	2.6	1 m 46 s	1 m 55 s
2012	14	98.61	97.12	94.15	2.8	1 m	1 m 50 s
2013	27	99.68	80.8	64.58	0.6	3 m 02 s	3 m 4 s
2014	33	94.63	80	82	10.6	2 m 34 s	3 m 49 s
2015	29	98.6	92.45	64.68	3	2 m 16 s	3 m 12 s
Avg.	19.82	98.21	92.82	87.79	3.57	1 m 48 s	2 m 29 s

2013 and 2015 sample, the recall is less at 63.4% and 64.23%, respectively. Therefore, the average recall is lower than the average accuracy and average precision.

Table 2 characterizes the results of implementing the KNN learning algorithm (with CFS) on the same set of samples from the Kyoto dataset. The average values of accuracy, precision and recall, respectively, are 98.21%, 92.82% and 87.79%. The average time taken is 2 min and 29 s when using all cores while using only one core takes 1 min 48 s on average. Like the previous case, in the 2013 and 2015 sample, the recall is less at 63.4% and 64.23%, respectively. Therefore, the average recall is lower than the average accuracy and average precision.

Adaptive Boost Model

Table 3 characterizes the results of implementing the adaptive boost model (without CFS) on the same set of samples from the Kyoto dataset [14]. The average values of accuracy, precision and recall, respectively, are 99.43%, 95.48% and 90.41%. This is significantly higher than the average values of KNN, both with and without CFS. The average time taken is 1 min and 36 s. The 2013 sample recall is 54.48% which brings down the recall average. The 2015 sample has recall and precision lesser than the average recall and precision.

AdaBoost with CFS

Table 4 characterizes the results of implementing the adaptive boost model (with CFS) on the same set of samples from the Kyoto dataset. Each sample is a day's worth of data used from a year. The average values of accuracy, precision and recall, respectively, are 99.39%, 96.24% and 91.18%. Compared to the previous table, the overall average values are similar. Using CFS has increased the average precision and average recall mildly, and there is a miniscule decrease in average accuracy. The average time taken is 1 min and 5 s. Similar to the AdaBoost without CFS, the 2013 sample has 60.34% recall which would reduce the average recall greatly.

Table 3 AdaBoost without CFS

Year	Size (in MB)	Accuracy (in % age)	Precision(% age)	Recall (% age)	Mean absolute error (% age)	Time
2006	18.5	99.13	99.05	89.16	0.8	1 m 39 s
2007	11.4	99.56	99.83	98.69	0.23	52 s
2008	9	99.75	99.78	99.77	0.5	36 s
2009	16.5	99.86	99.88	99.82	0.23	1 m 26 s
2011	20	99.81	98.98	99.13	0.38	1 m 32 s
2012	14	99.57	99.43	97.94	0.7	1 m 5 s
2013	27	99.61	74.16	54.48	0.7	2 m 5 s
2014	33	98.45	98.29	90.43	2.8	2 m 44 s
2015	29	99.16	89.9	84.28	1.7	2 m 24 s
Avg.	19.82	99.43%	95.48%	90.41%	0.89%	1 m 36 s

Table 4 AdaBoost with CFS

Year	Size (in MB)	Accuracy	Precision	Recall	Error metrics	Time
2006	18.5	99.5	98.79	94.32	2	38 s
2007	11.4	99.5	99.8	98.7	1.6	35 s
2008	9	99.21	99.24	99.33	3.1	32 s
2009	16.5	99.84	99.90	99.76	0.6	1 min 13 s
2011	20	99.7	98.4	98.8	1.0	1 min 11 s
2012	14	99.45	99.18	97.46	2	1 m 2 s
2013	27	99.65	77.97	60.34	0.13	1 m 30 s
2014	33	98.5	97.7	91.2	3	1 m 33 s
2015	29	99.19	95.18	80.76	1.9	1 m 31 s
Avg.	19.82	99.39%	96.24%	91.18%	1.70%	1 m 5 s

Bagging Model
Table 5 characterizes the results of implementing the bagging model (without CFS) on the same set of samples from the Kyoto dataset. Each sample is a day's worth of data used over a year. The average values of accuracy, precision and recall, respectively, are 99.28%, 95.17% and 92.48%. These values are better than the average values of KNN, with and without CFS and worse than the average values of AdaBoost. The average time required is 1 min 2 s; therefore, bagging is faster compared to AdaBoost and KNN. The 2006 sample has 61.48% recall and 70.19% precision, although the accuracy is 95%. In the 2006 sample, all values are lesser than the average.

Bagging with CFS Model
Table 6 characterizes the results of implementing the bagging model (with CFS) on the same set of samples from the Kyoto dataset. Each sample is a day's worth of data used from a year. The average values of accuracy, precision and recall, respectively,

Table 5 Bagging without CFS

Year	Size (in MB)	Accuracy (%age)	Precision (% age)	Recall (%age)	Mean absolute error (%age)	Time
2006	18.5	95.19	70.19	61.48	0.47	1 m 3 s
2007	11.4	99.57	99.73	99.94	0.4	32 s
2008	9	99.87	99.94	99.84	1	23 s
2009	16.5	99.91	99.94	99.85	0.17	54 s
2011	20	99.96	99.84	99.69	0.14	57 s
2012	14	99.85	99.72	99.4	0.22	48 s
2013	27	99.82	90.97	80.85	0.3	1 m 22 s
2014	33	99.67	99.36	98.26	0.6	1 m 42 s
2015	29	99.65	96.91	93.01	0.8	1 m 36 s
Avg.	19.82	99.28%	95.17%	92.48%	0.45%	1 m 2 s

Table 6 Bagging with CFS model

Year	Size (in MB)	Accuracy	Precision	Recall	Time
2006	18.5	99.68	98.66	97.01	53
2007	11.4	99.79	99.64	99.64	30
2008	9	99.70	99.73	99.73	24
2009	16.5	99.9	99.9	99.8	54
2011	20	99.8	99.3	99.4	52
2012	14	99.86	99.71	99.42	44
2013	27	99.8	88.1	77.7	1 m 16 s
2014	33	99.6	99.3	98.2	1 m 34 s
2015	29	99.59	96.22	91.88	1 m 32 s
Avg	19.82	99.74%	97.84%	95.86%	58 s

are 99.74%, 97.84% and 95.86%. The average values of bagging with CFS are significantly higher than bagging without CFS, AdaBoost and KNN. Bagging with CFS is also the fastest among the other models with 58 s. On modeling with CFS, the low values in the 2006 sample have been rectified. Evidently, this model bagging with CFS is the best among the lot.

6 Summary

Table 7 shows the summary of the accuracy, precision and recall of different models with and without CFS.

Table 7 Performance of different models

Model/Metrics	KNN	KNN + CFS	AdaBoost	AdaBoost + CFS	Bagging	Bagging + CFS
Accuracy	98.4	98.21	99.43	99.39	99.28	99.74
Precision	92.57	92.82	95.48	96.24	95.17	97.84
Recall	87.81	87.79	90.41	91.18	92.48	95.86

7 Conclusion

This paper presents an experimental framework for intrusion detection system. The three algorithms evaluated on the Kyoto dataset are AdaBoost, bagging and KNN. Each of them was modeled with and without CFS. CFS determines a subset of features which can improve the performance of bagging because it takes a few samples from the dataset. In addition, bagging tries to reduce the variance. However, KNN is a lazy learner which never utilizes the dataset in the training phase, but only for comparing the nearest neighbors. Hence, reducing the number of features will have little impact on performance. This is the reason why CFS helps bagging but not other models. Also, the results of these experiments, as tabulated above clearly indicate that Bagging + CFS with an average accuracy of 99.74 overpowers average accuracy of other models like bagging (99.27), KNN (98.5), KNN + CFS (98.03), AdaBoost (99.434) and AdaBoost + CFS (99.39). Thus, Bagging + CFS provides the best performance across all the metrics, accuracy, precision and recall.

References

1. Kemmerer RA, Vigna G (2002) Intrusion detection: a brief history and overview. Computer 35(4):supl27–supl30. https://doi.org/10.1109/mc.2002.1012428
2. https://hackernoon.com/what-is-machine-learning-and-why-is-it-important-introduction-to-machine-learning-part-1-30738cba5abd
3. Laskov P, Düssel P, Schäfer C, Rieck K (2005) Learning intrusion detection: supervised or unsupervised? In: Roli F, Vitulano S (eds) Image analysis and processing—ICIAP 2005. ICIAP 2005. Lecture notes in computer science, vol 3617. Springer, Berlin, Heidelberg
4. Bahrololum M, Salahi E, Khaleghi M (2009) Machine learning techniques for feature reduction in intrusion detection systems: a comparison. In: 2009 Fourth international conference on computer sciences and convergence information technology, Seoul, pp 1091–1095
5. Hui L, Yonghui C (2010) Research intrusion detection techniques from the perspective of machine learning. In: 2010 Second international conference on multimedia and information technology, Kaifeng, pp 166–168
6. Divyatmika, Sreekesh M (2016 A two-tier network based intrusion detection system architecture using machine learning approach. In: 2016 International conference on electrical, electronics, and optimization techniques (ICEEOT*)*, Chennai, pp 42–47
7. Almseidin M, Alzubi M, Kovacs S, Alkasassbeh M (2017) Evaluation of machine learning algorithms for intrusion detection system. In: 2017 IEEE 15th international symposium on intelligent systems and informatics (SISY), Subotica, pp 000277–000282

8. Das S, Nene MJ (2017) A survey on types of machine learning techniques in intrusion prevention systems. In: 2017 International conference on wireless communications, signal processing and networking (WiSPNET), Chennai, pp 2296–2299
9. Zaman M, Lung C (2018) Evaluation of machine learning techniques for network intrusion detection. In: NOMS 2018—2018 IEEE/IFIP network operations and management symposium, Taipei, pp 1–5
10. Çavuşoğlu Ü (2019) Appl Intell. https://doi.org/10.1007/s10489-018-01408-x
11. Tavallaee M, Bagheri E, Lu W, Ghorbani AA (2009) A detailed analysis of the KDD CUP 99 data set. In: 2009 IEEE symposium on computational intelligence for security and defense applications, Ottawa, ON, pp 1–6
12. Kumar A et al (2019a) An enhanced quantum key distribution protocol for security authentication. J Discr Math Sci Cryptogr 22(4):499–507
13. Kumar A et al (2019b) An improved quantum key distribution protocol for verification. J Discr Math Sci Cryptogr 22(4):491–498
14. Panwar D, Tomar P, Singh V (2018) Hybridization of Cuckoo-ACO algorithm for test case prioritization" J Stat Manage Syst 21(4): 539–546

Chapter 9
Color Image Encryption and Compression Using DCT in Joint Process

Umesh Maru, Gajendra Sujediya, and Yashika Saini

1 Introduction

A color image compression process is represented the original image without introducing the least bit of image data. The compressed image is a shirked form of the original image without losing original integrity. In the digital color image, each color component (R, G, B) contains 8 bits of data and also color image contains a lot of data redundancy and requires a large amount of data space. To save the storage space as well as transmission bandwidth, image compression plays a crucial role in the transmission and storage of data. To obtain the best possible compression ratio, DCT transform is applied and for each (n, n) DCT block a corresponding $(n \times n)$ vector containing the maximum possible run of zeroes at its end. The last step of this application of modified systematic lossless encoder.

Based on compression factors:

1.1 Compression Ratio

This is a process to calculate how much or to what extent a color image is shrunk, which is shown in Eq. (1).

$$\text{Compression Ratio} = \frac{\text{Size After Compression}}{\text{Size Before Compression}} \tag{1}$$

U. Maru (✉) · G. Sujediya · Y. Saini
Rajasthan Institute of Engineering and Technology, Bhankrota Ajmer Road, Jaipur, Rajasthan 302026, India
e-mail: umeshmaru12@gmail.com

© Springer Nature Singapore Pte Ltd. 2021

S. D. Purohit et al. (eds.), *Proceedings of International Conference on Communication and Computational Technologies*, Algorithms for Intelligent Systems,
https://doi.org/10.1007/978-981-15-5077-5_9

1.2 Compression Factor

Compression factor is an inverse of compression ratio. Compression factor is given in Eq. (2) below.

$$\text{Compression Factor} = \frac{\text{Size Before Compression}}{\text{Size After Compression}} \tag{2}$$

1.3 Compression Time

A time required by a compressor to compress a color image and measure execution time to improve performance by MATLAB.

1.4 Saving Percentage

Saving percentage is how much a color image compressed in a size as a percentage value shown in Eq. (3).

$$\text{Saving Percentage} = \frac{\text{size before compression} - \text{size after compression}}{\text{size before compression}}\% \tag{3}$$

There are some algorithms that perform compression in various ways some are lossless while others are lossy. In this paper, the proposed method for a color image compression balancing on compression ratio a color image quality by compressing the important parts of the image with high quality. Fusion implementation of the encryption process together with the color image compression process will be discussed in this paper. Encryption is a process of encoding and decoding of an image and information by using encryption a decryption keys. In this process, a secret image is divided into random parts.

Various confidential data in an image form are transmitted over the Internet. Thus, using secret image security should be taken into consideration because hackers/third parties may use information over the communication network, encryption is introduced as a process for secure communication and different methods developed to encrypt and decrypt data image.

2 Related Work

In line with a growing need for data and information transmission in a safe and quick manner, research work has been done by researchers in the area of image encryption and color image compression. Some of the encryption processes and color image compression processes are discussed below.

A comparative study has been carried out on image compression using DCT and discrete wavelet transform (DWT). A comparison is outlined to emphasize the results of this compression system between DCT and DWT using joint photographic experts group (JPEG) and portable network graphics (PNG) color images [1].

To reduce spatial correlation and concentrate the energy of the image, an iterative process based on the bisection method is used to determine the required threshold and control compression quality via achieving the prefixed peak signal-to-noise ratio (PSNR) [2].

An innovative method for lossless compression of discrete color images, such as map images, graphics, GIS, as well as binary images. This method comprises two main components. The first is a fixed-size codebook encompassing (8 × 8) bit blocks of two-tone data along with their corresponding Huffman codes and their relative probabilities of occurrence. The probabilities were obtained from a very large set of discrete color images which are also used for arithmetic coding. The second component is the row-column reduction coding [3].

Efficient color and texture feature-based adaptive color image compression. Color conversion from RGB to YCbCr is performed to extract color and texture features. The extracted features are used to select non-zero (significant) DCT coefficients. The storage space and bandwidth during transmission are efficiently utilized by encoding non-zero DCT coefficients, and thereby preserving texture and color information in the reconstructed image [4].

A fuzzy color-preserving rule-based system is introduced to express the dissimilarity between two regions. The compression part is based on the shape adaptive DCT with /SPL delta/dc correction method. The quantization matrices have been designed according to the properties of the used transform [5].

A hybrid DCT and DWT compression method are employed to capitalize on the advantages of both the techniques. The method involves generating color data of the white band and narrowband images in an intermediate format and then generating the decompressed image [6].

A method for encryption and decryption of color images using RGB pixel displacement. In the proposed method, the original plain image is split into its basic three components, which are the RGB components and the key image is also split into RGB components [7].

A hybrid image cryptographic encryption and digital watermarking technique are for the encryption and authentication of digital image content. The image encryption was done on the RGB channels and the spatial watermarking was successfully applied to the RGB color bands of the digital image and this rendered the watermark visibly subtle and difficult to detect under regular viewing. The digital watermarking technique was engaged to authenticate the image and detect modifications to the watermarked image [8].

Image steganography on gray and color image using discrete cosine transform (DCT) enhancement and RSA with the least significant bit (LSB) technique. Secret data is encrypted and decrypted using RSA. RSA is performed to increase security. Encryption is performed with a key which makes it difficult to get the covert message even if the presence of the hidden message is exposed. Now the stego image is enhanced using DCT [9].

GUI-based secure breast cancer identification system through a combination of cryptography and biomedical image watermarking. Through this scheme, second level of security can be applied for the embedded data in mammographic images. Several information related to patients (EPR), infected region (ROI), doctor's name and diagnosis from symptoms are encrypted in the first level using RSA encryption and then embedded in the mammographic image itself in the second level using bit plane slicing watermarking technique [10].

The encryption process involves three phases to encrypt the data. In the first phase, the bits of every eighth character in the text will be embedded in the most significant bit positions of preceding seven characters. This process creates an encrypted text data which will be further processed for second-level encryption [11]. During the second phase, every character of encrypted text is XORed with the secret key provided by the sender that authenticates the sender. In the third phase, final encryption is done using the Huffman encoding method. In the decryption process, the ciphertext passes through three phases to reconstruct the original text. The encrypted data is decrypted using the Huffman decoding method in the first phase. Then, the decrypted data and secret key that authenticates the receiver are used to reconstruct the original text in the last two phases of the decryption process [12, 13].

3 Proposed Work

The color image is taken as an input and this color image is first compressed by using DCT. Now the compressed image is encrypted by the encryption process. This encrypted image is transmitted over the communication network. At the receiver end, this encrypted image initially decrypted and this decrypted image is decompressed by using inverse DCT and finally the original image is revealed. The entire process is as shown in Fig. 1. In this process, color image is compressed using a DCT matrix. DCT converts the information contained by a color image in a block (8×8) of the pixel from special domain to the frequency domain. The original color image is divided into 8-8 block and two-dimensional DCT is computed for each block. Color image decompression is done by decoding each of the 8×8 blocks of the compressed image, the entire operation done by inverse DCT. Block schematic of the proposed approach to change the order of compression and encryption is shown in Fig. 4.

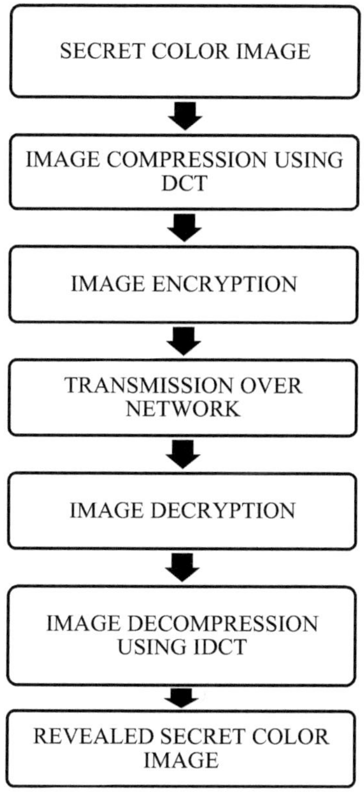

Fig. 1 Block schematic of the proposed approach

3.1 *Encoding Image Size*

The methodology assumes that an input image has size $2^n \times 2^n$ where $2 \leq n \leq 9$. If the image size is $2^n \times 2^n$, then it is encoded as a binary form of $n - 2$ using three bits. This encoding can be extended to larger image sizes.

In any segment, the first bit is either 0 or 1 and it is encoded as itself using one bit. The number of segments is encoded as its binary form using $2n$ bits if the image size is $2^n \times 2^n$. For example, let the image size be 8×8 and let 11111100000000000001100001110000000001111111000000 00111111111111000 be the bit sequence along the segment determined by the compression. The first bit of the segment is 1 and it is encoded as 1. The number of segments is 10 and it is encoded as 001010 using 6 bits.

3.2 Observation

For testing, we have taken two different images. One is 'lena.tif' and the second is 'bird.png'. The test images are chosen in such a way they can serve the purpose of thorough analysis of the algorithm. 'Lena.tif' is a 256 × 256 × 3 image and has homogeneous regions and gradual change of shades and intensities of different color components. 'bird.png' on the other hand is a 320 × 320 × 3 image. This image has more heterogeneous regions as compared to the previous image and the different color components are more prominent. Moreover, the images taken for testing the algorithm are of different formats, i.e., one is tagged image file (.TIF) format and other is portable network graphics (.PNG) format. So, the compatibility of the algorithm is also testified for different kinds of image formats. We have analyzed the above process and implemented coding in MATLAB.

3.3 Component of RGB

A color image is consists of three primary colors red, green and blue. So any system which displays a color image actually combines these three primary colors in different ratios to form any colors. The first operation that is performed on the image is to split the image into three different color component images. After that, the encryption and compression are done on each component image. After encryption and compression on three component images separately, the outputs again combined to form the final encrypted and compressed image.

3.4 Encryption and Decryption

A given image is first compressed, and then the compressed image is encrypted by rearranging the bits of the compressed image. Conversely, a compressed-encrypted image is decrypted by reversing the rearrangement of the bits, using the same set of encrypt key which is used for encryption. Encryption and decryption process illustration of a single image matrix is shown in Fig. 2. After the decryption, the compressed image can be decompressed to obtain the original image. The method of generation of encryption keys by the encryption algorithm used in MATLAB coding.

3.4.1 Encryption

This is done twice to reach a higher order of security and to achieve a higher compression as well. The first encryption is done by keeping the image pattern for compression and secret [14]. This image pattern is the first key of the encryption and after the

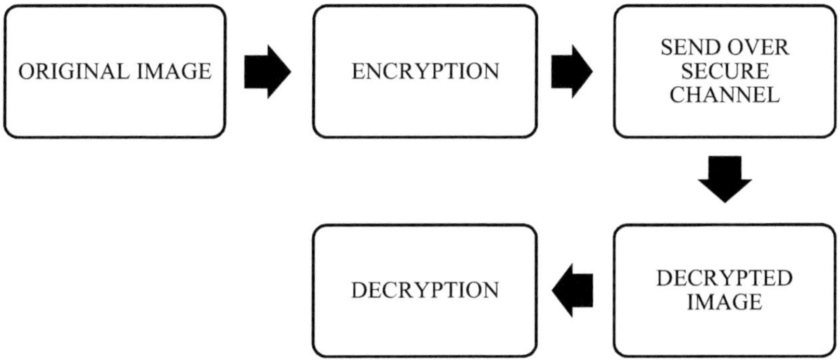

Fig. 2 Encryption and decryption process illustration of a single image matrix

Fig. 3 Illustration of encryption

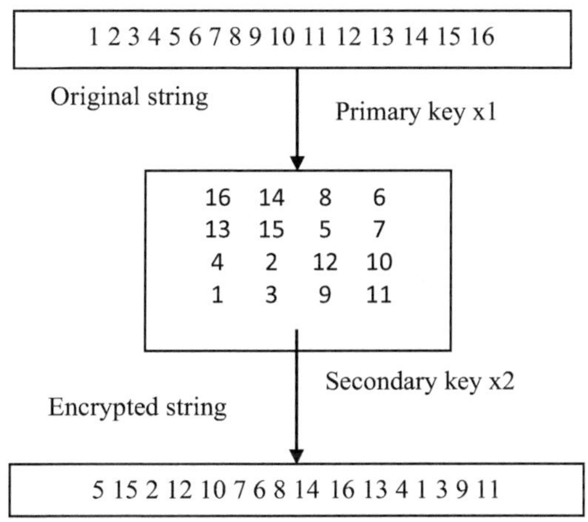

compression, another image pattern is applied to rearrange the bit sequence again and this image pattern also kept secret. This encryption key is applied after merging all the compressed bit planes. The second image pattern is an another key for encryption. The receiver should have the knowledge of both the keys to decrypt and decompress the received image. As encryption and decryption are done by the same set of keys, the encryption is known as symmetric key encryption that is shown in Fig. 3.

3.4.2 Decryption

Once run-length decoding is performed on each bit plane, the output bit stream is put back to the matrix form of its original size. Then, the image pattern used prior to the image compression is again applied on the output image to rearrange the bits in the correct order [15]. This is done on each bit plane and then all bit planes are merged to form the color component (red/green/blue) image. Then, in the final step,

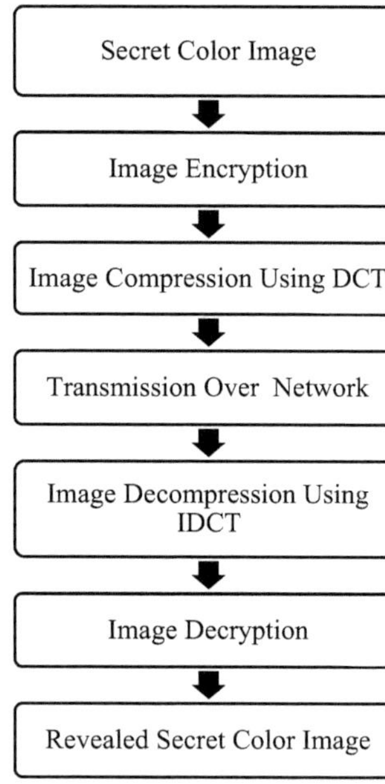

Fig. 4 Block schematic of the proposed approach to change the order of compression and encryption

all these color component images are put together to form the retrieved RGB color image (Fig. 4).

4 Result and Analysis

The proposed color image is implemented according to the description in section III, and here, we tested with different sizes of the color images. Initially, we take the first color image as an input that is the ".tif" format shown in Fig. 5. This image is compressed by using DCT. This compressed image shown is shown in Fig. 6. Then, this compressed image is encrypted by the encryption process and this encrypted image is shown in Fig. 7, and the encrypted image is sanded over the communication channel. At the receiver end, now the encrypted image is first decrypted by the decryption process. The decrypted image is shown in Fig. 8. Now, this decrypted image is decompressed by using the inverse DCT process and reconstructed the original image is shown in Fig. 9, and now we take the second color image as an input that is in the ".jpg" format is shown in Fig. 10. 'Bird.jpg' image is compressed by using the same process that used in the first image and the compressed image is shown in Fig. 11. Now this compressed image is encrypted for the secure transformation that is shown in Fig. 12. Again this image receives at the receiver end. The encrypted image is first decrypted and then decompressed by the decompression process is shown in Figs. 13 and 14, respectively.

Fig. 5 Original image

Fig. 6 Compressed image

The description of results for lena.tif and bird.jpg images is shown in the form of a table. Table 1 is containing the compression ratio and compression factor based on original image size that is calculated by Eqs. (1) and (2). Compressed image size and the corresponding values of saving percentage from Eq. (3) and compression time, that is, time required by a compressor to compress the image and measure execution time to improve performance by MATLAB are shown in Table 2.

Fig. 7 Encrypted image

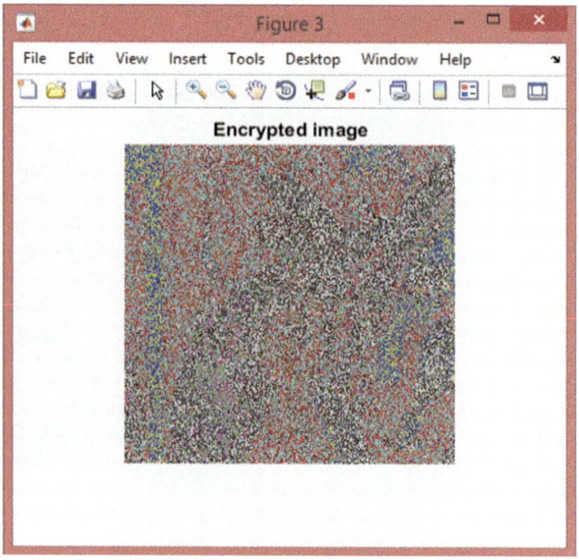

Fig. 8 Decrypted image

5 Conclusion

This paper shows a new methodology which performs both compression and encryption processes for color images. The encryption and compression processes are based on DCT. The first and foremost advantage of this algorithm is that it performs compression and encryption simultaneously. So there is no need to go through a

Fig. 9 Reconstructed original image

Fig. 10 Original image

separate algorithm to perform compression and encryption of a single image. As we know that least significant bit (LSB) has the least information about the image, while using this algorithm, we can achieve a great amount of compression. This will cost minimum information loss and the compression ratio will much better. Also, it takes

Fig. 11 Compressed image

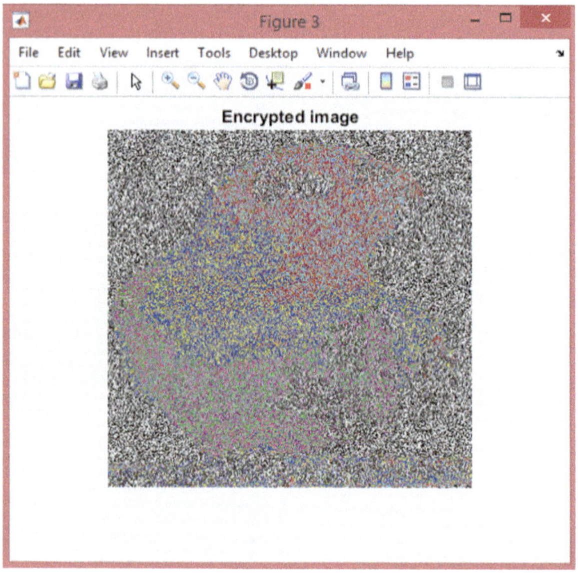

Fig. 12 Encrypted image

lesser compression time as compared to other techniques. The proposed compression process with encryption ensures more data security as compared to the existing image compression techniques while using transmission.

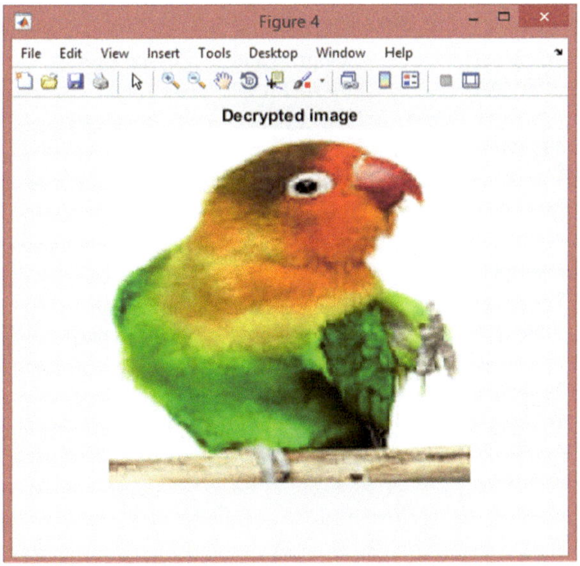

Fig. 13 Decrypted image

Fig. 14 Reconstructed original image

Table 1 Compression ratio and compression factor

Name of input image	Original image size (bytes)	Compress and encrypt image size (bytes)	Compression ratio (%)	Compression factor
Lena.tif	12,346	6850	55.48	1.8
Bird.jpg	15,939	9166	57.50	1.7

Table 2 Saving percentage and compression time

Name of input image	Original image size (bytes)	Compressed and encrypted image size (bytes)	Saving percentage	Compression time (ms)
Lena.tif	12,346	6850	44.51	11.48
Bird.jpg	15,939	9166	42.49	18.25

References

1. Barbhuiya AHMJI, Laskar TA, Hemachandran K (2014) An Approach for color image compression of JPEG and PNG Images Using DCT and DWT. In: International conference on computational intelligence and communication networks (2014)
2. Zhou Y, Wang C, Zhou X (2018) DCT-based color image compression algorithm using an efficient lossless encoder. In: 14th IEEE International conference on signal processing (ICSP)
3. Alzahir S, Borici A (2015) An innovative lossless compression method for discrete-color images. IEEE Trans Image Process 24(1)
4. Patil NK, Murgod SF, Boregowda L, Udupi VR (2013) Adaptive texture and color feature based color image compression. In: International conference on smart structures and systems—ICSSS'13
5. Makrogiannis S, Schelkens P, Fotopoulos S, Cornelis J (2001) Region-oriented compression of color images using fuzzy inference and shape adaptive DCT. In: Proceedings 2001 international conference on image processing (Cat. No. 01CH37205)
6. Sathi Raju Ch, Rama Koti Reddy DV (2015) On compression characteristics of white band and narrow band images using hybrid DCT and DWT. In: 2nd international conference on electronics and communication systems (ICECS)
7. Somaraj S, Hussain MA (2016) A novel image encryption technique using RGB pixel displacement for color images. In: IEEE 6th international conference on advanced computing (IACC)
8. Kester Q-A, Nana L, Pascu AC, Gire S, Eghan JM, Quaynor NN (2015) A Hybrid image cryptographic and spatial digital watermarking encryption technique for security and authentication of digital images. In: 17th UKSim-AMSS international conference on modelling and simulation (UKSim)
9. Singh YK, Sharma S (2016) Image steganography on gray and color image using DCT enhancement and RSA with LSB method. In: International conference on inventive computation technologies (ICICT)
10. Pal K, Ghosh G, Bhattacharya M (2014) GUI based smart breast cancer identification system for mammographic images through 2nd level secured combined Crypto-watermarking. In: IEEE international conference on bioinformatics and biomedicine (BIBM)
11. Kodabagi MM, Jerabandi MV, Gadagin N (2015) Multilevel security and compression of text data using bit stuffing and Huffman coding. IEEE (iCATccT)
12. Gupta P, Kumar S (2014) A comparative analysis of SHA and MD5 algorithm. Int J Comput Sci Inf Technol 5(3):4492–4495

13. Sharma A, Chaturvedi R, Dwivedi UK, Kumar S, Reddy S (2018) Firefly algorithm based effective gray scale image segmentation using multilevel thresholding and entropy function. Int J Pure Appl Math 118(5):437–443
14. Kumar A et al (2019) An improved quantum key distribution protocol for verification. J Discrete Math Sci Cryptogr 22(4):491–498
15. Kumar A et al An enhanced quantum key distribution protocol for security authentication. J Discrete Math Sci Crypt 22(4): 499–507

Chapter 10
Efficient Method for Prediction Accuracy of Heart Diseases Using Machine Learning

Manoj Raman, Vijay Kumar Sharma, Saroj Hiranwal, and Amit Kumar Bairwa

1 Introduction

1.1 Human Heart Structure

Typically, Heart or We can say cardiovascular disease (CVD) is Couse of several illnesses, disability, and death. Working on heart disease patients' databases can be compared to real-life applications. Doctor's knowledge to assign the weight to each attribute. More weight is assigned to the attribute having high impact on disease prediction. It also provides healthcare professionals with an extra source of knowledge for making decisions. In simple words, heart is a hollow muscle that pumps blood throughout the blood vessels by repeated, rhythmic contraction. The term cardiac (as in cardiology) means "related to the heat" and comes from the Greek Kupétd, kardia, for "heart". The adult human heart has a mass of between 250 and 350 g [1].

Heart has four chambers, two superior atria and two inferior ventricles. The atria are the receiving chambers and the ventricles are the discharging chambers. The figure below shows the diagram of human heart. The vertebrate heart is principally composed of cardiac muscle and connective tissue. Cardiac muscle is an involuntary striated muscle tissue specific to the heart and is responsible for the heart's ability to pump blood. The average human heart, beating at 72 beats per minute, will beat approximately 2.5 billion times during an average 66-year lifespan, and pumps approximately 4.7–5.7 L of blood per minute. It weighs approximately 250 to 300 g (9 to 11 oz.) in females and 300 to 350 g (11 to 12 oz.) in males.

Data mining offers great potential benefits for medical- based applied decision-making. The data mining medical application area has been receiving quite a big

M. Raman (✉) · V. K. Sharma · S. Hiranwal · A. K. Bairwa
Rajasthan Institute of Engineering and Technology, Bhankrota Ajmer Road, Jaipur, Rajasthan 302026, India
e-mail: manoj1985_raman@rediffmail.com

© Springer Nature Singapore Pte Ltd. 2021
S. D. Purohit et al. (eds.), *Proceedings of International Conference on Communication and Computational Technologies*, Algorithms for Intelligent Systems,
https://doi.org/10.1007/978-981-15-5077-5_10

attention. However, working with very large data sets with many attributes is difficult. Medical field experts use heavily advanced statistical analysis. The use of data mining techniques is fairly new and advanced. Data mining technology expands the possibilities of medical data mining that opens the door for a large source of medical data analysis. Data mining provides a user-oriented approach method that is used to find out the novel and hidden patterns in the medical datasets. Healthcare administrators can improve the quality of diagnosis by using this hidden knowledge. One of the important applications of data mining includes anticipating the patient's behavior from the given data. In medical science, most of the decisions taken by the doctors are based on intuition and experience rather than the knowledge-rich data hidden in the medical datasets (Fig. 1).

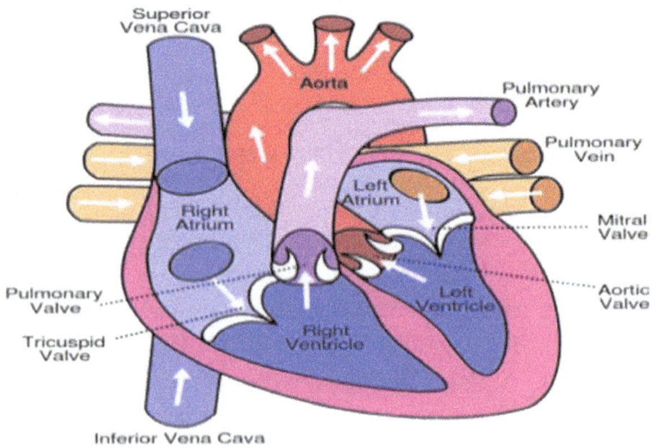

Fig. 1. Human heart structure

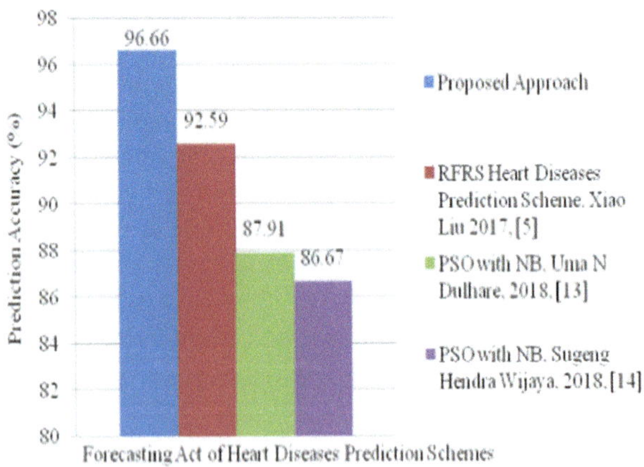

Fig. 2. Performance of proposed technique over standalone prediction scheme with statlog dataset

2 Key Factors of Heart Diseases

Some risk factors for heart disease can be controlled, and some can't. According to the American Heart Association (AHA), these are the leading factors that put you at risk for coronary artery disease.

Age More than 83% of people who die from coronary heart disease are 65 or older. Older women are more likely to die of heart attacks within a few weeks of the attack than older men.

Being Male Men have a greater risk of having heart attacks than women and they have attacks earlier in life. Even after menopause, when women's death rate from heart disease increases, it's not as great as men's

Family History Those with parents or close relatives with heart disease are more likely to develop it themselves.

Race Heart disease risk is higher among African Americans, Mexican Americans, American Indians, native Hawaiians, and some Asian Americans compared to Caucasians.

Smoking Cigarette smoking increases your risk of developing heart disease by two to four times.

High Cholesterol As blood cholesterol rises, so does risk of coronary heart disease.

High Blood Pressure High blood pressure increases the heart's workload, causing the heart to thicken and become stiffer. It also increases your risk of stroke, heart attack, kidney failure, and congestive heart failure. Smoking, high blood cholesterol levels, or diabetes, the risk of heart attack or stroke increases heart disease several times.

3 Related Work

"Associative Classification Approach for Diagnosing Cardiovascular Disease", was proposed by Noh et al. [1] uses a classification method for the extraction of multi-parametric features by assessing HRV (Heart Rate Variability) from ECG, data preprocessing and heart disease pattern. The dataset consisting of 670 peoples, distributed into two groups, namely normal people and patients with heart disease, were employed to carry out the experiment for the associative classifier.

"The problem of identifying constrained association rules for heart disease prediction" was studied by Ordonez [2]. The assessed data set encompassed medical records of people having heart disease with attributes for risk factors, heart perfusion measurements, and artery narrowing. Three constraints were introduced to decrease the number of patterns. First one necessitates the attributes to appear on only one side

of the rule. The second one segregates attributes into uninteresting groups. The ultimate constraint restricts the number of attributes in a rule. Experiments illustrated that the constraints reduced the number of discovered rules remarkably besides decreasing the running time. Two groups of rules envisaged the presence or absence of heart disease in four specific heart arteries. Data mining methods may aid the clinicians in the prediction of the survival of patients and in the adaptation of the practices consequently.

John Peter proposed a method "An Empirical Study On Prediction Of Heart Disease Using Classification Data Mining Techniques" [3] to investigate the performance of different classification algorithms such as DT, NB, K-NN, and NN on heart disease dataset. The dataset have a large volume of data that consumes more time for classification. Thereby it has reduced the dimensionality of data using the attribute selection methods. Then the reduced data is classified using various classification algorithms. And the result found that NB classifier gives better accuracy for heart disease prediction after applying the CFS attribute selection method.

Maryam Tayef et al. has presented a predictive model for coronary heart disease using a decision tree algorithm. Typically this study was based on a case-control study of patients referred to Ghaem Hospital, Mashhad–Iran for coronary angiography, between September 2011 and May 2013. The authors built a decision tree with 11 traditional related factors of CHD including age, gender, PAL, BMI, SBP, DBP, FBG, TC, TG, LDL, HDL and hs-CRP on the training group of 1640 records and evaluate their designed model with testing group of 706 different records from training datasets. The algorithm used the Gini index for selecting the variables, and the final tree was pruned. Author demonstrated that with using serum hs-CRP and other traditional CHD risk factors through decision tree algorithm they have obtained an accuracy rate of 94% [4].

Xiao Liu et al. have demonstrated a hybrid classification system for diagnose of heart disease using ReliefF and Rough Set (RFRS) method. Typically, designed system contains two subsystems: the RFRS feature selection system and a classification system with an ensemble classifier. The first system includes three stages: (i) data discretization, (ii) feature extraction using the ReliefF algorithm, and (iii) feature reduction using the heuristic Rough Set reduction algorithm. In the second system, they have proposed a classifier based on the C4.5 algorithm. For the experimental work, they obtain statlog (Heart) disease dataset from the UCI machine learning database which contains observation of 270 patients with two classes, namely presence and absence of heart disease. The author preprocesses the dataset to remove the missing value error of dataset and employ a feature reduction methodology. They include only 13 condition features and for the performance analysis, they have used four parameters as Confusion Matrix, Sensitivity, Specificity, and Accuracy. The authors have indicates that they achieve maximum classification accuracy of 92.59% which is superior to the performances of previously reported classification techniques [5].

A number of different investigator groups have intended different gaps in the research on heart disease diagnosis and treatment and propose a model to systematically close those gaps to discover heart disease treatment with reliable performance by using data mining techniques [6–11].

4 Proposed Work

A number of researchers have introduced huge number of algorithms for enhancing the prediction accuracy of heart diseases with low false alarms but each and every obtainable algorithm associates his unique limitation. With the motivation from literature that most of the researchers have considered neural network technique for building appropriates prediction approaches hence a hybrid technique has introduced in this paper by using the group of two most popular classification techniques of data mining, Naive Bayes and Neural Network. Naive Bayes algorithm employs a simplified version of Bayes formula to decide which class a feature belongs to. The posterior probability of each feature, given the feature values, presents in the instance; the instance is assigned the class with the highest probability.

The Bayes formula, which makes the assumption that feature values are statistically independent within each class. Neural networks are network structures consisting of a number of nodes connected through directional links. Each node represents a processing unit, and the links between nodes specify the causal relationship between connected nodes [12]. Due to high learning efficiency, the neural network technique produces effective results in comparison with the other classifier algorithms when the majority of variables are weakly relevant. Several neural network frameworks can be categories as Feed-forward & Recurrent neural networks. The Feed-forward neural network is a simple network where information travels in one direction only, from input to the out neuron by using intermediate neurons. On the other hand in recurrent neural networks established connections between the units shape a directed cycle.

To improve the prediction accuracy of heart disease the intended approach integrating the prediction mechanism of two popular classification techniques namely Naïve Bayes and neural network scheme into a single form. For enhancing the accuracy ratio, selected approaches are placed in a layered form in which if one technique will fail to predict data accurately then the second technique enhances the chance of correct prediction. The following are the pseudo steps of the proposed approach.

1. Gather medical dataset of heart disease and normal person from UCI repository.
2. Apply preprocessing technique.
   i. 10 Fold cross-validation method for dividing dataset into train and test part.
   ii. Employ CFS with BFS methodology for selecting optimized feature set.
3. Investigate data with the prediction mechanism of Naïve Bayes method.
4. If data predicted correctly update dataset with removing its existence from processed dataset.

5. If data predicted incorrectly it supplies to the next technique, i.e., MLP for prediction.
6. Repeat steps 3 to 5 until complete dataset has not analyzed.
7. Calculate the accuracy of employed method.

5 Experimental Setup and Result Analysis

To analyze the performance of designed approach over traditional as well as other most current proposed technique a good number of dissimilar simulations have done in incorporated work of this investigation. In order to examine actual performance and superiority of designed prediction system over the other accessible techniques, a number of experiments have been carried out with different data sets and parameters such as different attributes based on the opted dataset. For finding an effective and fair comparison on the base of prediction accuracy the proposed technique has been executed with the same parameters as used by separate traditional methodologies. The entire algorithms, i.e., proposed and standalone prediction technique has been analyzed on the base of their prediction accuracy. For fair comparison between the introduced technique and other accessible data prediction algorithm performance a number of experiments have done with the same parameters, use same datasets with other parameters such as number of attributes and instances. For the evaluation purpose, a number of dataset has been collected from the online available data pool, UCI data repository.

At the first phase of experiment statlog dataset has employed with 14 attributes and 270 instances. Typically, a single instance is a record of a person who has heart disease or a normal. The dataset consists the record of 120 persons who are having heart disease along with 150 normal person's entry. Following Table 1 and Fig. 3 present comparative results in tabular and graph form.

However, the statistics present in the above table have demonstrated the effectiveness of proposed approach over the other accessible heart disease prediction methods but in order to explain efficiency of designed approach in a better way, the following figure presents such comparative results in a graph form (Fig. 2).

Above figure evidently shows that the proposed approach has produced higher comparative results in comparison of other accessible heart disease prediction methods. As literature of prediction techniques states that an approach that produces

Table 1. Comparative performances over statlog dataset

Forecasting practice	Attain Accuracy (%)
Proposed approach	96.66
RFRS Heart Diseases Prediction Scheme, Liu [5]	92.59
PSO with NB, Dulhare [13]	87.91
PSO with NB, Wijaya [14]	86.67

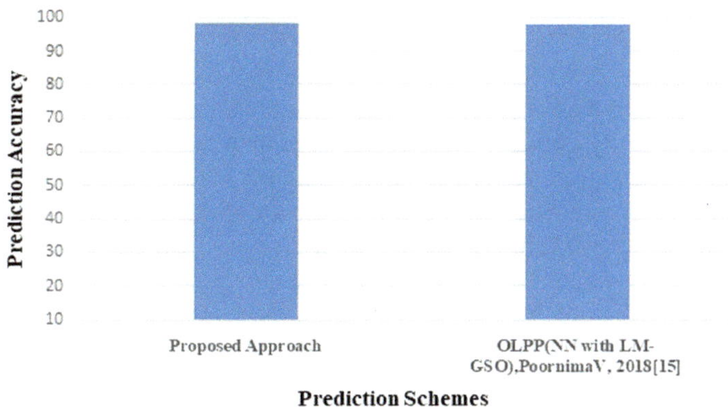

Fig. 3. Proposed approach versus modern heart diseases prediction schemes over hungarian dataset

Table 2. Prediction act of proposed and modern schemes over Hungarian dataset

Forecasting Practice	Attain Accuracy (%)
Proposed approach	98.29
Orthogonal local preserving projection (OLPP), (NN with LM-GSO), Poornima V, 2018 [15]	98

significant results with a dataset may perform poor with other statistics or with other parameters. Therefore, to examine the actual performance of designed approach different experiments have been done with other datasets that have been used by a number of researchers into their investigations. Another dataset with name of Hungarian and Cleveland dataset has opted for next experiment.

The evaluated fallouts depicted in the following Table 2 has evidently shown the efficiency of proposed approach in terms to attain high prediction accuracy of heart diseases. The assessment depicts that an act of proposed approach has remained unchanged even parameters of evaluation have modernized. The following table demonstrates a comprehensible difference image of proposed scheme in comparison to other modern heart disease prediction schemes.

The following figure illustrates the difference in prediction accuracy among proposed and other modern accessible approaches.

The statistics present in the above table and figure has shown that presents proposed prediction scheme has outperformed over other simulated technique as same as it performs in previous experiments. However, the difference in prediction accuracy does not have a huge gap but in the medical field little amount of dissimilarity may participate as a key role.

6 Conclusions

This paper introduced a novel hybrid data prediction technique by incorporating the prediction functionality of two weak algorithms of data mining named Naïve Bayes and Multi-layer perceptron technique into a single practice. The proposed approach of this investigation has not simple integrate these algorithms for the purpose of data prediction but also has taken attributes selection process into consideration at the time of building technique. For the selection of only relative and efficient attributes, a correlation-based feature selection technique has used with the concepts of best-first search method. The designed technique works in a layered format at which each layer has predicted entity separately with the use of optimized feature set that has been selected by the integrated attributes selection process. For the evaluation purpose, a number of datasets have been collected from the online available data pool, UCI data repository. Each and every evaluation result have demonstrated the effectiveness and efficiency of the proposed data prediction practice of this investigation.

References

1. Noh K, Lee HG, Shon HS, Lee BJ, Ryu KH (2006) Associative classification approach for diagnosing cardiovascular disease. vol 345. Springer, Berlin, pp 721–727
2. Ordonez C (2004) Improving heart disease prediction using constrained association rules. Seminar Presentation at University of Tokyo
3. Peter J (2012) An empirical study on prediction of heart disease using classification data mining techniques. In: IEEE-international conference on advances in engineering, science and management (ICAESM–2012), 30–31 Mar 2012, pp 514–517
4. Tayefi M, Tajfard M, Saffar S, Hanachi P, Amirabadizadeh AR, Esmaeily H, Taghipour A, Ferns GA, Moohebati M, Ghayour-Mobarhan M (2017) hs-CRP is strongly associated with coronary heart disease (CHD): a data mining approach using decision tree algorithm. Comput Methods Programs Biomed 141:105–109
5. Liu X, Wang X, Su Q, Zhang M, Zhu Y, Wang Q, Wang Q (2017) A hybrid classification system for heart disease diagnosis based on the RFRS method. Hindawi Comput Math Methods Med 2017:1–11
6. Krishnaiah V, Narsimha G, Subhash Chandra N (2016) Heart disease prediction system using data mining techniques and intelligent fuzzy approach: a review. Int J Comput Appl 136(2):43–51
7. Badbe VC Londhe V (2016) Heart disease analysis using LVQ. Int J Adv Found Res Sci Eng (IJAFRSE) vol 2 Special Issue Vivruti 2016 pp 44–50
8. Wiharto W, Kusnanto H, Herianto H (2016) Interpretation of clinical data based on C4.5 algorithm for the diagnosis of coronary heart disease. Healthc Inform Res 22(3):186–195
9. Arabasadi Z, Alizadehsani R, Roshanzamir M, Moosaei H, Yarifard AA (2017) Computer aided decision making for heart disease detection using hybrid neural network-genetic algorithm. Comput Methods Programs Biomed 141:19–26. https://doi.org/10.1016/j.cmpb.2017.01.004
10. Florence S, Amma NB, Annapoorani G, Malathi K (2014) Predicting the risk of heart attacks using neural network and decision tree. Int J Innovative Res Comput Commun Eng 2(11):7025–7030
11. Bharti S Singh SN (2015) Analytical study of heart disease prediction comparing with different algorithms. In: International Conference on Computing, Communication and Automation (ICCCA2015), IEEE, pp 78–82

12. Vadicherla D, Sonawane S (2013) Classification of heart disease using SVM and ANN. Int J Res Comput Commun Technol 2(9):694–701
13. Dulhare UN (2018) Prediction system for heart disease using Naive Bayes and particle swarm optimization. Biomed. Res. 29(12):2646–2649
14. Wijaya SH Pamungkas GT Sulthan M.B (2018) Improving classifier performance using particle swarm optimization on heart disease detection. IEEE International seminar on application for technology of information and communication (iSemantic), pp 603–608
15. Poornima V, Gladis D (2018) A novel approach for diagnosing heart disease with hybrid classifier. Biomed Res 29(11):2274–2280

Chapter 11
Neighbourhood-Inspired Grey Wolf Optimizer

Priyanka Meiwal, Harish Sharma, and Nirmala Sharma

1 Introduction

Nature-inspired algorithms (NIAs) are a significant source of motivation to both engineering and scientific calculations. NIAs provide reliability in solving the optimization problems [1]. Fundamentally, NIAs are classified into two broad categories: algorithms based on swarm intelligence (SI) and evolutionary algorithms (EAs). The nature of the solution gets convoluted as complexity of problem increases [6, 8]. All SI-based algorithms use multi-agents, inspired by the composite behaviour of communal insects like ants, bees and other animal societies. Grey wolf optimizer (GWO) is an SI-based algorithm. The GWO stipulates the hunting behaviour and social hierarchy of grey wolves. For incorporating this behaviour in GWO, Mirjalili et al. [5] define the positioning approach of each wolf by taking mean of positions of the alpha, beta and delta wolves. In GWO alpha, beta and delta are the first fittest, second fittest and third fittest solution of the search space, respectively.

In this paper, for enhancing the exploration capacity, an innovative alternate of GWO is introduced. In the proposed work, the searching procedure is supervised by an arbitrarily selected solution of the search space along with the first fittest, second fittest and third fittest solution. The proposed variant is known as neighbourhood-inspired grey wolf optimizer (NIGWO). In NIGWO, a new randomly selected agent is added for increasing the exploration capacity. A set of 16 benchmark functions are studied for experiments, to check the competitiveness of NIGWO. The state-of-

P. Meiwal (✉) · H. Sharma · N. Sharma
Rajasthan Technical University, Kota, India
e-mail: meiwalpriya500@gmail.com

H. Sharma
e-mail: harish.sharma0107@gmail.com

N. Sharma
e-mail: nsharma@rtu.ac.in

© Springer Nature Singapore Pte Ltd. 2021

S. D. Purohit et al. (eds.), *Proceedings of International Conference on Communication and Computational Technologies*, Algorithms for Intelligent Systems,
https://doi.org/10.1007/978-981-15-5077-5_11

the art algorithms are used to compare the obtained results. The obtained outcomes validate the authenticity of the proposed method.

The remainder part of the paper is organized as follows: Section 2 briefly introduces the GWO algorithm. The proposed NIGWO is described during the Sect. 3. The performance of the proposed algorithm is tested over benchmarks functions in Sect. 4. Finally, the conclusion is explained in Sect. 5.

2 Grey Wolf Optimizer Algorithm

The GWO is introduced by Mirjalili et al. [5] which specifies the social behaviour of grey wolves. The social behaviour includes approaching, tracking, pursuit, encircling and exploiting (attacking) the prey adopted by grey wolves. Grey wolves belong to the Canidae species. Wolves are classified by strong teeth, fluffy tassel and have shooting behaviour in a group which has medium size bounded within 5–12. The hierarchy of GWO is characterized by levels.

In level 1, **alpha wolf** is the leader of the group, which can be either male or female. The alpha wolf is mostly liable for decision making like how to perform the tracking task, hunting, pursuit, sleeping place, exploration, exploitation, etc.

In level 2, **beta wolf** is the subordinating wolf whose responsibility is to assist the alpha wolf in decision making and leadership. The beta wolf is treated the second fittest wolf to whom the leadership is handover if the alpha wolf is accidentally injured or dead because of some reasons. The beta wolf is an advisor to alpha and maintains the discipline for the group. Beta wolf performs the responsibilities to ensure that other subordinate wolves should obey the order or decisions made by the alpha wolf and gives feedback as an acknowledgement to the alpha wolf.

In level 3, **delta wolf** alike the beta wolf is responsible for assisting the alpha wolf. They are senior to omega wolves, and it reports to alpha and beta wolves. Further, the delta wolf can be categorized into the following:
Scout wolf: responsible for watching the boundaries.
Sentinels wolf: In this category, wolves are responsible for protecting the group.
Older or senior wolf: those who were alpha or beta sometimes.
Hunter wolf: helps alpha and beta in hunting.
Caretaker wolf: takes care of ill, weak and injured or wounded wolves.
Figure 1 specifies the social behaviour of grey wolves:

2.1 Calculation Modelling and Algorithm

In this subsidiary, the calculation models of the social ranking, exploration, encircling and exploitation phases are described. Then, the basic algorithm of GWO is defined as follows:

1. Social ranking:

   In procedure to describe the mathematical calculation, the social ranking of wolves when outlining GWO, we study according to the fitness value; the best fitness solution is the α_{wolf}, later the second fittest is β_{wolf} and the third fittest is the δ_{wolf}, respectively. These three solutions can be fixed as key group of wolves [12].

2. Encircle the prey:

   Wolves start with circumscribe or encircle the prey at the time of hunting. (To illustrate the mathematic equations for the encircling process is as follows [5]). The following equation describes the process of hunting the prey as:

$$\vec{D} = |\vec{C}_{\text{coef}} * \vec{w}_{\text{prey}}(\text{iter}) - \vec{w}_{gw}(\text{iter})| \tag{1}$$

$$\vec{w}(\text{iter}+1) = \vec{w}_{\text{prey}}(\text{iter}) - \vec{A}_{\text{coef}} * \vec{D} \tag{2}$$

   Here, $w_{gw}(\text{iter})$ is the position vector of a grey wolf and iter is the current iteration. $w_{\text{prey}}(\text{iter})$ indicates the position of the prey. As presented in Eq. (2), wolves shorten their distance from the prey.
   Their distance depends on A and D in which A is decreased gently, and the distance from the prey's position is presented by D. As the iteration number of the algorithm increases, the wolves reach closer to the prey. In other words, the wolves encircle the prey since their initial locations are determined randomly [3]. The coefficient vectors A_{coef} and C_{coef} can be evaluated by the following equation:

$$\vec{A}_{\text{coef}} = 2\vec{a} * \vec{r}_{1_A} - \vec{a} \tag{3}$$

$$\vec{C}_{\text{coef}} = 2 * \vec{r}_{2_C} \tag{4}$$

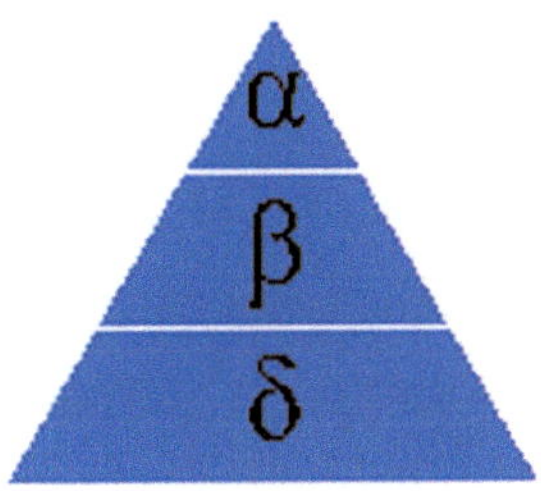

Fig. 1 Social behaviour of grey wolves

here, r_{1_A} and r_{2_C} are random vectors in [0,1] with the range and component $|a|$ decreased linearly from 2 to 0.
To attack again the prey, the grey wolf searches for the prey in the next iteration, wherein they will repeatedly be finding the next fittest solution alpha among the other wolves. This manner will be repeated until the terminus criterion is accomplished [7].

3. Hunting:

When the location of the prey is discovered by the grey wolf group then, the α_{wolf}, the β_{wolf} and the δ_{wolf} lead the group of the wolf to enclose the prey [12]. In this proposed article, we will be assuming that the group of wolves knows the location of the prey. Thus, as key group, save the fittest three possible solutions that are gained so far and according to the fundamentals of the group, will be updating the position of every wolf in the group. The equation for position refurbishing is shown as follows:

$$\vec{w}_{(\text{iter}+1)} = \frac{\vec{w}_1 + \vec{w}_2 + \vec{w}_3}{3} \tag{5}$$

$$\vec{w}_1 = \vec{w}_{\alpha,\text{wolf}} - \vec{A}_{\text{coef}_1} * \vec{D}_{\alpha,\text{wolf}} \tag{6}$$

$$\vec{w}_2 = \vec{w}_{\beta,\text{wolf}} - \vec{A}_{\text{coef}_2} * \vec{D}_{\beta,\text{wolf}} \tag{7}$$

$$\vec{w}_3 = \vec{w}_{\delta,\text{wolf}} - \vec{A}_{\text{coef}_3} * \vec{D}_{\delta,\text{wolf}} \tag{8}$$

Here, $w_{\alpha,\text{wolf}}$, $w_{\beta,\text{wolf}}$ and $w_{\delta,\text{wolf}}$ represent the position vectors of fittest three solutions and $A_{\text{coef}_1}, A_{\text{coef}_2}, A_{\text{coef}_3} \in A_{\text{coef}}$ are coefficient vectors, in Eqs. (6), (7), (8). Other parameters are given by the following equation:

$$\vec{D}_{\alpha,\text{wolf}} = |\vec{C}_{\text{coef1}} * \vec{w}_{\alpha,\text{wolf}} - \vec{w}\,| \tag{9}$$

$$\vec{D}_{\beta,\text{wolf}} = |\vec{C}_{\text{coef2}} * \vec{w}_{\beta,\text{wolf}} - \vec{w}\,| \tag{10}$$

$$\vec{D}_{\delta,\text{wolf}} = |\vec{C}_{\text{coef3}} * \vec{w}_{\delta,\text{wolf}} - \vec{w}\,| \tag{11}$$

Here, $C_{\text{coef}_1}, C_{\text{coef}_2}, C_{\text{coef}_3} \in C_{\text{coef}}$ is coefficient vectors, in Eqs. (9), (10), (11).

4. Exploitation (Attack towards the prey):

Usually, the grey wolf will be attacking the prey when the prey stops moving. To surrounding the prey, we will be decreasing the value of $|a|$. Notice that the

variation range of $|A|$ is also declined by $|a|$.
Therefore, if the value of $|A| < 1$, then alpha wolf attacks on prey; while if $|A| > 1$, then wolf far away from prey or maybe assumed that alpha wolf is injured and tired. Here, major parameters a and C_{coef} are to be adjusted. The value of a can be calculated by this formula: $|a| = [2 - 2 * ((1)/\text{Maxiteration})]$, which linearly decreases from 2 to 0.

The basic of grey wolf optimizer is depicted in basic algorithm 1 is as follow:

Initialize the population of grey wolf W_i , (i = 1 to n)
Initialize a, A_{coef} and C_{coef}
Evaluate the fitness of each search agent
$w_{\alpha,wolf}$ = the first fittest wolf
$w_{\beta,wolf}$ = the second fittest wolf
$w_{\delta,wolf}$ = the third fittest wolf
while iter < maximum number of iteration **do**
  for every search agent **do**
    Renovate the ongoing location of search agent
    Best position can be obtained by calculating average of result
    $\vec{w}_{(iter+1)} = \frac{\vec{w}_1+\vec{w}_2+\vec{w}_3}{3}$
  end for
  Renovate a, A_{coef} and C_{coef}
  Evaluate the fitness of all search agents
  Renovate $w_{\alpha,wolf}$, $w_{\beta,wolf}$ and $w_{\delta,wolf}$
  $iter = (iter + 1)$
end while
return $w_{\alpha,wolf}$

Algorithm 1: Basic Algorithm:

3 Proposed Neighbourhood-Inspired Grey Wolf Optimizer

In, the basic version of GWO, the position of a solution is modified based upon the first fittest solution (alpha), second fittest solution (beta) and third fittest solution (delta) of the search space. It means the searching ability is depended on the best three solutions. So when the position of alpha, beta and delta wolves are stuck, then the performance of the algorithm is degraded [5].

So to overcome the above problem and to enrich the GWO's exploitation capability, the search process is modified as per the following Eq. 12:

$$\vec{w}_{(\text{iter}+1)} = \frac{\vec{w}_1 + \vec{w}_2 + \vec{w}_3 + \vec{w}_4}{4} \tag{12}$$

where $\vec{w}_4$ is the position of any arbitrarily selected agent of the search space that is not equal to the three best solutions which are $w_{\alpha,\text{wolf}}$, $w_{\beta,\text{wolf}}$ and $w_{\delta,\text{wolf}}$. The value

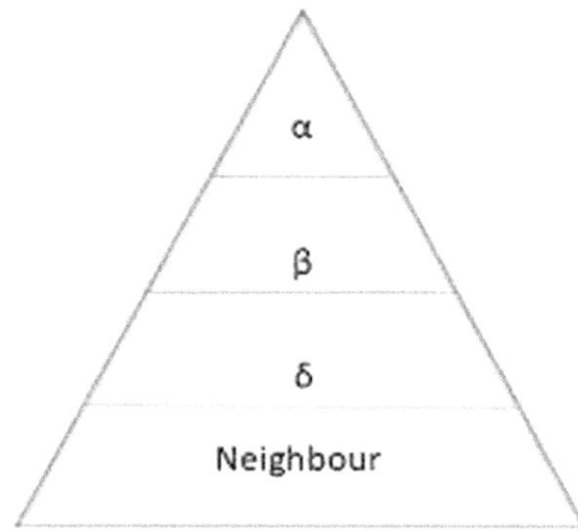

Fig. 2 Social behaviour of NIGWO

of w_4 is computed using the equations as follows 13 and 14.

$$\overrightarrow{w}_4 = \overrightarrow{w}_{\text{neigh,wolf}} - \overrightarrow{A} * \overrightarrow{D}_{\text{neigh,wolf}} \quad (13)$$

$$\overrightarrow{D}_{\text{neigh,wolf}} = |\overrightarrow{C}_{\text{coef}_4} * \overrightarrow{w}_{\text{neigh,wolf}} - \overrightarrow{w}_{\text{neigh,wolf}}| \quad (14)$$

Here, $C_{\text{coef}_4} \in C_{\text{coef}}$ and $w_{\text{neigh,wolf}}$ are the position vectors of the randomly selected forth fittest solution.

The proposed strategy is known as neighbourhood-inspired GWO (NIGWO). By adding this new neighbouring solution, the exploitation capacity increases. The principal contribution of the proposed work is to enhance exploitation capacity and efficiency of GWO. The social behaviour of NIGWO can be depicted, as shown in Fig. 2. As per the above consideration, the pseudocode of the proposed NIGWO is present in the Algorithm 2.

As per the above description, the pseudocode of the proposed NIGWO is depicted Algorithm 2:

4 Experimental Outcomes

To demonstrate the competence of the proposed NIGWO, 16 different global optimization problems f_{b1} to f_{b16} have been selected as depicted in Table 1. These problems have various complexity levels. A comparative analysis is carried out among neighbourhood-inspired grey wolf optimizer (NIGWO),grey wolf optimizer (GWO) [5], artificial bee colony (ABC) [11], power law-based local search in spider monkey optimization (PLSMO) [10], differential evolution (DE) [9] and particle swarm optimization (PSO) [2] to approve the compatibility of the proposed algorithm. To test them over the considered problems, the experimental set-up is carried out as follows:

- The total number of runs (TR) = 30
- Population size of grey wolves (PSGW) =50

Assume the W_p is the population grey wolf here p is 1 to n
while iter < maximum number of iteration **do**
  for every searching agent **do**
    Generate the current position of searching agent(grey wolf) by equation
    $\vec{D} = |C_{coef} * \vec{w}_{prey}(iter) - \vec{w}_{gw}(iter)|$
    $\vec{w}(iter+1) = \vec{w}_{prey}(iter) - \vec{A}_{coef} * \vec{D}$
    The coefficient vectors A_{coef} and C_{coef} can be calculated by the following equation
    $A_{coef} = 2a * r1_A - a$
    $C_{coef} = 2 * r2_C$
    here, $r1_A$ and $r2_C$ are random vectors between [0,1]
    Select a neighbour using randomly selected solution ($w_{neigh,wolf}$) from the search agent in such a way that it is not equal to alpha, β and delta. Most scrimping position for the encircle and hunt can be prevailed by the following equation.
    $\vec{w}_{(iter+1)} = \frac{\vec{w}_1 + \vec{w}_2 + \vec{w}_3 + \vec{w}_4}{4}$
  end for
  Renovate a, A_{coef} and C_{coef}
  Evaluate the fitness of all search agents
  Renovate $w_{\alpha,wolf}$, $w_{\beta,wolf}$, $w_{\delta,wolf}$ and $w_{neigh,wolf}$
  $iter = (iter + 1)$
end while
return $w_{\alpha,wolf}$

Algorithm 2: Neighbourhood-Inspired Grey Wolf Optimizer (NIGWO)

- Total number of emphasis or repitation (TE)=5000
- Maximum function evaluation (MFE)= 200000
- The parameter settings for GWO, ABC, DE, PLSMO and PSO have been taken from their underlying research paper respectively.

Table 2 furnishes the complete comparison between the contemplated algorithms. The arithmetical conclusions are presented in the form of success rate (SR), average function evaluation (AFE) and standard deviation (SD). It is fair from the outcomes shown in table 2 that NIGWO performs better as compared to GWO [5], ABC [11], DE [9], PLSMO [10] and PSO [2] in terms of accuracy, performance and efficiency.

Table 2 clearly specifies the efficiency and reliability of NIGWO in comparison with GWO [5], ABC [11], DE [9], PLSMO [10] and PSO [2].

Some other analytical tests like Mann–Whitney U rank sum test [4], AR test and boxplot are constructed to analyse the result more reliable.

To compare the overall observation in terms of AFE, boxplot is generated for the average function evaluation. The empirical distribution of the AFE can be graphically and efficiently understood by a boxplot. The boxplot for NIGWO, GWO [5], ABC [11], DE [9], PLSMO [10] and PSO [2] is represented in Fig. 3 which clearly shows that NIGWO is much efficient in terms of AFE because the median and range are very less for NIGWO as compared to GWO [5], ABC [11], DE [9], PLSMO [10] and PSO [2].

To compare the acceleration rate (AR) of NIGWO with GWO [5], ABC [11], DE [9], PLSMO [10] and PSO [2] , result of table 3 is analysed and through Eq. 15. The AR of GWO, ABC, DE, PLSMO and PSO is evaluated. AR is calculated as:

Table 1 Test problems

S. No.	Objective *function*	Search range	Dm	AE
1	$f_{b1}(x) = -\cos x_1 \cos x_2 e^{((-(x_1-\pi)^2-(x_2-\pi)^2))}$	[-100, 100]	2	$1.0E-13$
2	$f_{b2}(x) = 10D + \sum_{i=1}^{D}[x_i^2 - 10\cos(2\pi x_i)]$	[-5.12, 5.12]	30	$1.0E-05$
3	$f_{b3}(x) = \sum_{i=1}^{n}[\lvert x_{i\sin} x_i + 0.1x_i\rvert]$	[-10,10]	30	$1.0E-05$
4	$f_{b4}(x) = \sum_{i=1}^{D} x_i{}^2 + (\sum_{i=1}^{D} \frac{ix_i}{2})^2 + (\sum_{i=1}^{D} \frac{ix_1}{2})^4$	[-5.12,5.12]	30	$1.0E-02$
5	$f_{b5}(x) = x_0{}^2 + 100000\sum_{i=1}^{D} x_i{}^2$	[-10, 10]	30	$1.0E-05$
6	$f_{b6}(x) = \sum_{i=1}^{D-1}(x_i{}^{2(x_{i+1})^2+1} + x_{i+1}{}^{2x_i{}^2+1})$	[-1,4]	30	$1.0e-5$
7	$f_{b7}(x) = \sum_{i=1}^{n}\lvert x_i\rvert + \prod_{i=1}^{n}\lvert x_i\rvert$	[-10,10]	30	$1.0e-5$
8	$f_{b8}(x) = \sum_{i=1}^{D} i.x_i^2$	[-5.12,5.12]	30	$1.0e-5$
9	$f_{b9}(x) = \sum_{i=1}^{n-1}\left(\frac{\sin^2\left(\sqrt{x_{i+1}^2+100x_i^2}\right)-0.5}{0.001(x_{i+1}^2-2x_{i+1}x_i+x_i^2)^2+1.0} + 0.5\right)$	[-1,1]	30	$1.0E-01$
10	$f_{b10}(x) = \sum_{i=1}^{D}\lvert x_i\rvert^{i+1}$	[-1,1]	30	$1.0E-01$
11	$f_{b11}(x) = \sum_{i=1}^{n}(\lfloor x_i + 0.5\rfloor)^2$	[-100, 100]	30	$1.0e-5$
12	$f_{b12}(x) = \sum_{i=1}^{D} i.(x_i)^4$	[-1.28,1.28]	30	$1.0e-5$
13	$f_{b13}(x) = -\sum_{i=1}^{D-1}\left(\exp\left(\frac{-(x_i^2+x_{i+1}^2+0.5x_ix_{i+1})}{8}\right)\times \mathrm{I}\right)$	[-5,5]	10	$1.0e-5$
14	$f_{b14}(x) = \sum_{i=1}^{D}\frac{z_i^2}{4000} - \prod_{i=1}^{D}\cos(\frac{z_i}{\sqrt{i}}) + 1 + f_{\text{bias}}, z = (x-o), x = [x_1, x_2, \ldots x_D], o = [o_1, o_2, \ldots o_D]$	[-600,600]	10	$1.0e-5$
15	$f_{b15}(x) = (1 + (x_1 + x_2 + 1)^2 \cdot (19 - 14x_1 + 3x_1^2 - 14x_2 + 6x_1x_2 + 3x_2^2)) \cdot (30 + (2x_1 - 3x_2)^2 \cdot (18 - 32x_1 + 12x_1^2 + 48x_2 - 36x_1x_2 + 27x_2^2))$	[-2,2]	2	$1.0e-14$
16	$f_{b16}(x) = \sum_{i=1}^{D} 5i \times x_i^2$	[-5.12,5.12]	30	$1.0e-15$

Table 2 Comparison of outcomes of NIGWO, GWO, ABC, PLSMO, DE and PSO algorithms; TF:Test Function

TP	Algorithm	SR	AFE	SD
f_{b1}	NIGWO	30	88.33	0.03
	GWO	30	95.00	0.03
	ABC	30	17256.66	3.30E-06
	PLSMO	30	11198.90	1.15E-06
	DE	29	21062.06	1.73E-05
	PSO	30	31871.66	6.46E-07
f_{b2}	NIGWO	30	246.66	1.93796E-06
	GWO	30	2713.33	1.89684E-06
	ABC	30	70340.00	2.54E-06
	PLSMO	16	117022.10	8.23E-03
	DE	21	32988.09	4.99E-03
	PSO	29	220863.79	1.33E-03
f_{b3}	NIGWO	30	1723.33	7.42496E-07
	GWO	28	21898.33	2.40955E-05
	ABC	30	43613.33	4.08E-06
	PLSMO	30	50570.86	3.19E-06
	DE	4	72550.00	5.58E-02
	PSO	3	871966.66	1.55E-01
f_{b4}	NIGWO	30	1405.00	0.01
	GWO	30	7518.33	0.00
	ABC	30	63050.00	2.31E-06
	PLSMO	30	23150.20	7.49E-07
	DE	30	40451.66	1.34E-06
	PSO	30	69401.66	4.43E-07
f_{b5}	NIGWO	30	1858.33	2.07987E-06
	GWO	30	5973.33	1.15374E-06
	ABC	30	39670.00	1.72E-06
	PLSMO	30	13073.36	7.04E-07
	DE	30	22470.00	1.28E-06
	PSO	30	35786.66	5.03E-07
f_{b6}	NIGWO	30	1178.33	2.03782E-06
	GWO	30	3366.66	1.21913E-06
	ABC	30	80860.00	1.44E-06
	PLSMO	30	23699.60	5.87E-07
	DE	30	45693.33	6.92E-07
	PSO	30	71213.33	4.66E-07

(continued)

Table 2 (continued)

TP	Algorithm	SR	AFE	SD
f_{b7}	NIGWO	30	895.00	0.04
	GWO	30	3996.66	0.01
	ABC	30	42050.00	1.92E-06
	PLSMO	30	15136.86	7.52E-07
	DE	30	26296.66	1.58E-06
	PSO	30	44191.66	7.43E-07
f_{b8}	NIGWO	30	1275.00	2.3255E-06
	GWO	30	3895.00	9.97944E-07
	ABC	0.00	0.00	3.34E-01
	PLSMO	0	200052.10	8.97E-01
	DE	2	22775.00	1.68E+00
	PSO	0	0.00	1.04E-01
f_{b9}	NIGWO	30	1960.00	0.024864745
	GWO	2	189610.00	2.43
	ABC	30	28566.66	2.73E-06
	PLSMO	30	5334.63	1.40E-06
	DE	30	8666.66	1.99E-06
	PSO	30	9801.66	1.14E-06
f_{b10}	NIGWO	30	613.33	2.75054E-06
	GWO	30	1270.00	2.21962E-06
	ABC	30	20976.66	0.00E+00
	PLSMO	30	16922.16	0.00E+00
	DE	29	116579.31	1.80E-01
	PSO	30	35756.66	0.00E+00
f_{b11}	NIGWO	30	755.00	0
	GWO	30	1968.33	0
	ABC	0	0.00	3.28E-01
	PLSMO	0	200048.70	5.16E-01
	DE	0	0.00	3.36E-01
	PSO	0	0.00	4.45E-01
f_{b12}	NIGWO	23	103456.66	6.52083E-06
	GWO	1	198185.00	5.73624E-05
	ABC	12	126212.20	2.19E-06
	PLSMO	30	78461.16	2.30E-06
	DE	1	98200.00	6.11E-01
	PSO	6	533741.66	8.15E-01

(continued)

Table 2 (continued)

TP	Algorithm	SR	AFE	SD
f_{b13}	NIGWO	30	1401.66	2.44862E-06
	GWO	29	24970.00	0.492238821
	ABC	30	1174841.96	2.01E-07
	PLSMO	30	103017.80	7.34E-07
	DE	30	17758.33	1.75E-06
	PSO	30	67420.00	4.05E-07
f_{b14}	NIGWO	30	50.00	24242.26
	GWO	30	50.00	174.0798879
	ABC	12	1683.33	1.42E-05
	PLSMO	12	120365.20	1.41E-05
	DE	23	2260.86	8.15E-07
	PSO	17	2635.29	1.13E-05
f_{b15}	NIGWO	30	2698.33	2.08962E-16
	GWO	30	8963.33	1.29E-16
	ABC	30	121640.00	6.30E-17
	PLSMO	30	35040.03	6.92E-17
	DE	30	60066.66	1.29E-16
	PSO	30	104681.66	7.50E-17
f_{b16}	NIGWO	30	390.00	3.3921E-06
	GWO	30	390.00	2.73772E-06
	ABC	30	110686.66	1.28E-04
	PLSMO	30	37050.87	1.39E-04
	DE	30	68796.66	1.63E-04
	PSO	26	330305.76	3.35E-03

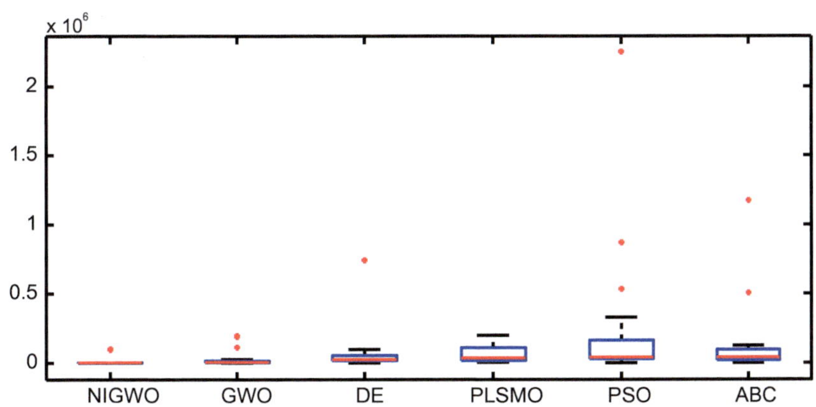

Fig. 3 Boxplot of average function evaluation

Table 3 Acceleration rate (AR) of NIGWO compared to the GWO, ABC, PLSMO, DE and PSO

TP	GWO	ABC	PLSMO	DE	PSO
f_{b1}	1.07	238.44	126.78	360.82	195.36
f_{b2}	30.71	373.46	1324.82	2500.43	796.33
f_{b3}	247.91	821.35	572.52	9871.69	493.75
f_{b4}	85.11	457.96	262.08	785.70	713.80
f_{b5}	67.62	254.38	148.00	405.13	449.11
f_{b6}	38.11	517.29	268.38	6.21	915.43
f_{b7}	45.24	297.70	171.36	500.20	476.00
f_{b8}	44.09	257.83	2264.8 0	0.00	0.00
f_{b9}	2146.60	98.11	60.39	110.90	323.40
f_{b10}	14.30	187.60	191.50	404.80	237.40
f_{b11}	22.24	0.00	2248.20	0.00	0.00
f_{b12}	2243.60	1111.74	888.27	6042.70	1428.80
f_{b13}	2926.83	1.46	1.44	1.46	1.46
f_{b14}	0.56	25.59	1362.60	29.83	19.05
f_{b15}	101.47	680.01	396.60	1185.12	1377.10
f_{b16}	4.41	778.80	419.44	3739.40	1253.09

$$AR = \frac{AFE_{ALGO}}{AFE_{NIGWO}} \tag{15}$$

Mann–Whitney U rank sum based on AFE at coefficient = 0.05 significance level is also done as shown in Table 4. This proposed algorithm performs outstanding for most of the standard problems as shown in Table 1. In comparison with GWO [5], NIGWO performs outstanding for 15 standard problems (f_{b1} to f_{b13}, f_{b15} and f_{b16}). In case of ABC [11], NIGWO performs much better for 15 distinct problems (f_{b1} to f_{b10} and f_{b12} to f_{b16}). In comparison with DE [9], it gives better performance for 13 standard problems (f_{b1} to f_{b7}, f_{b10} and f_{b12} to f_{b16}). In case of PLSMO [10], NIGWO performs much better for 15 distinct problems (f_{b1} to f_{b8} and $fb10$ to f_{b16}), while in case of PSO [2], NIGWO gives better performance for 13 distinct problems (f_{b1} to f_{b7}, f_{b10} and f_{b12} to f_{b16}).

5 Conclusion

This article presents a modified version of GWO, distinguished as neighbourhood-inspired grey wolf optimizer (NIGWO). In the proposed work, the solution is updated by incorporating the information from an arbitrarily selected neighbouring solution along with alpha, beta and delta solutions. The proposed NIGWO enhances the exploration capability of GWO. To certify the authenticity of NIGWO, 16 benchmark

Table 4 Comparison based on AFE and the Mann–Whitney U rank sum test at ('+' illustrate NIGWO is better, '−' illustrate NIGWO is worse and '=' illustrate that there is no noticeable difference)

TP	GWO	ABC	PLSMO	DE	PSO
f_{b1}	+	+	+	+	+
f_{b2}	+	+	+	+	+
f_{b3}	+	+	+	+	+
f_{b4}	+	+	+	+	+
f_{b5}	+	+	+	+	+
f_{b6}	+	+	+	+	+
f_{b7}	+	+	+	+	+
f_{b8}	+	+	+	−	−
f_{b9}	+	+	−	−	−
f_{b10}	+	+	+	+	+
f_{b11}	+	−	+	−	−
f_{b12}	+	+	+	+	+
f_{b13}	+	+	+	+	+
f_{b14}	−	+	+	+	+
f_{b15}	+	+	+	+	+
f_{b16}	+	+	+	+	+

functions of various levels and complexity are considered. The acquired outcomes are analysed with state-of-the art algorithms accessible in the literature. The listed outcomes validate the authenticity of the proposed NIGWO. In future, this proposed approach may be applied to solve some natural world optimization issues.

References

1. Coit DW, Smith AE (1996) Reliability optimization of series-parallel systems using a genetic algorithm. IEEE Trans Reliab 45(2):254–260
2. Jeyakumar DN, Jayabarathi T, Raghunathan T (2014) Particle swarm optimization for various types of economic dispatch problems. Int J Elect Power Energy Syst 28(1):2
3. Komaki GM, Kayvanfar V (2015) Grey wolf optimizer algorithm for the two-stage assembly flow shop scheduling problem with release time. J Comput Sci 8:109–120
4. McKnight PE, Najab J (2010) Mann-whitney u test. In: The Corsini encyclopedia of psychology, pp 1–1
5. Mirjalili S, Mirjalili SM, Lewis A (2014) Grey wolf optimizer. Adv Eng Soft 69:46–61
6. Mittal H, Pal R, Kulhari A, Saraswat M (2016) Chaotic kbest gravitational search algorithm (ckgsa). In: 2016 9th international conference on contemporary computing (IC3). IEEE, pp 1–6
7. Mittal N, Singh U, Sohi BS (2016) Modified grey wolf optimizer for global engineering optimization. Appl Comput Intell Soft Comput 2016:8

8. Pal R, Mittal H, Pandey A, Saraswat M (2016) Beecp: biogeography optimization-based energy efficient clustering protocol for hwsns. In: 2016 9th international conference on contemporary computing (IC3), IEEE, pp 1–6
9. Qin AK, Vicky LH, Suganthan PN (2009) Differential evolution algorithm with strategy adaptation for global numerical optimization. IEEE Trans Evol Comput 13(2):398–417
10. Sharma Ajay, Sharma Harish, Bhargava Annapurna, Sharma Nirmala (2017) Power law-based local search in spider monkey optimisation for lower order system modelling. Int J Syst Sci 48(1):150–160
11. TSai P-W, Pan J-S, Liao B-Y, Chu S-C (2009) Enhanced artificial bee colony optimization. Int J Inn Comput Inf Cont 5(12):5081–5092
12. Xu H, Liu X, Su J (2017) An improved grey wolf optimizer algorithm integrated with cuckoo search. In: 2017 9th IEEE international conference on intelligent data acquisition and advanced computing systems: technology and applications (IDAACS)

Chapter 12
The Analysis of Machine Learning Techniques for Heart Disease Prediction

Surendra Singh and Vishal Shrivastava

1 Introduction

There are large amount of data being stored in files, databases, and many other appliances, as the confidential and the private data should not be stored anywhere else. So, it is very important to discover and propose a system in which all the data and information can be stored safely and securely. Sometimes it becomes very difficult for the users to extract and use only relevant data from this large data. So, in order to overcome this situation, data mining is used. Data Mining is the process of selecting, choosing, and extracting only that data that is useful and relevant for that particular instant of time. It allows the user to access their data anytime and from anywhere [1]. Data mining is commonly known as Knowledge Discovery in Databases (KDD), which is non-trivial extraction of implicit, previously stored, and most popularly used information from the stored data. There are large amount of relevant and irrelevant data being stored on the databases and many other areas. It has given rise to the term data mining, which can be further useful in decision-making process. It is the phenomenon of extracting useful and important data from a large amount of data being stored almost everywhere on the internet. It mostly deals with the already stored and gathered data for any other purpose instead of data mining purposes [2]. This shows that the main objective of data mining is to collect and choose the relevant data from the previously stored data. There are enormous types of data sets being used in data mining. Heart diseases are one of the major causes of death nowadays. Smoking, consumption of alcohol in large quantities, cholesterol, and pulse rate are the reason for heart diseases. The heart is the operating

S. Singh (✉) · V. Shrivastava
Department of Computer Science and Engineering, Arya College of Engineering and IT, Jaipur, Rajasthan, India
e-mail: Er.Surendrait@gmail.com

V. Shrivastava
e-mail: Vishalshrivastava.cs@aryacollege.in

© Springer Nature Singapore Pte Ltd. 2021

S. D. Purohit et al. (eds.), *Proceedings of International Conference on Communication and Computational Technologies*, Algorithms for Intelligent Systems,
https://doi.org/10.1007/978-981-15-5077-5_12

system of human body, if it will not function properly then it will directly affect the functioning of the other body parts. Some of the major factors that lead to heart diseases are family history, high blood pressure, high rate of cholesterol, age, poor diet, and many more. The stretching of blood vessels will increase the blood pressure which will further cause the cardiac rest. Smoking is one of the major causes of heart diseases; almost 40% of the population is dying because of this. Because it limits the supply of oxygen in our body and prevents the proper flow of blood and tightens the blood vessels [3]. Different types of data mining techniques are employed for the prediction of data mining like Naïve Bayes, KNN algorithm, Decision tree, Neural Network. In KNN algorithm K user-defined value is used to find the values of factors that lead to heart diseases. Decision tree is used to deploy a classified report on the heart suffering patients. The Naïve Bayes is employed to predict the probability of heart diseases. Last but not the least, the neural networks are used to minimize the errors occurred at the time of prediction. By using all these techniques, the records are classified as well as maintained regularly. The activity of every patient is properly checked, if there is any change, and then the level of risk is informed to the patients. With the help of all these classifiers, the doctors are able to predict the heart diseases at the very initial stage [4]. There are various classifiers used for the prediction of heart diseases like decision tree, SVM, neural networks, Naïve Bayes, and KNN. These classifiers are employed and used so that the risk of heart diseases can be predicted at an early stage. This will help the doctor as well to the patient to have the proper treatment on time. This will help to diagnose the diseases at its early stages. Decision tree frames the classification in the form of tree. It divides the datasets into smaller and smaller subsets while on the other hand decision tree is incrementally developed. The final structure is the tree along with the decision nodes and leaf nodes. A decision node has two or more than two branches; leaf denotes the classification or decision. Root node is the top node which is the best predictor [5]. It can handle both the numerical as well as categorical data. Naïve Bayes Classification Algorithm is denoted by the supervised learning method as well as the statistical method used for the classification. It allows capturing uncertain conditions about the model in a fundamental way by determining the possibilities of the outcomes. It is useful in solving diagnostic and predictive issues. It is used to solve practical learning algorithms in which prior knowledge and observed data is combined. It underlines the perspective used for the evaluation of the learning algorithms. It evaluates the explicit probabilities and its robust noise in the input data [6]. The classifier that utilizes training samples in order to perform learning in it is known as K-nearest neighbor classifier. A point within an n-dimensional space is represented by each sample. Further, within this n-dimensional pattern space, all the training samples are stored. The pattern space for k-training samples which are nearest to the unknown sample is identified by the K-nearest neighbor classifier in the case when an unknown sample is given. In terms of Euclidean distance, the closeness is defined here. Support Vector Machine (SVM) is an effective classifier that is used [7] to perform regression, classification as well as general pattern recognition and hence it is called SVM classifier. It highly generalized the performance without including any prior knowledge even in the case when the input space is very high;

this classifier is one of the best classifier amongst all. It is used for binary classification and for solving multiclass issues.

2 Literature Review

Babu et al. introduced [8] a technique called data mining, which is the mechanism of discovering new set of information from huge data. It is used to analyze large volume of data and the patterns were extracted to convert the irrelevant information into useful information. This collected information is fed into several classifiers like KNN, Naïve Bayes, SVM, and decision trees, each of them perform some specific tasks. These techniques are used to predict heart diseases at an early stage. It is shows very effective performance in order to achieve the correct and perfect diagnose for the heart-related diseases. There are certain advantages of this approach such as the diseases that can be predicted at their very initial stages and can be diagnosed correctly and properly on time. Therefore, the researcher concluded that this method is very useful in preventing heart-related issues.

Islam et al. proposed a non-invasive technique to detect the heart rate from Photoplethysmography (PPG) signal. [9] This proposed technique plays a very major role in the diagnosis of heart-related disease. A suitable heart disease detection mechanism is used with the help of PPG and extracted from human facial videos. This can overcome the expensive ECG machine for heart-related disease detection. The FFT can detect the heart rate and re-check with the conventional heart rate using ECG machines. The contactless approach provides the solution and identifies the heart-related disease it could be diagnosed by the doctor. Therefore, the result of this approach concludes that it ensures safety by diagnosing the diseases by its early symptoms. From the study and the experimental results it also concludes that green channel predicts more accurate heart rate than any other. It has been found 97.7% accurate results from green channel.

KarayÕlan and KÕlÕç proposed backpropagation algorithm for the prediction of heart diseases with the help of artificial neural networks [10]. It has some clinical features in which neural networks are used as input and are trained along with this proposed back propagation technique. It can predict the heart-related diseases with an accuracy of 95%. The already proposed methods were not sufficient for the early prediction of heart diseases. The advancement of technology will lead to the use of machine learning techniques for the prediction of cardiac diseases at their initial stages. Therefore, the researcher draws the conclusion that the proposed approach has almost 100% accuracy in prediction heart-related diseases at their early stages. It gives better results in comparison to the other techniques.

Mahboob et al. introduced various learning practices that assist the detection of innumerable heart disease [11]. There are certain techniques were used like Hidden Markov Models, Support Vector Machine, Computational intelligent classifier, and data mining, and so on. As cardiac disease treatment is very expensive and unaffordable to any normal individual, these types of advanced technology are developed to overcome this problem. These techniques are also useful in early stage predictions. It avoids any other future sufferings by making slight changes in daily routine. Hence, the author concludes that the predicted approach has several advantages and is very useful.

Nag et al. proposed [12] a very effective technique that is very useful in the prediction of heart diseases at the initial stage. The researcher has developed a prototype called Acute Myocardial Infarction (AMI). Heart attack have various symptoms like chest pain, breathing problems, palpitation, vomiting, and continuous sweating. Therefore, the researcher draws the conclusion that the advancement of computer technology in medical and health region provides useful aids and people are becoming more dependent on these technologies. The results of data mining are very beneficial and are used for better assistance to many physicians as lot of data is related to diseases.

Pouriyeh et al. proposed a machine learning method for the detection of heart diseases [13]. The data related to the heart diseases consist of 303 instances which are used as control house of the database for the training and development of the system. For the increment in the amount of data, 10-Fold Cross-Validation has been used, otherwise, there would be a very limited amount of data available. There are several approaches have used like; Decision Tree, Naïve Bayes, Multilayer Perceptron, Support Vector Machine, and so on. The efficiency of each of these classifier employing bagging, boosting, and stacking techniques have been accurately calculated. The main objective of this approach was to compare the proposed algorithm with different machine learning algorithms. Therefore, the researcher draws the conclusion that the proposed method improves the accuracy of the already existed techniques just to have a better comparison between them. The author also concludes that the SVM technique boosts and outperforms as compared to any other.

Priyanga and Naveen proposed an intelligent and efficient technique called Naïve Bayes for the prediction of heart-related issues [14]. The data is collected from the given attributes and then they are implemented as web-based applications. All the methods which are already been discovered and used don not show effective and satisfactory results for the prediction of heart diseases. But all the techniques are introduced for the very same purpose that is for predicting severe diseases like a cardiac crest, cancer, brain tumor. But they were failed to discover the diseases at their initial stage. Therefore, the researcher concludes that the approach classifies that it has low cost and extensively tested by experienced cardiologists. The research mainly focuses on detection of heart disease using UCI dataset.

Priyanka and RaviKumar proposed [15] two methods of data mining that is Naïve Bayes and decision tree for the early prediction of severe diseases like brain tumor, cancer, and heart attack. The results and accuracy of both the introduced methods are compared and then decides which one is better and most widely used. Due to deficient resources available in the medical and health region, it was quite difficult to predict the severe diseases on time. Appropriate and suitable technologies should be used to get better and effective results. It is very important to provide good facilities for the hospitals and patients. Data mining is specially designed to overcome this issue. Therefore, the study of the respective paper concludes that both the methods were compared and results that both are suitable and best at different fields, also provides a more accurate and efficient result for heart diseases detection.

Authors' names	Year	Description	Outcomes	Advantages	Disadvantages
Sarath Babu	2017	There are certain advantages of this approach such as the diseases that can be predicted at their very initial stages and can be diagnosed correctly and properly on time	The researcher concluded that this method is very useful in preventing heart-related issues	The accuracy of the proposed model is very high approx to 90%	Due to use of many models, the execution time is increased
Monira Islam	2017	A non-invasive technique was proposed to detect the heart rate from Photoplethysmography (PPG) signal	From the study and the experimental results it also concludes that green channel predicts more accurate heart rate than any other. It has been found 97.7% accurate results from green channel	The proposed model is less complex and also features are extracted efficiently	The proposed model is good but the number of attributes considered is very less
Tülay KarayÕlan	2017	Backpropagation algorithm was proposed for the prediction of heart diseases with the help of artificial neural networks	The proposed approach has almost 100% accuracy in prediction heart-related diseases at their early stages. It gives better results in comparison to the other techniques	The system is designed which give very high accuracy	The complexity and time is very high due to training of system again and again
Tahira Mahboob	2017	As the cardiac diseases treatment is very expensive and unaffordable to any normal individual so, these types of advanced technology are developed to overcome this problem	The author concludes that the predicted approach has several advantages and is very useful	The author reviews various technique which is already proposed for the disease prediction	The model are just reviewed not implemented for the result analysis

(continued)

(continued)

Authors' names	Year	Description	Outcomes	Advantages	Disadvantages
Procheta Nag	2017	The researcher has developed a prototype called Acute Myocardial Infarction (AMI). Heart attack having various symptoms	The results of data mining are very beneficial and are used for better assistance to many physicians as a lot of data is related to diseases		
Seyedamin Pouriyeh	2017	The data related to the heart diseases consist of 303 instances which are used as control house of the database for the training and development of the system. For the increment in the amount of data 10-Fold Cross-Validation has been used, otherwise there would be very limited amount of data available	The proposed method improves the accuracy of the already existed techniques just to have a better comparison between them. The author also concludes that the SVM technique boosts and outperforms as compared to any other	The proposed model gives result on a large number of dataset	The proposed model can be validated over certain number of instances
Priyanga	2017	An intelligent and efficient technique called Naïve Bayes was used by the proposed technique for the prediction of heart-related issues	The approach classifies that it has low cost and extensively tested by experienced cardiologists	The proposed model is applied using Naïve Bayes classifier	The complexity of the system can be further reduced
Priyanka N	2017	Two methods were used that are Naïve Bayes and decision tree for the early prediction of severe diseases like brain tumor, cancer, and heart attack	Both the methods were compared and result that both are suitable and best in different fields, also provides a more accurate and efficient result for heart diseases detection	The proposed is the combination of more than two classifiers which increase accuracy	The classifiers which are combined are complex which increase execution time

3 Conclusion

In this paper, it is concluded that machine learning is the technique which can learn from previous experiences and drive new values. The prediction analysis is the approach of machine learning [16]. The heart disease prediction is the problem which is solved with the machine learning techniques. The various machine learning techniques are reviewed for heart disease prediction. In the future, the novel approach of machine learning will be proposed which gives high accuracy for heart disease prediction.

References

1. Disease prediction using different data mining techniques (2012) IJERT 1(8)
2. Amin SU, Agarwal K, Beg R (2013) Genetic neural network based data mining in prediction of heart disease using risk factors. IEEE
3. Chen AH, Huang SY, Hong PS, Cheng CH, Lin EJ (2011) HDPS: heart disease prediction system. IEEE
4. Jabbar MA, Deekshatulu BL, Chandra P (2013) Heart disease prediction using lazy associative classification. IEEE
5. Dangare CS, Apte SS (2012) Improved study of heart disease prediction system using data mining classification techniques. IJCA 47(10)
6. Bhandari P, Yadav S, Mote S, Rankhambe D (2016) Predictive system for medical diagnosis with expertise analysis. IJESC 6:4652–4656
7. Nishara Banu G (2014) Disease forecasting system using data mining methods. IEEE Trans Intell Comput Appl
8. Babu S, Vivek EM, Famina KP, Fida K, Aswathi P, Shanid M, Hena M (2017) Heart disease diagnosis using data mining technique. IEEE
9. Islam M, Ashikuzzaman M, Tabassum T, Yusuf MSU (2017) A non-invasive technique of early heart diseases prediction from photoplethysmography signal. IEEE
10. KarayÕlan T, KÕlÕç Ö (2017) Prediction of heart disease using neural network. IEEE
11. Mahboob T, Irfan R, Ghaffar B (2017) Evaluating ensemble prediction of coronary heart disease using receiver operating characteristics. IEEE
12. Nag P, Mondal S, Ahmed F, More A, Raihan M (2017) A simple acute myocardial infarction (heart attack) prediction system using clinical data and data mining techniques. IEEE
13. Pouriyeh S, Vahid S, Sannino G, De Pietro G, Arabnia H, Gutierrez J (2017) A comprehensive investigation and comparison of machine learning techniques in the domain of heart disease. IEEE
14. Priyanga P, Naveen NC (2017) Web analytics support system for prediction of heart disease using Naïve Bayes weighted approach (NBwa). IEEE
15. Priyanka N, RaviKumar P (2017) Usage of data mining techniques in predicting the heart diseases—Naïve Bayes & decision tree. IEEE
16. Panwar D, Tomar P, Singh V (2018) Hybridization of Cuckoo-ACO algorithm for test case prioritization. J Stat Manag Syst 21(4):539–546

Chapter 13
The Analysis of Stock Market Prediction Techniques

Swati Chedwal and Vishal Shrivastava

1 Introduction

Data mining can be described as "making better use of data." Every human being is increasingly faced with unmanageable amounts of data; hence, data mining or knowledge discovery apparently affects all of us. It is therefore recognized as one of the key research areas. Ideally, we would like to develop techniques for "making better use of any kind of data for any purpose." However, we argue that this goal is too demanding yet [1]. Over the last three decades, increasingly large amounts of historical data have been stored electronically and this volume is expected to continue to grow considerably in the future. Yet despite this wealth of data, many fund managers have been unable to fully capitalize on their value. Recently, data mining techniques and artificial intelligence techniques like decision trees, rough set approach, and artificial neural networks have been applied to this area. Data mining refers to extracting or mining knowledge from large data stores or sets [2]. Some of its functionalities are the discovery of concept or class descriptions, associations and correlations, classification, prediction, clustering, trend analysis, outlier and deviation analysis, and similarity analysis. Data classification can be done in many different methods; one of those methods is the classification by using decision tree. It is a graphical representation of all possible outcomes and the paths by which they may be reached. Following the assumption of technical analysis that patterns exist in price data, it is possible in principle to use data mining techniques to discover these patterns in an automated manner. Once these patterns have been discovered, future prices can be predicted [3]. Today, the grand challenge of using a database is

S. Chedwal (✉) · V. Shrivastava
Department of Computer Science and Engineering, Arya College of Engineering and IT, Jaipur, Rajasthan, India
e-mail: swatichedwal@gmail.com

V. Shrivastava
e-mail: vishal500371@yahoo.co.in

© Springer Nature Singapore Pte Ltd. 2021
S. D. Purohit et al. (eds.), *Proceedings of International Conference on Communication and Computational Technologies*, Algorithms for Intelligent Systems,
https://doi.org/10.1007/978-981-15-5077-5_13

to generate useful rules from raw data in a database for users to make decisions, and these rules may be hidden deeply in the raw data of the database. Traditionally, the method of turning data into knowledge relies on manual analysis; this is becoming impractical in many domains as data volumes grow exponentially. The problem with predicting stock prices is that the volume of data is too large and huge. Public companies raise funds by issuing shares to public and institutions. By purchasing shares, an investor acquires partial ownership of company. A stock market is a place where public listed company stocks are traded. Stocks are exchanged among buyer and seller which generate transaction data. Prices are changing as per demand and supply of stocks [4]. All trading data is captured by stock exchange where stock companies are listed. Stock trading data is nonlinear, fluctuating hence highly time variant. A lot of information is hiding in this huge data captured by stock exchange is difficult and time-consuming for human being to extract without powerful tools. Thus, predicting future price of stock is highly challenging. Stock prices are changing due to demand and supply of stocks in stock market. Based on past history and current information, investors are buying or selling stocks of listed companies. Investors, stock analyst, and stock brokers are predicting demand and supply of stocks after studying fundamentals and technical information of stocks [5]. In order to study and analyze stocks data, investor needs to go through vast amount of data to discover hidden patterns which is very cumbersome and tiring task. Usually stock analysts analyze stocks based on three vital aspects—external environmental factors, fundamental analysis, and technical analysis. Several stock market prediction theories have been proposed by various researchers. Efficient market hypothesis was an idea developed in 1965. EMH states that the price of a security will reflect the whole market information. As soon as there is any information indicating that a security is underpriced and therefore offers a profit opportunity, investors will immediately buy it and its price will rise up to a fair value. There were three forms of financial market efficiency introduced: weak, semi-strong, and strong [6]. In weak-form efficiency, future prices cannot be predicted by analyzing prices from the past; hence, the technical analysis was unreliable. In semi-strong-form efficiency, the price will rapidly adjust to publicly available new information; neither fundamental analysis nor technical analysis techniques are able to reliably produce excess returns. And in strong-form efficiency, the price will reflect all information including insider information; no one can earn excess returns. Behavioral finance explains why and how markets might be inefficient. It is to study the influence of psychological, social, cognitive, and emotional factors on the security and the markets. The historical patterns and the current market circumstance can psychologically affect the investor, and their respond will lead to a predictable market trend. Classification consists of predicting a certain outcome based on a given input. In order to predict the outcome, the algorithm processes a training set containing a set of attributes and the respective outcome, usually called goal or prediction attribute [7]. The algorithm tries to discover relationships between the attributes that would make it possible to predict the outcome. Next, the algorithm is given a data set not seen before, called prediction set, which contains the same set of attributes, except for the prediction attribute—not yet known. The algorithm analyzes the input and produces a prediction. The prediction accuracy defines how "good" the algorithm is.

Decision tree is a classification method which yields output as a flowchart-like tree structure. The result from D-Tree is highly interpretable, but the outcome must be represented in categorical data [8]. Mostly, D-Tree algorithm called "J48" is applied to classify future stock market direction. Naïve Bayes is a simple probabilistic classifier based on Bayes' theorem, with a naive assumption of independence between every pair of features. K-nearest neighbor (k-NN) is a non-parametric lazy learning algorithm that predicts class of the object based on the k closest training examples in the feature space. An object is classified by a majority vote of its neighbors; the object will be assigned to the class most common among its k-nearest neighbors ($k = 15$ in this study).

2 Literature Review

Zhang et al. [9] proposed the utilization of Elman network to predict the opening price of stock market. Typically, Elman neural network is a local recurrent neural network, having one context layer that memorizes the past states, which is quite fit for resolving time series issues [9]. Afterward, the optimized data, regarded as initial weight and threshold value, is given to Elman network for training, accordingly the prediction model for opening price of stock market based on self-adapting variant PSO-Elman network is formed. Finally, this paper verifies that model by some stock prices and compares with BP network and Elman network, so as to draw the result that shows the precision and stability of this predication model both are superior to the traditional neural network.

Nonita Sharma et al. [10] emphasized on the prediction of future stock market index values based on historical data. The experimental evaluation is based on historical data of 10 years of two indices, namely CNX Nifty and S&P Bombay Stock Exchange (BSE) Sensex from Indian stock markets [10]. The prediction performance of the proposed model is compared with that of well-known support vector regression. Technical indicators are selected as inputs to each of the prediction models. The closing value of the stock price is the predicted variable. Results show that the proposed scheme outperforms support vector regression and can be applied successfully for building predictive models for stock prices prediction.

Yaojun Wang et al. [11] used the social media mining technology to quantitative evaluation market segment, and in combination with other factors to predict the stock price trend in short term. In this paper, starting from the efficient market hypothesis, the authors fetch the stock comments information from social media and then preprocessing the data to emotion vectors [11]. By calculating the segment value of each stock's emotion vector, we found that the segment value is very sensitivity to the stock price movement. In addition, through testing our social media mining algorithm we discovered that the SVM model contains segment index and has higher prediction accuracy than SVM model not combined segment index. Experiment results show that by using social media mining combined with other information, the stock prices prediction model can forecast more accurate.

Mustain Billah et al. proposed an improved Levenberg–Marquardt (LM) training algorithm of artificial neural network [12]. Improved Levenberg–Marquardt algorithm of neural network can predict the possible day-end closing stock price with less memory and time needed, provided previous historical stock market data of Dhaka Stock Exchange such as opening price, highest price, lowest price, and total share traded. Moreover, improved LM algorithm can predict day-end stock price with 53% less error than ANFIS and traditional LM algorithm. It also requires 30% less time, 54% less memory than traditional LM and 47% less time, 59% less memory than ANFIS. Moreover, it can be used for any prediction purposes. In future, we will use this concept in predicting network traffic prediction purpose.

R. Yamini Nivetha et al. [13] analyzed various prediction algorithms to build a prediction model. The prediction model will be based on monthly prediction and daily prediction to forecast the next-day market price [13]. This model estimates the open value of the next day in the market. A comparative study of these three algorithms which are multiple linear regression, support vector machine, and artificial neural network is done. The stock price is predicted by sentiment analysis with the best forecasting algorithm. The result exhibits that the deep learning algorithm performs better than the MLR and SVM. In deep learning algorithm, the hidden layer neuron learns in every prediction. Hence, the output layer neuron produces the best outcome. Artificial neural network is the best predicting algorithm.

Harun Ercan used artificial neural networks to predict OMX Baltic Benchmark GI (OMXBBGI) market index value [14]. The research performed on the Baltic stock market uses index value and EUR/USD exchange rate to predict the index. After the inputs have been defined, the data has been gathered for the period of January 1, 2013, through January 1, 2017. The training set is determined to include about 70% of the data set, 15% will be used for testing, and the rest of 15% will be used for validation purposes. MATLAB ANN tool is used to complete the tests. The results showed that Baltic Market values are successfully predicted by NARX methodology with the variables: exchange rate and previous days' index values. The uniqueness of the paper especially comes from variables that have never been tried to forecast Baltic Stock Market values by ANN.

Muhammad Waqar et al. investigated the problem of high dimensionality of stock exchange to predict the market trends by applying the principal component analysis (PCA) with linear regression [15]. PCA can help to improve the predictive performance of machine learning methods while reducing the redundancy among the data. Experiments are carried out on a high-dimensional spectral of 3 stock exchanges such as: New York Stock Exchange, London Stock Exchange, and Karachi Stock Exchange. The accuracy of linear regression classification model is compared before and after applying PCA. The experiments show that PCA can improve the performance of machine learning in general if and only if relative correlation among input features is investigated and careful selection is done while choosing principal components. Root mean square error (RMSE) is used as an evaluation metric to evaluate the classification model.

Vivek Rajput et al. constructed a model to predict stock value movement using the opinion mining and clustering method to predict National Stock Exchange (NSE)

[16]. The author used domain specific approach to predict the stocks from each domain by taking some stock with maximum capitalization. The proposed methodology will give us two outputs set, i.e., one from sentiment analysis and another from clustering-based prediction with respect to some specialized parameters of stock exchange. By examining both the results, an efficient prediction is produced. In this paper, stocks with maximum capitalization within all the important sectors are taken into consideration for empirical analysis [17] (Table 1).

Table 1 Comparison analysis

Authors' names	Year	Description	Outcomes
Ze Zhang, Yongjun Shen, Guidong Zhang, Yongqiang Song, Yan Zhu	2017	The utilization of Elman network was done in the proposed research to predict the opening price of stock market	Results show that the precision and stability of this predication model both are superior to the traditional neural network
Nonita Sharma, Akanksha Juneja	2017	The prediction performance of the proposed model is compared with that of well-known support vector regression	Results show that the proposed scheme outperforms support vector regression and can be applied successfully for building predictive models for stock prices prediction
Yaojun Wang, Yaoqing Wang	2016	In this paper, starting from the efficient market hypothesis the authors fetch the stock comments information from social media and then preprocessing the data to emotion vectors	Experiment results show that by using social media mining combined with other information, the stock prices prediction model can forecast more accurate
Mustain Billah, Sajjad Waheed, Abu Hanifa	2016	Improved Levenberg–Marquardt algorithm of neural network can predict the possible day-end closing stock price with less memory and time needed	Improved LM algorithm can predict day-end stock price with 53% less error than ANFIS and traditional LM algorithm. It also requires 30% less time, 54% less memory than traditional LM and 47% less time, 59% less memory than ANFIS
R. Yamini Nivetha, Dr. C. Dhaya	2017	A comparative study of these three algorithms which are multiple linear regression, support vector machine, and artificial neural network is done	The output layer neuron produces the best outcome. Artificial neural network is the best predicting algorithm

(continued)

Table 1 (continued)

Authors' names	Year	Description	Outcomes
Harun Ercan	2017	The research was performed on the Baltic stock market uses index value and EUR/USD exchange rate to predict the index	The uniqueness of the paper especially comes from variables that have never been tried to forecast Baltic Stock Market values by ANN
Muhammad Waqar, Hassan Dawood, Muhammad Bilal Shahnawaz, Mustansar Ali Ghazanfar, Ping Guo	2017	The problem of high dimensionality of stock exchange was investigated to predict the market trends by applying the principal component analysis (PCA) with linear regression	The experiments show that PCA can improve the performance of machine learning in general if and only if relative correlation among input features is investigated and careful selection is done while choosing principal components
Vivek Rajput, Sarika Bobde	2016	A model was constructed to predict stock value movement using the opinion mining and clustering method to predict National Stock Exchange (NSE)	In this paper, stocks with maximum capitalization within all the important sectors are taken into consideration for empirical analysis

3 Conclusion

In this paper, it is concluded that stock market prediction is the problem which can solved with the machine learning techniques. The stock market prediction techniques has the three various steps which are pre-processing, feature extraction and classification. It is analyzed that machine learning techniques are best performing for the stock market prediction. In future, novel approach will be proposed for the stock market prediction which gives high accuracy.

References

1. Qamar S, Adil SH (2009) Comparative analysis of data mining techniques for financial data using parallel processing. In: Proceedings of the 7th international conference on frontiers of information technology
2. Adil SH, Qamar S (2009) Implementation of association rule mining using CUDA. In: International conference on emerging technologies (ICET)
3. Ali SSA, Moinuddin M, Raza K, Adil SH (2014) An adaptive learning rate for RBFNN using time-domain feedback analysis. Sci World J
4. Asur S, Huberman, BA (2010) Predicting the future with social media. In: International conference on web intelligence and intelligent agent technology (WI-IAT)

5. Aurangzeb CDR (2012) Factors affecting performance of stock market: evidence from South Asian countries. Int J Acad Res Business Social Sci 2(9)
6. Shen S, Jiang H, Zhang T (2012) Stock market forecasting using machine learning algorithms
7. Ali I, Rehman KU, Yilmaz AK, Khan MA, Afzal H (2010) Causal relationship between macroeconomic indicators and stock exchange prices in Pakistan. Afr J Bus Manag 4(3):312
8. Bollen J, Mao H, Zeng X (2011) Twitter mood predicts the stock market. J Comput Sci 2(1):1–8
9. Zhang Z, Shen Y, Zhang G, Song Y, Zhu Y (2017) Short-term prediction for opening price of stock market based on self-adapting variant pso-elman neural network. IEEE
10. Sharma N, Juneja A (2017) Combining of random forest estimates using lsboost for stock market index prediction. In: 2017 2nd International conference for convergence in technology (I2CT)
11. Wang Y, Wang Y (2016) Using social media mining technology to assist in price prediction of stock market. In: 2016 IEEE International conference on big data analysis (ICBDA)
12. Billah M, Waheed S, Hanifa A (2016) Stock market prediction using an improved training algorithm of neural network. In: 2nd International conference on electrical, computer & telecommunication engineering (ICECTE), 8–10 Dec 2016
13. Yamini Nivetha R, Dhaya C (2017) Developing a prediction model for stock analysis. In: 2017 International conference on technical advancements in computers and communications
14. Ercan H (2017) Baltic stock market prediction by using NARX. In: 12th International Scientific and technical conference on computer sciences and information technologies (CSIT)
15. Waqar M, Dawood H, Shahnawaz M,, Ghazanfar MA, Guo P (2017) Prediction of stock market by principal component analysis. In: 2017 13th International conference on computational intelligence and security
16. Rajput V, Bobde S (2016) Stock market prediction using hybrid approach. In: International conference on computing, communication and automation (ICCCA2016)
17. Agarwal P (2019) Sustainable smart-farming framework: smart farming. In: Smart farming technologies for sustainable agricultural development. IGI Global, pp 147–173

Chapter 14
Mitigating the Impact of DDoS Attack on Upsurge Network Performance in MANET

Chour Singh Rajpoot, Amit Kumar Bairwa, and Vijay Kumar Sharma

1 Introduction

A mobile ad hoc network (MANET) is a set of wireless mobile nodes which may form a temporary network, without the use of any infrastructure or centralized administration [1]. Nodes calculate on multi-hop routing protocols to forward data packets sent from a source node to a destination node which is out of its transmission range. Every node may function as both a data source and a router that forward data for other nodes [2].

1.1 MANET Features

1. Autonomous terminal: In MANET, every mobile node can be a self-determining node, which can work as both sender and receiver. In other words, besides the basic process ability as a bunch, the mobile nodes can also perform change functions as a router. Therefore, usually endpoints and switches are identical in MANET [3].
2. Distributed operation: There is no central control over the network for the central management of the network operations. In wireless network, the terminal nodes work as central control of management in the networks [4]. The nodes concerned during a MANET ought to collaborate among themselves, and every node acts as a propagation as required to implement functions, e.g., security and routing.

C. S. Rajpoot (✉)
Rajasthan Institute of Engineering and Technology, Jaipur, Rajasthan, India
e-mail: csrajput22@gmail.com

A. K. Bairwa · V. K. Sharma
Manipal University Jaipur, Jaipur, Rajasthan, India

© Springer Nature Singapore Pte Ltd. 2021

S. D. Purohit et al. (eds.), *Proceedings of International Conference on Communication and Computational Technologies*, Algorithms for Intelligent Systems,
https://doi.org/10.1007/978-981-15-5077-5_14

3. Multi-hop routing: It is based on single-hop and multi-hop concept that is known as accidental routing algorithms and totally supported by different link-layer attributes and routing protocols. Single-hop MANET is better than multi-hop in terms of structure and implementation [5], with the value of lesser practicality and relevance. Once delivering information packets from a supply to its destination out of the direct wireless transmission space, the packets ought to be forwarded via one or a lot of intermediate nodes.
4. Dynamic network topology: Since the nodes are mobile, the configuration might modify quickly and erratically and therefore the characteristics among the terminals may change with time. MANET ought to accept the traffic and propagation conditions in addition because of the quality arrangement of the mobile network nodes. Nodes inside the MANET dynamically establish the connections among them, as mobility of the node is depending on the setup of the network [6]. Moreover, in MANET a single user cannot operate inside the accidental network; therefore, user has to access a public connected network (e.g., Internet).
5. Unsteady link capacity: The essence of high bit error rates of wireless association may be a lot of profound during a MANET. One end-to-end path will be shared by many sessions [7–9]. The channel over that the terminals communicate relies on noise, fading, and interference, and has less information measure than a wired network. In some situations, the trail between any combinations of users will traverse multiple wireless links and therefore the link themselves will be dissimilar [10].
6. Lightweight terminals: In most cases, the MANET nodes are mobile devices with less processing capability, tiny memory size, and low power storage. Such devices required more precious algorithms and tools to implement computing and execute functions in the network [11–13].

2 Survey on Different Types of Attacks

Different types of security attacks are discussed here [14].

2.1 Physical Layer Attacks

Eavesdropping is that prevents reading of messages and conversations by accidental receivers. The mobile hosts in mobile unplanned networks share a wireless intermediate [15]. The wireless communications mostly use the RF spectrum and broadcast naturally. Signals broadcast over airwaves may be simply intercepted with receivers tuned to the right frequency. Thus, messages unfold may be eavesdropped, and faux messages may be injected into network.

2.2 Link-Layer Attacks

The MANET is an open multi-point peer-to-peer specification. Exactly, one-hop characteristics of neighbors are maintained by the link-layer protocols, and also the network layer protocols expand the property to other nodes within the network. Attacks might target the link layer by disturbing the cooperation of the layer protocols [16].

2.3 Network Layer Attacks

Network layer protocols extend property from neighboring one-hop nodes to any or all different nodes in MANET [17]. The connectedness between mobile hosts over a doubtless multi-hop wireless link powerfully depends on cooperative reactions among all network nodes. The numbers of attacks focus on the network layer, which required detailed analysis in this paper [18]. By assaulting the routing protocols, attackers will absorb network traffic, add themselves into the trail between the supply and destination, and so curb the network traffic, as shown in the figure a pair of (a) and (b), wherever a malicious node M will add itself into the routing path among sender S and receiver D. The traffic packets might be forwarded to a non-optimal path, which may be called vital delay. In extension, the packets might be forwarded to a nonexistent path and found lost. The attackers will create routing loops; introduce strict network holdup; and channel competition into assured areas. Multiple colluding attackers might even stop a supply node from finding any route to the destination, infecting the network to partition, that triggers redundant network management traffic and more intensifies network congestion and performance degradation [19].

2.4 Transport Layer Attacks

The motive of TCP-like transport layer protocols in MANET embodies framework of end-to-end connection, end-to-end reliable delivery of packets, management, congestion management, clearing of end-to-end association. The same as transmission control protocols within the net, the mobile node is accessible to the classic SYN flooding attack or session hijacking attacks [20]. However, a MANET includes a higher channel error rate once related to wired networks. As a result, transmission control protocol doesn't have any technique to tell apart, whether the loss is due to congestion, random error, or malicious attacks. Transmission control protocol multiplicative loss its congestion window upon encountering losses, that degrades network performance considerably.

2.5 *Application Layer Attacks*

Application layer attacks can be mobile viruses, worm attacks, and repudiation attacks. In wormhole attacks, the attacker connects two distant parts of the ad hoc network using an extra communication channel (e.g., a fast LAN connection) as a mine. As a result, two distant nodes assume they are neighbors and send data using the mine. The attacker has the possibility of conducting a traffic analysis or selective forwarding attack [21].

3 Multiple Layer Attacks

Some security attacks are often launched from multiple layers rather than a selected layer. Examples of multilayer attacks are denial of service (DoS), man-in-the-middle, and different ways of attacks.

3.1 *Denial of Service (DoS)*

It aims to crab the provision of bound node or even the services of the complete ad hoc networks. In the traditional wired network, the DDoS attacks are implemented by flooding therefore considerate of network traffic to the target by disable the processing power of the target and denied the services provided by the target system [22].

3.2 *Impersonation*

Impersonation attack may be severely strict to the protection of mobile impromptu network. If there is not such a correct authentication mechanism the nodes, the opponent will capture some nodes within the network and create them seem like delicate nodes. During this manner, the agree nodes will be part of the network because the traditional nodes start to conduct the malicious behaviors like propagate invented routing information and gain inappropriate priority to access some hint [23].

3.3 *Sinkhole Attack*

The offensive node tries to supply a really engaging link, e.g., to a gateway. Therefore, lots of congestion avoid that node. Besides easy traffic analysis, different attacks like selective forwarding or denial of service are often merged with the sinkhole attack [24].

3.4 Sybil Attack

The Sybil attack particularly aims at distributed system environments. The offender plays multiple roles. It tries to act as many totally different identities/nodes instead of one. This grants to forge the results of a selection used for threshold security strategies [25] .

3.5 Traffic Analysis

It is a passive attack that wants to gain info on that node which connects to each other and the way large information is processed.

3.6 Black Hole Attack

The black hole attack has two properties. First, the node avails the mobile ad hoc routing protocol, like AODV, to show itself as having a certain route to a destination node, while the route is fake, with the intention of preventing packets. Second, the offender consumes the ambush packets with none forwarding. However, the wrongdoer runs the chance that neighboring nodes can monitor and expose the continued attacks. There is a lot of slight style of this charge once a wrongdoer by selection forwards packets. Modify packets activated from some nodes, whereas move the info from the opposite nodes unaffected that limits the suspicion of its wrongdoing [26, 27].

3.7 Gray Hole Attack

A variation of the part attack is that the gray hole attack. The Gray hole attack is one of the types of black hole attack. This attack once launched by the intermediate nodes by selection overhang drops the packets, rather than forwarding all [28].

4 Problem Statement

The work focuses on AODV-based MANETs. In this, each unit is an independent wireless mobile node that has both networking and computing capacities. Each node is designed to comply with the AODV protocol and cooperate on routing and data forwarding. Each node arbitrarily joins and leaves the network, and the nodes and the wireless connections are fragile and unstable. The DDoS attackers could be any combination of nodes, even including some compromised security system nodes. This work focuses on the DDoS attacks of flooding and black hole. An operating AODV-based MANET would require a security strategy to defend against existing and potential DDoS attacks. Because the MANETs are in a dispersal pattern, and the nodes may be individually controlled, some or all nodes are out of reach or even out of control of the network administrators. It is very difficult to apply to network-wide security. Not only operating MANETs but also any upcoming or planned MANET would meet this problem. Though an about to be deployed MANET can apply an updated defense strategy, any unpredictable, unforeseen DDoS attack technique in the future can threaten the network and put it in the same situation of those operating unsafe MANETs. A proper solution should require no change of the existing network system. It should not require a large-scale upgrade. It should not depend on cooperation from the individuals. It should be backward compatible and transparent to the other components of the network.

5 Proposed Methodology

Based on survey and analysis of current MANET DDoS attacks and defense strategies, we conclude the fundamental principles of designing the secure MANET routing protocols and the DDoS defense strategies, which should also apply on AODV defense.

1. The DDoS attacks abuse the definition of the network protocols by consuming network or system resources, so that other legitimate requests cannot be served. A good DDoS defense should reactively stop only the malicious traffic and ensure that legitimate traffic is passed. There are two opposing methods of defense. One is to reinforce the tolerance and resilience capability of the system by applying encryption protection, evading the attacks, allowing redundancy. On the other hand, significant efforts actively stop the attacks by monitoring and filtering the malicious traffic, and/or tracing back to the attackers, such as Mayday, hop count filtering, and hash-based trackback. The methods under this category require large-scale deployment of the defense routers and a necessary modification on the network layer protocols, but they attempt to solve the specific problems effectively from the source.
2. Distributed detecting architecture efficiently responds to attack. The defense initiated by a victim is much less practical or efficient. Because of the nature of the attack traffic, the overwhelming traffic could have already congested and crashed the intermediate nodes before it reaches the victim. Victim initiated intrusion detection could not meet the attack traffic and fail in this scenario. So, an efficient detection and defense strategy should be based on a distributed architecture and is initiated by the intermediate nodes.

3. The defense system should be dynamic, distributed, adaptable, and effective. Dynamic means a filter or a defense node can join and exit as needed. There should be no constraint on the deployed topology; distributive means the security workload is shared and balanced among the defense system nodes; adaptability means the defense system does self-recovery on the network change or under the DDoS attacks; effectiveness requires the system to be functional under DDoS attacks, which includes detecting, logging and reporting, tracing back the attack path, isolating the attackers, and, ultimately, stopping the attack. There are many external restrictions for a security system to fulfill all the functionalities listed above. Some functionality is infeasible for some specific network architectures or under some specific scenarios.
4. If a defense system is required for an operational network, the practicality of the strategy will be a crucial issue. In this situation, the system tries to avoid modifying the current network infrastructure, tries to impact as fewer nodes as possible, and tries to build the anti-DDoS in the application layer of the participating routers and nodes. Less modification means less cost or more practicality to the public or commercial networks.
5. The defense system should work for all types of network traffic, which means it can fully control the target bandwidth (in contrast to the strategy that works only on a type of traffic, such as a Web browser service, and for all types of customer, which agree to the model of a public service. The defense system requires generalization of the system to fit in the practical environment and defend against generic DDoS traffic.
6. Tracking down a specific packet or source is not very useful in defending against a DDoS attack because many attack sources are only randomly compromised nodes. The tracking procedure consumes resources too. This is especially true in MANETs, where either the links or the location of nodes is volatile. The only good way to track back the malicious traffic is to have a guard located as close to the attacker as possible.
7. The defense system should provide good performance with the least overhead. QoS capability is required for some commercial scenarios. A defense system may protect only one specific network service. Several individual defense systems may work together to provide an overlay-based protection. The owner of the defense system can be the ISP, a public security organization, or a security service provider. Therefore, different scenarios may have different QoS demands.
8. The defense system should assume that the defense mechanism and structure may be fully known by all network members including the attackers, and still have self-protection. So, it should be durable toward attacks on the system itself. Some protection tactics should be taken. For instance, the system should be able to detect suspicious or malicious scanning from the attackers and return phony or misleading information, or evade the attack by address hopping.

6 Results

6.1 Simulation Parameters

Simulations have been carried out in order to evaluate routing protocol. We focused our attention on the evaluation of network performance in terms of routing overhead, throughput, packet delivery ratio, and normalized routing load of a mobile ad hoc network where a number of nodes are varying. The proposed method is compared with the existing algorithm of efficient route method based on the proposed method by packet delivery ratio (PDR), throughput, end-to-end delay (E2E), etc. The results of the simulation are detailed below:

6.2 Throughput

It is defined as the ratio of number of bytes received per second at the destination end. As we can see in Fig. 1, the performance of the network is decreasing while black hole attack is performed. Simulation outcomes reflect the comparison between normal AODV, AODV with DDoS attack, and AODV based on proposed method.

6.3 Packet Delivery Ratio

Figure 2 shows that the packet delivery ratio is decreasing while DDoS attack is performed. If describing about the original AODV working, it decreases delivery of packet with an increase in the number of nodes.

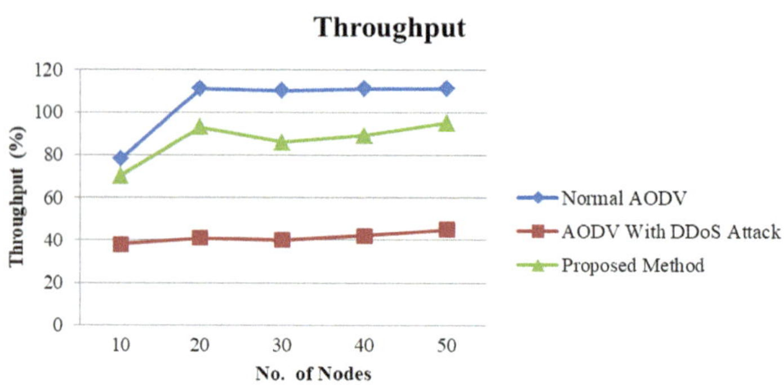

Fig. 1 Throughput

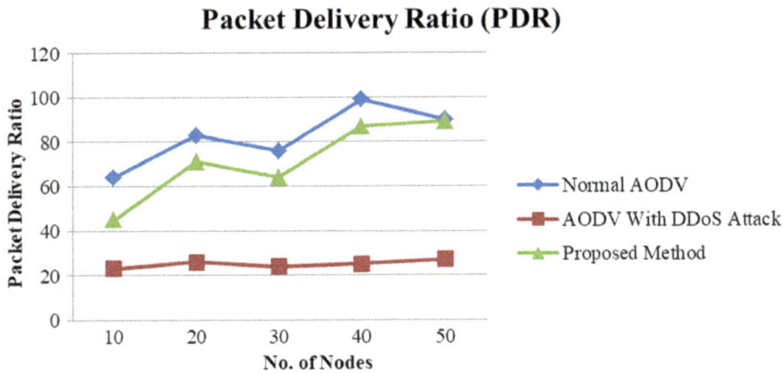

Fig. 2 Packet delivery ratio

6.4 *End-to-End Delay*

It represents the time required to move the packet from the source node to the destination node as in Fig. 3.

6.5 *Dropped Packets*

Dropped packets show in Fig. 4 the number of nodes that sent from the source and fail to reach to the destination.

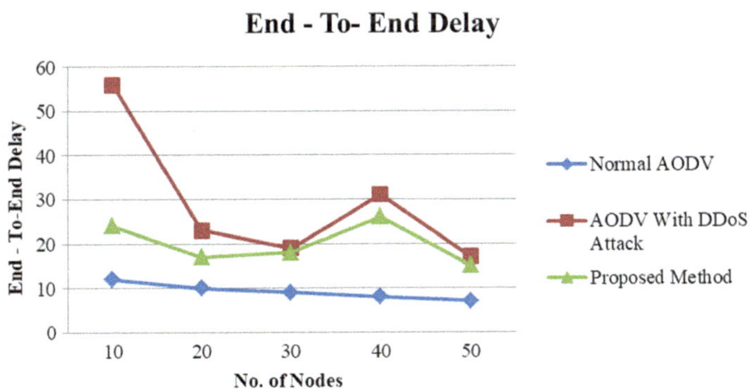

Fig. 3 End-to-end delay

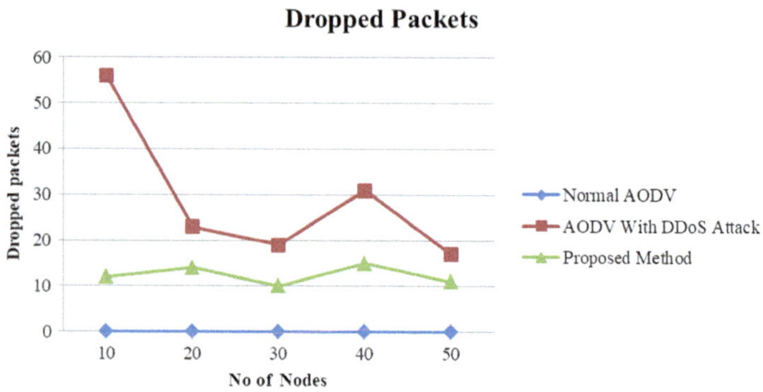

Fig. 4 Dropped packets

7 Conclusion

The implementation of this proposed method is more efficient and can be able to detect the attack in the network. It clearly shows the performance of the AODV-based mobile ad hoc network which is effectively improved. The overall summary of performance is throughput, packet delivery ratio, and end-to-end delay improved. In the proposed method, two main problems are improved such as performance and security, in one solution. This algorithm is useful for the black hole and able to detect only one attack.

8 Future Work

In this work, the simulation of Static and Dynamic routing protocols is done using Bayesian Filtering and Collaborative Message Passing Interface methods. Future work involves the study of specific attacks on the network under stochastic modeling for nodes participating in the routing path and its effect on routing protocol by comparing various network parameters. It is also aimed to find the analytical expression for the same.

References

1. Yan Q, Gong Q, Deng F-A (2016) Detection of DDoS attacks against wireless SDN controllers based on the fuzzy synthetic evaluation decision-making model. Ad Hoc Sens Wirel Netw 33:275299
2. Zargar ST, Joshi J, Tipper D (2013) A survey of defense mechanisms against distributed denial of service (DDoS) flooding attacks. IEEE Commun Surv Tutor 15(4):2046–2068

3. Karthik S, Bhavadharini RM, Arunachanlam VP (2008) Analyzing interaction between denial of service attack and threats. In: International conference on computing, communication and networking (ICCCN), 2008; Chang RKC (2002) Defending against flooding-based distributed denial-of-service attacks: a tutorial. IEEE Commun Magaz 42–51
4. Spyridopoulos T, Karanikas G, Tryfonas T, Oikonomou G (2013) A game theoretic defense framework against DoS / DDoS cyber-attacks. Comput Sec 38:39–50
5. Srivatsa M, Iyengar A, Yin J, Liu L (2008) Mitigating application-level denial of service attacks on Web servers: a client-transparent approach, ACM Trans Web (TWEB) 2(3)
6. Singh V, Poonia RC, Raja L, Sharma G, Trivedi NK (2019) Source redundancy management and host intrusion detection in wireless sensor networks. Recent Patents Comput Sci 12(1)
7. Singh V, Poonia RC, Raja L, Sharma G, Trivedi NK, Mathur GN (2020) Redundancy management and host intrusion detection in WSN. In: Forensic investigations and risk management in mobile and wireless communications. IGI Global, pp 153–167
8. Shastri A, Dadhich R, Poonia RC (2011) Performance analysis of on-demand Routing protocols for vehicular ad-hoc Networks. Int J Wirel Mobile Netw (IJWMN) 3(4):103–111
9. Kambourakis G, Moschos T, Geneiatakis D, Gritzalis S (2008) Detecting DNS amplification attacks, critical information infrastructures security. Lecture Notes Comput Sci 5141:185–196
10. Aldaej A, Ahamad T (2016) AAODV (Aggrandized Ad Hoc on Demand Vector): a Detection and prevention technique for MANETs. Int J Adv Comput Sci Appl (IJACSA) 7(10)
11. Kumar S, Saini ML, Kumar S (2020) Improved DYMO-based ACO for MANET using distance and density of nodes. In: Microservices in big data analytics. Springer, Singapore, pp 29–38
12. Kumar S, Saini ML, Kumar S (2019) Improvement in routing in MANET using spider monkey optimization. Sandeep, improvement in routing in MANET using spider monkey optimization (April 2019)
13. Kumar S, Saini ML, Kumar S (2018) A survey: swarm based routing algorithm toward improved quality of service in MANET. Int J Manage Technol Eng 8(5):311–322
14. Monika G, Poonia SK, Goyal D (2017) Attacks finding and prevention techniques in MANET: a survey. Adv Wirel Mobile Commun 10(5):1185–1195
15. Aljumah A, Ahamad T (2016) Black hole and mobile Ad hoc network (MANET) a simple logical solution. In: International conference on CyberWarfare and security, March 2016
16. Aljumah A, Ahamad T (2016) A novel approach for detecting DDoS using artificial neural networks. Int J Comput Sci Netw Sec 16(12):132
17. Piyush Agrawal R, Ghosh K, Das SK (2008) Cooperative black and gray hole attacks in mobile Ad hoc networks. In: 2nd international conference on ubiquitous information management and communication, pp 310–314
18. Kumar S, Goyal M, Goyal D, Poonia RC (2017) Routing protocols and security issues in MANET. In: 2017 international conference on infocom technologies and unmanned systems (trends and future directions)(ICTUS), IEEE, pp 818–824
19. Goyal M, Kumar S, Sharma VK, Goyal D (2020) Modified Dragon-Aodv for efficient secure routing. In: Advances in computing and intelligent systems. Springer, Singapore, pp 539–546
20. Goyal M, Goyal D, Kumar S (2020) Dragon-AODV: efficient Ad Hoc on-demand distance vector routing protocol using dragon fly algorithm. In: Soft computing: theories and applications. Springer, Singapore, pp 181–191
21. Chen W, Long X, Bai Y, Gao X (2007) A new solution for resisting gray hole attack in mobile Ad-Hoc networks. In: 2nd international conference on communications and networking in China, pp 366–370
22. Arshad J, Nathan G, Anthony D, Richard M (2008) An outlook on the impact of trust models on routing in mobile
23. Su M-Y (2010) Prevention of selective black hole attacks on mobile ad hoc networks through intrusion detection systems. Comput Commun 34(1):107–17
24. Gonzalez O, Ansa G, Howarth MP, Pavlou G (2008) Detection and accusation of packet forwarding misbehaviour in mobile ad-hoc networks. J Inter Eng 2(1):181–192
25. Yan Q, Yu FR, Gong Q, Li J (2016) Software defined networking (SDN) and distributed denial of service (DDoS) attacks in cloud computing environments: a survey, some research issues, and challenges. IEEE Commun Surv Tutor 18(1):602–22

26. Shin S, Gu G (2013) Attacking software defined networks: a first feasibility study. In: Proceedings Of ACM Hot SDN, Hong Kong, China, pp 165–166
27. Li H, Li P, Guo S, Nayak A (2014) Byzantine-resilient secure software-defined networks with multiple controllers in cloud. IEEE Trans Cloud Comput 2:436447
28. Dao NN, Park J, Park M, Cho S (2015) A feasible method to combat against DDoS attack in SDN network. In: 2015 International conference on information networking (ICOIN). IEEE, pp 309–311

Chapter 15
Protection of Wind Farm-Integrated Series Capacitor-Compensated Three-Phase Transmission Line Using Daubechies-7 Wavelet Transform

Gaurav Kapoor

1 Introduction

The series-compensated transmission lines (SCTLs) carry massive quantity of electrical power. The fault incidence and successive tripping of a SCTL would outcome in extensive interruption of electrical power. Thus, an accurate detection of fault in a SCTL turns out to be very crucial for reducing the loss of profit and providing fast preservation. In the latest years, lots of investigations have been devoted for the protection of SCTLs. The short reviews of recent investigations have been discussed hereafter. In [1], authors presented the DNN-based fault location technique for the SCTL. In [2], WT has been applied for the protection of TPTL. An impedance-based technique has been employed in [3] for SCCTL. In [4–6], authors used WT for the different configurations of transmission lines. In [7], the HHT has been used. In [8], a technique based on mathematical-morphological gradients has been implemented on six-phase transmission line. In [9], a WHT technique has been proposed. Further in [10], WT-based algorithm has been suggested for six-phase transmission line with series compensation. An EMD-HHT-based fault recognition technique has been proposed in [11]. In [12], discrete Walsh-Hadamard transform has been used as a tool for transmission line protection. In [13], wavelet-based fault detector has been designed for transmission line protection.

In this work, the Db7WT is executed for the detection of faults in the WFISCCTPTL. Such type of investigation has not been depicted so far to the best of the information of the author. The results show that the Db7WT competently detects the faults, and the uniformity of the Db7WT is not susceptible to the deviations in fault factors.

G. Kapoor (✉)
Department of Electrical Engineering, Modi Institute of Technology, Kota, India
e-mail: gaurav.kapoor019@gmail.com

© Springer Nature Singapore Pte Ltd. 2021
S. D. Purohit et al. (eds.), *Proceedings of International Conference on Communication and Computational Technologies*, Algorithms for Intelligent Systems,
https://doi.org/10.1007/978-981-15-5077-5_15

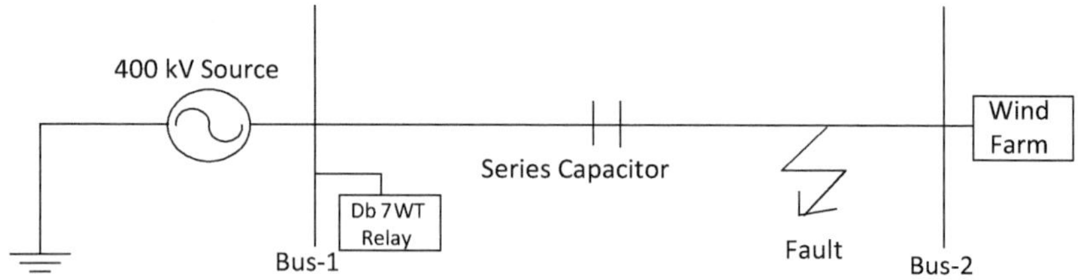

Fig. 1 Schematic of WFISCCTPTL

This chapter is structured as follows: Sect. 2 presents the specifications of WFISCCTPTL. Section 3 describes the process of Db7WT. Section 4 presents the performance appraisal of the investigations which are carried out in this study. Section 5 concludes the chapter.

2 The Specifications of WFISCCTPTL

The simulation model (Fig. 1) consists of WFISCCTPTL of 400 kV, divided into two parts each of length 100 km. The WFISCCTPTL is fed from a 400-kV source at the one end and integrated with two wind turbines at the other end. The current measurement blocks are installed at the bus-1 for the measurement of fault currents.

3 The Flowchart for Db7WT

Figure 2 exemplifies the procedure for Db7WT. The steps are shown below.

1. Simulate the WFISCCTPTL and generate fault currents.
2. Analyze the currents using Db7WT for characteristics retrieval.
3. The phase will be declared as the faulted phase if its Db7WT coefficient has the larger amplitude as compared to the healthy phase.

4 Performance Appraisal

The Db7WT has been tested for the converting faults, boundary faults, faults at two different positions, and varying wind turbine units. The results are revealed below.

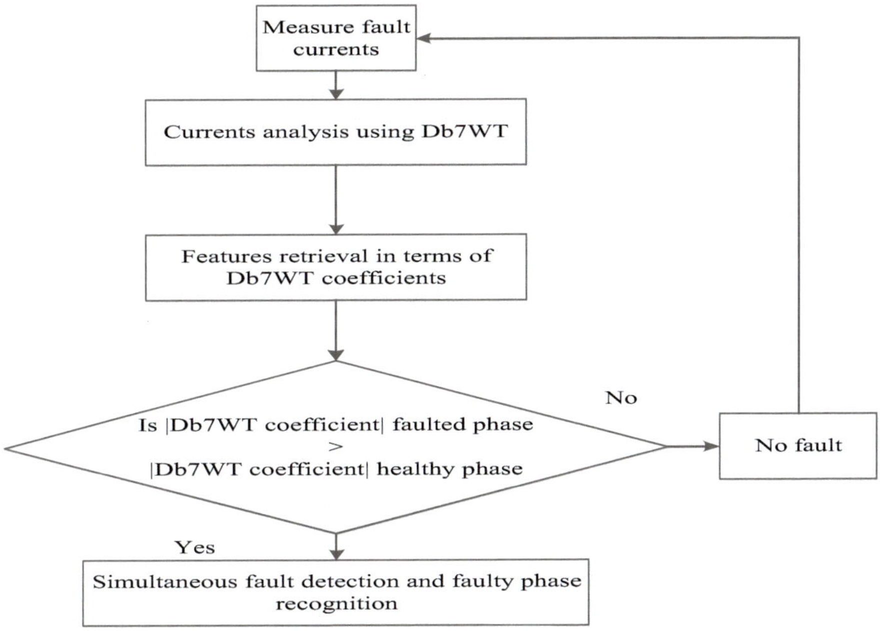

Fig. 2 Procedure for Db7WT

4.1 *The Effectiveness of Db7WT for Healthy Condition*

Figure 3 exemplifies the currents and voltages for no fault. Figure 4 shows the Db7WT coefficients for no-fault. Table 1 reports the results of Db7WT for no-fault.

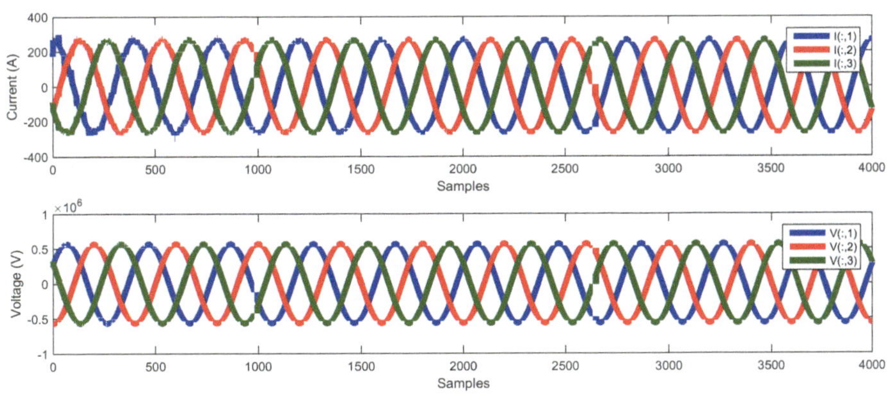

Fig. 3 Currents and voltages for no-fault

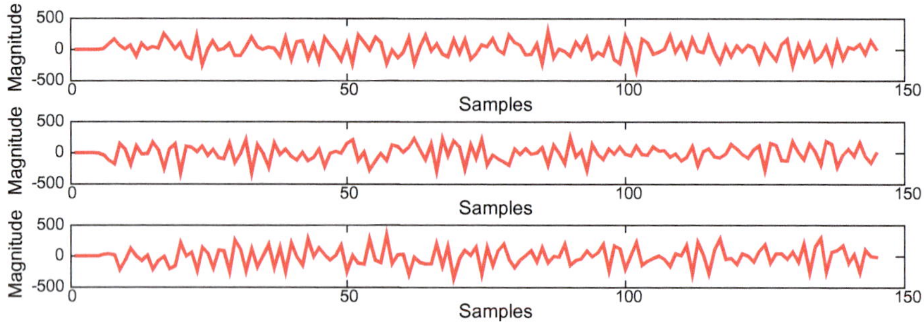

Fig. 4 Db7WT coefficients of currents for no-fault

Table 1 Results of Db7WT for no-fault

Db7WT coefficients		
Phase-A	Phase-B	Phase-C
297.3026	246.5708	345.3360

4.2 *The Effectiveness of Db7WT for Converting Faults*

The Db7WT has been explored for the converting faults. Figure 5 depicts the plot when the ABCG fault at 100 km at 0.05 s is converted into the AG fault at 100 km at 0.25 s among $R_F = 0.5\ \Omega$ and $R_G = 0.25\ \Omega$. Figure 6 exemplifies the Db7WT coefficients when the ABCG fault is converted into the AG fault. The fault factors selected for the rest of the fault cases are: $F_L = 100$ km, $R_F = 0.5\ \Omega$, and $R_G = 0.25\ \Omega$. Table 2 reports the results for different types of converting faults. It is evident from Table 2 that the Db7WT is insensitive to the converting faults.

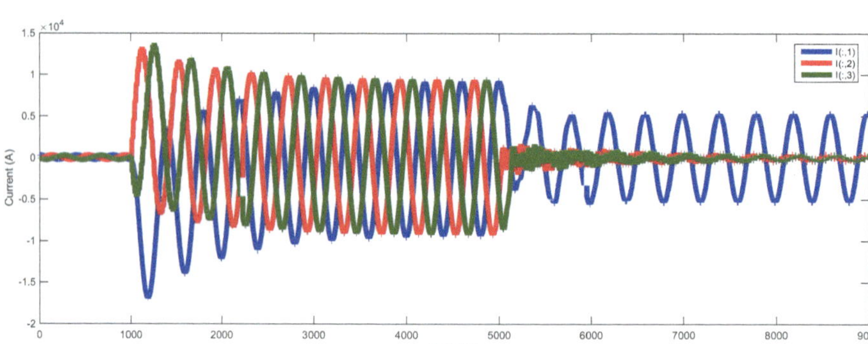

Fig. 5 Waveform when ABCG fault at 100 km at 0.05 s is converted into AG fault at 100 km at 0.25 s among $R_F = 0.5\ \Omega$ and $R_G = 0.25\ \Omega$

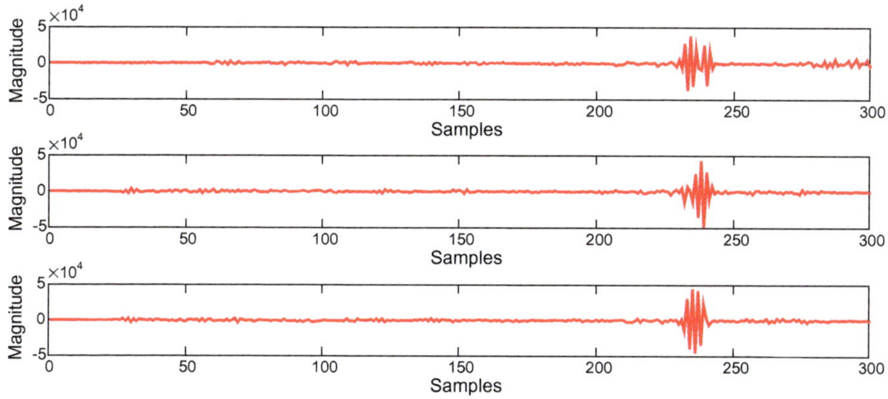

Fig. 6 Db7WT coefficients when ABCG fault at 100 km at 0.05 s is converted into AG fault at 100 km at 0.25 s among $R_F = 0.5\ \Omega$ and $R_G = 0.25\ \Omega$

Table 2 Results of Db7WT for converting faults

Fault type with time (Sec)	F_L (km)	R_F (Ω)	R_G (Ω)	Converted fault type with time (Sec)	Db7WT coefficients		
					Phase-A	Phase-B	Phase-C
ABCG (0.05)	100	0.5	0.25	AG (0.25)	$3.7966 * 10^4$	$4.2689 * 10^4$	$4.4554 * 10^4$
BG (0.04)	100	0.5	0.25	BCG (0.1225)	722.8298	$1.6901 * 10^4$	$1.4367 * 10^4$
ABG (0.06)	100	0.5	0.25	AG (0.25)	$1.3875 * 10^4$	$1.1162 * 10^4$	$3.4353 * 10^3$
ACG (0.07)	100	0.5	0.25	ABCG (0.2)	$3.6909 * 10^4$	$4.2370 * 10^4$	$4.5346 * 10^4$
CG (0.0545)	100	0.5	0.25	ACG (0.1445)	$1.4811 * 10^4$	826.4624	$1.7549 * 10^4$

4.3 *The Effectiveness of Db7WT for Near-in Relay Faults*

The WFISCCTPTL is simulated for various near-in relay faults. Figure 7 illustrates the BG near-in relay fault current at 5 km at 0.0865 s among $R_F = 2\ \Omega$ and $R_G = 3\ \Omega$. For different near-in relay faults, the fault factors selected are: $R_F = 2\ \Omega$, $R_G = 3\ \Omega$, and $T = 0.0865$ s. It is investigated from Table 3 that the near-in relay faults do not affect the results of Db7WT (Fig. 8).

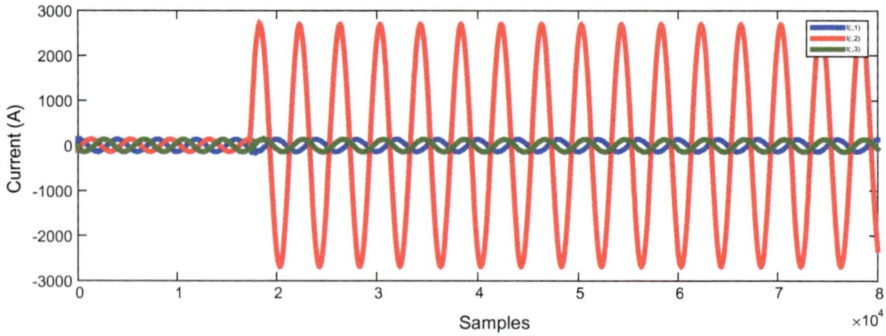

Fig. 7 BG near-in relay fault current at 5 km at 0.0865 s among $R_F = 2\ \Omega$ and $R_G = 3\ \Omega$

Table 3 Results of Db7WT for near-in relay faults

Fault type	F_L (km)	FIT (S)	Db7WT coefficients		
			Phase-A	Phase-B	Phase-C
BG	5	0.0865	106.4144	$1.2730 * 10^3$	113.3509
ACG	10	0.0865	$4.9843 * 10^3$	131.2176	$7.7471 * 10^3$
ABCG	15	0.0865	$8.8963 * 10^3$	$1.1626 * 10^4$	$8.0301 * 10^3$
BCG	20	0.0865	$1.1955 * 10^3$	$7.8041 * 10^3$	$6.5644 * 10^3$
AG	25	0.0865	$8.1469 * 10^3$	761.2906	773.6441

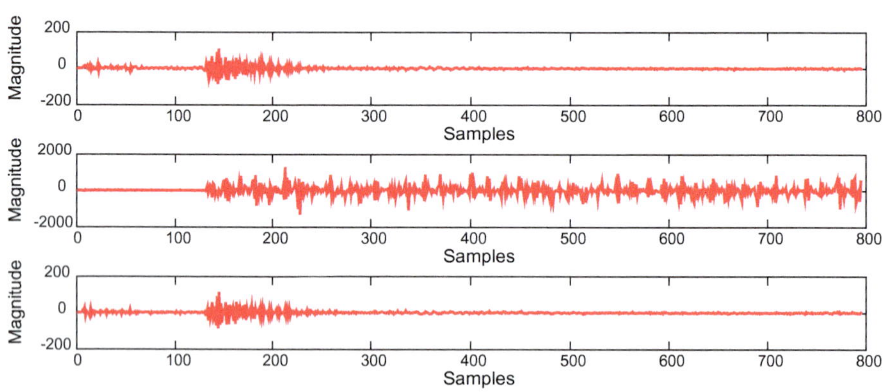

Fig. 8 Db7WT coefficients for BG near-in relay fault at 5 km at 0.0865 s

4.4 *The Effectiveness of Db7WT for Far-End Relay Faults*

The Db7WT is tested for various far-end relay faults. Figure 9 depicts the ACG fault current at 180 km at 0.0565 s among $R_F = 4.5\ \Omega$ and $R_G = 2.5\ \Omega$. Figure 10 exemplifies the Db7WT coefficients for the ACG far-end relay fault. The fault factors

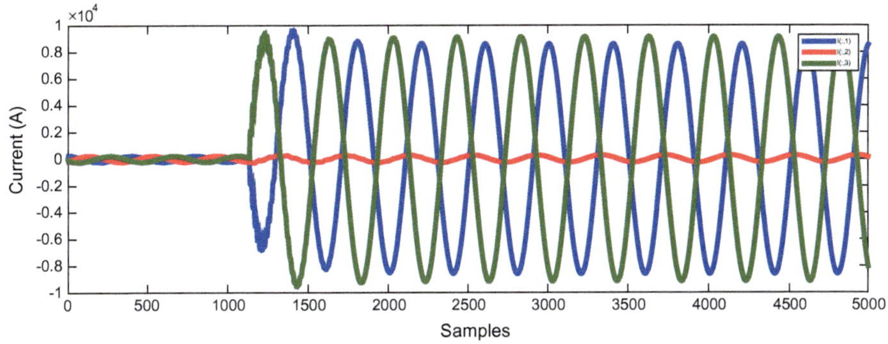

Fig. 9 ACG far-end relay fault at 180 km at 0.0565 s among $R_F = 4.5\ \Omega$ and $R_G = 2.5\ \Omega$

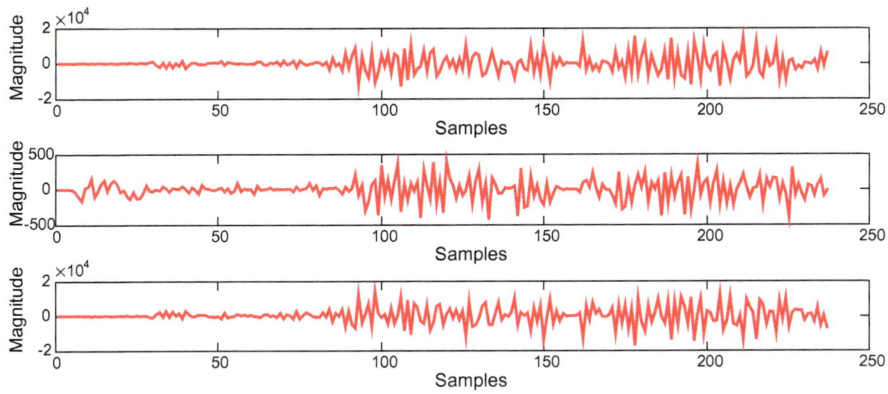

Fig. 10 Db7WT coefficients for ACG far-end relay fault at 180 km at 0.0565 s

chosen for the further fault cases are: T = 0.0565 s, $R_F = 4.5\ \Omega$ and $R_G = 2.5\ \Omega$. It is explored from Table 4 that the Db7WT well detects the far-end relay faults.

Table 4 Results of Db7WT for far-end relay faults

Fault type	F_L (km)	FIT (S)	Db7WT coefficients		
			Phase-A	Phase-B	Phase-C
ACG	180	0.0565	$1.5392 * 10^4$	402.5338	$1.4138 * 10^4$
ABG	185	0.0565	$1.4749 * 10^4$	$1.7782 * 10^4$	613.0042
ABCG	190	0.0565	$1.1931 * 10^4$	$1.5385 * 10^4$	$1.3321 * 10^4$
CG	195	0.0565	775.0358	780.4825	$4.3690 * 10^3$
AG	200	0.0565	$3.6352 * 10^3$	511.7331	549.1708

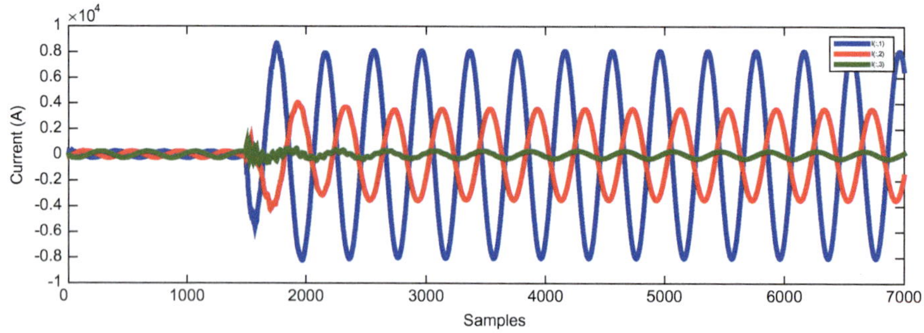

Fig. 11 Waveform of AG fault at 60 km and BG fault at 140 km at 0.075 s

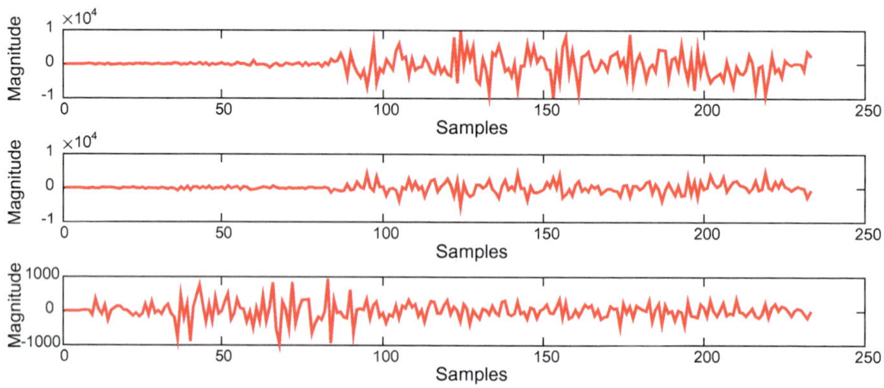

Fig. 12 Db7WT coefficients for AG fault at 60 km and BG fault at 140 km at 0.075 s

4.5 The Effectiveness of Db7WT for Faults at Two Different Positions

The Db7WT has been investigated for the faults which occur at two different positions on WFISCCTPTL. Figure 11 exemplifies the currents for AG fault simulated at 60 km before series capacitors and BG fault simulated at 140 km after series capacitors at 0.075 s among $R_F = 5\ \Omega$ and $R_G = 7\ \Omega$. Figure 12 depicts the Db7WT coefficients for the AG and BG fault currents. It is explored from Table 5 that the Db7WT performs efficiently for the detection of two different position faults.

4.6 The Effectiveness of Db7WT for Variation in R_F and R_G

The Db7WT has been inspected for fault resistances from 0.25 Ω to 150 Ω and ground resistances from 0.5 to 100 Ω. Figure 13 exemplifies the ABCG fault current

Table 5 Results of Db7WT for two different position faults

Fault-1 before capacitor	Fault-2 after capacitor	FIT (S)	R_F (Ω)	R_G (Ω)	Db7WT coefficients		
					Phase-A	Phase-B	Phase-C
AG (60 km)	BG (140 km)	0.075	5	7	$9.6029 * 10^3$	$4.5373 * 10^3$	930.2501
BG (75 km)	CG (125 km)	0.1	5	7	$1.1673 * 10^3$	$7.8862 * 10^3$	$3.6477 * 10^3$
ABG (90 km)	CG (110 km)	0.2	5	7	$9.7396 * 10^3$	$1.3877 * 10^4$	$5.7543 * 10^3$
CG (45 km)	AG (155 km)	0.05	5	7	$2.7883 * 10^3$	829.3423	$8.1682 * 10^3$
BG (125 km)	AG (75 km)	0.25	5	7	$5.6218 * 10^3$	$6.5287 * 10^3$	$1.3760 * 10^3$

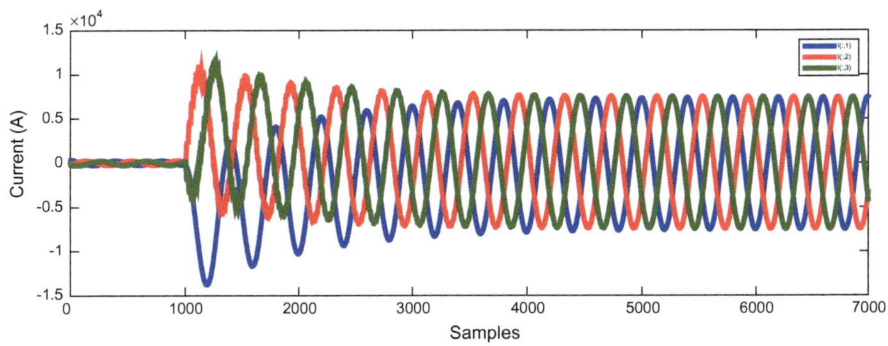

Fig. 13 ABCG fault current at 130 km at 0.05 s among $R_F = 0.25$ Ω and $R_G = 0.5$ Ω

at 130 km at 0.05 s among $R_F = 0.25$ Ω and $R_G = 0.5$ Ω. Figure 14 shows the Db7WT coefficients for the ABCG fault. It is explored from Table 6 that the Db7WT works well for different R_F and R_G.

4.7 *The Effectiveness of Db7WT for Variation in Wind Turbine Units*

The Db7WT is investigated for variation in the wind turbine units in the wind farm. Figure 15 exemplifies the AG fault current at 100 km at 0.1 s among one wind turbine, $R_F = 1.75$ Ω and $R_G = 2.75$ Ω. Figure 16 shows the Db7WT coefficients for the AG fault. The fault factors for other fault cases are: T = 0.1 s, $F_L = 100$ km, $R_F = 1.75$ Ω, and $R_G = 2.75$ Ω. It is examined from Table 7 that varying the wind turbine units does not affect the results of Db7WT.

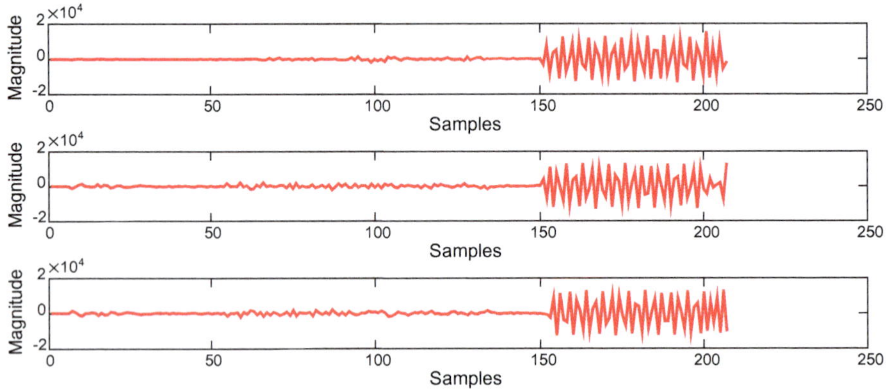

Fig. 14 Db7WT coefficients for ABCG fault at 130 km among $R_F = 0.25\ \Omega$ and $R_G = 0.5\ \Omega$

Table 6 Results of Db7WT for variation in fault and ground resistances

Fault type	R_F (Ω)	R_G (Ω)	Db7WT coefficients		
			Phase-A	Phase-B	Phase-C
ABCG	0.25	0.5	$1.5182 * 10^4$	$1.2969 * 10^4$	$1.3021 * 10^4$
ABG	25	15	$5.9860 * 10^3$	$6.3688 * 10^3$	385.9512
ACG	55	30	$3.8879 * 10^3$	295.6817	$5.0738 * 10^3$
BG	85	60	307.6703	$1.8991 * 10^3$	361.2466
CG	150	100	345.7419	291.7594	$1.1086 * 10^3$

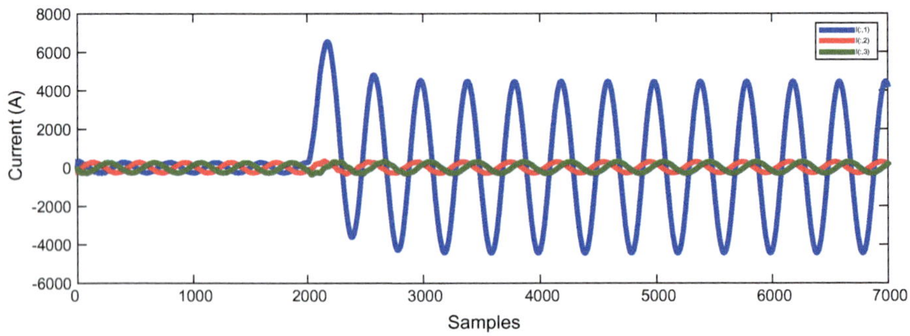

Fig. 15 AG fault current at 100 km at 0.1 s with one wind turbine

5 Conclusion

A Db7WT-based fault recognition method has been presented in this work for the protection of WFISCCTPTL. The Db7WT is applied which decomposes the fault currents to evaluate the Db7WT coefficients. The fault factors of the WFISCCTPTL

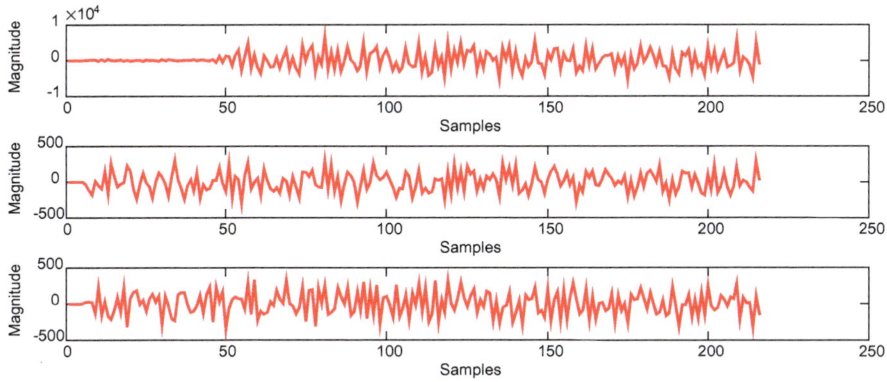

Fig. 16 Db7WT coefficients for AG fault at 100 km at 0.1 s with one wind turbine

Table 7 Results of Db7WT for variation in the wind turbine units

Fault type	No. of WT units	F_L (Km)	Db7WT coefficients		
			Phase-A	Phase-B	Phase-C
AG	1	100	$6.6862 * 10^3$	307.2280	329.8145
ABG	3	100	$1.8425 * 10^4$	$1.6408 * 10^4$	669.3002
BG	5	100	529.0671	$6.5886 * 10^3$	514.3646
BCG	2	100	468.4600	$1.9942 * 10^4$	$1.3249 * 10^4$
CG	4	100	682.8242	684.7961	$5.6651 * 10^3$

are varied. According to the performance appraisal, it is discovered that the Db7WT detects the faults efficiently.

References

1. Mirzaei M, Vahidi B, Hosseinian SH (2018) Fault location on a series-compensated three terminal transmission line using deep neural networks. IET Sci Meas Technol 12(6):746–754
2. Kapoor G (2018) Wavelet transform based detection and classification of multi-location three phase to ground faults in twelve phase transmission line. Majlesi J Mechatron Syst 7(4):47–60
3. Bains TPS, Sidhu TS, Xu Z, Voloh I, Zadeh MRD (2018) Impedance-based fault location algorithm for ground faults in series capacitor compensated transmission lines. IEEE Trans Power Deliv 33(1):189–199
4. Kapoor G (2018) Six phase transmission line boundary protection using wavelet transform. In: Proceedings of the 8th IEEE india international conference on power electronics (IICPE). IEEE, Jaipur, India
5. Gautam N, Ali S, Kapoor G (2018) Detection of fault in series capacitor compensated double circuit transmission line using wavelet transform. In: Proceedings of the IEEE international conference on computing, power and communication technologies (GUCON). IEEE Greater Noida, India, pp 769–773

6. Kapoor G (2018) Fault detection of phase to phase fault in series capacitor compensated six phase transmission line using wavelet transform. Jordan J Electr Eng 4(3):151–164
7. Sharma N, Ali S, Kapoor G (2018) Fault detection in wind farm integrated series capacitor compensated transmission line using Hilbert Huang transform. In: Proceedings of the IEEE international conference on computing, power and communication technologies (GUCON). IEEE Greater Noida, India, pp 774–778
8. Kapoor G (2018) Six phase transmission line boundary protection using mathematical morphology. In: Proceedings of the IEEE international conference on computing, power and communication technologies (GUCON). IEEE Greater Noida, India, pp 857–861
9. Sharma P, Kapoor G, Ali S (2018) Fault detection on series capacitor compensated transmission line using Walsh hadamard transform. In: Proceedings of IEEE international conference on computing, power and communication technologies (GUCON). IEEE Greater Noida, India, pp 763–768
10. Kapoor G (2018) Detection of phase to phase faults and identification of faulty phases in series capacitor compensated six phase transmission line using the norm of wavelet transform. i-Manager's J Digit Signal Process 6(1):10–20
11. Kapoor G (2019) Detection and classification of single line to ground boundary faults in a 138 kV six phase transmission line using Hilbert Huang transform. i-Manager's J Electr Eng 12(3):28–41
12. Kapoor G (2019) Protection technique for series capacitor compensated three phase transmission line connected with distributed generation using discrete Walsh Hadamard transform. Int J Eng Sci Technol (Africa) 11(3):1–10
13. Kapoor G (2018) Wavelet transform based fault detector for protection of series capacitor compensated three phase transmission line. Int J Eng Sci Technol (Africa) 10(4):29–49

Chapter 16
Smart Wireless Reconfigurable Device Using LoRa Technology

Ronak Jain, Rekha Mithal, and Naveen Kumar Kedia

1 Introduction

Management of streets using LoRa technology [1] is growing rapidly in Smart cities. So nowadays, we are controlling all the lights through particular areas. And it is easier to control with one station. Furthermore, the energy-efficient light system [2] and intensity-based technology [3] have been well studied [4]. But it is not enough as lights should be reconfigurable itself that can change the behaviour itself on basics of power availability. And we have access and information about every single light. So in this research paper, a machine named Light Guard is introduced which gives notification of unforeseen shutdown and illuminates the emergency street lights at night. The mesh network of LoRa can be set up near power station, and a LoRa gateway will collect the data to server where we can see the notification and information about every single Light Guard. An overview is described in Fig. 1. In today's world, power cut-off is a major threat, and currently, India faces daily power outage of 30,000 MW [5]. So we can also use this technology to reduce power outage by notifying in real time.

R. Jain
Department of Electronics and Communication Engineering, JECRC Foundation, Jaipur, India
e-mail: ronakjainsocial@gmail.com

R. Mithal
Department of Chemistry, JECRC Foundation, Jaipur, India
e-mail: rekhamithal.chem@jecrc.ac.in

N. K. Kedia (✉)
Department of Information Technology, JECRC Foundation, Jaipur, India
e-mail: naveenkedia.it@jecrc.ac.in

© Springer Nature Singapore Pte Ltd. 2021
S. D. Purohit et al. (eds.), *Proceedings of International Conference on Communication and Computational Technologies*, Algorithms for Intelligent Systems,
https://doi.org/10.1007/978-981-15-5077-5_16

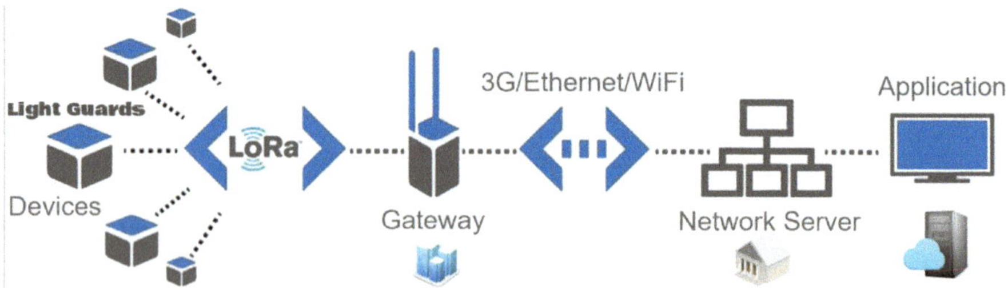

Fig. 1 Setup of smart reconfigurable device using LoRa technology

2 LoRa Technology

A Semtech innovation, LoRa devices work on wireless radio frequency technology which is long-range, low-power wireless platform that has become a revolution in the Internet of things (IoT) networks. The Lora SX1278 433 M Wireless Spread Spectrum Module RA-02 which is acting as a transceiver in Light Guard; it works on 433 MHz frequency for India. They cost only 450 INR ($6.5) which is a one-time investment. It is low power consumption module that enables low-cost battery-operated IoT applications. Furthermore, at power station LoRa gateway network has been set up which transfers the data of the all connected devices to the server. A personalized gateway is shown in Fig. 2. LoRa gateways are connected to the network server via standard IP connections. Tata Communications has announced it has successfully completed trials for connected devices and another Internet of things applications based on LoRa technology. This is a significant step taken by an Indian operator which will greatly help the Internet of things ecosystem in India. LoRa technology is the DNA of IoT, creating a smarter planet.

3 Smart Light Guard

Light Guard is a machine based on LoRa Technology run by Arduino controller. It detects the unforeseen breakdown and sends a notification to the LoRa gateway. Lights are reconfigurable as there are different modes. The first mode is normal mode which glows a light on the night and shuts off during the day time. The second mode is the emergency mode in which light will glow only when government light is not coming and send a notification of cutout to the power station; it acts as an emergency light for a better service. The third mode is shutdown mode in which the system will shut down itself when the battery health is not good and send a notification for repairing to the power station. They are reconfigurable by online applications. The machine has itself modes like active mode, deactivate mode, power saving mode based on time. Smart Light Guard can also have different models for other applications like residential colony facility power management, detection of

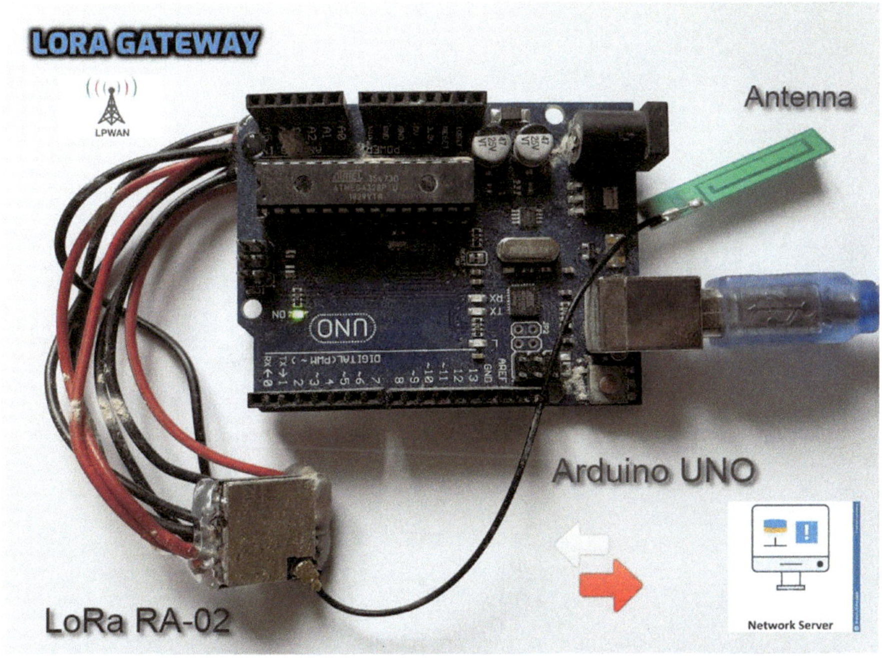

Fig. 2 A simple gateway using LoRa RA-02 and Arduino to network server.

faulty power lines, mobile tower supply detection, personalized power operation projects, etc. One of them is Smart Street Light Guard as shown in Fig. 3 which is specialized in street light timer, line fault detection and emergency light service [6].

3.1 Components of Light Guard

Arduino microcontroller (Atmega328p) to run system.

LoRa is a transceiver.

Real-time clock (RTC) module for a self-clock-in system.

Relay to switch on/off the light.

SMPS as a power supply to the system.

Relay system to detect power supply.

Step-down buck converter.

12 V Battery to operate system and light.

Voltage divider circuit for battery health.

Overcharge protection circuit using Op-amp.

A prototype handmade PCB is shown in Fig. 4 which can be assembled with a street light and a battery. Further, we can redesign this prototype into a product and an online application can be developed for whole project. These all are basic concepts

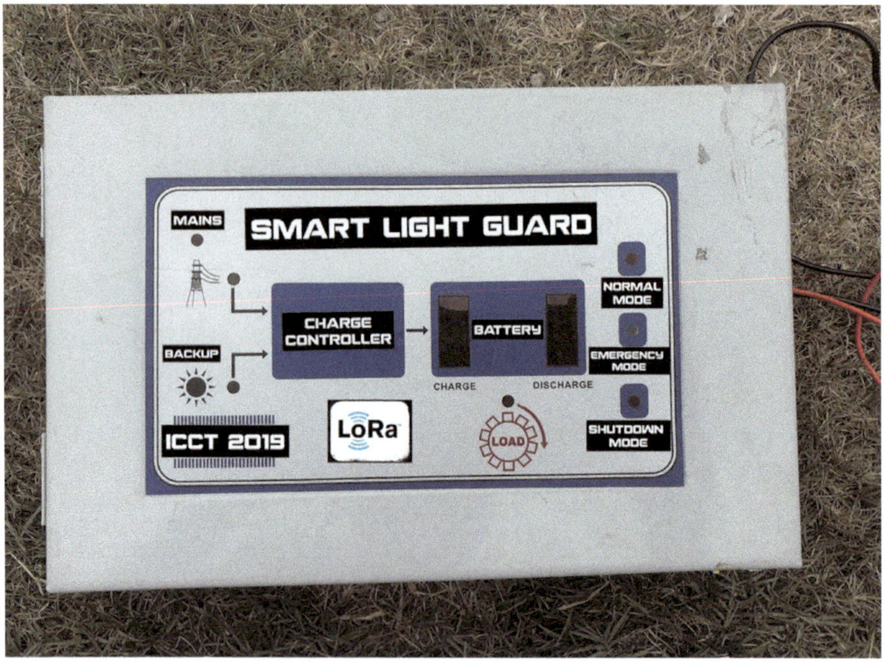

Fig. 3 Smart street Light Guard panel

which are introduced. Further, we can explore more things about it according to implementation.

3.2 *Testing Between Street Light Guard and Power Station*

Figure 5 shows the connection between the LoRa receiver and LoRa transmitter. Whenever the unforeseen shutdown happens, it will send a notification with activity status like online, offline, time when a shutdown occurs, and location of Light Guard. All Light Guards have unique address which has already uploaded on server with all information.

In Fig. 6, real-time testing is shown in which the JECRC foundation is power station where LoRa gateway has been set up. There are 3 Light Guards located in different areas. When a line is disconnected, they send an alert to JECRC foundation.

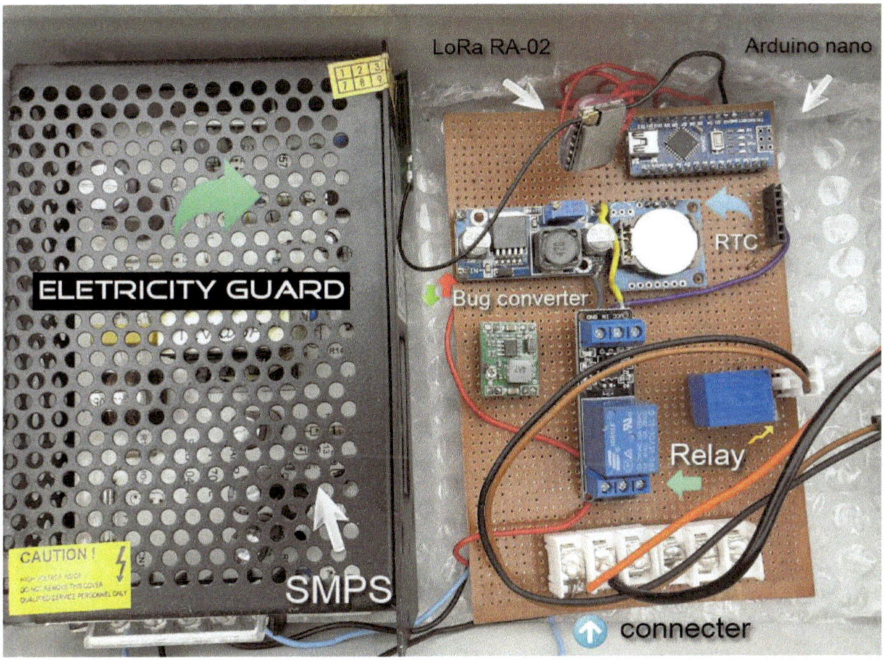

Fig. 4 A PCB prototype of street light guard

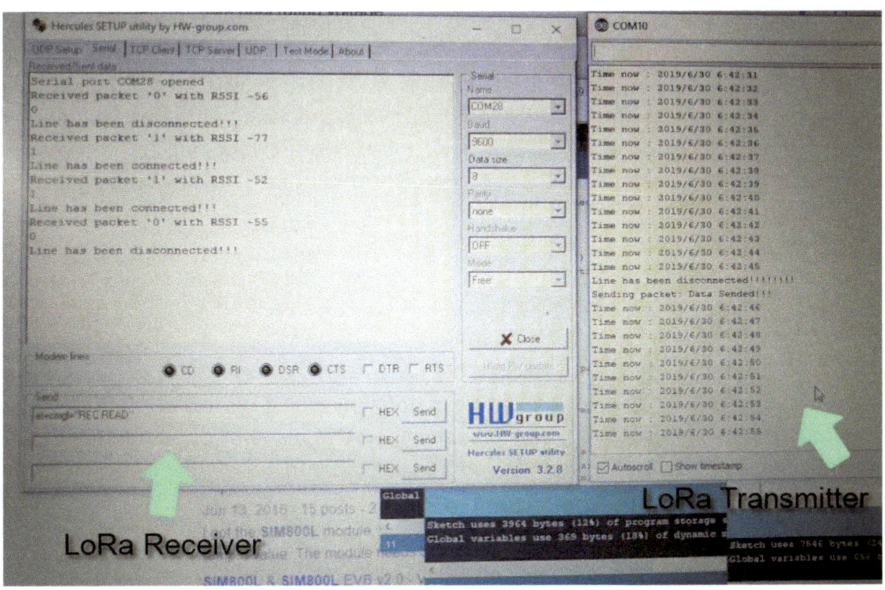

Fig. 5 Computer testing between Light Guard and power station

Fig. 6 Real-time testing site

4 Implementation

In every Smart City project by the Government of India, we can implement this technology in a cost-effective way. Moreover, we are working with this technology in the emergency light system. Besides all these, we are focusing on unforeseen shutdown, so it will work perfectly in emergency situations. It can be redesigned for power distributed lines.

5 Conclusion

From this technology, we also reduce the maintenance cost of every light as we can monitor the health of the battery and deactivate system using bi-directional communication. The unforeseen breakdown can be detectable instantly. We can establish a worthwhile LoRa network near any power station, so any shutdown can be notified and boost the maintenance service.

Reference

1. Jha A, Babab A, Ranjan I (2017) Smart street light management system using LoRa technology. Int J Sci Res(IJSR) 6(3):2349–2352
2. Anand S, Jain N (2015) Intelligent street light system using wireless transmission: review. Int J Adv Res Electron Commun Eng (IJARECE) 4(2):271–275
3. Ankalkote T, Shere VB (2016) Modern LED street lighting system with intensity control based on vehicle movements and atmospheric conditions using WSN. Int J Innovative Res Adv Eng (IJIRAE) 3(05)
4. Vidhate B, Patil S, Somani S (2018) Lora based street light automation for smart-cities. Int J Electrical Eng Technol (IJEET) 9(5)
5. Abrar MdM (2016) Power cut off and power blackout in india a major threat. Int J Adv Res Technol 5:8–15
6. Mishra CS, Nayyar A, Kumar S, Mahapatra B, Palai G (2019) FDTD approach to photonic based angular waveguide for wide range of sensing application. Optik 176:56–59

Chapter 17
Microstrip Aerial for 2.45/5.2 GHz by Pin Loading

Rajesh Kumar Raj and Himanshu Khandelwal

1 Introduction

During the past few years, demand of multifunctional small wireless communication systems with higher mobility is generated/increased. Compact, small, microstrip patch aerial at frequency of 2.45 and 5.2 GHz has various demanding applications in the wireless (Analog/Digital) communication system.

Microstrip antenna is being designed and studied since 1953 [1]. In the recent methodologies for designing integrable microstrip patch antenna with a smaller device, the latest emerging and encouraging method of shorting pin shall be used. Pin loading technique with diode was introduced in 1977 [2]. To suppress surface wave propagation, the demonstration is suggested in [3] for shorting post in contiguity to the feed point. Recently, circularly-polarized patch aerial by pin loading technique is demonstrated with sharing of gain roll-off rate and 3-dB widening of axial ratio beamwidth [4], higher and improved gain and increased bandwidth in [5–9] patch antennas. Different positions/numbers of pins loading are tried [17] for modification in the current distribution, and consequently manipulate impedance of aerial (to attune the resonance frequency and broaden the bandwidth of aerial).

But, it is observed that such kind of structures don't depict electrically small size aerials. Studies of these works show that a shorting pin will result in the resonant frequencies above the fundamental mode of the patch antenna (without pin loading). On the other hand of these designs, major focus is on miniaturizing the patch aerial by use of the pin loading technique; but, as per the studies [10–13], the results are having disadvantages due to their low gain.

This paper has the main goal to present a small, efficient, miniaturized, planar, thin aerial with a medium gain, and good impedance matching. The physical perspective shows that the proposed aerial structure could be integrated with various types of

R. K. Raj · H. Khandelwal (✉)
Government Engineering College Ajmer, Ajmer, Rajasthan, India
e-mail: maiet.himanshu@gmail.com

© Springer Nature Singapore Pte Ltd. 2021

S. D. Purohit et al. (eds.), *Proceedings of International Conference on Communication and Computational Technologies*, Algorithms for Intelligent Systems,
https://doi.org/10.1007/978-981-15-5077-5_17

movable devices, where the antenna has to be thin and easy to integrate. In such kinds of applications, the antennas like dipole might be preferred because of the existence of ground plane that is normally introduced for supporting Radio Frequency (RF), digital circuits and its radiation pattern (omnidirectional), aerial like a wire is not applicative. In order to determine the feasibility of reducing the width of patch, the aerial is studied with shorting pin technique and concurrently increasing the antenna's resonant frequency. It's depicted that the physical limitations (low gain and low-efficiency) of conventional microstrip patch antennas could be overcome by shorting pin. Second, an approach by introducing extra shorting pins is proposed to mitigate the drawbacks of low-efficiency. By utilizing the proposed method, the matching as well as radiation of a pin-loaded single patch antenna are increased.

Dominant mode's resonant frequency can be tuned by effect of shunt inductance due to shorting pins. Width of the patch is directly related to resonant frequencies.

In addition to this, a greatly enhanced matching is also noted after comparison with an equivalent simple patch antenna with loaded pins and without loaded pins at equivalent resonant frequency. The latest structures are described in [14–17] are suffering from its big electrical length. A miniaturized, one layer, single feed aerial results in dual-band behaviour. Furthermore, increasing the shorting pins doesn't help, but, properly positioned shorting pins help in changing and improving resonant frequency, behaviour of aerial as a function of pin loading and positioning of pins are also analyzed and the results are demonstrated in [17]. Moreover, the pins location variation in the perpendicular direction has a dramatic and drastic effect on the dominant resonant frequency and gain of aerial. We can analyze from the latest studies [14–17] that position of loading of pins beside the edge of patch has an impressive effect on the antenna's resonant frequency. Resonant frequency shifts up to 2 GHz by aiding of a centered, offset, and side pin, which is an important factor to improve efficiency and gain. This also helps in changing the direction of surface current and preventing opposite-current cancelation. Offset pin is used for tuning the low-frequency band to the required application, although, it does not affect the upper band much remarkably.

The proposed aerial structure is designed using the CST and different cases have been studied, the design schematic and results of the simulation are shown in Figs. 1, 2, 3, 4, 5, 6, 7, 8, 9, 10, 11, and 12.

2 Subject Definition

Shorting pins can be loaded in a miniaturized, slim patch antenna. Loading of pin results in the enhancement of electrical length of patch and miniaturization in aerial, significantly, it decreases the radiating edge, due to which, aerial gain drops. The proposed design in this paper has the main target to study, design, and bring a solution to understand the effective tuning as well as the matching of patch antenna by pin loading technique for 2.45/5.2 GHz resonant frequency.

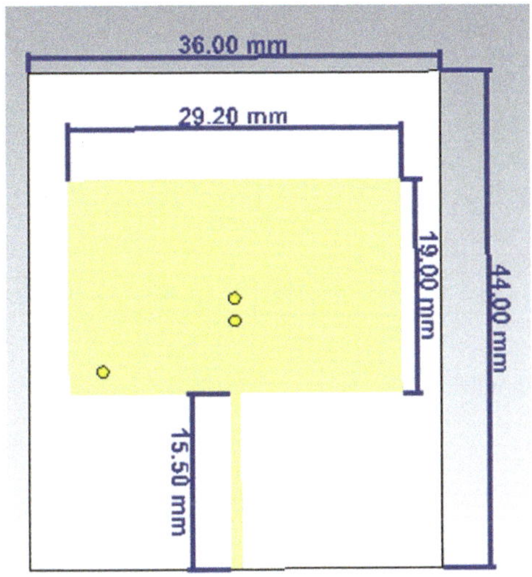

Fig. 1 Design-1: top view of patch antenna loaded with three pins (side, centre offset) resonating at 2.45 GHz

Fig. 2 Design-1: side view of patch antenna loaded with three pins (side, centre offset) resonating at 2.45 GHz

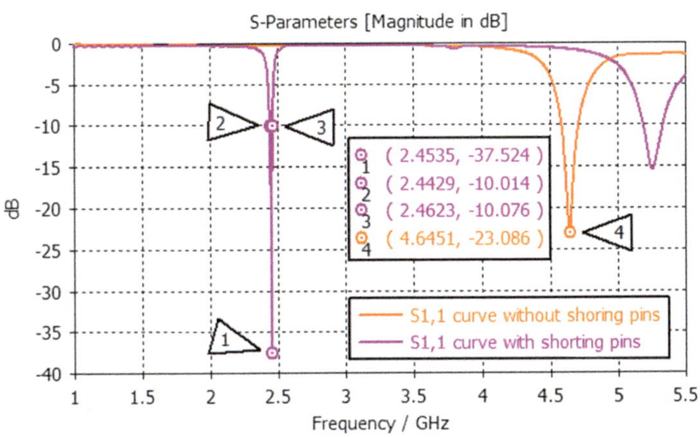

Fig. 3 Design-1: comparison of S-Parameter of patch antenna between without and with shorting/loading three pins (side, centre offset)

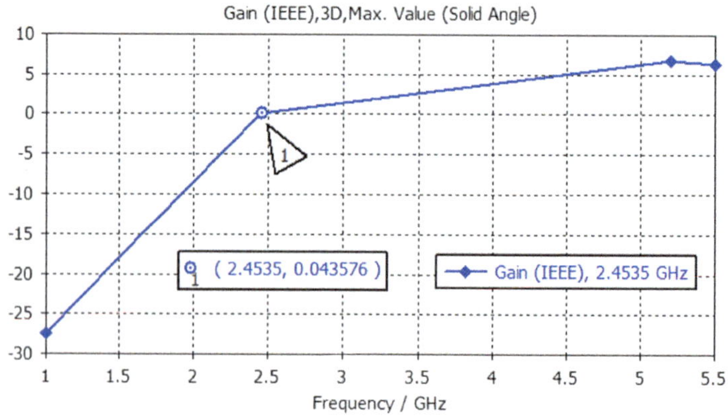

Fig. 4 Design-1: gain (IEEE) of antenna (resonant frequency—2.45 GHz)

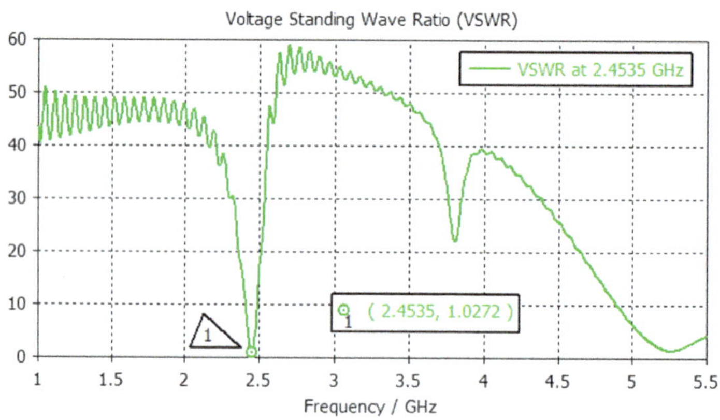

Fig. 5 Design-1: VSWR of antenna (resonant frequency—2.45 GHz)

To excite the patch antenna, Microstrip feed is used from the edge of substrate at a distance of F_l (which is the distance between substrate edge and patch edge). To examine the behaviour of antenna, an effective and latest version simulator is used.

After the use of a shorting pin having a diameter of 1 mm, the distribution of field will definitely to be changed because of the effect of introduced shorting pin. Figures 1, 2, and 3 depict the design and results after simulation of patch antenna without pin loading effect on S-Parameter. Simulation results demonstrate that patch aerial without pin loading in dominant mode has a resonant frequency about to 4.6 GHz (Fig. 3).

Resonant frequency of aerial (loaded without/with three pins) is also shown in the Figs. 3 and 9, which, distinctly show the resonant frequency of a conventional (without shorting pins) patch antenna and shorted-pin microstrip patch antenna.

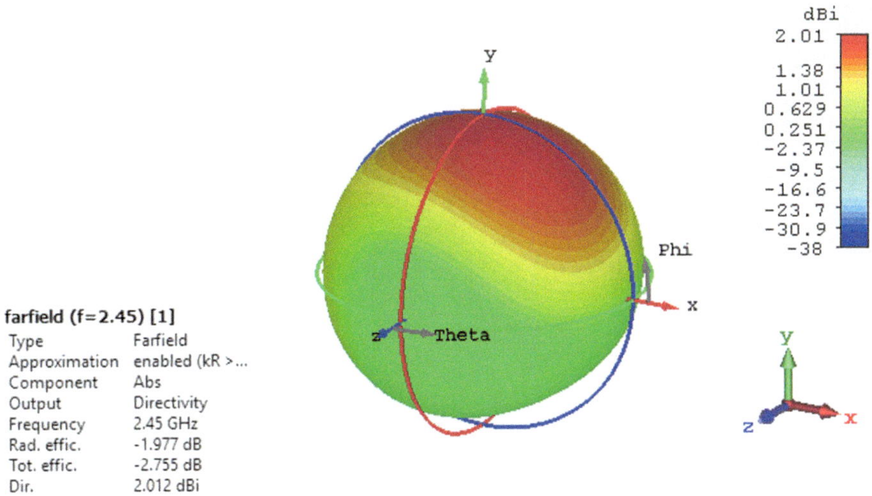

Fig. 6 Design-1: radiation pattern of antenna (resonant frequency—2.45 GHz)

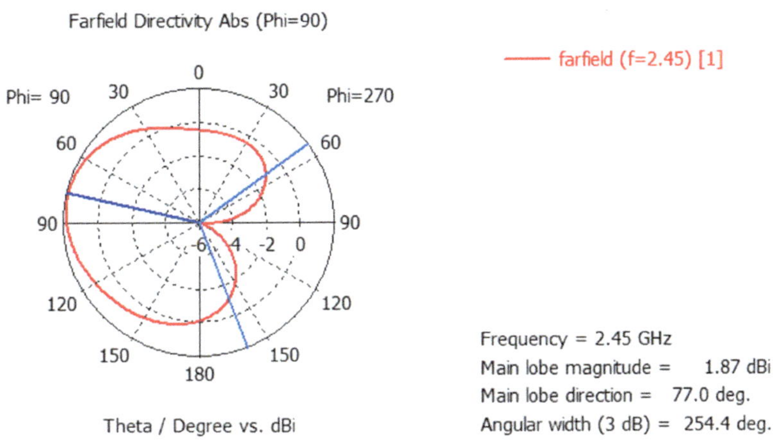

Fig. 7 Design-1: polar plot of antenna (resonant frequency—2.45 GHz)

3 Improvement in Design and Results

To validate the mentioned concept, simple microstrip patch aerial has been picked; the parametric diagram of an aerial with pin loading is shown in Fig. 1 (further, extended in Fig. 8) and parameters are indicated in Table 1.

The aerial is designed on a single-layer Rogers RT-5880 substrate having dimensions/area 36 × 44 mm^2, whose dielectric constant is 2.2, and substrate height is 1.57 mm, and size of patch is 19 × 29.2 mm^2. The side pin is loaded along the diagonal of antenna and beside the edge for showing the miniaturization.

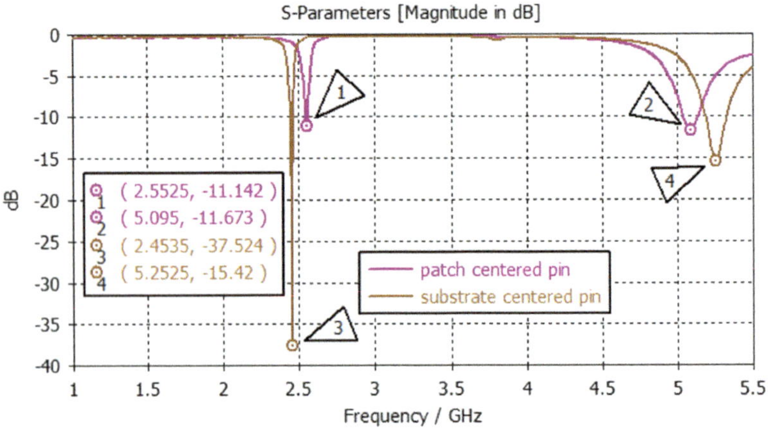

Fig. 8 Design-1: comparison of S-Parameter of patch antenna loaded with three pins [side, centre (patch/substrate), offset]

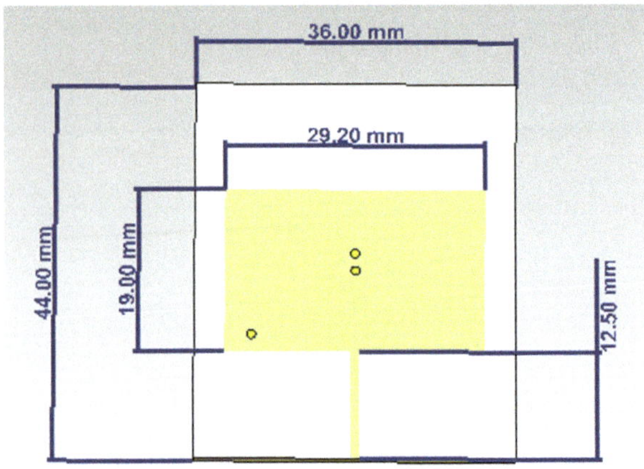

Fig. 9 Design-2: top view of patch antenna loaded with three pins (side, centre offset) resonating at 5.2 GHz

Shorting pins are implemented between the ground plane and patch of aerial. A small microstrip line is used to feed the aerial. To examine the proposed design, the latest version of CST Studio Suite simulator is used. The parameters of aerial are mentioned in Figs. 1 and 8 and Table 1. The aerial is designed so that it can easily integrate and work in the portable/movable/handheld applications.

Figures 2, 3 and 9 depict the S-Parameter (S_{11}) results after simulation of the various designs and the effects of pins and its position on it. When a patch antenna is operating in its resonant mode, it can be comparable with a lossy shunt resonator. For obtaining better results and understanding, more than 30 designs were previously studied (which is not shown here) and designed on different substrate material in [17]. The few fruitful design schematic and results are shown in this paper in Figs. 1, 2, 3, 4, 5, 6, 7, 8, 9, 10, 11, and 12.

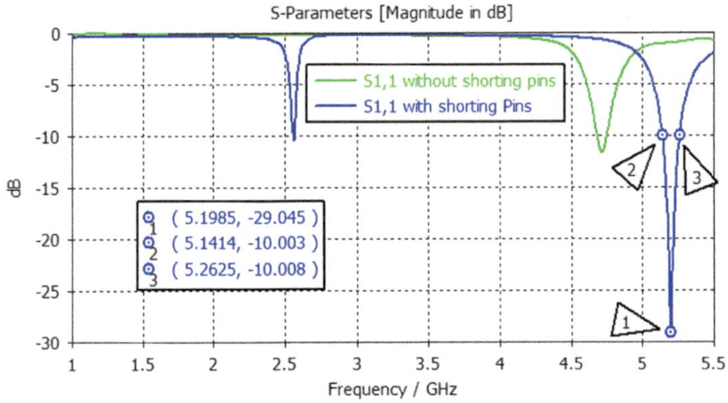

Fig. 10 Design-2: comparison of S-Parameter of patch antenna between without and with shorting/loading three pins (side, centre offset)

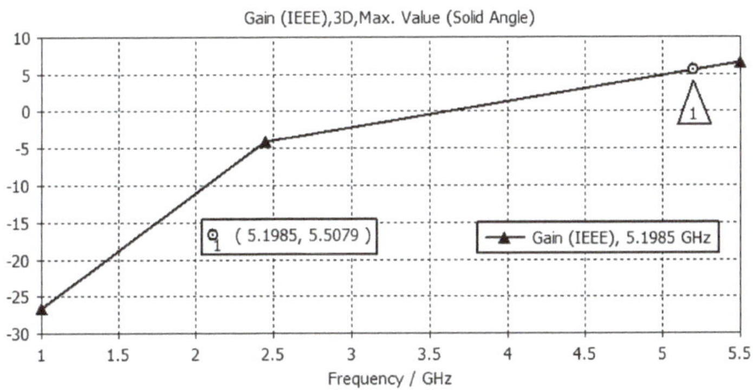

Fig. 11 Design-2: gain (IEEE) of antenna resonating at 5.2 GHz

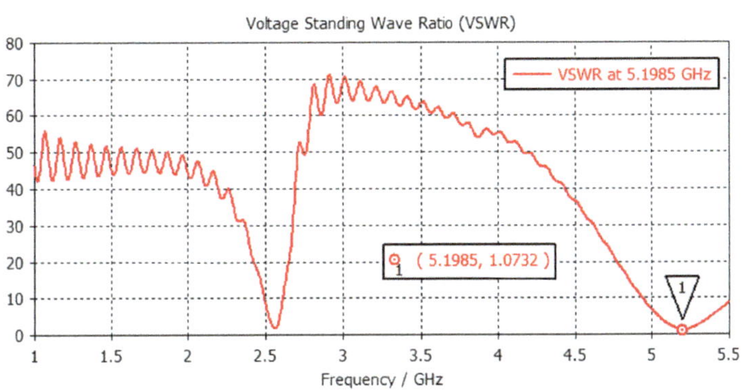

Fig. 12 Design-2: VSWR of antenna (resonant frequency—5.2 GHz)

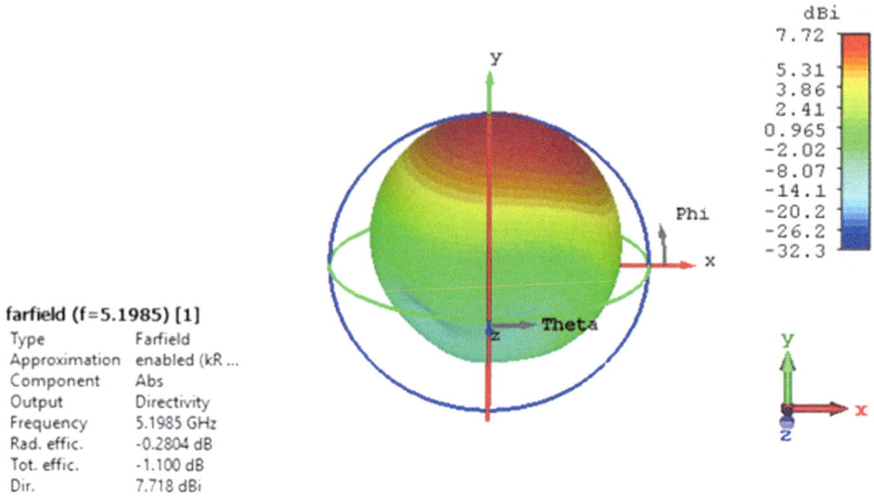

Fig. 13 Design-2: radiation pattern of antenna (resonant frequency—5.2 GHz)

Table 1 Parameters and results for antennas (design-1 and design-2)

Particulars of patch, feeding & shorted pins (copper lossy) and substrate (rogers RT 5880 lossy, $\in_r = 2.2$, $S_h = 1.57$ mm, tan $\delta = 0.0009$) details	Resonant frequency (fr) (GHz)	S_{11} (dB)	Gain (dBi)
Design-1: $P_w = 29.2$ mm, $P_l = 19$ mm, $F_w = 1$ mm, $F_l = 15.5$ mm, $S_w = 36$ mm, $S_l = 44$ mm, $D_{cp} = D_{op} = D_{sp} = 1$ mm	2.4535	−37.52	2.01
Design-2: $P_w = 29.2$ mm, $P_l = 19$ mm, $F_w = 1$ mm, $F_l = 12.5$ mm, $S_w = 36$ mm, $S_l = 44$ mm, $D_{cp} = D_{op} = D_{sp} = 1$ mm	5.1985	−29.05	7.72

According to numerical calculations and as compared with structure shown at Fig. 1, the recommended structure gives more than 70% patch area contraction. The overall dimension of aerial is $0.294\lambda r \times 0.36\lambda r \times 0.013\lambda r$, and the overall dimension of patch is $0.156\lambda r \times 0.239\lambda r \times 0.013\lambda r$ [fr = 2.45 GHz and λr is the free space wavelength analogous to resonant frequency]. The proposed aerial resolves the issues of matching and also provides medium gain and efficiency.

As per the numerical simulation analysis (not shown in this paper), the various cases shows that increasing the pins results in decreasing resonant frequency, but, could not be matched to requirement, due to which, overall performance of aerial decreases.

In this paper, patch antenna designed, simulated, and results obtained using CST software. Study of shorting pins on the dimension of conventional patch antenna has been done and various cases have been designed, studied, and illustrated for three pins shorting and without shorting, which can be seen in Figs. 1 and 8. The analysis of the case is also studied and demonstrated, as under.

3.1 Design-1

Initially, the patch antenna was designed of same dimensions and parameters (as shown in Fig. 1) without shorting pins and simulated results obtained. Further, three shorting pins [side, offset and center (in the center of substrate)] introduced, results obtained, and compared with the S_{11} parameter results of initial design, i.e., illustrated in Fig. 3.

In addition to the above, IEEE Gain, Radiation pattern, VSWR(Voltage Standing Wave Ratio), and polar plot results were also obtained and depicted in Figs. 4, 5, 6, and 7. Comparative results and design parameters are illustrated in Table 1.

As per obtained result in Fig. 3, it can be easily seen that resonant frequency shifted from 4.6 GHz to desired resonant frequency 2.45 GHz due to shorting pins positioned at proper place and bandwidth (−10 dB) is around 20 MHz.

In Figs. 4 and 5, gain (IEEE) and VSWR is shown vide which gain and VSWR is marked at 2.45 GHz respectively. Here, VSWR is good enough and reflections would be minimum.

In Fig. 6, 3D radiation pattern is shown, which shows the isotropic gain up to 2.01 dBi in a particular direction, and in Fig. 7, polar plot is shown at 2.45 GHz.

Further, patch antenna was designed of same dimensions and parameters (as shown in Fig. 1), but, the position of center pin is taken in the center of patch in place of substrate and results obtained and compared with initial design, i.e., illustrated in Fig. 7.

In Fig. 8, comparative results have been shown for different position of centre shorting pin, which shows that if centre pin is shorted at the centre of patch then desired results may not be obtained but when the centre pin is shorted at centre of substrate the desired results obtained, which concludes that centre pin positioning also has important/drastic role to get the desired results in patch antenna.

3.2 Design-2

Furthermore, patch size was kept the same, but, shifted towards the feed side and feed length reduced to 12.50 mm (as per Fig. 8) in place of 15.50 mm (as per Fig. 1). Now, the center pin is located at the center of patch as well as substrate.

Comparative result of S-Parameter is obtained for the design-2 without shorting pins and with shorting pins and shown in Fig. 10. In this case, we find the desired resonant frequency around 5.2 GHz.

IEEE Gain, VSWR, Radiation pattern and Polar plot results for design-2 (with three shoring pins) were also obtained and shown in Figs. 11, 12, 13, and 14, respec-

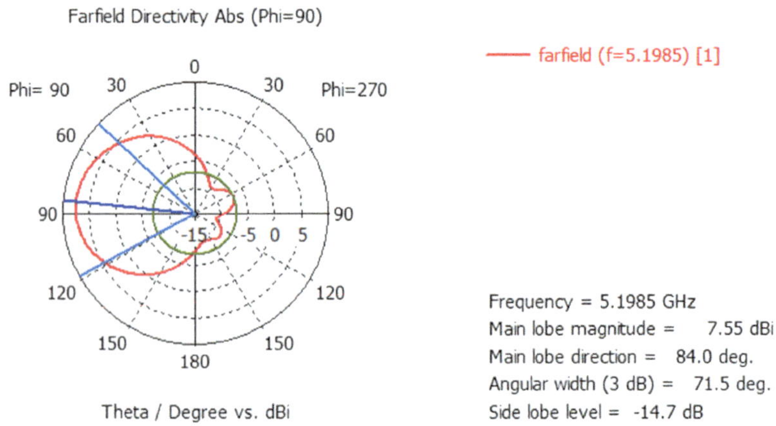

Fig. 14 Design-2: polar plot of antenna (resonant frequency—5.2 GHz)

tively. Here it may be clearly noted that IEEE gain at 5.2 GHz has good value and VSWR is also have good value, which will help in avoiding the reflecting waves.

In the design-1 and 2, the offset pin is 2 mm distant(in opposite side of feed) from the center pin; side pin is 3 mm distant from left edge of patch and 2 mm distant from feed side edge of patch. Substrate material is Rogers RT 5880; material of ground, feed, patch, and shorting pins is copper (lossy). Other parameters and comparative results are indicated in Table 1, as under.

In above Table 1, relative permittivity of substrate is represented by $\in_r$, tan δ is loss tangent of substrate, S_h is height of substrate, S_w is substrate width, S_l is substrate length, P_w is patch width, P_l is length of patch, F_w is width of feed, F_l is length of feed, D_{cp} is diameter of center pin, D_{op} is diameter of offset pin, D_{sp} is diameter of side pin.

4 Conclusion

We have proposed two novel aerials, whose resonant frequency are 5.2 and 2.45 GHz, for feeding simple microstrip line is used. Use of centered, offset pin with side pin helps in impedance matching and miniaturization of a patch aerial at the lower frequency. Simultaneously, it helps in improving radiation. The shorting pin technique reduces the patch size. Two compact, thin/slim, and single-band patch antennas are presented [18]. Use of shorting pins leads to correct matching and desired resonant frequency.

The overall dimensions of both the aerials is $0.294\lambda \times 0.36\lambda \times 0.013\lambda$ and maybe fabricated as per case results shown in Fig. 1 and 8. The aerial shows a proper resonant frequency of 2.45/5.2 GHz with bandwidth around 20/121 MHz. The aerials at above resonant frequency may have its applications in Wireless Local Area

Network (WLAN), Zigbee Bluetooth, Wi-fi, Industrial Scientific Medical (ISM), Industrial Automation, Internet of Things (IoT) devices, Satellite, Military, Aircraft, Worldwide Interoperability for Microwave Access (WiMAX), and Missiles.

References

1. Deschamps GA (1953) Microstrip microwave antennas. In: Presented at the 3rd USAF symposium on antennas
2. Malagisi CS (1977) Electronically scanned microstrip antenna array. U.S. Patent 4053895
3. Bhattacharyya AK, Jackson DR, Williams JT, Smith R (1991) Microstrip patch designs which do not excite surface waves. In: Antennas and propagation society symposium 1991 digest. IEEE, pp 68–71
4. Zhang X, Zhu L, Liu NW, Xie DP (2018) Pin-loaded circularly-polarised patch antenna with sharpened gain roll-off rate and widened 3-dB axial ratio beamwidth. IET Microwaves Antennas Propag 12(8):1247–1254
5. Zhang X, Zhu L (2016) Gain-enhanced patch antennas with loading of shorting pins. IEEE Trans Antennas Propag 64(8):3310–3318
6. Liu N-W, Zhu L, Choi W-W, Zhang X (2017) Wideband shorted patch antenna under radiation of dual-resonant modes. IEEE Trans Antennas Propag 65(6):2789–2796
7. Liu N-W, Zhu L, Choi W-W (2017) A differential-fed microstrip patch antenna with bandwidth enhancement under operation of TM10 and TM30 modes. IEEE Trans Antennas Propag 65(4):1607–1614
8. Liu N-W, Zhu L, Choi W-W, Zhang X (2017) A low-profile aperture coupled microstrip antenna with enhanced bandwidth under dual resonance. IEEE Trans Antennas Propag 65(3):1055–1062
9. Waterhouse R (1995) Small microstrip patch antenna. Electron Lett 31(8):604–605
10. Waterhouse RB, Targonski SD, Kokotoff DM (1998) Design and performance of small printed antennas. IEEE Trans Antennas Propag 46(11):1629–1633
11. Porath R (2000) Theory of miniaturized shorting-post microstrip antennas. IEEE Trans Antennas Propag 48(1):41–47
12. Hirasawa K, Haneishi M (1992) Analysis, design and measurement of small and low-profile antennas. Artech House, Norwood
13. Sievenpiper DF et al (2012) Experimental validation of performance limits and design guidelines for small antennas. IEEE Trans Antennas Propag 60(1):8–19
14. Amani N, Jafargholi A (2015) Internal uni-planar antenna for LTE/WWAN/GPS/GLONASS applications in tablet/laptop computers. IEEE Antennas Wirel Propag Lett 14:1654–1657
15. Jafargholi A, Jafargholi A (2018) Miniaturisation of printed slot antennas using artificial magnetic conductors. IET Microwave Antenna Propag. 12(7):1054–1059
16. Jafargholi A, Jafargholi A, Ghalamkari B (2018) Dual-band slim microstrip patch antennas. IEEE Trans Antennas Propag 66(12):6818–6825
17. Raj RK, Khandelwal H (2018) Pin loaded microstrip antenna for 2.45 GHz. J Basic Appl Eng Res (JBAER) 5(7):596–602
18. Raj RK et al (2017) A small monopole W shaped antenna for UWB and X band application with band notched for WiMAX. In 2018 Springer Nature Singapure Pte Ltd. 2018

Chapter 18
Decision Tree Using Artificial Neural Network: A Proposed Model

Monika Rathore and Shruti Gupta

1 Introduction

Artificial neural networks and symbolic learning algorithms are the two approaches through which we can go for machine learning. Decision tree proves to be a quite efficient approach, but when the large amount of data is encountered, [2] it leads to poor statistical efficiency. Artificial neural network always produces the accurate result, but they explicitly reveal the reasoning behind their decisions [1]. This paper proposes a model in which we combine the explicit explanation of the decision tree along with the predictive accuracy offered by artificial neural network.

1.1 Decision Tree

A decision tree is a decision support tool where each node represents a "test" on a particular attribute. Each branch of decision tree represents the result of the analysis, and each leaf node of the tree signifies the decision taken after computing all attributes, that is also known as class label [2].

There are three types of nodes in a decision tree:

1. Decision nodes, be a symbol of squares.
2. Chance nodes are represented by circles.
3. End nodes are symbolized by triangles (Fig. 1).

M. Rathore (✉) · S. Gupta
International School of Informatics and Management, Jaipur, Rajasthan, India
e-mail: monikadrathore@gmail.com

S. Gupta
e-mail: guptashr2712@gmail.com

© Springer Nature Singapore Pte Ltd. 2021

S. D. Purohit et al. (eds.), *Proceedings of International Conference on Communication and Computational Technologies*, Algorithms for Intelligent Systems,
https://doi.org/10.1007/978-981-15-5077-5_18

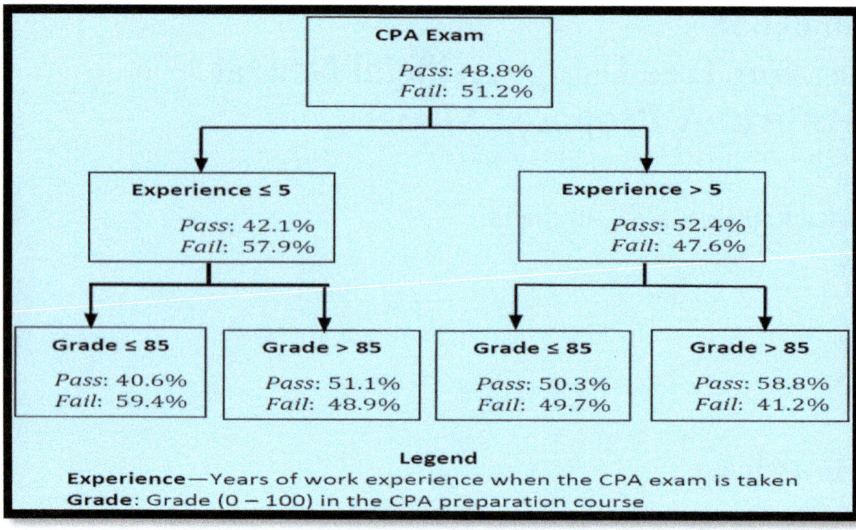

Fig. 1 Decision tree [1]

Decision Rules

Decision rules are the if-then statements that are used to predict the results. A decision rule comprises of single or a combination of multiple rules. On the basis of the conditions, the outcome is generated by the decision tree [3].

$$if \ < condition1 > \ and \ < condition2 > \ and \ < condition\ N > \ then\ outcome.$$

1.2 Artificial Neural Network

An artificial neural network is a set of associated units called artificial neurons, similarly the neurons in a biological brain. Each association of these neurons can send out a signal from one artificial neuron to another neuron. An artificial neuron first receives a signal (Fig. 2).

After processing it, and send out this to other neurons. They have distributed representation in their hidden layers which is hard to understand. There are two types of topologies used in artificial neural network [4]:

- **Feedforward:** In this type of topology, the flow of the information is unidirectional. There is no system of feedback.
- **Feedback:** Feedback loops are allowed in the feedback topology of ANN [5].

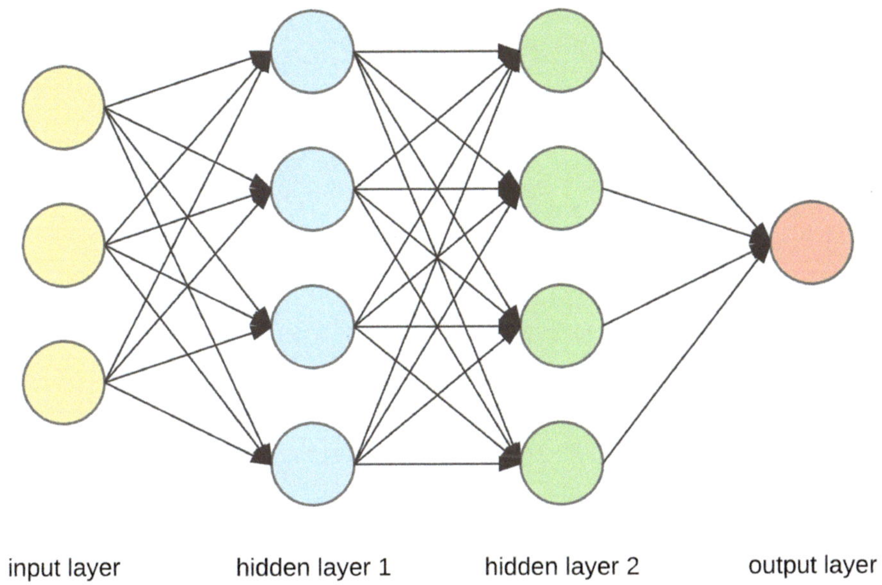

Fig. 2 Artificial neural network [4]

2 Construction of Decision Tree

2.1 Experiment Setup

RStudio 3.4.4 version is used for the decision tree construction. This is done with the help of R programming language.

2.2 Dataset Used

Cardiotocography dataset of UCI Machine Learning Repository is used for constructing the decision tree. This CTG dataset comprises the measurements of uterine contraction (UC) as well as the fetal heart rate (FHR).

The structure of the dataset is as follows (Fig. 3).

The decision tree is created on the cardiotocography dataset, which consists of 22 columns with 2129 rows of data. The dataset is classified by the three expert obstetricians and features on cardiotocograms. A consensus classification label was also allotted to each of them. Classification is prepared on the basis of a fatal state (N, S, and P), for that reason the dataset can be used for 3-class experiments.

```
> str(data)
'data.frame':	2129 obs. of  23 variables:
 $ LB      : int  120 132 133 134 132 134 134 122 122 122 ...
 $ AC      : int  0 4 2 2 4 1 1 0 0 0 ...
 $ FM      : int  0 0 0 0 0 0 0 0 0 0 ...
 $ UC      : int  0 4 5 6 5 10 9 0 1 3 ...
 $ ASTV    : int  73 17 16 16 16 26 29 83 84 86 ...
 $ MSTV    : num  0.5 2.1 2.1 2.4 2.4 5.9 6.3 0.5 0.5 0.3 ...
 $ ALTV    : int  43 0 0 0 0 0 0 6 5 6 ...
 $ MLTV    : num  2.4 10.4 13.4 23 19.9 0 0 15.6 13.6 10.6 ...
 $ DL      : int  0 2 2 2 0 9 6 0 0 0 ...
 $ DS      : int  0 0 0 0 0 0 0 0 0 0 ...
 $ DP      : int  0 0 0 0 0 2 2 0 0 0 ...
 $ Width   : int  64 130 130 117 117 150 150 68 68 68 ...
 $ Min     : int  62 68 68 53 53 50 50 62 62 62 ...
 $ Max     : int  126 198 198 170 170 200 200 130 130 130 ...
 $ Nmax    : int  2 6 5 11 9 5 6 0 0 1 ...
 $ Nzeros  : int  0 1 1 0 0 3 3 0 0 0 ...
 $ Mode    : int  120 141 141 137 137 76 71 122 122 122 ...
 $ Mean    : int  137 136 135 134 136 107 107 122 122 122 ...
 $ Median  : int  121 140 138 137 138 107 106 123 123 123 ...
 $ Variance: int  73 12 13 13 11 170 215 3 3 1 ...
 $ Tendency: int  1 0 0 1 1 0 0 1 1 1 ...
 $ NSP     : int  2 1 1 1 1 3 3 3 3 3 ...
 $ NSPF    : Factor w/ 3 levels "1","2","3": 2 1 1 1 1 3 3 3 3 3 ...
>
```

Fig. 3 Dataset structure

2.3 *Decision Tree Creation*

To construct the decision tree, we make the use of CTG data, which consists of 21 features and a response variable NSP. NSP represents the fetal state class code. The classification is done by experts into normal state, suspect state, and pathologic state. To construct the decision tree, the following steps are taken into consideration:

- The CTG dataset imported into RStudio (Fig. 4).
- The response variable NSP is in integer format, and it is converted into the categorical or factor variable (Fig. 5).
- Data is partitioned into two parts, that is, training data and the validation data (Figs. 6 and 7).

```
$ NSP     : int  2 1 1 1 1 3 3 3 3 3 ...
```

Fig. 4 Response variable

```
$ NSP     : int  2 1 1 1 1 3 3 3 3 3 ...
$ NSPF    : Factor w/ 3 levels "1","2","3": 2 1 1 1 1 3 3 3 3 3 ...
```

Fig. 5 Categorical variable

```
> train<-data[pd==1,]
> validate<-data[pd==2,]
```

Fig. 6 Data partition

```
pd                    int [1:2129] 2 1 1 1 1 1 1 1 2 2 ...
```

Fig. 7 Variable

```
data_train            1708 obs. of 23 variables
data_validate         421 obs. of 23 variables
```

Fig. 8 Probability assignment

- The datasets are assigned probabilities of 70–30%. The training dataset is given with the value 0.07 and the validation data with 0.03 (Fig. 8).
- "party" package installation (Fig. 9).
- The decision tree is created using "ctree()" function.
- Data used = train data
- Variables used = LB, AC and FM (Fig. 10).
- Plot the tree, using the plot() function. This function is used to plot the decision tree. The parameters used is the ctree_control with the following attributes—mincriterion = 0.99 and minsplit = 500 (Fig. 11).

2.4 Tree Explanation

After the construction of decision tree, we can easily predict a new patient that whether it is normal, suspect, and pathogenic. If a patient has an AC of 0.0009, that is, less than 0.001, then we traverse the left part of the tree. After coming toward the left side, suppose the value of LB is less than 136, say 45, then the tree will traverse toward the left side. According to the generated tree structure, the response for 1, that is, normal is 0.07 (70%). So the patient is having more probability of being normal. This is how decision tree is used to predict the outcomes. These probabilities can be predicted with the help of the predict() function. For the complete dataset, the model gives the prediction that whether they belong to class 1, 2, or 3 (normal, suspect, or pathogenic).

The image shows the probabilities for each given row of data, on the basis of the three respective classes (Fig. 12).

```
1) AC <= 0.000834028; criterion = 1, statistic = 263.403
  2) LB <= 136; criterion = 1, statistic = 131.511
    3) FM <= 0.111898; criterion = 1, statistic = 35.729
      4)* weights = 405
    3) FM > 0.111898
      5)* weights = 11
  2) LB > 136
    6)* weights = 314
1) AC > 0.000834028
  7) AC <= 0.002209945; criterion = 1, statistic = 52.155
    8) LB <= 136; criterion = 0.999, statistic = 17.292
      9) FM <= 0.0121396; criterion = 1, statistic = 42.826
        10)* weights = 103
      9) FM > 0.0121396
        11)* weights = 7
    8) LB > 136
      12)* weights = 78
  7) AC > 0.002209945
    13) LB <= 110; criterion = 1, statistic = 18.889
      14)* weights = 18
    13) LB > 110
      15) LB <= 147; criterion = 0.965, statistic = 8.877
        16) FM <= 0.2354892; criterion = 0.986, statistic = 10.725
          17)* weights = 742
        16) FM > 0.2354892
          18)* weights = 12
      15) LB > 147
        19)* weights = 28
>
```

Fig. 9 Values

```
library(party)
tree <- ctree(NSPF~LB+AC+FM,data = train,
controls = ctree_control(mincriterion = 0.99, minsplit = 500))
tree
plot(tree)
```

Fig. 10 Use of ctree() function

2.5 *Problems Associated with Decision Tree*

After the experiment is done on CTG dataset, many limitations and problems are observed. Some of the issues related to the decision tree construction can be seen in Fig. 4. This is a decision tree made up of the CTG dataset, but using different parameter values.

1. The decision tree generated through CTG dataset is quite perfect, but it seems to be highly unstable. Even a small change that is done in the input dataset may change the complete tree structure. Smaller changes may lead to redrawing of the tree.

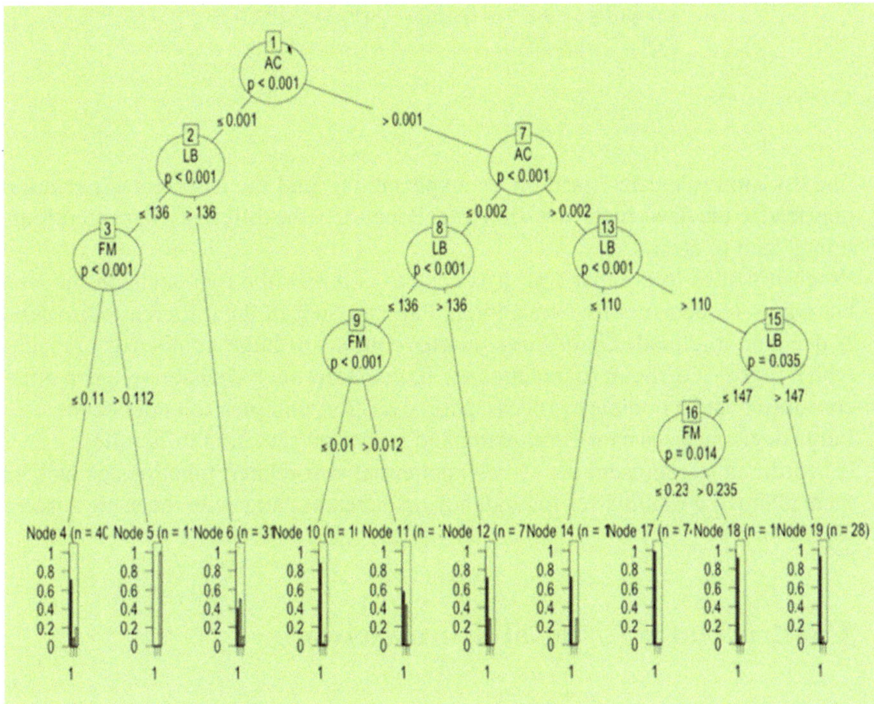

Fig. 11 After plotting the tree structure will appear like this

```
[[400]]
[1] 0.985933504 0.007672634 0.006393862

[[401]]
[1] 0.69230769 0.08894231 0.21875000

[[402]]
[1] 0.69230769 0.08894231 0.21875000

[[403]]
[1] 0.69230769 0.08894231 0.21875000

[[404]]
[1] 0.69230769 0.08894231 0.21875000
```

Fig. 12 Probabilities

```
samplesize = 0.60 * nrow(data)
set.seed(80)
```

Fig. 13 Sample size

2. The decision rules are based on the expectations, and the incorrect expectation may lead to the flaws in the decision tree. Because of the difference in perceptions, it may lead to error.
3. When it comes to comparison, it is easy to compare the two decision models, but when the decision tree encounters huge amount of data, it create disasters. A decision tree made up of large dataset consists of huge number of branches which makes it difficult to understand. To construct such decision tree is a time-consuming and a complex process. Moreover, it is complex to understand.
4. Large decision trees cause confusion and lots of presentation difficulties.
5. It is difficult to represent parity or exponential size-related functions as well as there is much possibility of redundancy on different paths with the same subtree (Fig. 13).

3 Construction of Artificial Neural Network

In order to construct an artificial neural network, we generally follow three steps.

1. BUILD IT
2. TRAIN IT
3. TEST IT.

After these steps, a trained artificial neural network will be constructed with least errors and perfect accuracy.

3.1 Creating Index Variable

Read the dataset and store it into another variable, data. This will work as a reference variable for the dataset. The next step to perform is the random sampling of the dataset.

Random Sampling

The purpose of random sampling is to eliminate the unnecessary parameters from the dataset. It is efficient and easy to implement. In this, each dataset has equal probability for selection. As the sampling of the data is done on a random basis, the model performance is less biased. Here, in this experiment random sampling of data is done with the help of sample() function and the setseed() function is used to maintain consistency by generating the same random samples always.

samplesize	1275.6

Fig. 14 Output

```
index = sample( seq_len ( nrow ( data ) ), size = samplesize )
```

Fig. 15 Index variable

index	int [1:1275] 2008 1690 2071 742 1066 1719 16 32 1448 1969 ...

Fig. 16 Output of Fig. 16

```
# Create training and test set
datatrain = data[ index, ]
datatest = data[ -index, ]
max = apply(data , 2 , max)

min = apply(data, 2 , min)
scaled = as.data.frame(scale(data, center = min, scale = max - min))
```

Fig. 17 Creating sets

Code 1: See Fig. 14.

Output: See Fig. 15.

Code 2: See Fig. 16.

Output: See Fig. 17.

3.2 Create Training and Test Set

After random sampling, the data is partitioned into two parts: First is the training dataset, and another is the testing dataset (Fig. 18).

This code is used to partition the data. There are two types of variables—dependent and independent variables. Our first partition that is training set is used to find the relationship between both the types of variables, while the performance of the model is checked by the test set (Fig. 19).

datatest	851 obs. of 4 variables
datatrain	1275 obs. of 4 variables

Fig. 18 Output of set creation

```
max        Named int [1:4] 160 26 564 3
min        Named int [1:4] 106 0 0 1
```

Fig. 19 Scaling

According to the partition done in the experiment, training set is the 60% of the dataset and the assignment of the data is purely done on the basis of random sampling. Index variable is used to fit the dataset into two partitions.

3.3 Scale Data for Neural Network

In this step, we will scale the CTG dataset. Unscaled data may lead to the meaningless results; that's why, scaling of dataset is considered to be quite important because if the scale is too large, then it has a large impact on the prediction variable of the dataset. MAD, tan-h estimators, min-max normalization, Z-score normalization, median, etc., are some of the techniques used for scaling the data. The technique used here is the min-max normalization. This technique removes the scaling effect from all the variables and transforms the data into a common range. It retains the original distribution of the variables. The scaling is done as follows (Figs. 20 and 21).

```
scaled        2126 obs. of 4 variables
```

Fig. 20 Scaled output

```
# install library
install.packages("neuralnet ")

# load library
library(neuralnet)

# creating training and test set
trainNN = scaled[index , ]
testNN = scaled[-index , ]

# fit neural network
set.seed(2)
NN = neuralnet(NSP ~ LB + AC + FM, trainNN, hidden = 1, linear.output = T )
# plot neural network
plot(NN)
```

Fig. 21 Fitting network

3.4 *Fitting the Neural Network*

In this step, we will finally fit the neural network. The artificial neural network can be fit using the scaled data. After that, neural network can be visualized using the weights with each variable (Fig. 22).

Install Library: To create the neural network, it is important to install the associated libraries. The library used here for the experiment is the "neuralnet". First of all, this library should be installed, and after that, rest of the experiment will take place.

Load Library: The next step is to load the installed library. Library "neuralnet" is then loaded into the RStudio software.

Creating training and test set: We already partitioned the data into two parts, that is, training data and the test data using the index variable. The next step here is to use the scaled values. For that purpose, the data is partitioned using the scaled values.

Fitting neural network: After the completion of the installing and load, here we will fit the neural network.

Plot neural Network: After fitting the neural network, the last step is to plot it. The plotting of the neural network will be done using the plot() function of the neural network.

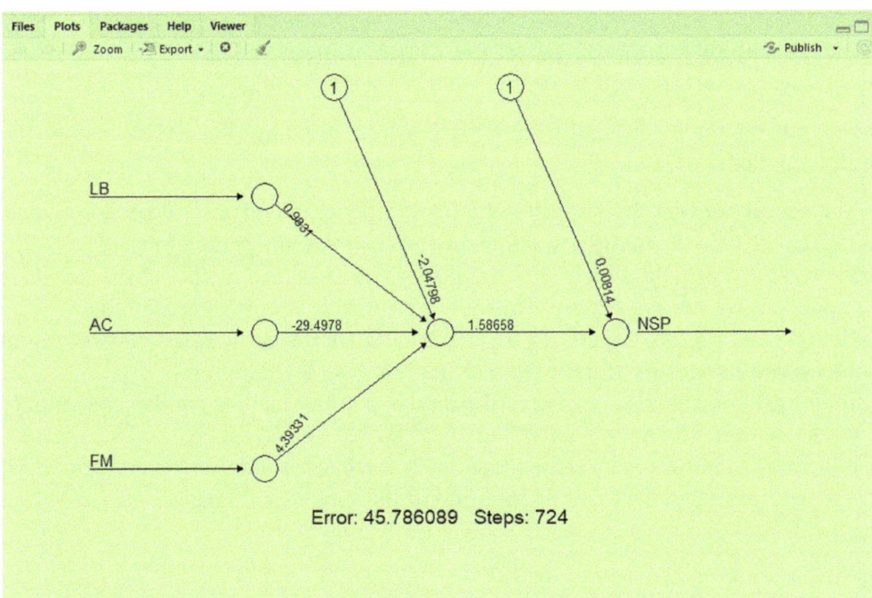

Fig. 22 Snapshot of artificial neural network created using cardiotocography dataset

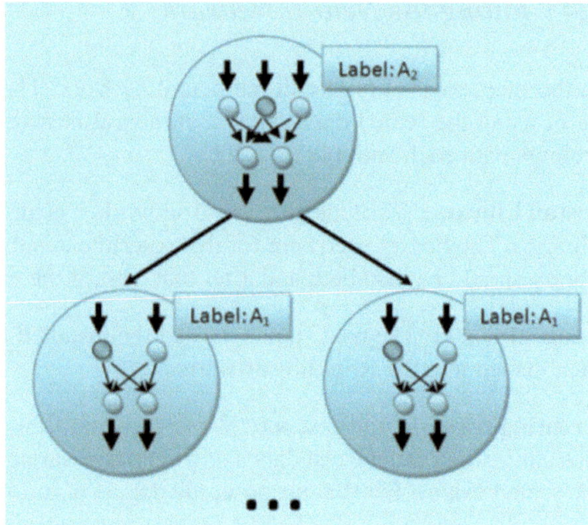

Fig. 23 Decision tree using ANN

3.5 *Artificial Neural Network Created Using Cardiotocography Dataset*

See Fig. 23.

3.6 *Observed Limitations of Artificial Neural Network*

The most common limitation observed while creating the ANN using the cardiotocography dataset is:

- The neural network created using the cardiotocography dataset does not give us a clue that why the things are going on and how the things are going on.
- There is no rule, by which we can specify the structure of the neural net. It is dependent on various factors. Thus, the structure is simply unpredictable.
- Neural networks are dependent on errors and experience. It keeps on working in a loop and iterates itself until the number of errors is minimum.
- To construct the ANN, I gathered the data in a numerical format, because it only works on that. The data in an alphabetical format has no scope.
- It is found that it is totally dependent on the number of iterations it performed, so it is hard to determine the time required or the time it will take.

4 Conclusion

After conducting the experiment using R Studio and analyzing all the limitations associated with decision tree and artificial neural network, this research came to the following conclusions.

4.1 It was analyzed that there are several limitations of a decision tree.
4.2 When it comes to large dataset, decision tree gives a complex output, which seems to be quite confusing and difficult to understand.
4.3 In case of artificial neural network, it takes direct input and gives a direct output.
4.4 It works same, irrespective of the size of data, and provides an accurate answer of the problem.
4.5 Hence, artificial neural network overcomes the problem associated with the decision tree.
4.6 But the problem here is artificial neural network does not show the way how it comes to a particular outcome.
4.7 The procedure to obtain a particular result is not defined because of the hidden layers in neural network.
4.8 Direct output is given.
4.9 This problem associated with the artificial neural network can be overcome by using decision tree, because decision tree shows the complete procedure before coming to a particular decision.
4.10 This is done through its branches.

5 Proposed Model

This paper focuses on combining a decision tree with artificial neural network. So, the conclusion here is, decision tree and artificial neural network, both can be combined to work in an efficient manner.

5.1 The problem associated with each of them can be overcome by using them interchangeably.
5.2 The problem will come to an end, when we combine the decision tree with an artificial neural network.
5.3 The proposed model will work well and overcome the limitations of each other.
5.4 Artificial neural network algorithm can be used to avoid pruning difficulty, and apart from making it easier, it contains better efficiency and accurateness during the phase of training the datasets. And on the other hand, decision tree can be used because they are less complex and can be easily interpreted, can be easily comprehended, and also holds nonlinear characteristics among values. These features make decision tree capable to provide better outputs. To employ of this kind of learning algorithm in that way can do good to the medical field.

6 Future Scope

Further, using the outcome of this paper, a new code for the proposed idea of combining the decision tree with artificial neural network can be developed using R Programming.

References

1. Frosst N, Hinton G (2016) Distilling a neural network into a soft decision tree. Google Brain Team, pp 5–7
2. Simplilearn (2018) Decision tree algorithm. Data science tutorial. Machine learning. Simplilearn
3. Kabari LG, Nwachukwu EO (2011) Neural networks and decision trees for eye diseases diagnosis, pp 3–6
4. Great Learning (2017) Building artificial neural network using R. Machine learning tutorial. Great Learning
5. Kontschieder P, Fiterau M, Criminisi A, Rota Bulo, S (2001) Deep neural decision forests, pp 4–5 (2001)

Chapter 19
A Comparative Analysis of R and Weka for Market Basket Analysis Using Apriori Algorithm

Monika Rathore and Shruti Gupta

1 Introduction

In today's era, competition in retail market is at its peak, as high completion customers are tend to move where they get better deals; due to this problem, every organization wants to retain their customers so that it will not affect their business. To retain customers, organizations are using different technologies so that their customers keep on buying from them and to attract new customer also. One of the methods is association rule mining (ARM) which can help business organization in two ways after using the market basket analysis [1].

1.1 This facilitates buyers to know all the associated items that belong to same place; it helps customer where they can have their wish list to one place instead of visiting different places of the store.
1.2 This provides a solution to organization for extra promotion of items when putting items nearer which are sold collectively. Various business houses across the world are using market basket analysis technique; for example, big organization like *Walmart* is using this technique to increase sales.

Apriori algorithm is the association rules of mining frequent item sets; it is first proposed by R. Agrawal and R. Srikan in 1994 [1]. The main principles of this theory are the subsets of frequent item sets are frequent item sets and the supersets of infrequent item sets are infrequent item sets [2]. First, the static data association rules mining was given by Agrawal in 1994 [1]; then, frequent closed item sets mining was proposed by Pasquier in year 1991. Then, maximum frequent item sets mining was given by Bayardo 1998, then fuzzy association rules mining was proposed by

M. Rathore (✉) · S. Gupta
International School of Informatics and Management, Jaipur, Rajasthan, India
e-mail: monikadrathore@gmail.com

S. Gupta
e-mail: guptashr2712@gmail.com

© Springer Nature Singapore Pte Ltd. 2021

S. D. Purohit et al. (eds.), *Proceedings of International Conference on Communication and Computational Technologies*, Algorithms for Intelligent Systems,
https://doi.org/10.1007/978-981-15-5077-5_19

Srikant, in 1996, and then, uncertain data mining based on frequent item sets mining was proposed by Agarwal 2009.

This paper compares the outcomes of two data mining tools, i.e., R programming and Weka for association rule mining. This paper includes the result of association rule algorithm for R and Weka with day-by-day transaction data from point of sale system of super store sales. This paper tested the apriori for both Weka and R on data of grocery store (Bharatpuria derpartmental Store, Jaipur). By making changes in the parameters like support and confidence for this algorithms and very remarkable results given by Association rule mining which are discussed later. The buyers are grouped based on their purchasing power, time and purchasing habits, most repeated buyers, and amount of transactions that helps in paying attention for advertisement based on customer appearance time and their purchasing behavior [3]. Other sections of the paper are structured as follows: Sect. 2 comprises preprocessing of sales data which will be used by the algorithm. Section 3 contains experimental results after applying apriori algorithm in R [5] (Sect. 3.1) and Weka (Sect. 3.2), Sect. 4 comprises comparative analysis between Weka and R, Sect. 5 will conclude with discussion and conclusion, and the last section comprises references.

2 Preprocessing of Sales Data

- Here taking a very common example of grocery.
- There are only two variables in the data sets.

(2.1) Customer Id—which shows the unique identification of customers in form of number

(2.2) Products—Products bought by the customers.

This paper discusses in detail the comparative analysis of both the tools (Weka and R) for association rule mining over the transaction data normalized. 83 records were taken for experiment before importing the excel file into Weka and R; first, we have changed file into .csv extension; following is the screen shot of input data file "MBA_data (name of file)".

3 Experiment Results

The following sections will provide the results of apriori algorithm after implementation of association rules in R and Weka using the above-mentioned sales data.

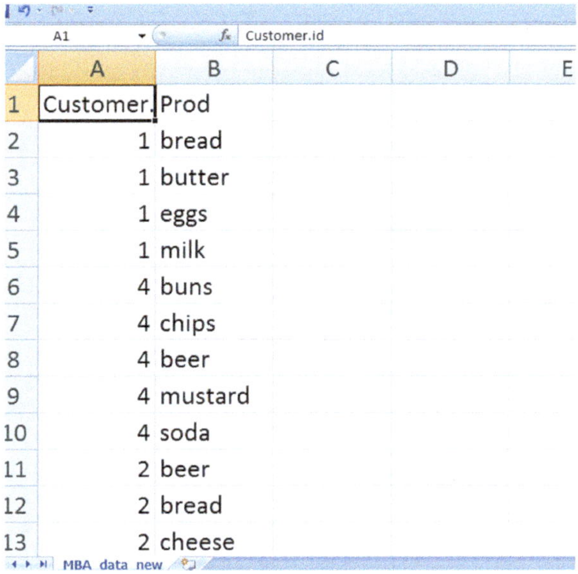

	A	B
1	Customer.	Prod
2	1	bread
3	1	butter
4	1	eggs
5	1	milk
6	4	buns
7	4	chips
8	4	beer
9	4	mustard
10	4	soda
11	2	beer
12	2	bread
13	2	cheese

Fig. 1 Data set used for experiment

3.1 Experiment Using R

After importing the file into R, this paper applies the following association rules like support should be at least 10%, confidence should be equal or more than 50%, maximum products can be associated would be 2, and same in case of minimum products to be associated would be 2 (Fig. 2).

```
> rules = apriori(trans, parameter=list(support=0.10, confidence=0.5,maxlen=2,minlen=2))

Parameter specification:
 confidence minval smax arem  aval originalSupport support minlen maxlen target   ext
        0.5    0.1    1 none FALSE            TRUE     0.1      2      2  rules FALSE

Algorithmic control:
 filter tree heap memopt load sort verbose
    0.1 TRUE TRUE  FALSE TRUE    2    TRUE

Warning in apriori(trans, parameter = list(support = 0.1, confidence = 0.5,  :
  You chose a very low absolute support count of 1. You might run out of memory! Increase minimum support
.

apriori - find association rules with the apriori algorithm
version 4.21 (2004.05.09)        (c) 1996-2004   Christian Borgelt
set item appearances ...[0 item(s)] done [0.00s].
set transactions ...[15 item(s), 15 transaction(s)] done [0.00s].
sorting and recoding items ... [14 item(s)] done [0.00s].
creating transaction tree ... done [0.00s].
checking subsets of size 1 2 done [0.01s].
writing ... [66 rule(s)] done [0.00s].
creating S4 object  ... done [0.00s].
```

Fig. 2 Screen shot of codes applied in R experiment

```
> inspect(rules)
   lhs             rhs           support   confidence lift
1  {buns}       => {mustard}     0.2000000 1.0000000  3.0000000
2  {mustard}    => {buns}        0.2000000 0.6000000  3.0000000
3  {buns}       => {chips}       0.2000000 1.0000000  2.5000000
4  {chips}      => {buns}        0.2000000 0.5000000  2.5000000
5  {buns}       => {soda}        0.2000000 1.0000000  2.1428571
6  {buns}       => {beer}        0.2000000 1.0000000  2.1428571
7  {mayo}       => {mustard}     0.1333333 0.6666667  2.0000000
8  {mayo}       => {chips}       0.2000000 1.0000000  2.5000000
9  {chips}      => {mayo}        0.2000000 0.5000000  2.5000000
10 {mayo}       => {cheese}      0.2000000 1.0000000  2.5000000
11 {cheese}     => {mayo}        0.2000000 0.5000000  2.5000000
12 {mayo}       => {soda}        0.2000000 1.0000000  2.1428571
13 {mayo}       => {beer}        0.2000000 1.0000000  2.1428571
14 {mayo}       => {bread}       0.2000000 1.0000000  1.3636364
15 {oranges}    => {banana}      0.2000000 0.7500000  2.2500000
16 {banana}     => {oranges}     0.2000000 0.6000000  2.2500000
17 {oranges}    => {eggs}        0.1333333 0.5000000  1.2500000
18 {oranges}    => {butter}      0.1333333 0.5000000  1.0714286
19 {oranges}    => {milk}        0.2000000 0.7500000  1.6071429
20 {oranges}    => {cheese}      0.1333333 0.5000000  1.2500000
21 {oranges}    => {bread}       0.2000000 0.7500000  1.0227273
22 {chocolate}  => {eggs}        0.2000000 0.7500000  1.8750000
23 {eggs}       => {chocolate}   0.2000000 0.5000000  1.8750000
24 {chocolate}  => {butter}      0.1333333 0.5000000  1.0714286
25 {chocolate}  => {milk}        0.2000000 0.7500000  1.6071429
26 {chocolate}  => {bread}       0.2000000 0.7500000  1.0227273
27 {banana}     => {cheese}      0.2000000 0.6000000  1.5000000
28 {cheese}     => {banana}      0.2000000 0.5000000  1.5000000
29 {banana}     => {bread}       0.2000000 0.6000000  0.8181818
```

Fig. 3 Result of experiment in R

The above screen shot shows that after applying above parameters R gives 66 association rule for given input data (second last line of screen shot); following is the screen shot of given association rules by R (Fig. 3).

Here, by observing these results, the snapshot says that for 83 records and when the parameters are like support 10%, confidence 50%, and minimum and maximum association of products are 2, it gives 66 rules.

3.2 Experiment Using Weka

Like in R before importing data file into Weka, first there is a requirement to change the extension into.csv; following is the screen shot of Weka after importing the same data file which in this paper used is R (Fig. 4).

After importing the data file in weka it provides the option association rule as menu bar user has to just use that provided option wherein R for applying association rule user has to import the association rule packages by applying command before installing the packages user has to set the cran mirrors too but in case of Weka the

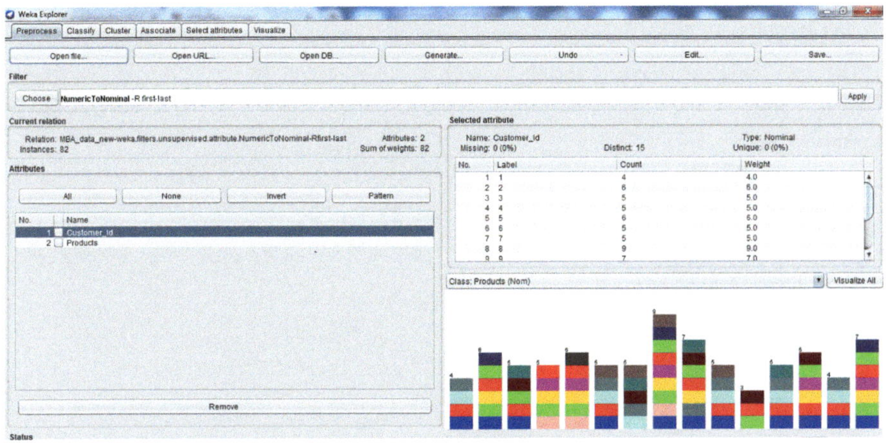

Fig. 4 Imported MBA_data file in Weka

association rule packages is already installed user has to just click on the option and has to provide the required parameters. By applying the default parameters, the Weka was not providing the association rules given records. Following is the screen shot of providing default parameters (Fig. 5).

After applying default parameters, the Weka did not provide any association rules (Fig. 6)

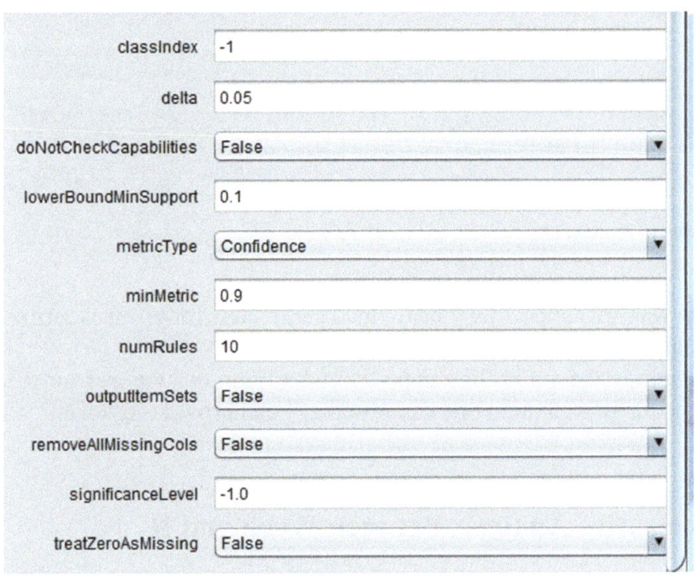

Fig. 5 Parameters setting as per experiment

```
=== Run information ===

Scheme:       weka.associations.Apriori -N 5 -T 0 -C 0.2 -D 0.01 -U 0.1 -M 0.5 -S -1.0 -c -1
Relation:     MBA_data_new-weka.filters.unsupervised.attribute.NumericToNominal-Rfirst-last
Instances:    82
Attributes:   2
              Customer_Id
              Products
=== Associator model (full training set) ===


No large itemsets and rules found!
```

Fig. 6 Weka not supporting given parameters

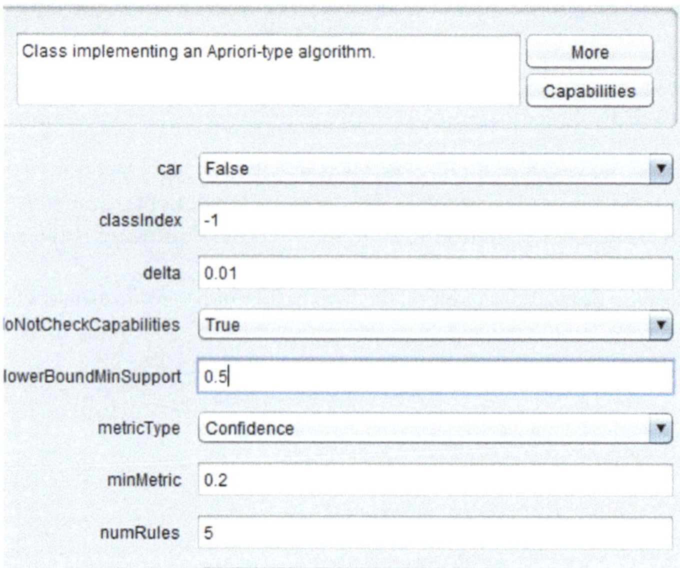

Fig. 7 Change in parameters

As Weka was not supporting the provided parameters for given experimental data, we have changed the rules as follows:

After changing the values like upper bound minimum support from 1 to 0.5 and no of rules from 10 to 5, here one can observe Weka provides association rules.

4 Comparative Analysis Between Weka and R

Weka, a data mining tool, is a research tool which is based on java; it consists of large compilation of data mining algorithms like lazy, Bayesian classification

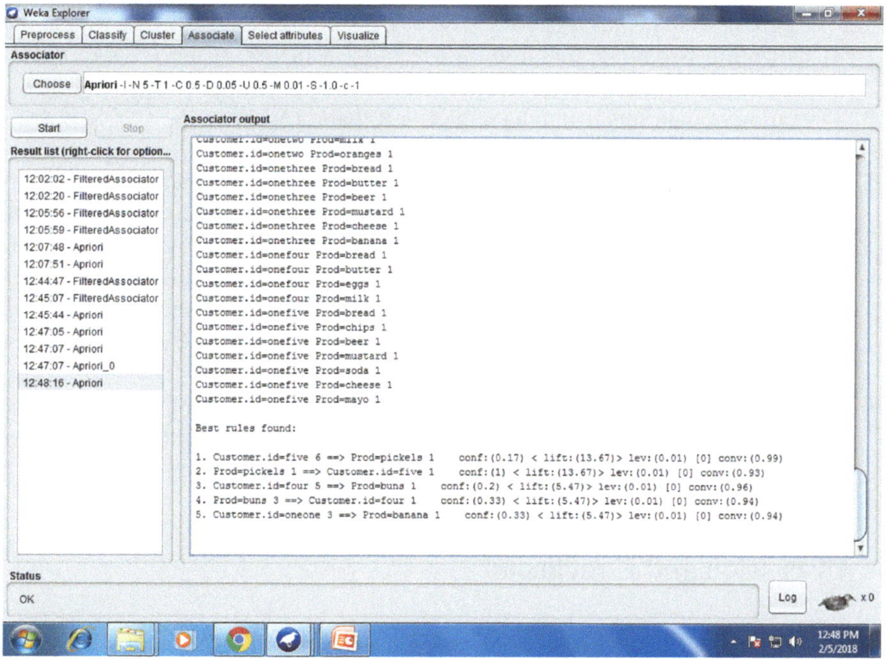

Fig. 8 Results in Weka

rules, decision trees, and so on. Weka contains four options (Explorer, Experimenter, Knowledge Flow, and Simple CLI). Among these four options, Explorer and Experimenter are widely used for data mining. When a researcher wants to compare multiple algorithms, Experimenter should be used; when a specific result is required of data mining, then Explorer should be used. When Weka is opened, the Explorer starts with a screen of data preprocessing. The problem with Weka is opening file because Weka supports only csv files, and mostly the data is stored in excel files, so there is a requirement to change the excel file into csv which is again a time-consuming process as there is a requirement to change the format from semicolon to comma. On the other hand, Weka provides knowledge on algorithms and multiple user choices are the finest features about Weka; any user can use Weka without any training because it is simple to learn and user can implement its own algorithm too.

Working on R programming is somehow similar to working in MATLAB, and it uses same logic in R; user can load CSV format file with no trouble and work with functions. But in case of R, default functions are very less; user has to install packages by applying particular code.

4.1 As compared to Weka, R is much simpler to learn due to its GUI and it is easy to get started too.
4.2 The Weka GUI offers numerous integral "visualization" sections, but these are very inadequate, particularly when judge against to what R packages can provide

like ggplot. Visualization of data is one of the strong points of R. In Weka, the visualization is paying attention on accepting behavior of the AI algorithms, instead of data sets.

4.3 Operation of data sets is much simpler in R than in Weka.

4.4 Help documentation is more purposeful in R.

4.5 In case of large data set and in big data environment, R has capability to support high volume of data as it offers RHadoop which can be used for big data processing. RHadoop works as a supporter which connects R to Hadoop environment and runs and helps in process R programs to the entire Hadoop nodes and clusters. But R does not provide visual features which can help in making producing workflows challenging, particularly for a novice. But R Studio can provide many graphical utilities.

4.6 In case of Weka for big data processing, Weka contains packages for MapReduce programming to keep autonomy over platform and also offers wrappers for Hadoop. Weka delivers workflow support using its Knowledge Flow utility.

5 Discussion and Conclusion

For grocery data set this paper uses parameters like support 10%, confidence more than 50% maximum products can be associated would be 2 and same in case of minimum products to be associated would be 2 and after applying the above parameters the R was providing 66 rules (Fig. 3) but in case of Weka for given default parameters (Fig. 7) it was not providing any rules but after altering the parameters like when support changed from 1 to 0.5 and minimum number of rules changed from 10 to 5 (Fig. 8) Weka was providing only 5 rules here by observing both result this paper can conclude then for given grocery dataset after applying almost similar parameters R provides 66 rules and on the other hand Weka provides only 5 so this paper conclude that though Weka is simpler to apply test and it has lot of inbuilt applications but as compared to R it provides lesser rules so associate data items amongst several data items it is preferable to use R than Weka.

References

1. Agrawal R, Srikant R (1994) Fast algorithms for mining association rules. In: Proceedings of the 20th VLDB conference, pp 487–499
2. Annie MCLC, Kumar DA (2012) Market basket analysis for a supermarket based on frequent itemset mining. Int J Comput Sci 9(5):257
3. Annie MCLC, Kumar DA (2011) Frequent item set mining for market basket data using K-Apriori algorithm. Int J Comput Intell Inform 1(1):14–18
4. Bhandari A, Gupta A, Das D (2015) Improvised apriori algorithm using frequent. In: Proceedings in Computing, 9th International Research Conference-KDU, Sri Lanka, vol 2016, p 133

5. Han J, Pei H, Yin Y (2000) Mining frequent patterns without candidate generation. In: Proceedings of the Conference on the management of data (SIGMOD 2000). ACM Press, New York

Chapter 20
An Energy-Efficient and Secured Hop-by-Hop Authentication Protocol for Wireless Sensor Networks

S. V. N. Santhosh Kumar, M. Selvi, A. Kannan, and S. Ganapathy

1 Introduction

Wireless Sensor Network is a wireless technology which has tiny devices which can sense the events in the deployed environment and disseminate the sensed events to the base station (BS) using radio frequency through the wireless channel [1]. WSNs are applicable in various domains such as military, disaster management system, and transportation system [2]. In WSN, providing security and privacy [3, 4] to the nodes is a major challenge. Considering the inherent nature where the nodes are deployed, the attacker can compromise the nodes of the network during data dissemination. The attackers [5] can steal the data or eavesdrop the sensed information with false data, and attacker can send the malicious data along with sensed information to make the nodes compromise and prevent them to perform the intended services. On the other hand, the attackers can exhaust the resources of sensor nodes by launching Denial of Service attack [6], where nodes fail to forward the data to other nodes.

S. V. N. Santhosh Kumar (✉)
School of Information Technology and Engineering, Vellore Institute of Technology University, Vellore, Tamil Nadu, India
e-mail: Santhoshkumar.svn@vit.ac.in

M. Selvi · A. Kannan
School of Computer Science and Engineering, Vellore Institute of Technology University, Vellore, Tamil Nadu, India
e-mail: selvi.m@vit.ac.in

A. Kannan
e-mail: kannan.a@vit.ac.in

S. Ganapathy
School of Computing Science and Engineering, Vellore Institute of Technology, Chennai campus, Tamil Nadu, India
e-mail: ganapathy.S@vit.ac.in

© Springer Nature Singapore Pte Ltd. 2021

S. D. Purohit et al. (eds.), *Proceedings of International Conference on Communication and Computational Technologies*, Algorithms for Intelligent Systems,
https://doi.org/10.1007/978-981-15-5077-5_20

Considering all these scenarios, it is necessary to provide secure authentication to ensure privacy and security. In most of the existing systems, secure authentication is provided by employing different cryptography mechanisms like hash functions and public key cryptography. In these mechanisms, the authentication protocols mainly depend on the integration of cryptographic primitives with exchange of messages between the protocols. The limitations in existing authentication protocols for WSNs are that most of the existing schemes use either hashing or public key cryptography for secure authentication [3]. By doing so, they fail to provide adequate security with better energy efficiency. Motivated from all these observations, it is necessary to provide energy-efficient authentication protocol, which ensures effective secure authentication with better energy efficiency.

2 Related Works

There are various mechanisms which have been proposed to provide security by means of effective authentication in WSN. Some of the mechanisms available in the literature for providing data authentication are discussed in this section. Spin and Brosk [7] is a lightweight authentication protocol based on symmetric key cryptography. The advantage of this protocol is low computational overhead. Limitations are it is vulnerable to key capture attack. TESLA and EMSS [8] are the protocols which are meant for lossy multicast streams. TELSA provides low overhead with delay-tolerant authentication. EMSS provides stream signature which ensures non-repudiation. The limitations of these protocols are they are vulnerable against DOS attack [6] during authentication delay. Interleaved hop-by-hop authentication [9] provides the detection of false injection attack. This scheme ensures each node validates the correctness of data transmitted with their keys. Limitations are there exits computational and communication overhead during data authentication. Statistical en route filtering (SRF) [10] is designed to detect false report during data forwarding. SFR employs Message Authentication Code (MAC) to ensure the integrity of data transmitted. Limitations are it provides only limited authentication during data forwarding by the nodes.

Robust authentication scheme (RAS) [11] is designed to filter out false data. In this scheme, data are subdivided into smaller items called chunks. The limitations are this scheme suffers from scalability problem. Perturbation polynomial authentication [12] is a novel message authentication scheme which ensures intermediate authentication during data forwarding. The limitations are scalability and there exits the computational and communication overhead during data transmission. Singh et al. have proposed the collaborative lightweight trust-based routing protocol (CLT) [13]. This scheme consists of three steps. In the first step, the trust is to be calculated as a positive integer of range 0 to 100. In the second step, the computed trust value is stored in transaction table, and in the third phase, routing takes place based on trust values. The advantages of this scheme are that it significantly reduces memory consumption and increases the packet delivery ratio. Since this scheme focuses only on reducing the memory consumption, security is not considered in their design. Raja Rajeswari et al. have proposed a secured energy conserving slot-based topology maintenance protocol [14] for Wireless Sensor Networks. In this scheme, the sleep/wake-up schedules of nodes are synchronized and authenticated based on the symmetric key to increase the lifetime and efficiency of the network. The advantages of this scheme are that it reduces the communication and computational overhead to increase the lifetime of the network. However, this protocol is vulnerable to node compromise attack. Santhosh Kumar et al. have proposed Secured Selective Design Relay Inquiry Protocol (S-SELDRIP) [5] which provides hop-by-hop authentication by using distributed key management technique. The advantage of this protocol is better security. Limitations are this protocol can be applied only for distributed systems. The observations from the literature survey are the most of the existing authentication protocols focus either only on energy efficiency or in security. They fail to provide trade-off between energy efficiency and security. Motivated from all these observation, in this paper a novel energy-efficient authentication is proposed which can provide better security with energy efficiency.

3 System Model and Assumptions

The proposed protocol is designed by considering the following assumptions

1. Base station is a resourceful device and controls the entire network
2. Random static deployment is used for deploying N sensor nodes.
3. The deployed sensor nodes have a same initial energy as 100 J
4. Though sensor nodes are resource constraint in nature, it has ability to perform basic cryptographic operations.

4 Proposed Authentication Protocol

In this paper, a novel energy-efficient authentication protocol is proposed which provides efficient hop-by-hop authentication for the nodes of WSN. The hop-by-hop authentication is ensured by elliptical curve cryptography [1]. The proposed protocol consists of two phases, namely system setup and key pair generation phase and hop-by-hop authentication phase. In system setup and key pair generation phase, the nodes are deployed by the BS and key pairs of public key and private key is generated by BS to all the nodes. In hop-by-hop authentication phase, each hop of nodes are authenticated and then forwarded to other nodes until it reaches BS. By doing so, the proposed protocol provides hop-by-hop authentication and ensures security during data transmission.

4.1 System Setup and Key Pair Generation Phase

In system setup and key pair generation phase, first the nodes are deployed using random method by BS. After the deployment of nodes, the next is the generation of private key and public key by the BS. The proposed authentication protocol uses homomorphic encryption to ensure security and privacy of the sensed information until it reaches BS. The procedure for generation of public key and private key is explained as follows.

Algorithm1: Generation of public key by BS

1. Choose the two large random prime numbers p, q which are generated from pseudo random integer function
2. BS selects a random prime number RX € to Z*(P), where Z*(P) is the set of invertible prime elements from the group of prime numbers
3. BS selects another random point YX € to Z*(q), where Z*(q) is the set of invertible prime elements from the group of prime numbers.
4. Public key generated by BS is $PUK = (RX)^{(p-1)}.\ (YX)^{q-1}$ (1)

Equation (1) gives the public key generated by the BS

Algorithm 2: Generation of private key by BS

1. For private key generation, ECC cure E is set up and its points are generated based on point multiplication
2. BS chooses the random number (RR) from the point (X1) on E as a unique identifier for each node
3. Then BS chooses a random prime number (RP) € to Z*(q) from the generator (G1)
4. BS chooses another random prime number (Rq) € to Z*(p) from the generator (G2).
5. The generated private $PRK = (RP)^{(pq)}.\ (RP)^{(E)}$ (2)

Equation (2) gives the private key generated by BS. The generated public and private keys are preloaded into the nodes during pre-deployment state.

4.2 Hop-by-Hop Authentication Phase

After the generation of public key and private key, the next step is hop-by-hop authentication phase. In hop-by-hop authentication phase, initially the source nodes have to establish the optimal route to reach the destination nodes using S-SELDRIP protocol. The procedure for hop-by-hop authentication is explained as follows

1. for all the nodes present in a Route RN
   Do

2. BS identifies the source node (SN) and destination node (DN) in a Route RN, and each route in RN is stored in a array [SN DN]
3. Initialize array [SN DN] = 0
   End For
4. From all the given Route RN, source node (SN) encrypts the sensed data with its public key and sends to all the nodes which are present in a given array
5. The encrypted message format is EM = SD || SN (PUK) || TN, where SD is the sensed data, SN (PUK) is the source node public key, and TN is the time stamp
6. for the array elements of nodes to 1 to N do
7. The intermediate nodes in the Route RN verifies the correctness of senders public key by using Eq. (3)

8. Node N = SD || Node N || PRK (3)

9. If the validation of sender's public key is correct
10. Then
11. Node N forwards the message to next node in a given Route RN
12. Else
13. The message from the nodes will be discarded and information regarding the validation failure is informed to the BS by the corresponding nodes.
14. For all the nodes present in the Route RN repeat steps 7–13.
15. End For

By using this procedure, the hop-by-hop message authentication is carried out until it reaches the destination nodes. By doing so, the proposed protocol provides effective message authentication and ensures message integrity during data transmission in WSN.

5 Proposed Authentication Protocol

The proposed protocol is tested and implemented with NS2 Simulator. Table 1 gives the simulation parameters used in proposed system.

Table 1 Simulation parameters

Network simulator	NS2
Transmission range	20–25 ms
Physical layer	Phy/wirelessphy-mica2
Radio propagation model	Two ray model
Environment	Urban
Initial energy	100 J
Packet size	25 bytes

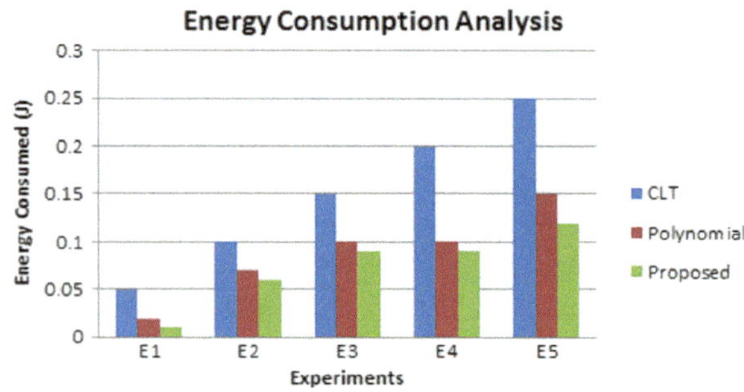

Fig. 1 Energy consumption analysis

6 Results and Discussions

The proposed protocol is compared with other existing protocols in terms of energy consumption analysis, packet delivery ratio (PDR), and network lifetime. The security level can be observed from these evaluation parameters.

6.1 *Energy Consumption Analysis*

Figure 1 shows the energy consumption analysis for the proposed system and the existing models such as CLT and polynomial. From the graph, it is clear that the proposed protocol has better average spend energy compared with other protocols because proposed protocol provides efficient hop-by-hop authentication and reduces the impact of malicious nodes in terms of packet dropping and also prevents data forwarding attacks. Hence, it has average spend energy compared with other existing protocols.

6.2 *Packet Delivery Ratio*

The proposed protocol is compared with other existing protocols such as CLT and the polynomial scheme in terms of packet delivery ratio.

From Fig. 2, it is clear the PDR of proposed protocol is better than other existing protocols because proposed protocol provides efficient hop-by-hop authentication and identifies the nodes which drops the data packets. Hence, it has better PDR when compared with other existing related protocols.

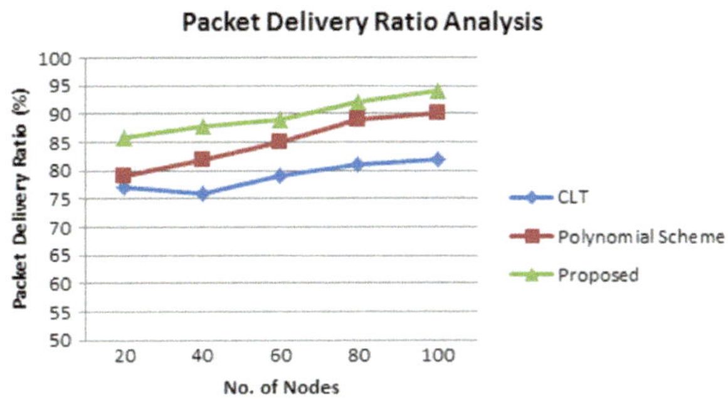

Fig. 2 Packet delivery ratio

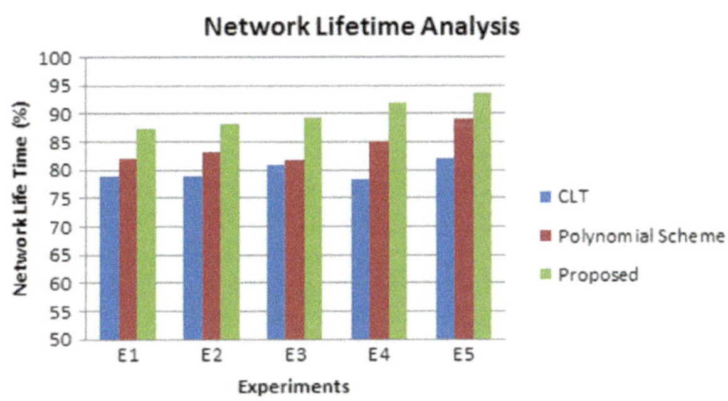

Fig. 3 Nodes lifetime

6.3 Nodes Lifetime

The proposed protocol is compared with other existing protocols in terms of node lifetime in Fig. 3. Five experiments have been conducted for performing comparative analysis. Here, the various numbers of nodes have been involved in the different experiments.

From Fig. 3, it can be seen that the performance of the proposed system performs well than the existing works such as CLT and the polynomial scheme. This is due to the incorporation of hop-by-hop authentication scheme. The proposed system secured the network communication efficiently by detecting the malicious nodes.

7 Conclusion and Future Work

In this paper, a novel authentication protocol is proposed which can provide an energy-efficient effective hop-by-hop authentication in WSN using ECC. The proposed protocol is implemented using NS2 simulator. Simulation results show that the proposed protocol can be able to provide the efficient secured data transmission. The future work of the proposed scheme is to improve its security and energy efficiency ability in mobile environment.

References

1. Dinarvand N, Barati H (2019) An efficient and secure RFID authentication protocol using elliptic curve cryptography. Wireless Netw. 25(1):415–428
2. Heinzelman WR, Chandrakasan A, Balakrishnan H (2000) Energy-efficient communication protocol for wireless microsensor networks. In: Proceedings of IEEE 33rd annual Hawaii international conference on system sciences, pp 1–10
3. He D (2015) Accountable and privacy-enhanced access control in wireless sensor networks. IEEE Trans Wirel Commun 14(1):387–398
4. Santhosh Kumar SVN, Yogesh P (2016) Privacy preserving with enhanced access control for distributed data dissemination in WSN. Asian J Res Soc Sci Hum 6(12):561–577
5. Santhosh Kumar SVN, Palanichamy Y (2018) Energy efficient and secured distributed data dissemination using hop by hop authentication in WSN. Wirel Netw 24:1343–1360
6. Rajendran R, Santhosh Kumar S, Palanichamy Y, Arputharaj K (2019) Detection of DoS attacks in cloud networks using intelligent rule based classification system. Cluster Comput 22(1):423–434
7. Ruj S, Nayak A, Stojmenovic I (2011) Distributed fine grained access control in wireless sensor networks. In: Proceedings of IEEE IPDPS, pp 352–362
8. He D, Bu J, Zhu S, Chan S, Chen C (2011) Distributed access control with privacy support in wireless sensor networks. IEEE Trans Wirel Commun 10(10):3472–3481
9. Zhu S (2004) An interleaved hop-by-hop authentication scheme for filtering of injected false data in sensor networks. In: IEEE symposium on security and privacy
10. Lee HW (2007) A method to control the probability of attempts to verify a report in statistical en-route filtering networks. In: 15th IEEE international conference on networks
11. Hu Y (2007) RAS: a robust authentication scheme for filtering false data in wireless sensor networks. In: 15th IEEE international conference on networks
12. Li J, Ren J (2014) Hop by hop message authentication and source privacy inn WSN. IEE Trans Parallel Distrib Syst 25:1223–1232
13. Singh M, Sardar AR, Sahoo RR., Majumder K, Ray S, Sarkar SK Lightweight trust model for clustered WSN, In: Proceedings of the 3rd international conference on frontiers of intelligent computing: theory and applications, pp 765–773
14. Raja Rajeswari S, Seenivasagam V (2015) Secured energy conserving slot-based topology maintenance protocol for wireless sensor networks. Wirel Pers Commun 87(2):527–550

Chapter 21
A Study of Gujarati Character Recognition

Arpit A. Jain and Harshal A. Arolkar

1 Introduction

Today, the core part of any organization is data; without data, we cannot work. Data can be kept in many forms like soft copy or hard copy. Hard copy can be a handwritten document or printout of soft copy. Printing of a soft document and scanning of a hard document are easy tasks nowadays. The most common way of creating data is the use of keyboard and scanner. The text entered by using keyboard can be easily edited although for the documents which have huge data set, the task can be tedious and time-consuming. By using scanner, the data can be stored in editable and non-editable format. If the data is stored in non-editable format, then we need an OCR to convert the data in editable format. The conversion of printed or handwritten scanned document into ASCII or Unicode character set is possible nowadays. This technique really helps in the area of text recognition. OCR has been using this approach to convert scanned documents into editable formats. The non-Indic scripts do not have 'Matras,' so it is an easy task to recognize and find out the equivalent ASCII or Unicode, but for Indic script that has matra's the task is difficult. Gujarati is one such language where building a handwritten character recognizer is bit difficult.

2 The Guajarati Language

The Gujarati script is one of the oldest scripts which has been designed from an ancient Devanagari script. Devanagari has a huge character set, and it is considered one of the most used parent scripts for many Indian regional languages. It has total

A. A. Jain (✉) · H. A. Arolkar
GLS University, Ahmedabad, Gujarat, India
e-mail: jain.arpit03@gmail.com

© Springer Nature Singapore Pte Ltd. 2021

S. D. Purohit et al. (eds.), *Proceedings of International Conference on Communication and Computational Technologies*, Algorithms for Intelligent Systems,
https://doi.org/10.1007/978-981-15-5077-5_21

forty-four characters which further can be divided into thirty-three consonants and eleven vowels. The Devanagari also has modifiers which can be associated with the consonants and vowels to make a different articulation. The modifiers in the Devanagari script are matras and pure consonants. Matras are the modifiers which can be placed with the consonants and vowels in several ways. Pure consonants are generally termed as half written consonants. A pure consonant and consonant written together is called conjunct consonants. The difference between Gujarati and Devanagari script is that Devanagari words and characters have a upper line called Shirorekha (a header line), while Gujarati words and characters do not have Shirorekha. Gujarati language also has a huge character set with fifty-nine characters and sixteen diacritics. The language Gujarati has thirty-four consonants, two compound consonants and thirteen vowels. Any of these written together can form a word of Gujarati language. It also has modifiers like matras and compound consonants, which can be used with consonants. The compound consonants and consonants written together is called conjunct consonants. Sixteen diacritics are separated into thirteen vowels and three other characters [1–3]. The Gujarati language also has ten numerals.

The character set of Gujarati is almost double than the character set of English language. Recognizing the characters in Gujarati at time is little bit confusing, as they have similar looks for example numeral five (૫) and alphabet 'Pa' (પ), numeral two (૨) and alphabet 'Ra' (ર) can be easily miss interpreted.

Furthermore, the conjunct consonants create a character by combining one and half character. This adds to another problem of identifying additional permutation and combination of an existing character. These are the few characteristics of this script which can be considered as a reason for the slow progress in development of Gujarati Character Recognition.

2.1 Handwritten Character Recognition (HCR)

Handwritten Character Recognition (HCR) is a software that allows us to identify characters from an image. The HCR can be classified based on the type of images that are used for recognition. It can generally work on two types of images online and offline. An offline image is created by capturing a picture of contents written using a pen or pencil whereas an online image is usually created by using a Light pen. Figures 1 and 2 show a sample offline and online image. The online image is generally smooth, and all characters have same attributes as the light pen is used. On the other hand, offline images vary often as it depends on the type of pen used, ink used and pressure applied by writer.

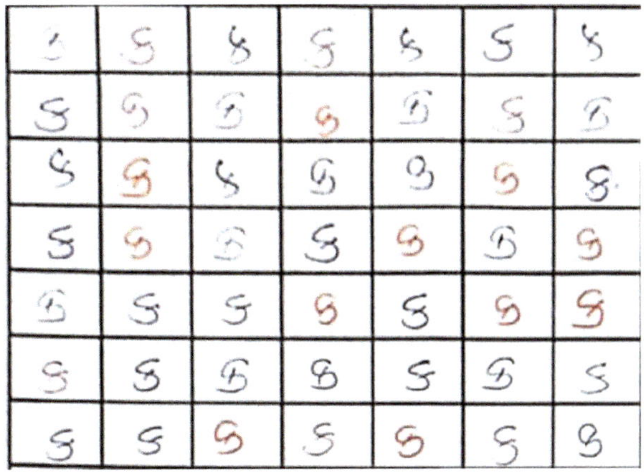

Fig. 1 Offline image for the character ‘ ક’

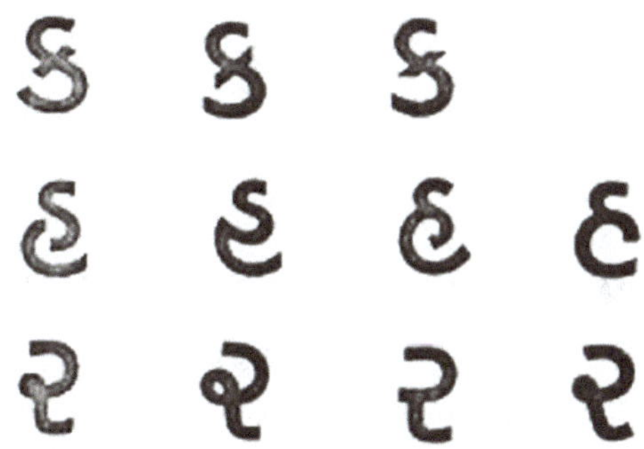

Fig. 2 Online image for the characters ‘ ક, ડ, ર’

3 Processing of Characters

The characters are processed using six step method. The first step is image acquisition followed by pre-processing if required. Once pre-processing is taken care of we do segmentation followed by feature extraction and then after classification and recognition as a final steps. All the steps are discussed below in detail.

3.1 Image Acquisition

Image acquisition is the preliminary phase for the process of character recognition. It is a process where the image of the document to be processed needs to be acquired by using a scanner or camera. The image can be stored as jpeg or png format and is considered to be final input image. This input image can be a colored image or a grayscale image. After the image acquisition phase, the image is given for the further processing.

3.2 Pre-processing

In Pre-Processing phase multiple operations are performed on an image. The Pre-Processing phase includes noise removal, binarization, edge detection and thinning of a scanned document.

- **Noise Removal**
  There are many possibilities by which the noise is introduced in an image while scanning. Noise can also come as a form of damaged image. Even scanner can at times introduce noise due to wrong placement of paper. There are variety of noises which are basically termed as errors in the image acquisition process. Removing such signals is a part of noise removal mechanism which involves several filtering techniques [4]. Use of proper filters can minimize the noise occurrence while scanning. The filters which can be used in this process are mean filter, median filter, and Gaussian filter. Mean filtering is one of the simplest method to implement smoothing images operation. Its main task is to reduce the intensity between two consecutive pixels [5]. Median filter is a non-linear filter and comparatively generate better results as compared to Mean filter, by replacing the pixel value to median of those values, whereas Gaussian filter comes in the category of linear filter. The main task of Gaussian filter is to blur the image and to reduce the noise.
- **Binarization**
  It is one of the most basic and important steps in document processing. It is the conversion of a grayscale image into binary form, i.e., 0, 1. A scanned image can be divided into two different sections, i.e., background and foreground. The task here is to separate both the sections from the scanned image and the process is called Binarization. Thresholding is one of the most used techniques to perform Binarization [6]. It is the process of separating the image into foreground and background. The pixel intensity allocated to foreground is 1 and for background is 0 [6]. Thresholding can be understood by a function T with f(x,y) as the gray level and l(x,y) is the local property, i.e.

  $$T = T[x, y, l(x, y), f(x, y)]$$

  If $f(x, y) < T$ then $f(x, y)$ is considered to be a point in the background of the image otherwise it is considered a point on foreground. The threshold T can be classified as global, local and dynamic or adaptive based on its dependency on $f(x, y)$, $l(x, y)$ and use of spatial coordinates. T is global if T depends only on $f(x, y)$, local if T depends on $f(x, y)$ and $l(x, y)$. If T depends on spatial coordinates, it is dynamic or adaptive [7]. The three methods popularly used to achieve thresholding are adaptive thresholding, global thresholding, and local thresholding.
- **Edge Detection**
  Edges are generally considered as boundaries of the object, i.e., an image. Identifying the boundaries in an image is good choice before the segmentation and identification phases. It helps further phases to get a filtered image which is free from all anomalies. It also gives images which do not have any unnecessary parts

to be considered. It is one of the useful sub phase under the pre-processing stage. Gradient and Laplacian are the most common used method for the process of edge detection.
The gradient-based edge detection method is a search-based method which concentrates on the first-order derivative of the image for the edge detection by observing for maximum and minimum value, while the Laplacian-based edge detection method is a zero-crossing method which concentrates on the second-order derivative of the image [8].

- **Thinning**
Image thinning is the process of reducing the memory requirement for information storage. A digital image can be represented in terms of pixel matrix. The pixel is either Zero (white) or One (black). A digital image at times consists of few additional pixels. It is possible to delete these additional pixels without affecting the contents within the image. The thinning process is repetitive manner so that in each iteration the image is thinned without losses.
The final phase of pre-processing is Skeletonization. Here skeleton of an image is extracted such that no loss of topological properties happen.

3.3 Segmentation

Processing an entire image at times is not possible, so it is necessary to break the image into multiple parts. Segmentation is a process that separates an input image into multiple subsets. The input digital image can be divided in three ways, namely line-based segmentation, word-based segmentation and character-based segmentation.

In, line-based segmentation the input image is separated into the number of lines which image comprised. Here lines are treated as small block and the segmentation algorithm is applied on these blocks. In word-based segmentation, the image is first separated into lines and then into words. Here each word is treated as block and in character-based segmentation, the image is first separated into lines and words. It is then additionally separated into characters. Here each character is treated as small block.

3.4 Feature Extraction

Feature Extraction is a methodology of retrieving meaningful data from raw data. Each character has some distinct and special features that can be used to define and identify it. The characters that have similar features like numeral '૫' (five) and 'પ' (pa) at times create conflicting results. To avoid such conflicts, we need to define subset of features, by which each part of character can be uniquely identified. The process of identifying such a subset is called feature selection.

Statistical features, Syntactical structural features and Hybrid features are the three used process for the process of feature selection. [9].

Template Matching, n-tuples, Zoning, Binary Transformation and other methods can be used for feature extraction. We have worked at how template matching is used for feature extraction.

- **Template Matching**

  We need a repository of characters for working with template matching algorithm. The repository stores a template of each and every character that needs to be recognized. Each and every extracted character is compared with all the templates in the record/file till the adjoining template is found. One of the techniques to identify the adjoining template is Mean Squared Distance.

$$W_i = \sum_{i=1}^{N} \left(X(a_i, b_i) - T_j(a_i, b_i)\right)^2$$

where, X is the Input Character Image and T_j is the jth template in record/file. To Classify the character in a unique way structural feature identification approach can be used. In this approach, it is possible to differentiate the character irrespective of its size and writer's style. The feature called connected and disconnected components can be examined within the character of different geometrical structure and shapes.

To perform structural feature identification following steps are used:

Step 1: Find out the character is connected or disconnected, if character can be written as a single component, which do not have multiple division is considered as connected otherwise disconnected. Figure 3 shows binary representation of alphabet 'ણ'

Step 2: Now examine the total number of separate components, if the number of separate component is greater than 1, then the character in consideration might be disconnected.

Step 3: Now examine the number of end points.

1	ણ	ણ	ણ	ણ	ણ	ણ	1	ણ
1	ણ	ણ	ણ	1	ણ	ણ	1	ણ
1	ણ	ણ	ણ	ણ	1	ણ	1	ણ
1	ણ	ણ	ણ	ણ	1	ણ	1	ણ
1	ણ	ણ	ણ	ણ	1	ણ	1	ણ
ણ	1	ણ	ણ	1	ણ	ણ	1	ણ
ણ	ણ	1	ણ	ણ	ણ	ણ	1	ણ
ણ	ણ	ણ	1	ણ	ણ	ણ	ણ	1
ણ	ણ	ણ	ણ	1	ણ	ણ	ણ	1

Fig. 3 Character 'ણ' with three components and its binary representation

Step 4: Further find out the number of closed loops available within the character along with the number of end points.
Step 5: Draw a feature table which will examine on few specific points.
Zoning
Zoning is a process that divide a character into multiple regions that is it converts cells within specific boundary. The area covered by specified boundary consists of several overlapping or non-overlapping regions where the black densities are considered as features. Once zoning is done we may apply further processing.

3.5 Classification and Recognition

Once the features have been finalized and selected the next step performed is of classification where we try to recognize an individual character and transform into machine readable format by applying classifiers. The general classifiers which are used for the recognition of a character are Template Matching method, Fuzzy Classifiers and Neural Network. These further are categorized as syntactic method and statistical method. Syntactic method works on the concept of similarity between structural components. Here, different structural components are classified into class that may be represented by a string or pattern. The characters available as input are then matched against the class.

The statistical classifiers are further divided into parametric and non-parametric. (KNN) Nearest Neighbor is a non-parametric lazy learning method, which is widely used in character recognition. It provides more appropriate results as compared to other methods. It follows instance-based approach. To classify a characteristic the system uses the K-nearest neighbor method. If the value of k is equal to 1, then the object is simply assigned to the classification of its nearest neighbor. It is a lazy algorithm, it doesn't use the training data points to do any generalization. The equation for the KNN algorithm is as follows [10]:

$$Y(d_i) = \arg\ \mathrm{MAX}_k \sum_{x_i < \mathrm{KNN}} \mathrm{Sim}\big(d_i, X_j\big) Y\big(X_j, C_K\big)$$

where d_i is a test character, Sim(d_i, X_j) is a similarity function for d_i, $Y(X_j, C_K)$ € {0,1} indicates whether X_j belongs to class $C_{K,}$ X_j is one of the neighbors in the training set [10].

Few more methods like weighted Euclidean Distance, Hierarchical Classification, Asymmetric Mahala Nobis's Distance and Support Vector Machines are used for the non-parametric recognition.

3.6 Post-processing

After the recognition of a character, the final stage is to find and recover the errors and correct them. In handwritten characters, pattern of writing the characters are not uniform. The distance in between two characters is variable, i.e., not fixed. The post processing part has to solve such problems [11]. The major concern is to resolve the problem of skewed and slanted characters.

Post Processing part is the final phase in the process of character Identification.

4 Comparative Analysis of Gujarati Character Recognition

Table 1 shows the comparative analysis of the major work done in the area of Handwritten Gujarati Character Recognition.

5 Conclusion

In this paper, various phases involved in character recognition have been discussed. The paper also gave basic knowledge and understanding of Gujarati language and its character set. After reviewing the phases, it is observed that every phase in character recognition is having its importance in identifying a character. The study also gives the understanding that for better accuracy the data sample size should be very large. If the data sample size is not appropriate, then the accuracy level of identifying characters will be always low. Though character recognition techniques have been around from over more than five decades, the accuracy level required for handwritten character recognition does not seem to have been achieved properly. The study thus shows that there is scope of research in the area of handwritten character recognition.

Table 1 Comparative analysis of the major work done by different researchers in the area of Gujarati HCR

Title of paper	Year	Researcher	Sample data	Sample data size	Technique used	Accuracy	Avg. accuracy
Features fusion based approach for handwritten Gujarati character recognition	2016	Sharma et al. [12]	All consonants	88,000	Support vector machine, Naive Bayes classifier, chain code, zone based, projection profile	99.47% using chain code with SVM, 99.80% using chain code with zone based feature	99.47 and 99.80%
Structural feature extraction to recognize some of the offline isolated handwritten Gujarati characters using decision tree classifier	2014	Thaker and Kumbharana [3]	Only 5 specific characters namely ણ, શ, ગ, લ, ષ	750	Structural Feature extraction, classification based on decision tree	98% for ણ, 80.6% for શ, 94.66% for ગ, 87.33% for લ, 83.33% for ષ	88.78%
Extraction of characters and modifiers from handwritten Gujarati words	2013	Patel and Desai [13]	Four Gujarati characters, ગ, ણ, શ, લ	500 words	Connected components labeling, zone identification	–	Average 60%

(continued)

Table 1 (continued)

Title of paper	Year	Researcher	Sample data	Sample data size	Technique used	Accuracy	Avg. accuracy
Recognition of Gujarati numerals using hybrid approach and neural networks	2013	Baheti and Kale [1]	Gujarati numeral 0–9	1800 Training set is 600 Sample data is 1200	Feature extraction by hybrid approach, neural network, mean squared error, gradient descent back propagation	96% for 0 91% for 1 88% for 2 84% for 3 83% for 4 82% for 5 79% for 6 78% for 7 81% for 8 94% for 9	Average 86%
Gujarati numeral recognition: affine invariant moments approach, soft computing	201	Baheti and Kale [14]	Gujarati numeral 0–9	1600 Training set is 800 Sample data is 800	Affine invariant moment, principal component analysis, support vector machine, K-NN, Gaussian distribution	Support vector machine: 92.28% Gaussian distribution: 87.2% K-NN classifier: 90.04% Principal component analysis: 84.1%	Average 84.1%
Gujarati handwritten numeral optical character recognition through neural network	2010	Desai [2]	Gujarati numeral 0–9	Training set is 610 Sample data is 2650 digits	Feed forward, back propagation neural network, CLAHE, median filter	–	81.6%

References

1. Baheti MJ, Kale KV (2013) Recognition of Gujarati numerals using hybrid approach and neural networks. In: International conference on recent trends in engineering and technology (ICRTET 2013), pp 12–17
2. Desai AA (2010) Gujarati handwritten numeral optical character reorganization through neural network. Pattern Recogn 43(7):2582–2589
3. Thaker H, Kumbharana CK (2014) Structural feature extraction to recognize some of the offline isolated handwritten Gujarati characters using decision tree classifier. Int J Comput Appl 99(15):46–50 ISSN 0975–8887
4. Kasturi R, O'gorman L, Govindaraju V (2002) Document image analysis: a primer. Sadhana 27:3–22 Part 1
5. Rajesh Kumar N, Uday Kumar J (2015) A spatial mean and median filter for noise removal in digital images. Int J Adv Res Electr Electron Instrum Eng. ISSN (Online) 2278–8875, (Print) 2320–3765. http://www.rroij.com/open-access/a-spatial-mean-and-median-filter-for-noiseremoval-in-digital-images.php?aid=43137
6. Modi M, Macwan F, Prajapati R (2014) Gujarati character identification: a survey. Int J Innov Res Electr Electron Instrum Control Eng 2(2):939–943 ISSN(online) 2321–2004, ISSN (print) 2321–5526
7. https://www.coursehero.com/file/p7h9hhn/11-What-is-global-Local-and-dynamic-or-adaptive-threshold-When-Threshold-T/
8. Kaur S, Singh I (2016) Comparison between edge detection techniques. Int J Comput Appl 145(15):15–18 0975–8887
9. Maloo M, Kale KV (2011) Gujarati script recognition: a review. Int J Comput Sci Issues (IJCSI) 8(4):480–489 ISSN (Online) 1694–0814
10. Patel C, Desai A (2013) Gujarati handwritten character recognition using hybrid method based on binary tree-classifier and k-nearest neighbour. Int J Eng Res Technol (IJERT) 2(6):2337–2345 ISSN 2278–0181
11. Kumar A et al (2019) An improved quantum key distribution protocol for verification. J Discrete Math Sci Cryptogr 22(4):491–498
12. Sharma A et al (2017) Features fusion based approach for handwritten Gujarati character recognition. Nirma Univ J Eng Technol (NUJET) 5(2):13–19
13. Patel C, Desai A (2013) Extraction of characters and modifiers from handwritten Gujarati words. Int J Comput Appl 73(3):7–12 0975–8887
14. Baheti MJ, Kale KV (2012) Gujarati numeral recognition: affine invariant moments approach. In: 1st international conference on recent trends in engineering and technology, special issue of international journal of electronics, communication and soft computing science and engineering, pp 140–146 (ISSN 2277–9477)

Chapter 22
Security-Aware Workflow Allocation Strategy for IaaS Cloud Environment

Mohammad Shahid, Mahfooz Alam, Faraz Hasan, and Mohammad Imran

1 Introduction

Cloud platform has got more attention and established itself as a new form of computing model. Cloud computing integrates various resources, viz. storage, software, executing platforms, machines for processing to provide resources as a reliable services effectively by implementing multi-level virtualization and abstraction [1]. There are some typical cloud system used for storage applications, Web search and social networking, namely Google App Engine [2], Amazon EC2, Microsoft Azure and IBM Cloud. Cloud system (CS) is mainly used to satisfy on demand requirements of high computation power for solving complex problem by resource sharing with increased efficiency and rapid growing network infrastructure. Virtualization technique is used in cloud system which allows sharing a single physical host among various users by creating more virtual machines (VMs) as VMs are easier to manage than physical host [3, 4].

M. Shahid
Department of Commerce, Aligarh Muslim University, Aligarh, India
e-mail: mdshahid.cs@gmail.com

M. Alam (✉)
Department of Computer Science, Al-Barkaat College of Graduate Studies, Aligarh, India
e-mail: mahfoozalam.amu@gmail.com

F. Hasan
Department of Computer Applications, BSA Crescent Institute of Science and Technology, Chennai, India
e-mail: faraz.hasan@live.in

M. Imran
Department of Computer Science Engineering, Polytechnic MANUU, Hyderabad, India
e-mail: khanimran.cs@gmail.com

© Springer Nature Singapore Pte Ltd. 2021

S. D. Purohit et al. (eds.), *Proceedings of International Conference on Communication and Computational Technologies*, Algorithms for Intelligent Systems,
https://doi.org/10.1007/978-981-15-5077-5_22

In heterogeneous cloud environment, scientific workflow applications are presented by directed acyclic graph (DAG) to describe interrelated tasks relationships. Allocation problem in CS is verified to be NP-complete. The emerging use of IaaS service model of CS is to process large numbers of compute intensive workflow applications with optimized QoS parameters. This problem of allocation of workflows focused on optimum mapping of tasks/applications onto by preserving the execution order so that task independence/precedence conditions are fulfilled with optimized QoS parameters [5–7]. Yet, to fully exploit and efficiently use of resources in cloud system, resource management layer is needed to cover the complication of different physical architectures, communication and allocation a batch of tasks in workflow, reducing complexity arises in managing synchronization, dealing with exploitation of parallelism and preserving precedence constraints in workflow applications. Moreover, the allocation of workflow applications is more difficult to achieve needed performance in cloud systems.

Various security applications such as E-commerce services, transaction processing system, digital computer for government and medical electronics are using cloud system for their purpose [8, 9]. Therefore, security-aware allocation models come into picture to assure the applications to fulfill their security demands on machines. Yet, fulfilling the security demand, some extra security overhead is tolerated. Moreover, security heterogeneity existed in CS makes solving more complex as security overhead is machine dependent, which indicates for the identical security level of service different computing machines acquire separate security overhead. Due to this reason, it is needed to develop a set of security-aware allocation strategy to prevent security appliances in CS. Security-aware workflow allocation plays a vital role on cloud system. When the scheduler considers the security requirements are exerted by the task in heterogeneous system, then such scheduler is known as security-aware (SA) scheduler [10]. SA scheduler assigns the task on the machines that match the task security constraint in a security-aware (SA) environment, and the similar time achieves few performance scheduling parameters such as resource utilization, response time and turnaround time. The very important parameter for security-aware scheduling algorithm optimizes the risk probability or failure rate/probability.

In this paper, a security-aware workflow allocation (SAWA) strategy has been proposed for workflow application by integrating security overheads of tasks to virtual machines (VMs) mapping on heterogeneous cloud system. Model considers the security demand of tasks in the workflows and negotiating cipher suite between the tasks requirements and trust level offered by VM. The proposed strategy has been evaluated with comparative performance analysis to find the suitable place for it in the literature. Performance simulation results clearly show that our proposed security-aware workflow allocation model betters in terms of number of failure tasks (NFT) and average failure probability (AFT). Security-aware concerns are incorporated for single DAG in LBSIR [23] is base model for this work.

The remaining paper is categorized as following: Sect. 2 describes the related work of workflow allocation and security-aware task scheduling for this domain. In Sect. 3, the problem formulation has been presented with parameter estimation for this work. In Sect. 4, proposed security-aware workflow allocation model (SAWA) with illustrative example is explained. Section 5 discusses the simulation results for the work done. Section 6 concludes the paper and suggests possible avenues for future scope.

2 Related Work

Over past years, the workflow allocation schemes have been designed extensively focusing on parallel and distributed systems (clusters, grids and clouds). The workflow allocation is a kind of task scheduling problem. The aim of workflow allocation is to allocate tasks on machines and sorts their assignment order so that precedence demands are fulfilled with a minimum scheduling length/turnaround time. Workflow allocation algorithm is divided into static and dynamic. Static allocation algorithm assumes that the overall information like structure, the priority constraints and estimated running time of the workflow is known prior, while dynamic allocation algorithms dynamically allocate resources based on the changes in the workflow process [11]. The key advantage of the static workflow allocation is that it is simpler to develop code for scheduler's site. In dynamic scheduling, it is too difficult to approximate the costs, jobs and which are in coming. This model regarding task execution times and processing resources cannot be assumed at run time. Dynamic workflow allocation is depended on the relocation of processes among the nodes during run time. So, it is not simpler to develop code to scheduler point.

Workflow allocation (e.g., DAG) has focused on determining the suboptimal results in an acceptably degree. The problem of DAG scheduling is very important for efficient resource utilization of heterogeneous CSs, and a polynomial time solution was found only for simple cases [12–14]. There are many DAG scheduling list algorithms were proposed. This scheme consists of two phases: the priority phase for giving rank of each tasks and selection phase for choosing an appropriate node that reduces the completion time and cost function. Various list-based DAG scheduling algorithms have been reported to deal with heterogeneous distributed systems (HDS) such as ETF [15], LMT [16], DLS [18], CPOP [17], HEFT [17], SDS [19] and LBSIR [23].

LMT algorithm is very simple and a set of tasks based their precedence constraints on the processor having minimum finish time for HDS. CPOP algorithm attains better schedules than LMT as well as DLS and other similar schedules with lower time complexity. CPOP assigns all the tasks belonging to the critical path in DAG are assigned on single best node. HEFT algorithm executes in two phases as follows: In priority phase, it determines the upward and downward rank for maintaining the dependency of the tasks, and in node selection phase, it selects the nodes which have EFT of the selected tasks. The aim of HEFT is to reduce the schedule length and time complexity.

SDS algorithm is incorporated to security into inter task dependency. SDS algorithm is to improve security of HDS with minimum computational overheads, and it can achieve high-class of security applications. SDS is used to high security computed as trust level reputation of the tasks. Furthermore, some of the other security-aware scheduling algorithms are as follows, a secured scheduling model with risk involvement [20]. In [8], three scheduling methods were proposed based on dissimilar risk level and also modified the sufferage and MinMin algorithm. SP-MinMin is an enhancement of the secure mode. SP-MinMin algorithm is an improvement over MinMin algorithm for certain task or grid environment and proposed for SA scheduling model then compared with MinMin-SA and MaxMin-SA scheduling model [21]. Moreover, security-aware scheduling models make HDSs more challenging due to the security overheads is processor-dependent scheduling algorithm. Zhu and Wang [22] were proposed a security-driven model for real-time applications using genetic algorithm for maximizing winner rate with superior security efficiency. This paper proposes security-aware workflow allocation model for heterogeneous CS of security as well as computations and allocates tasks to only guarantee the security overhead or demand of the tasks.

3 Problem Formulation

In this section, security-aware workflow allocation model has been presented to put the effort for generating schedules of a batch of tasks in a workflow to minimize failure probability in the heterogeneous cloud computing environment. Now, workflow and virtual machine models, problem statement with parameter estimation are given as follows:

At first, we consider a single workflow with a set of tasks represented by DAG which is denoted by set $w = \{T_i: 1 \leq i \leq m\}$. Workflow is further divided into sub-workflows/tasks with their depth/precedence level. The tasks of the workflow having equal depth level are allocated at same time without any communication requirement. Execution of each task in workflow with specific depth level depends on precedence order. Communication cost is required for dependent tasks in the workflow allocated on different machines. Although, dependent tasks allocated on same machine have negligible communication cost share.

The set of virtual machines $V = \{V_j: 1 \leq j \leq k\}$ and each virtual machine V_j have trust level offered for computing (TL_j), beginning time (BT_j^1) and computing capacity (C_k) of virtual machines. Each task T_i has security demand (SD_i) in the range [1, 2, 3, …, R] with 1 is the lowest and R is the highest security demand on virtual machines V_j. The measure of communication cost depends on two factor inter task communication (ITC_{xy}) between the tasks T_x and T_y assigned on virtual machines V_k and V_r with has been considered with the expected time to compute matrix ETC[T_i, V_j] means estimated completing time of task T_i on virtual machines V_j. Some assumptions are considered in the proposed strategy to develop the mathematical models, and estimation of performance parameters is as follows:

- Pre-processing has been done prior, and workflows application is in the proper representation DAG.

- All VMs are considered with single system image.
- The model is assumed for only one DAG.

Now, the problem is determining the mapping (p) for single workflow (w) allocated onto the set of machines (V) in heterogeneous cloud computing

$$p : w \rightarrow V \tag{1}$$

Such that the average failure probability has been minimized with the constraints as follows:

- Each task T_i is assigned only on a unique virtual machine.
- Each virtual machine can be allocated for more tasks.
- Each task follows the precedence constraints as in the workflow.

ETC[i][j] can be computed as the ratio of the coordinates of workload (WL$_i$) in millions of instructions and C_j vectors. The communication cost $\left(\mathrm{CC}_{xykr}^{l-s,l}\right)$ of $T_y \in p^{l}$ and $T_x \in p^{l-s}$ allocated on V_k and V_r, respectively, can be written as [23]:

$$\mathrm{CC}_{xykr}^{l-s,l} = z\left(\mathrm{ITC}_{xy} \times \mathrm{D}_{kr}\right) \tag{2}$$

where z is constant and CC_j^{l} is communication cost on V_j at depth level l is estimated as max ($\mathrm{CC}_{xykr}^{l-s,l}$). Failure probability (FP) is an important parameter for performance of security-aware system. The failure probability may be used as a matching means of quantifying the excellence of security. If TL$_j$ of V_j is greater than or equal from T_i of SD$_i$ referred to qualified node and failure probability will be zero. Moreover, the failure probability (FP$_{ij}$) of T_i satisfying security demand (SD$_i$) on V_j, with trust level (TL$_j$) can be estimated as follows:

$$\mathrm{FP}_{ij} = \begin{cases} 0; & \text{if } \mathrm{SD_i} \le \mathrm{TL_j} \\ 1 - \mathrm{e}^{-\lambda_j\left(\mathrm{SD}_i - \mathrm{TL}_j\right)}; & \text{if } \mathrm{SD_i} > \mathrm{TL_j} \end{cases} \tag{3}$$

where λ is coefficient of failure rate and is a global parameter. $\mathrm{ET}_j^{l,S}$ is the execution time with security demand of tasks at depth level l allocated on the virtual machine V_j. The computation of $\mathrm{ET}_j^{l,S}$ is sum of expected time value (E_{ij}) and the security overhead due to failure probability of the tasks. $\mathrm{ET}_j^{l,S}$ is estimated as follows [8]:

$$\mathrm{ET}_j^{l,S} = \sum_{\forall T_{i \rightarrow} V_j} E_{ij}\left(1 + \mathrm{FP}_{ij}\right) \tag{4}$$

$\mathrm{TET}_j^{l,S}$ is the total execution time at depth level (l), and it is sum of the communication cost $\left(\mathrm{CC}_j^{l}\right)$, beginning time $\left(\mathrm{BT}_j^{1}\right)$ and $\mathrm{ET}_j^{l,S}$ as follows:

$$\mathrm{TET}_j^{l,S} = \mathrm{BT}_j^1 + \mathrm{CC}_j^l + \mathrm{ET}_j^{l,S} \tag{5}$$

Average failure probability (AFP) of complete work flow can be calculated as follows:

$$\mathrm{AFP} = \frac{\sum_{\forall T_i} \mathrm{FP}_{ij}}{m} \tag{6}$$

4 The Algorithm

In this section, the algorithm of proposed security-aware workflow allocation (SAWA) model for a workflow represented using DAG has been presented. The allocation and idle slot reduction are same as in [23], and assignments of workflow tasks are carried out by considering security demand (SD_i) of tasks and trust level (TL_j) of VM. The objective of such allocation scheme is to allocate maximum tasks with satisfying security demand of tasks. The algorithmic steps for proposed model have been shown as template given in Fig. 1.

We are describing the working of SAWA by using single workflows having ten tasks have been considered for illustration as shown in Fig. 2. The depth level of this workflow is 5 [24]. The task information of the batch and virtual machines considered are presented in Tables 1 and 2. We assume that in Table 1, the beginning time is 50 s, 30 s and 70 s of the VMs V_1, V_2 and V_3, respectively. Each task has security demand such as 8, 12, 10, 6, 7, 11, 5, 12, 14 and 4 are the tasks T_1, T_2, T_3, T_4, T_5, T_6, T_7, T_8, T_9 and T_{10}, respectively. Inter task communications associated with tasks are given with edge labels in the DAG.

At first, batch is divided into five depth levels and sorted in larger to smaller tasks order as $p^1 = \{T_1\}$, $p^2 = \{T_2, T_3\}$, $p^3 = \{T_6, T_5, T_4\}$, $p^4 = \{T_8, T_9, T_7\}$ and $p^5 = \{T_{10}\}$. The workflow allocation schedule by SAWA is represented by 5 tupple as $<T_i, V_j, \mathrm{BT}_j^1, \mathrm{FT}_{ij}, \mathrm{FP}_{ij}>$ for all the tasks assigned where FT_{ij} is the finish time of T_i on V_j and rest of the tupples already mentioned earlier. The allocation of tasks of ρ^1 is accomplished as T_1 could be assigned on V_1, V_2 and V_3, respectively, but T_1 is assigned on V_3 virtual machine due to SD requirement of T_1 is met on this machine. After execution of T_1, level l_1 is completed with finish time 95. The idle slots (IS) list is $\mathrm{IS}^1 = \{S_1{:}\ 45, S_2{:}\ 65\}$ where S_1 and S_2 are the idle slots on V_1 and V_2, respectively. The next level tasks are T_2 and T_3, and both have security demand 12 and 10, respectively, which is not fulfilled on V_2 and V_3. T_2 has execution time $(\mathrm{CC}_1^1 + E_{21} = 8 + 66.83 = 74.83)$ and $(\mathrm{CC}_2^1 + E_{22} = 10 + 51.927 = 61.927)$ on V_1 and V_2, respectively. The succeeding T_2 of T_1 is fitted into idle slots S_2 on V_2. So, T_2 is executed on V_2 with FP 0.90. Similarly, T_3 with execution time $(\mathrm{CC}_1^1 + E_{21} = 12 + 50.75 = 62.75)$ cannot be fitted into idle slots S_1. Now, at depth level l_2, T_3 has security demand 10 which is not fulfilling the any VM. So, T_3 will assign on the VM having minimum execution time, and execution time with communication cost

SAWA ( )
Input – *w*, WL_i, SD_i, BT_j^1, MIPS, TL_j, V, λ
Output – NFT, AFP

Begin

1. Partitioning batch as per depth levels
2. Generate ETC matrices
3. NFT=0
4. For each partition
5. Do
6. Each time get largest task from partition
7. If (SDi ≤ TL_k) //the security demand T_i is fulfill then two condition may occur
   i. Qualified VM is only one then execute Ti on it
   ii. Qualified VM more than one, then the VM fulfilling the SDi having minimum execution time is allocated to Ti
   iii. FPij is zero //as per equation (3)
8. Else
   i. Task will assigned on the VM that execute T_i with minimum execution time
   ii. FP_{ij} is computed //as per equation (4)
   iii. NFT=NFT+1

   End do
9. Task from next level are inserted into idle slots
   i. Largest selected task assigned in the slot at which it best fitted with satisfying SD_i //(NFT=NFT+1)
   ii. Otherwise allocated it on only best fitted VM slot.
   iii. FP_{ij} is computed // as per equation (4)
10. Compute **NFT** and **AFP** // as per equation (6)

End

Fig. 1 SAWA strategy

of T_3 is 12 + 50.75, 15 + 43.6271 and 0 + 31.80 on V_1, V_2 and V_3, respectively. So, T_3 will be assigned on V_3 with FP 0.09. The depth level l_2 is competed with finish time 126.8, and the idle slots list is created $\text{IS}^2 = \{S_1\text{: } 31.8, S_2\text{: } 31.8\}$. The succeeding tasks T_6, T_5 and T_4 having security demand 11, 7 and 6, respectively, cannot be fitted into idle slots S_1 and S_2 on V_1 and V_2, respectively, so we leave them and switch to next level assignment. Now, the security demand of T_6 is not fulfilling on any VMs, so T_6 will be assigned on V_3 with FP 0.18 having minimum execution time. In the same manner mentioned earlier, T_5 and T_4 will be assigned on V_1. The depth level l_3 is completed after adding communication cost ($l_3 = 232.8$) with idle slots list $\text{IS}^3 = \{S_2\text{: } 106, S_3\text{: } 40.1206\}$. T_7 with security demand 5 is fitted on S_3 without any FP. T_8 and T_9 are not fitted on ant slot at l_3. Again, at level l_4, T_8 and T_9 are assigned on V_3 and V_2 with FP 0.25 and 0.95, respectively. The depth level

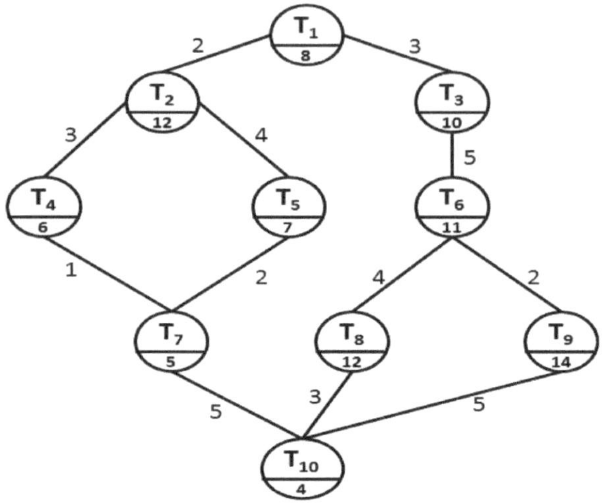

Fig. 2 Sample of single workflow for illustration

Table 1 Virtual machine information

VM	V_1	V_2	V_3
MIPS	10	15	12
BT_j^1 (μS)	50	30	70
TL_j	7	4	9
λ	0.2	0.3	0.1
VM distance (in hop counts)			
V_1	0	2	4
V_2	2	0	5
V_3	4	5	0

Table 2 Task information

Task	WL_i	SD_i	ETC			FP_{ij}		
			V_1	V_2	V_3	V_1	V_2	V_3
T_1	300	8	30	20	25	0.18	0.69	0
T_2	410	12	41	27.33	34.17	0.63	0.90	0.25
T_3	350	10	35	23.33	29.17	0.45	0.87	0.09
T_4	470	6	47	31.33	39.17	0	0.45	0
T_5	510	7	51	34	42.5	0	0.59	0
T_6	670	11	67	44.67	55.83	0.55	0.87	0.18
T_7	380	5	38	25.33	31.67	0	0.25	0
T_8	910	12	91	60.67	75.83	0.63	0.90	0.25
T_9	885	14	88.5	59	73.75	0.75	0.95	0.39
T_{10}	730	4	73	48.67	60.83	0	0	0

Table 3 Schedule by SAWA strategy

Tasks	Schedule	Status
T_1	$<T_1, V_3, 70, 95, 0>$	NF
T_2	$<T_2, V_2, 30, 91.927, 0.90>$	F
T_3	$<T_3, V_3, 95, 126.80, 0.09>$	F
T_6	$<T_6, V_3, 126.80, 192.6794, 0.18>$	F
T_5	$<T_5, V_1, 126.80, 177.8, 0>$	NF
T_4	$<T_4, V_1, 177.8, 224.8, 0>$	NF
T_7	$<T_7, V_3, 192.6794, 232.3494, 0>$	NF
T_8	$<T_8, V_3, 232.8, 327.5875, 0.25>$	F
T_9	$<T_9, V_2, 232.80, 357.85, 0.95>$	F
T_{10}	$<T_{10}, V_1, 232.8, 325.8, 0>$	NF

l_4 is completed with adding communication cost ($l_4 = 357.85$), and idle slots list is $\text{IS}^4 = \{S_1\text{: } 125.05, S_3\text{: } 30.2922\}$. Finally, T_{10} having security demand 4 is fitted on S_1 into level 4.

As per SAWA, the task-wise allocation schedule is presented in Table 3.

Total number of failure tasks (NFT) = 5.

AFP = ((0 + 0.90 + 0.09 + 0.18 + 0+0 + 0+0.25 + 0.95 + 0)/10) * 100 = 23.7%.

On same data, allocation schedule produced by LBSIR as presented in Table 4.

Total number of failure tasks (NFT) = 6.

AFP = ((0.69 + 0.90 + 0.09 + 0.87 + 0+0 + 0+0.90 + 0.39 + 0)/10) * 100 = 38.4%.

In Tables 3 and 4, **F** and **NF** stand for failure and non-failure task, respectively.

Table 4 Schedule by LBSIR

Tasks	Schedule	Status
T_1	$<T_1, V_2, 30, 63.8, 0.69>$	F
T_2	$<T_2, V_2, 70, 121.927, 0.90>$	F
T_3	$<T_3, V_3, 70, 116.7953, 0.09>$	F
T_6	$<T_6, V_2, 121.927, 230.45, 0.87>$	F
T_5	$<T_5, V_3, 121.927, 184.427, 0>$	NF
T_4	$<T_4, V_1, 121.927, 174.927, 0>$	NF
T_7	$<T_7, V_3, 184.427, 220.097, 0>$	NF
T_8	$<T_8, V_2, 230.4599, 345.7329, 0.90>$	F
T_9	$<T_9, V_3, 230.4599, 343.1114, 0.39>$	F
T_{10}	$<T_{10}, V_1, 230.4599, 323.4599, 0>$	NF

5 Experimental Results

Experimental result has been conducted to evaluate SAWA model by observing the performance parameters on allocation of the workflow on the heterogeneous virtual machines for different system configurations comparing with its base model LBSIR [23]. The cloud service providers provide the end user with different types of VMs. The experiments have been done by using ETC simulation approach. The results are taken for single workflow by varying number of tasks. For the case, experiments are reported to performance comparison on the QoS parameters such as number of failure tasks (NFT) and average failure probability (AFT). Here, number of tasks is varied by keeping remaining input parameters in the system fixed. The other input parameters are as follows: BT_k^1, V_j, MIPS, TL_j same as Table 1, and $SD_i = 1$–15, workload of tasks (WL_i) = 100–1000 (MIs).

The number of failure tasks (NFT) is increasing when the number is increased as shown in Fig. 3 for both SAWA and LBSIR. Yet SAWA performs better than LBSIR due to having approximately half of the failure tasks for fixed number of VMs and other input parameters.

The average failure probability (AFP) is directly proportional to NFT. As shown in Fig. 4, the average failure probability is almost increasing from 31.8 to 40.61% when the number of tasks is increased from 5 to 160 tasks for fixed number of VMs for LBSIR. LBSIR has not been developed to satisfy the security demand. In contrast to LBSIR, SAWA performs better due to having security-aware allocation as expected. When the batch size increases from 5 to 160, then average failure probability is decreased from 23.7 to 15.34%.

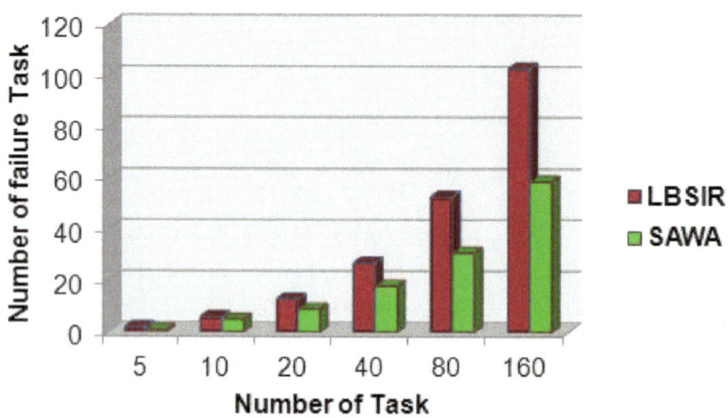

Fig. 3 Number of tasks versus NFT

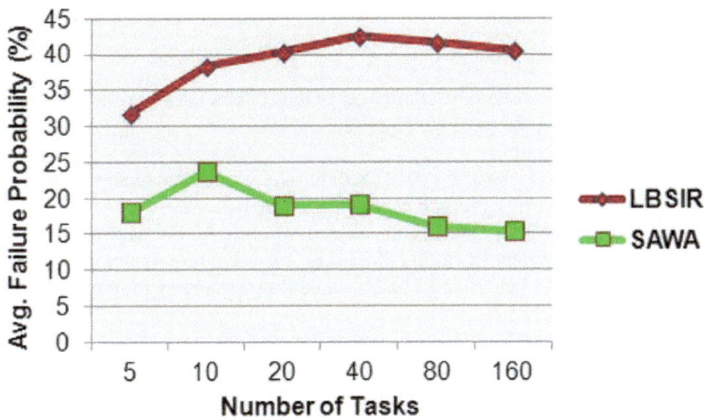

Fig. 4 Number of tasks versus AFP

6 Conclusions and Future Scope

In IaaS cloud, lack of trust is experienced in providing services to the users in the organization. Therefore, workflow allocation in cloud resources needs to satisfy security demand to fulfill the user request in real-world cloud. In this work, we present a security-aware workflow allocation (SAWA) strategy for cloud environments, which maps a workflow application to cloud resources, constrained to security demand and trust level values with the objective of minimizing number of failure tasks and average failure probability. The algorithm was compared with other/its base model in which security-aware allocation scheme is incorporated. The experimental study has been evaluated by varying the number of task in workflow. The study reveals that SAWA is performing better than LBSIR in terms of parameters considered for evaluation as suggested in the study.

In our expected future directions, the work can be extended for multiple workflow applications to provide the better system performance with satisfied user constraint. The proposed strategy further can be improved and compared to other state of art model (namely HEFT, DLS, etc.) for better suitability in the IaaS cloud. Also, we can further extend the work to estimate and optimize the other QoS parameters.

Acknowledgements This work is supported by the grant from UGC-BSR (Research Startup Grant with sanction No. F.30-377/2017 (BSR), Code-10057/B30528) to Dr. M. Shahid. Authors are also thankful to Dr. Zahid Raza, SC&SS, JNU, New Delhi, for his valuable suggestions and support needed to accomplish this work.

References

1. Alam M, Khan ZA (2017) Issues and challenges of load balancing algorithm in cloud computing environment. Indian J Sci Technol 10(25):1–2
2. http://www.googlecloud.com
3. Camati RS, Calsavara A, Lima Jr L (2014) Solving the virtual machine placement problem as a multiple multidimensional knapsack problem. ICN 2014 23:264
4. Suseela BB, Jeyakrishnan V (2014) A multi-objective hybrid ACO-PSO optimization algorithm for virtual machine placement in cloud computing. Int J Res Eng Technol 3(4):474–476
5. Shahid M, Raza Z (2014) Level-based batch scheduling strategies for computational grid. Int J Grid Util Comput 5(2):135–148
6. Forti A (2006) DAG scheduling for grid computing systems
7. Donoho G (2004) Building a web service to provide real-time stock quotes. MCAD Net 38–42
8. Song S, Kwok YK, Hwang K (2005) Trusted job scheduling in open computational grids: security-driven heuristics and a fast genetic algorithms. In: Proceedings of international symposium on parallel and distributed processing
9. Connelly K, Chien AA (2002) Breaking the barriers: high performance security for high performance computing. In: Proceedings of the 2002 workshop on new security paradigms. ACM, pp 36–42
10. Ahmad I, Kwok YK (1998) On exploiting task duplication in parallel program scheduling. IEEE Trans Parallel Distrib Syst 9(9):872–892
11. Barbosa JG, Moreira B (2011) Dynamic scheduling of a batch of parallel task jobs on heterogeneous clusters. Parallel Comput 37(8):428–438
12. Braun TD, Siegel HJ, Beck N, Bölöni LL, Maheswaran M, Reuther AI, Robertson JP, Theys MD, Yao B, Hensgen D, Freund RF (2001) A comparison of eleven static heuristics for mapping a class of independent tasks onto heterogeneous distributed computing systems. J Parallel Distrib Comput 61(6):810–837
13. Darbha S, Agrawal DP (1998) Optimal scheduling algorithm for distributed-memory machines. IEEE Trans Parallel Distrib Syst 9(1):87–95
14. Papadimitriou CH, Yannakakis M (1990) Towards an architecture-independent analysis of parallel algorithms. SIAM J Comput 19(2):322–328
15. Hwang JJ, Chow YC, Anger FD, Lee CY (1989) Scheduling precedence graphs in systems with interprocessor communication times. SIAM J Comput 18(2):244–257
16. Iverson MA, Özgüner F, Follen GJ (1995) Parallelizing existing applications in a distributed heterogeneous environment. In: 4th heterogeneous computing workshop (HCW'95)
17. Topcuoglu H, Hariri S, Wu MY (2002) Performance-effective and low-complexity task scheduling for heterogeneous computing. IEEE Trans Parallel Distrib Syst 13(3):260–274
18. Sih GC, Lee EA (1993) A compile-time scheduling heuristic for interconnection-constrained heterogeneous processor architectures. IEEE Trans Parallel Distrib Syst 4(2):175–187
19. Xiaoyong T, Li K, Zeng Z, Veeravalli B (2010) A novel security-driven scheduling algorithm for precedence-constrained tasks in heterogeneous distributed systems. IEEE Trans Comput 60(7):1017–1029
20. Song S, Hwang K, Kwok YK (2006) Risk-resilient heuristics and genetic algorithms for security-assured grid job scheduling. IEEE Trans Comput 55(6):703–719
21. Kashyap R, Vidyarthi DP (2012) Security-aware scheduling model for computational grid. Concurr Comput Pract Exp 24(12):1377–1391
22. Zhu H, Wang Y (2008) Security-driven task scheduling based on evolutionary algorithm. In: 2008 international conference on computational intelligence and security, vol 2. IEEE, pp 451–456
23. Shahid M, Raza Z, Sajid M (2015) Level based batch scheduling strategy with idle slot reduction under DAG constraints for computational grid. J Syst Softw 108:110–133
24. Gupta S et al (2019) Tier application in multi-cloud databases to improve security and service availability. In: Handbook of research on cloud computing and big data applications in IoT. IGI Global, pp 82–93

Chapter 23
Design Coverage and Connectivity Framework for Wireless Sensor Networks Using Optimum Energy

Sanchit Goyal, Shreshtha Goel, and Vinita Tiwari

1 Introduction

Wireless sensor network (WSN) offers a wide range of applications in areas such as traffic monitoring, medical care, inhospitable terrain, robotic exploration, and agriculture surveillance [1]. The advent of efficient wireless communications and advancement in electronics has enabled the development of low-power, low-cost, and multifunctional wireless sensor nodes that are characterized by miniaturization and integration.

In WSNs, thousands of physically embedded sensor nodes are deployed over a large covered area [2, 3]. The main hurdles to an efficient energy consumption and network lifetime of WSN are the network coverage and connectivity of nodes. Network coverage measures the degree of the area of interest covered by sensor nodes and connectivity measures the extent to which nodes are connected to each other in a network (possibly over multiple hops) [4].

Clustering is a useful mechanism in WSNs which helps to cope with scalability problems. So, it is combined with data aggregation to increase the energy efficiency of network. Data aggregation can be defined as the process of aggregating the data from multiple sensors to one sensor to eliminate redundant transmission and provide fused information to the base station (also called sink) [5]. One node can receive a huge amount of repeated data from different neighbors. Since the data processing consumes less power than data transmission, we can solve that by data aggregation and data fusion to remove the redundant data. Data aggregation together with the clustering algorithm can improve connectivity and efficiency. Many energy-efficient routing protocols such as the minimum transmission energy and [6, 7] low-energy adaptive clustering hierarchy (LEACH) protocol have been proposed, where the

S. Goyal (✉) · S. Goel · V. Tiwari
Electronics and Communication Department, Indian Institute of Information Technology Kota, Kota, Rajasthan, India
e-mail: Sanchitg14a3@gmail.com

© Springer Nature Singapore Pte Ltd. 2021

S. D. Purohit et al. (eds.), *Proceedings of International Conference on Communication and Computational Technologies*, Algorithms for Intelligent Systems,
https://doi.org/10.1007/978-981-15-5077-5_23

concept of clustering, cluster head, and techniques like data aggregation and data fusion are used which help in increasing the efficiency of the network.

The received signal strength at a distance d from the transmitter varies as the inverse nth power of the distance d. The received power $P_r(d)$ is related to distance d through the Friis equation.

$$P_r(d) = \frac{P_t G_t G_r \lambda^2}{4\pi (d)^n}$$

$P_r(d)$—Received Power at distance d
P_t—Transmitted Power
G_t—Gain of transmitting antenna
G_r—Gain of receiving antenna
λ—Wavelength
d—Distance between transmitter and receiver
n—Path loss exponent ($2 <= n < =6$), ($n = 2$ for free space).

To improve the efficiency of energy consumption, power is calculated using optimum internodal distance keeping other factors constant. Energy (E) dissipated by a sensor node is directly related to power transmitted by it which in turn is directly related to the nth power of internodal distance.

$$E \, \alpha \, d^n$$

So, to save energy, it becomes very important to reduce the distance between the node and the sink or the next cluster head. The different metrics to calculate distance between nodes are Euclidian, Manhattan, and Chebyshev out of which Chebyshev metric gives the least distance.

There are two parameters, namely the sensing range and transmission range, which plays an important role in determining coverage and connectivity of WSN. The sensing area of a sensor is the geographic area covered by a sensor. It is given by a circle with the radius of the circle equal to the sensing range. The value of an event taking place in this region is detected by the sensor [8]. The transmission range is the maximum distance between any two nodes such that the signal coming out from one node could directly reach the other node with strength appreciable enough to correctly extract the encoded information. It is well known that transmitting data consumes much more energy than collecting data [9]. So, to improve the energy efficiency for transmitting data, most of the existing energy-efficient routing protocols attempt to find the minimum energy path between a source and a sink to achieve optimal energy consumption [10, 11].

Since it is cumbersome to calculate the percentage of overlapped area using circular sensing regions, grid model is used which helps in determining not only the exact coverage area by considering the extent of overlapping but also the ideal number of nodes to be deployed in order to cover the entire field. Grid model is defined as the fashion in which the neighboring grids are covered by sensor nodes. It includes the number of grids and shape in which they are covered.

Multiple sink deployment [12] makes the WSN robust so that even if one of the sinks dies out, other sinks can take the charge and prevent the WSN from failing. Also, it enhances the coverage and connectivity by connecting the unique nodes. A unique node is the one which is not connected to any other cluster heads (CH). The deployment of sink can be random depending on the density of nodes or manual [13]. In the proposed method, manual deployment of sinks is considered.

The rest of this paper is organized as follows. Section 2 provides a brief overview of the different areas in WSN, coverage, and connectivity. Section 3 provides the system model to enhance the coverage and connectivity of WSN using the grid model. Section 4 provides the algorithm to calculate coverage for random deployment of nodes and improve connectivity for efficient energy consumption by using Chebyshev distance. Section 5 details the simulation and results. Finally, Sect. 6 concludes the paper and proposes the future work that can be incorporated to further enhance the network efficiency and lifetime.

2 Related Work

When we come across wireless sensor networks, the primary problems that we face are the accurate calculations of the degree of coverage of a field since the nodes are randomly distributed across the field and the extent to which the deployed nodes are connected to each other and ultimately to the sink. The degree of coverage also influences information processing algorithms. High coverage may be exploited to extend the network lifetime by switching redundant nodes to power-saving sleep modes [4].

The necessary and sufficient condition under which coverage implies connectivity states that the radio range is at least twice the sensing range [14]. This ensures that most of the nodes are connected to at least one other node and only then can they be considered as a part of coverage. Sriporamanont and Liming created a simulator to support the wireless sensor network simulation. They implemented the network simulator (NS-2) which supports both wire and wireless networks and provides extensive support for simulating TCP/IP [15]. Although NS-2 is the most widely used network simulator, it does not support directly the wireless sensor network, and to use this simulator with the sensor networks, some implemented modules are needed [15].

In this work, MATLAB has been used for network simulation due to its easy programming capability and appropriate platform for users to develop their own custom functions. The implementations done before have mostly calculated the coverage in terms of circular coverage regions [9]. This makes it difficult to calculate the coverage of the field with accuracy due to overlapping areas. In this work, 23-grid model has been used which closely resembles hexagonal coverage regions, which, in turn, is closely related to the circular region [16, 17]. Using the grid model not only

Table 1 Comparison between single sink and multiple sink

Reference number	Parameter	Single sink	Multiple sink
[12]	Energy consumption	Higher	Lower
[23]	End-to-end delay	Higher	Lower
[24]	Coverage	Lower	Higher
[24]	Connectivity	Lower	Higher
[23]	Data delivery	Lower	Higher
[12]	Scalability	Lower	Higher
[5]	Data aggregation	Required	Not required
[5]	Cost	Lower	Higher

simplifies our calculation but also helps in accurate calculations of total coverage area.

Based on topology, there are four different types of networks [12, 18]

1. Single hop single sink routing (SH-SS)
2. Single hop multiple sink routing (SH-MS)
3. Multiple hops single sink routing (MH-SS)
4. Multiple hops multiple sink routing (MH-MS).

In case of SH-SS and SH-MS, the coverage and the connectivity are too low as there is high probability of nodes not being lying in the transmission range of any sink. Also, since nodes are connected directly to the sink, optimum distance is not considered in these topologies. Hence, energy consumption is very high which makes them unfeasible to implement.

Table 1 gives the differences between using a single sink and a multi-sink. As clearly depicted, there are many advantages of using multiple sinks. Increasing the lifetime of network is a major challenge. Since in a multi-hop WSN a sensor spends most of its energy for relaying data packets, it is important to shorten the distance a packet has to travel until reaching the sink. The difference in energy consumption and end-to-end delay is also attributed to the lower overall distance of sensor nodes from the sink which can be reduced by deploying multiple sinks instead of one. So, every node communicates with the closest sink for transferring the sensed data. However, the sinks need to be deployed in a coordinated way [19]. This must be done to ensure maximum connectivity and avoid a particular node being connected to multiple sinks. Mobility of the sensor nodes also impacts the overall network performance.

3 System Model

The grid models have been used in the simulation to facilitate better calculation of areas and improvement in coverage.

3.1 *Different Grid Models*

The following grid models have been proposed to cover the field area as well as to connect nodes:

- 5-Grid Model
- 9-Grid Model
- 13-Grid Model
- 23-Grid Model
- 29-Grid Model.

Here, the number in the grid model represents the number of grids covered by each sensor node in its sensing range.

3.2 *Assumptions*

The following assumptions have been made for the proposed network.

(1) The nodes are randomly deployed. They are homogenous with similar and fixed sensing ranges and transmission ranges.
(2) The nodes transmit the data to sink or base station (BS) in multiple hop fashion.
(3) Sensor nodes and base station(s) are stationary.
(4) The sensors used have a transmitting range of 50–60 m (outdoor) and 10–20 m (indoor). If (16 m) and (2.83 m) represent transmission and sensing range, respectively, it is assumed that ≥ 2 as given in [14].
(5) If there is one more sensor within the transmission range of a sensor, then the sensed information is the same for both the sensors.

3.3 *Internodal Distance Calculation*

In order to connect nodes with each other and the sink, calculate the $D8$ or Chebyshev distance between the nodes. These distances are then represented in a distance matrix.

$D8$ distance is given by:

$$D_{i,j} = \text{maximum}\left(\left| x_i - x_j \right|, \left| y_i - y_j \right|\right) \quad i = 1,\ 2, \ldots,\ N, \quad j = 1,\ 2, \ldots,\ N; \tag{1}$$

where x_i is the x-coordinate of the ith node, y_i is the y-coordinate of the ith node, x_j is the x-coordinate of the jth node, and y_j is the y-coordinate of the jth node.

4 Proposed Algorithm

Step 1 Deploy N number of nodes randomly in the field to be monitored. Deploy BS(s) manually on the field.

$$N = \{n_1,\ n_2,\ n_3, \ldots, n_N\} \tag{2}$$

Step 2 Calculate the distance vector d_{ij} using Eq. (1). It represents the distance of ith node from all the jth nodes.

Step 3 Calculate the set of neighbor nodes Nb_i and the number of neighbor nodes count_i for the ith node n_i on the basis of the transmission range as specified.

$$\mathrm{Function}[\mathrm{count}_i,\ Nb_i] = \mathrm{neighbour\_inf\,o}(n_i) \tag{3}$$

The function neighbor_info returns all the neighboring nodes within the transmission range of a node n_i. Same happens for the BS.

Step 4 All the nodes are given ids from 1 to n as per their distances from the BS. Therefore, the node nearest to the BS is given the id 1.

$$\mathrm{Id} = f\left(d_{ij}\right) \tag{4}$$

Step 5 Connect all the nodes in Nb_i to n_i starting from id 1 to n. Maintain a P_{id} vector of zeros and give a P_{id} of 1 to all the connected nodes. Now, connect only those nodes which have a P_{id} of 0 in further iterations to avoid a node being connected to multiple nodes.

$$P_{\mathrm{id}} = \mathrm{zeros}(1,\ n) \tag{5}$$

In case of multi-sink, the above processes are repeated for all the sinks deployed on the field. A lot of nodes may get connected to multiple sinks depending on their distances from them. In such a case, maintain another vector x_{ij} which contains Chebyshev distance of ith sink from all the jth nodes where i varies from 1 to S

Table 2 Simulation table

Simulation parameters	Value
Deployment type	Random topology
Deployment area	64 m x 64 m
Grid model	23
Number of nodes	20
Number of sinks	1, 2, 4
MAC layer	IEEE 802.15.4
Transmission range	16 m
Sensing range	2.83 m
Simulation time	20 s

(number of sinks) and j varies from 1 to N. The nodes are, therefore, connected to the nearest sink using distances given by this vector.

Step 6 Transfer data in the form of packets to BS from all the nodes in multi-hop fashion. The nodes which do not lie in the transmission range of any node or those which have a P_{id} of 0 is called the unique nodes as they are isolated from the whole network and cannot transmit any data.

Step 7 Calculate the percentage area covered using stochastic geometry which involves Poisson point process. As per Poisson distribution, the probability of total area covered by random deployment of k nodes is given by:

$$P(X = k) = \frac{(\lambda A)^k e^{(-\lambda A)}}{k!} \tag{6}$$

where λ is the mean density of points to cover the entire region, A is the total area of the region, and k is the number of nodes to be deployed. Multiplying the probability obtained with 100 will give the percentage area covered by random deployment of nodes (Table 2).

5 Simulation

5.1 Simulation Setup

A random topology is built in a 64 m $\times$ 64 m area with 20 homogeneous sensor nodes, and one, two, and four sinks placed inside the field grid, where the radio transmission range is up to 16 m, which is considered an effective coverage area for sensor nodes. The IEEE 802.15.4 standard is used at the MAC and physical layers.

Table 3 Simulation results for 23-grid model

Nodes	Ideal area	Post simulation	Difference (in %)
10	230	221	3.913
50	1150	913	20.609
100	2300	1625	29.348
178	4096	2240	45.286

The sensing is calculated as 2.83 m considering 23-grid model having each grid of 1 m × 1 m (Table 3).

5.2 Field Coverage

The area covered is calculated for the different number of nodes deployed in the given area. Also, the ideal area and the simulation area covered by the deployment of nodes are calculated. Ideal area represents the area covered without overlapping of any grids, whereas simulation area represents the actual total area covered by sensor nodes. These two have been compared to find out the difference in the percentage area covered. As seen from the table, as the number of nodes increases, the difference is also increasing implying increase in overlapping of grids. Hence, the experimental area will always be less than or equal to the ideal area.

For 23-grid model, 178 is the ideal number of nodes for which entire field can be covered (considering no overlapping of grids). Here, 45.2% implies that grids are highly overlapping because of random deployment of nodes.

Figure 1 clearly depicts the area covered for random deployment of 20 nodes by 23-grid model. The area covered is coming out to be 10.693% of the total field area. The ideal area (without any overlapping of grids) covered by deploying 20 nodes will come out to be 11.23% of the total field area. This deviation (< 1%) of calculated area from the ideal area suggests overlapping of grids. 23-grid model is chosen as it resembles most closely to the hexagonal shape which is an ideal shape for calculations.

5.3 Connectivity of Nodes

The deployed nodes have to be connected to BS either directly or indirectly in order to transmit the data which can be further processed by BS and can be sent to user. As shown in Fig. 2, deployed nodes are connected to the sink (or BS) in multi-hop fashion. The *D*8 distance of all the nodes from the sink is calculated and as per the nodes which lie in the transmission range of sink is connected to sink directly. These directly connected nodes to the sink act as cluster heads. In this similar fashion, remaining nodes which lie in the transmission range of these cluster

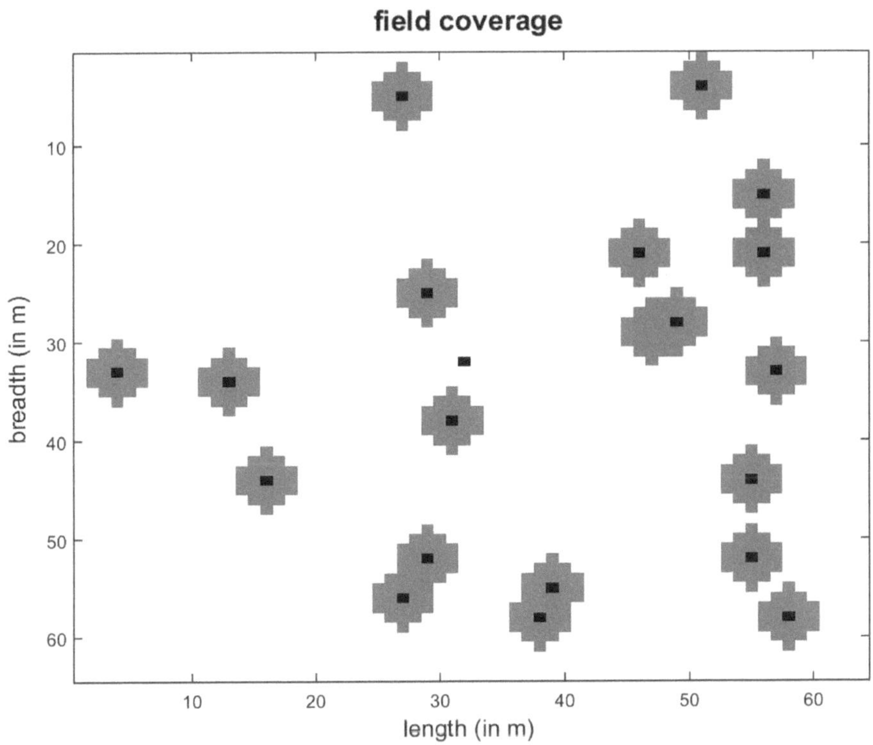

Fig. 1 Coverage by random deployment of nodes in 23-grid model

heads are connected. A node which lies in the transmission range of multiple nodes will be connected to the node from which its distance is minimum [20]. The nodes which do not lie in the transmission range of any other node are not connected so such nodes are not useful to us. Since these nodes are not connected to any cluster head, any node lying in the transmission range of these unconnected nodes will also remain unconnected, provided it does not lie in the transmission range of any other connected node. Also, the nodes are connected as per its id so a lower id node cannot be only connected to a higher id node [21]. The sink is directly connected to the user via Internet or fiber cable.

5.4 *Field Coverage Post-Connectivity*

As shown in Fig. 3, the nodes which are not connected to sink in multi-hop fashion will not be useful to the wireless sensor network as they will not be able to communicate the data sensed to the sink. So, the area covered by such nodes will not be included in our calculations of total area covered by the source nodes. Hence, as depicted in the

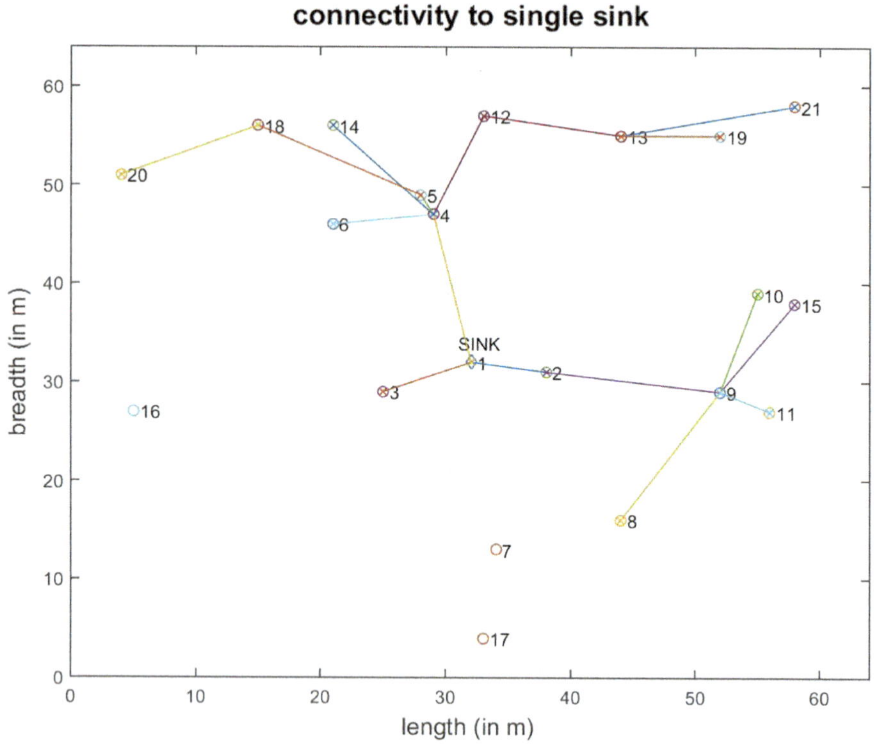

Fig. 2 Connectivity of nodes to single sink in 23-grid model

figure, the single points are the nodes which are not connected to any cluster head (except the one in the center as it the BS).

5.5 *Connectivity of Nodes to Multi-sink*

As shown in Fig. 4a, b, multiple sinks are deployed at the fixed positions. Figure 4a shows the connectivity of nodes to two sinks while Fig. 4b shows the connectivity to four sinks. All the BS(s) will be directly connected to the user via Internet or fiber cable. The advantage of using multiple sink is that the nodes which were not connected in the case of single sink are now connected to the sink. This will increase the network lifetime and the field area covered. Also, the distance between the sink and the nodes decreases so this will lead to efficient energy consumption.

Now, in this case, it is possible that some nodes might lie in the transmission range of multiple sinks so these nodes will be connected to only that sink for which the distance between the two is minimum. In case of equal distance of nodes from

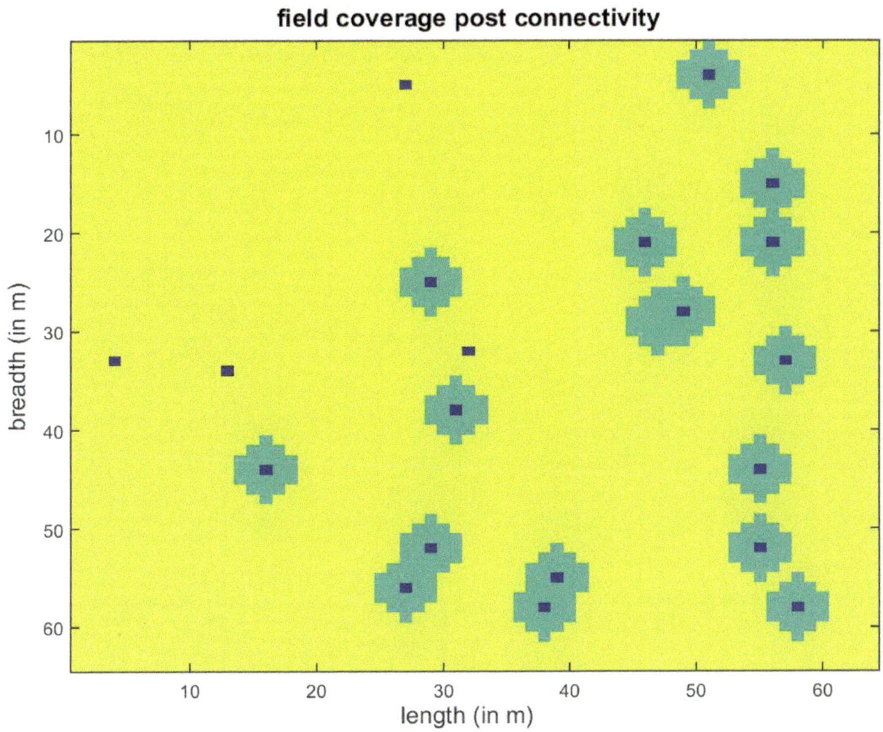

Fig. 3 Field coverage post-connectivity of nodes to single sink in 23-grid model

multi- sink, the node will be connected randomly to a sink in a stochastic scheme, similar to tossing a die [22].

5.6 Comparison of Different Grid Models

As shown in Fig. 5 and Table 4, as the number of nodes is increasing, the area covered is increasing. For 29-grid model, we get area coverage as high as 90% for 500 nodes. However, we prefer 23-grid model over 29-grid model since it provides efficient coverage even with lesser sensing ranges of sensors, thereby reducing the cost of the sensors.

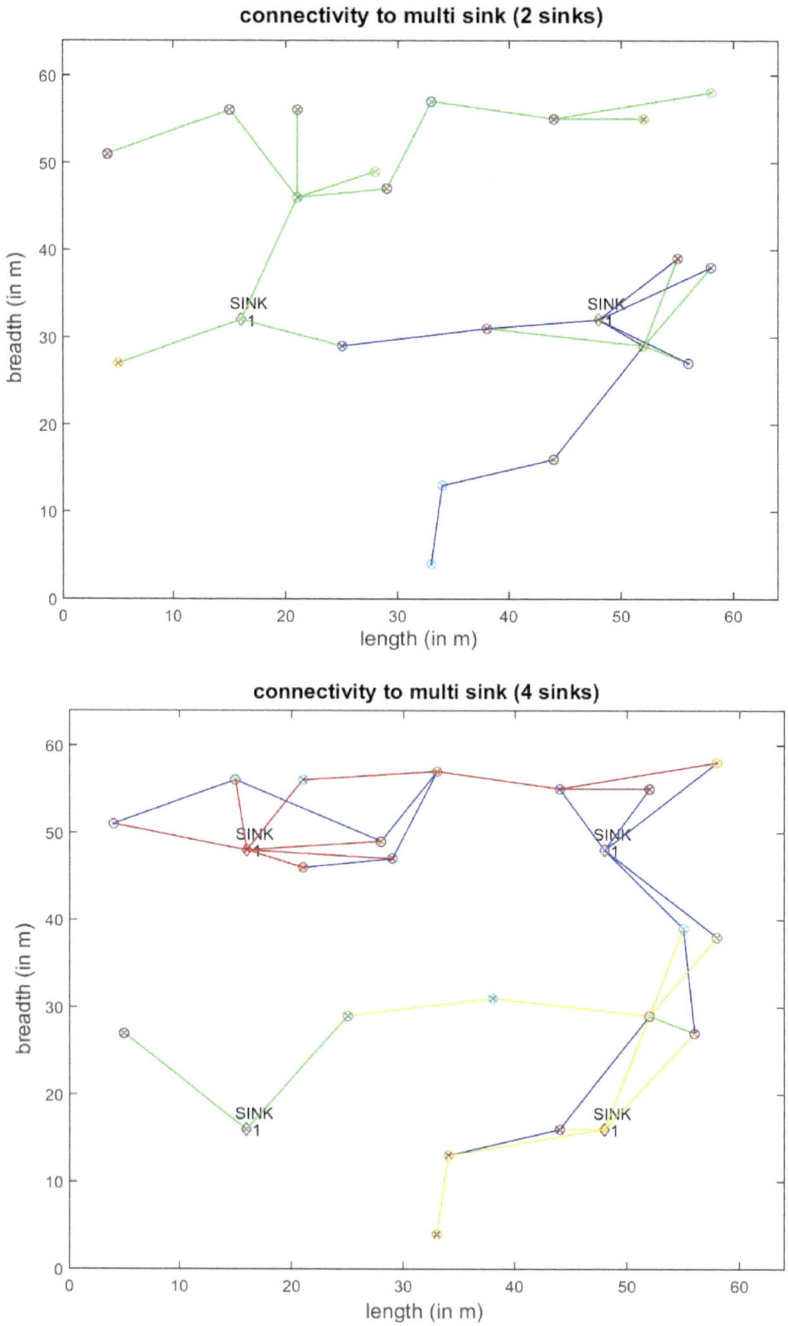

Fig. 4 **a** Connectivity of nodes to multi-sink (2 sinks). **b** Connectivity of nodes to multi-sink (4 sinks)

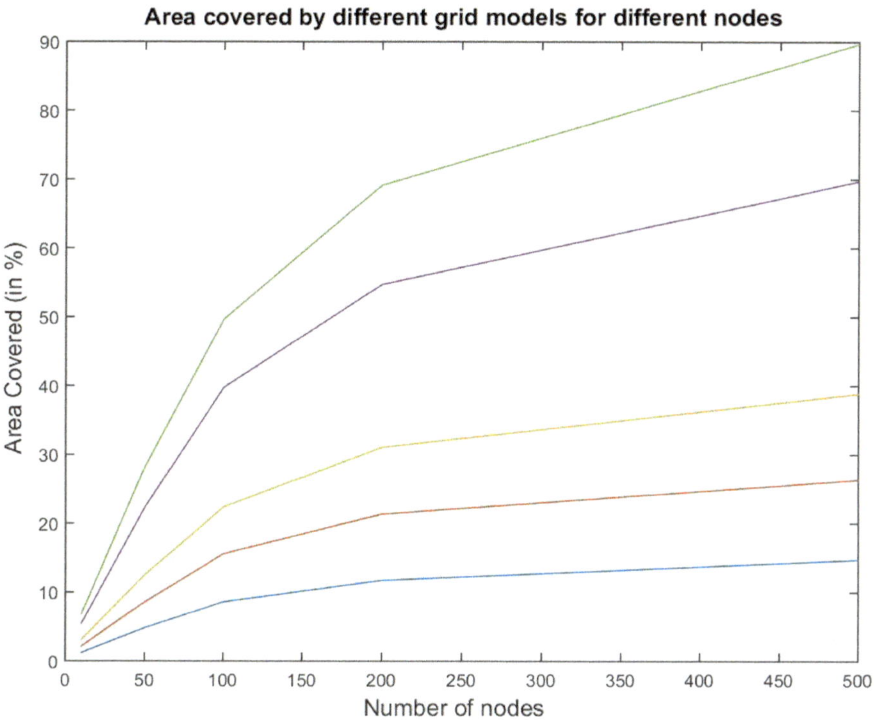

Fig. 5 Comparison of different grid models

Table 4 Area covered by different grid models (in %) for different number of nodes

Nodes	5 grid model	9 grid model	13 grid model	23 grid model	29 grid model
10	1.1963	2.124	3.0762	5.4199	6.8115
50	4.7852	8.5693	12.524	22.314	28.174
100	8.5938	15.576	22.363	39.697	49.634
200	11.743	21.338	31.055	54.712	69.141
500	14.722	26.392	38.818	69.727	89.673

6 Conclusion

The coverage and connectivity of these sensor nodes are very important for efficient energy consumption and increasing network lifetime. In order to reduce energy consumption, both the internodal difference and residual energy of nodes are considered. This paper has successfully calculated the area of the field covered using different grid models. As the number of nodes increase, the coverage of the field increases. However, the coverage can never be 100% in random node deployment. 23-grid model is coming out to be an optimal model for calculations as it resembles most closely to hexagonal shape which, in turn, resembles most closely to circular shape. The grid method also helped us to calculate the overlapped area of sensing regions of different nodes. As the information about the field coverage can now be obtained, we can divide the field into blocks and find out the block of field where the density of nodes is less. Hence, more nodes can be deployed there. The sensor nodes to sinks have been calculated using Chebyshev distance metric. The energy consumption for data transmission using this distance would be minimum which would reduce the cost and end-to-end delay. Multiple sinks have been deployed to increase the network coverage and increase the connectivity of remaining nodes.

Acknowledgements The authors are sincerely thankful to the Indian Institute of Information Technology, Kota, mentored by Malviya National Institute of Technology, Jaipur, for the provision of software facilities.

References

1. Bruckner D, Picus C, Velik R, Herzner W, Zucker G (2012) Hierarchical semantic processing architecture for smart sensors in surveillance networks. IEEE Trans Ind Informat 8(2):291–301
2. Torfs T, Sterken T, Brebels S, Santana J, van den Hoven R, Spiering V et al (2013) Low power wireless sensor network for building monitoring. IEEE Sens J 13(3):909–915
3. Yang J, Zhang C, Li X, Huang Y, Fu S, Acevedo M (2010) Integration of wireless sensor networks in environmental monitoring cyber infrastructure. Wireless Netw 16(4):1091–1108
4. Romer Kay, Mattern Friedemann (2004) The design space of wireless sensor networks. IEEE Wirel Commun 11(6):54–61
5. Rajagopalan R, Varshney PK (2006) Data aggregation techniques in sensor networks: a survey
6. Heinzelman WR, Chandrakasan A, Balakrishnan H (2000) Energy-efficient communication protocol for wireless microsensor networks. In: 2000 proceedings of the 33rd annual Hawaii international conference on System sciences, p 10, IEEE
7. Heinzelman W, Chandrakasan A, Balakrishnan H (2002)An application specific protocol architecture for wireless microsensor networks. IEEE Trans Wireless Commun 1(4)
8. Slijepcevic S, Potkonjak M (2001) Power efficient organization of wireless sensor networks. In: IEEE ICC 2001 international conference on communications, IEEE, vol 2. pp 472–476
9. Pottie GJ, Kaiser WJ (2000) Wireless integrated network sensors. Commun Assoc Comput Mach 43(5):51–58
10. LoBello L, Toscano E (2009) An adaptive approach to topology management in large and dense real-time wireless sensor networks. IEEE Trans Ind Informat 5(3):314–324
11. Zhang D, Li G, Zheng K, Ming X, Pan Z-H (2014) An energy-balanced routing method based on forward-aware factor for wireless sensor networks. IEEE Trans Ind Informat 10(1):766–773
12. Jain TK, Saini DS, Bhooshan SV (2014) Increasing lifetime of a wireless sensor network using multiple sinks. In: 2014 11th international conference on information technology: new generations (ITNG), IEEE, pp 616–619
13. Jain T (2012) Wireless environmental monitoring system (WEMS) using data aggregation in a bidirectional hybrid protocol. In: Proceedings of the 6th international conference on information systems, technology and management (ICISTM '12), pp 414–420, Grenoble, France
14. Zhang Honghai, Hou Jennifer C (2005) Maintaining sensing coverage and connectivity in large sensor networks. Ad Hoc Sens Wirel Netw 1(1–2):89–124
15. Sriporamanont T, Liming G (2006) Wireless sensor network simulator

16. Mateska A, Gavrilovska L (2011) Wsn coverage & connectivity improvement utilizing sensors mobility. In: 11th European wireless conference 2011-sustainable wireless technologies (European wireless), VDE, pp 1–8
17. Xiao J, Han S, Zhang Y, Xu G (2010) Hexagonal grid-based sensor deployment algorithm. In: 2010 Chinese control and decision conference (CCDC), IEEE, pp 4342–4346
18. Koutsopoulos Iordanis, Stanczak Slawomir (2012) The impact of transmit rate control on energy-efficient estimation in wireless sensor networks. IEEE Trans Wireless Commun 11(9):3261–3271
19. Wang X, Xing G, Zhang Y, Lu C, Pless R, Gill C (2003) Integrated coverage and connectivity configuration in wireless sensor networks. In: Proceedings of the 1st international conference on embedded networked sensor systems, ACM, pp 28–39
20. Bhushan S, Saroliya A, Singh V (2013) Implementation and evaluation of wireless mesh networks on MANET routing protocols. Int J Adv Res Comp Commun Eng 2(6(2013)
21. Singh V et al. (2019) Source redundancy management and host intrusion detection in wireless sensor networks. Recent Patents on Comp Sci 12(1)
22. Schurgers C, Srivastava M (2001) Energy efficient routing in wireless sensor networks. In: IEEE military communications conference, vol 1, pp 357–361
23. Tan DD, Kim DS (2014) Dynamic traffic-aware routing algorithm for multi-sink wireless sensor networks. Wireless Netw 20(6):1239–1250
24. Vincze Z, Vida R, Vidacs A (2007) Deploying multiple sinks in multi-hop wireless sensor networks. In: IEEE international conference on pervasive services, IEEE, pp 55–63

Chapter 24
Hybrid Forecasting Scheme for Enhance Prediction Accuracy of Spambase Dataset

Divyanjali Saini and Monalisa Meena

1 Introduction

Spam prediction with high accuracy using spam filtering technique and design machine learning model. In this research, we have used the classification learning methods, naïve Bayes and MLP [1–3]. Additionally with the stimulation from related field of literature that instead of the utilization of total feature set, an usage of only most appropriate features can significantly increase the quality of service of implemented algorithm; the proposed approach has integrate the feature reduction functionality of correlation-based feature selection (CFS) method [4]. This research observed spam-based features and found correlation feature between them with all spam data which classify users as spammer and non-spammers. The proposed solution defines higher accuracy which uses hybrid formation for filtering spam or non-spam content with the help of supervised learning methods.

Contributed in the literature of spam classification by the following steps:-

1. Observed exist spam datasets spambase and WEBSPAM-UK2011 have taken from UCI repository [5].
2. Developed and validated correlation feature selection as a feature selection method.
3. Discovered and validated naïve Bayes as the fast classifier machine learning model.
4. Observed neural network (MLP) as the best way to evaluate deep learning method.
5. Observed ten-fold cross validation method that developed efficient hybrid forecasting approach using naïve Bayes and MLP.
6. Proposed both classifiers as to remove low efficiency of feature selection method.

D. Saini (✉) · M. Meena
Department of Computer Science and Engineering, GWEC, Ajmer, Rajasthan, India
e-mail: divyanjalisaini24@gmail.com

© Springer Nature Singapore Pte Ltd. 2021

S. D. Purohit et al. (eds.), *Proceedings of International Conference on Communication and Computational Technologies*, Algorithms for Intelligent Systems,
https://doi.org/10.1007/978-981-15-5077-5_24

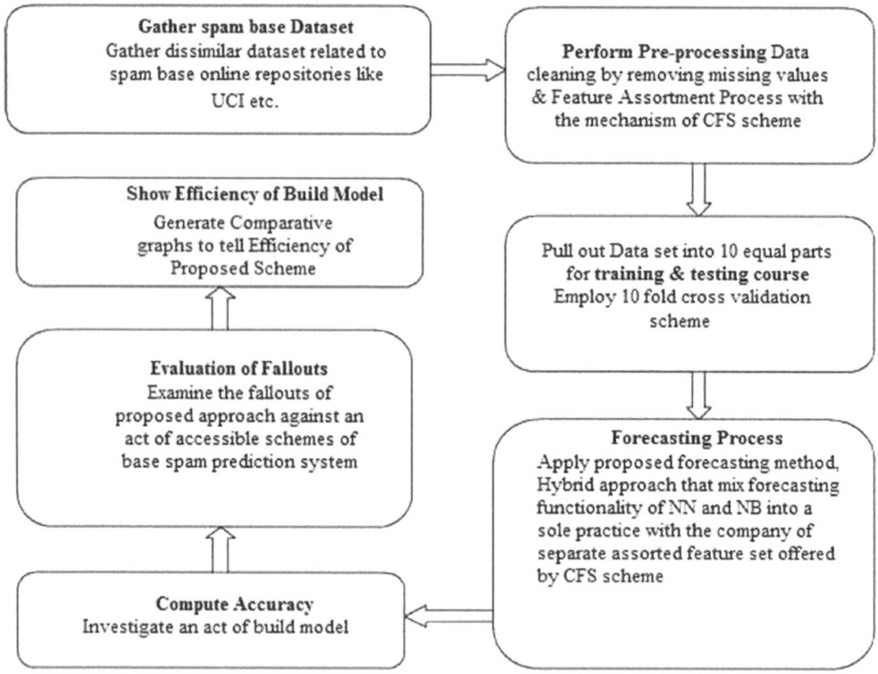

Fig. 1 Process flow diagram of proposed work

7. Proposed new model: hybrid approach using NB and NN (MLP) classifier to improve prediction accuracy.

2 Proposed Work

To defeat an inadequacy of accessible spam detection prediction mechanism in terms of fabricating high prediction accuracy of spam detection with low amount of false alarms, the proposed approach has exploit two of the most accepted and employed data forecasting methods in a ladder form with the utilization of feature assortment process (Fig. 1).

3 Implementation of Proposed Approach

In order to analysis data prediction recital of designed approach against a number of existing classification algorithms, a Java-based application has been designed in this investigation. Java is a very popular and freely accessible, portable, robust programming language that has been used in various areas to design numerous approaches.

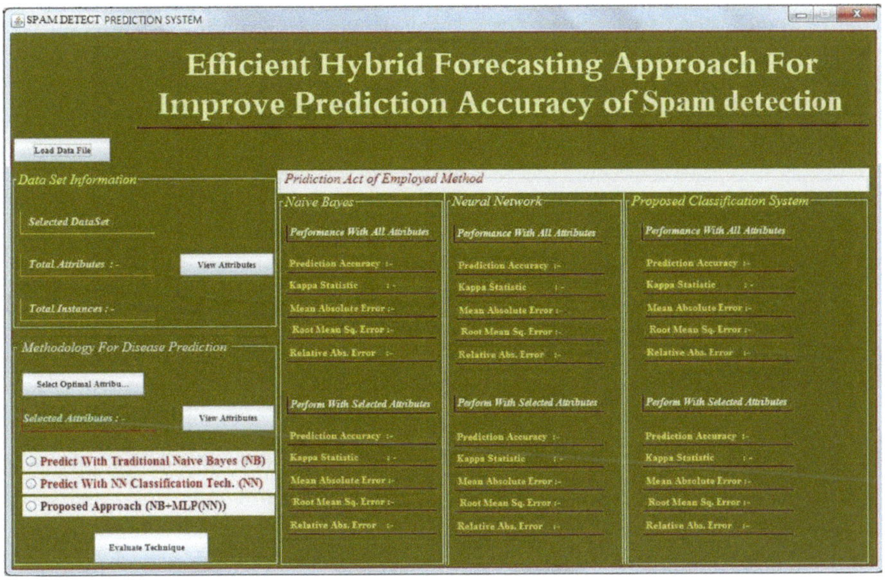

Fig. 2 Interfaces to asses an act of implemented approach

Apart from simple learning, this language has provided a number of unique functionality to its programmer that makes it the most popular and acceptable software developing tool around the world. With the strong and effective functionality of Java, the designed application provides a trouble-free and efficient GUI for its users. The application integrated into number of classification algorithms by incorporating the functionality of the most popular data simulation tool in the area of machine learning named as Weka. The designed techniques access the inbuilt algorithms of such tool in intended programming code to omit the requirement of redesign data classification process of such implemented algorithms. For the classification purposes, two different existing classification algorithms have been used for the recital comparison against the proposed prediction approach (Fig. 2).

4 Result Analysis

For fair evaluation, an act of proposed approach against traditional and modern spam prediction methods, a common spam dataset "spambase dataset" has taken in online access from UCI repository [6–8]; these are using in different methods for the prediction of spam and non-spam. All datasets enclose statistics of 4601 instances in spambase dataset [6]

Table 1 Opted features by proposed approach from spambase dataset for prediction of spam detection

Sl. No.	Feature	Sl. No.	Feature
1.	word_freq_ make	9.	word_freq_original
2.	word freq Address	10.	word_freq_conference
3.	word_freq_remove	11.	char_freq_;
4.	word_freq_internet	12.	char_freq_(
5.	word_freq_adresses	13.	char_freq_[
6.	word_freq_business	14.	char_freq_!
7.	word_freq_technology	15.	char_freq_$
8.	word_freq_meeting		

On the other hand, dataset associated with 58 attributes from that the proposed approach utilized only optimal set of 15 features that significantly improved act of proposed methods, as shown in Table 1.

Table 2 illustrates the list of complete feature set that are associated with spambase dataset.

Figure 3 illustrates the relative outcomes of evaluated schemes, i.e. proposed approach and sole version of traditional NB and NN, in the company of full and selected feature sets.

The evaluated fallouts depicted in above figure evidently demonstrate that proposed approach has significantly improve the accuracy level of spam prediction in comparison with sole practices of data mining, where NB method has to attain a prediction accuracy rate 89.85% with the inclusion of all feature and 92.59% with the optimal feature set, and MLP (NN) gets 89.39% with full and 92.2% in the company of reduced features set, and the proposed approach has attained 97.23% with full feature set and enhances accuracy with the execution of optimal feature set that is 97.71%. As the figure indicates that proposed approach has attained higher accuracy in comparison with classical forecasting method, as shown in Table 3 (Fig. 4).

The comparative fallout of the proposed approach shown in the above table has depicted its efficiency over the traditional forecasting mechanism of data mining in terms of high prediction accuracy of spam. The fallout has also compared against the one modern proposed algorithm of the [9–11], comparison have shown in following Table 4.

The above table show that the proposed approach has significantly attained high prediction accuracy of spambase not in comparison with the traditional sole practices of data mining but also against one of the modern intended algorithms [12]. The following figure presents the difference of attained accuracy in a graphical manner that explains the efficiency of proposed approach over the classical sole version of data mining forecasting method (Fig. 5).

The above figure clearly indicates the efficiency of the proposed approach in comparison with other evaluated algorithm. Experimental fallouts of the proposed

Table 2 Associated features of spambase dataset

Sl. No.	Features name	Sl. No	Features name	Sl. No.	Feature name
1.	word_freq_ make	20.	word_freq_credit	39.	word_freq_pm
2.	word freq Address	21.	word_freq_your	40.	word_freq_direct
3.	word freq All	22.	word_freq_font	41.	word_freq_cs
4.	word_freq_3d	23.	word_freq_000	42.	word_freq_meeting
5.	word_freq_our	24.	word_freq_money	43.	word_freq_original
6.	word_freq_over	25.	word_freq_hp	44.	word_freq_project
7.	word_freq_remove	26.	word_freq_hpl	45.	word_freq_re
8.	word_freq_internet	27.	word_freq_george	46.	word_freq_edu
9.	word_freq_order	28.	word_freq_650	47.	word_freq_table
10.	word_freq_mail	29.	word_freq_lab	48.	word_freq-conference
11.	word_freq_receive	30.	word_freq_labs	49.	char_freq_;
12.	word_freq_will	31.	word_freq_telnet	50.	char_freq_(
13.	word_freq_people	32.	word_freq_857	51.	char_freq_[
14.	word_freq_report	33.	word_freq_data	52.	char_freq_!
15.	word_freq_addresses	34.	word_freq_415	53.	char_freq_$
16.	word_freq_free	35.	word_freq_85	54.	char_freq_#
17.	word_freq_business	36.	word_freq_technology	55.	capital_run_length_aver age
18.	word_freq_email	37.	word_freq_1999	56.	capital_run_length_longest
19.	word_freq_you	38.	word_freq_parts	57.	capital_run_length_total
				58.	class

approach were depicted in above table show that it has outperformed in comparison with sole techniques of data mining.

5 Conclusion

To defeat inadequacies of obtainable spambase prediction schemes, the incorporated work of this investigation has proposed a naïve ensemble approach by utilizing the functionality of naïve Bayes, a form of Bayesian theorem and multilayer perceptron (MLP), a form of neural network into a sole practice. An act of build procedure the designed approach integrate the characteristic of feature assortment by exploit the mechanism of Correlation based Feature Selection (CFS) method. Typically, to crush the inefficiency of accessible models and aid to secure system professionals for diagnose such Spambase with good accuracy the implemented method robotically test the false predict statistics twice in a manner that each false predict fact of

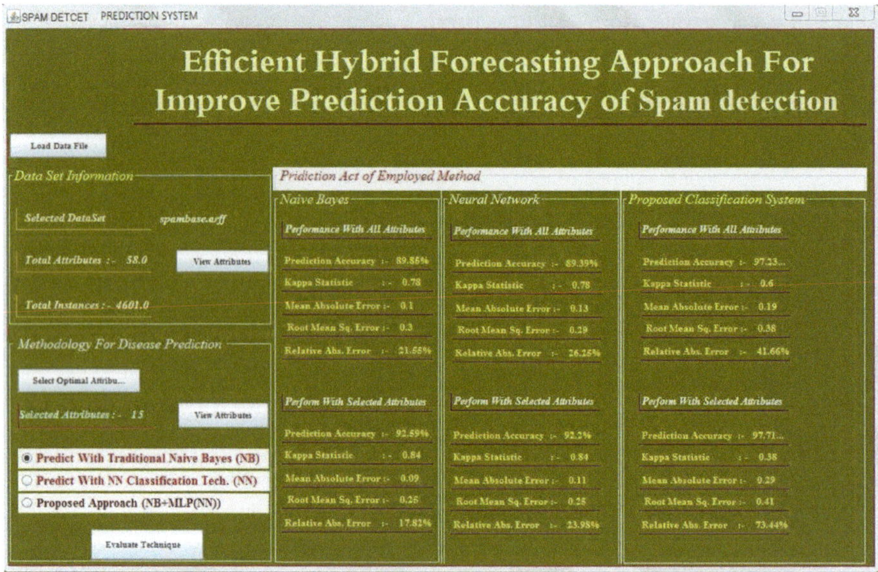

Fig. 3 Spam detection prediction act of proposed and traditional mechanism of data mining over spambase dataset

Table 3 Prediction act of proposed and classical schemes over spambase dataset

Forecasting Practice	Attain accuracy (%)	
	With full feature	With optimal feature
Proposed approach	**97.23**	**97.71**
Naïve Bayes	89.85	92.59
Multilayer perceptron	89.39	92.2

first applied approach has cross examine by second employed method that significantly enhance the precision rate of spam recognition. To examine and reveal the competence of proposed approach it has coded in a most popular and employed open source programming language JAVA. The weight you comes of proposed system have proven its efficiency over the separate version of prediction method as well as current innovative accessible Spambase prediction techniques.

6 Future Scope

However, it builds procedure that effectively outperformed in comparison with various obtainable spambase prediction algorithms but due to the inclusion of week

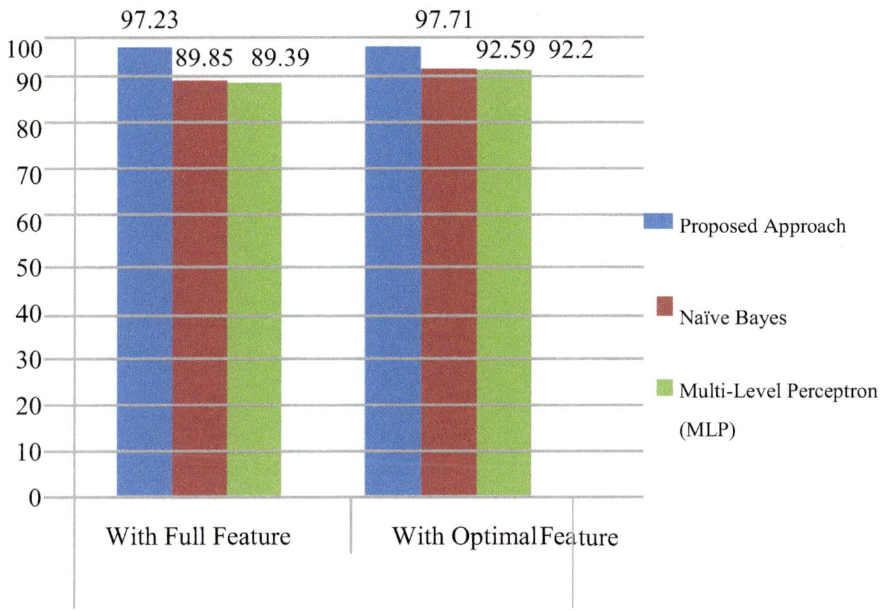

Fig. 4 Prediction act of proposed approach, NB and MLP mechanism over spambase dataset

Table 4 Act of proposed and other evaluated spambase dataset

Forecasting practice	Attain accuracy (%)
Proposed approach	**97.71**
Naïve Bayes	92.59
Multilayer perceptron	92.2
Machine learning tech. (NB), [9]	89
Machine learning tech. (MLP), [9]	93
Adaptive approach (NB), [10]	88.54
Adaptive approach (MLP), [10]	93.28
Naïve Bayes filtering, [11]	82.54

forecasting mechanism into a sole practice, the implemented approach has been associated with some restraint that can be eradicated with further optimization in future. Future work can be done in the direction to

1. Improve prediction accuracy with generating of low false alarms by discover and integrate the more powerful forecasting mechanisms in the company of efficient feature assortment procedure.
2. Reduce the requirement of manual efforts with the inclusion of mechanized functions that automatically explore the best method and utilize that to improve actions.

Fig. 5 Comparative fallouts of evaluated algorithms for prediction of over spambase dataset

References

1. Costa K, Ribeiro P, Camargo A, Rossi V, Martins H, Neves M, Fabris R, Imaisumi R, Paulo JP Comparison of the techniques decision tree and MLP for data mining in spams detection to computer networks. In: Third international conference on innovative computing
2. Hans K, Ahuja L, Muttoo S (2014) Approaches for web spam detection. Int J Comput Appl 101:975–8887
3. Sanghani G, Kotecha K (2019) Incremental personalized E-mail spam filter using novel TFDCR feature selection with dynamic feature update. Expert Syst Appl 115:287–299
4. Idris I, Selamat A, Omatu S (201) Hybrid email spam detection model with negative selection algorithm and differential evolution. Eng Appl Artif Intell
5. WEBSPAM-UK2011 Data set at https://www.ncbi.nlm.nih.gov/pmc/articles/PMC5795495/
6. Spambase dataset at https://archive.ics.uci.edu/ml/datasets/spambase
7. Rathi M, Pareek V (2013) Spam mail detection through data mining—a comparative performance analysis. Int J Mod Educ Comput Sci 5:31–39
8. Feng W, Sun J, Zhang L, Cao C, Yang Q (2016) A support vector machine based naive Bayes algorithm for spam filtering. In: 2016 IEEE 35th international performance computing and communications conference (IPCCC), Las Vegas, NV
9. Renuka DK, Hamsapriya T, Chakkaravarthi MR, Surya PL (2011) Spam classification based on supervised learning using machine learning techniques. ICTACT J Commun Technol

10. Rusland NF, Wahid N, Kasim S, Hafit H (2017) Analysis of Naïve Bayes algorithm for email spam filtering across multiple datasets. In: IOP conference series: materials science and engineering
11. Sharma S, Arora A (2013) Adaptive approach for spam detection. IJCSI Int J Comput Sci Issues 10(4), No 1
12. Panwar D, Tomar P, Singh V (2018) Hybridization of Cuckoo-ACO algorithm for test case prioritization. J Stat Manag Syst 21(4):539–546

Chapter 25
An Enhance Security of the Color Image Using Asymmetric RSA Algorithm

Shreya Agrawal, Mayank Patel, and Amit Sinhal

1 Introduction

The image is that the communication mode is most employed in completely different fields like medical field, analysis field, industry, military space, etc.. The vital transfer of pictures can crop up in an unsecured Web network. Therefore, there is a necessity for acceptable security in order that the image prevents access by the unauthorized person to special data. The advantage of the image is that it covers more multimedia data and needs protection [1]. In currently going on years with the high advancement of digital technology and also the scope of rights in the online broadens. Interests also end up in a series of negatives effects, as an example data larceny, piracy, etc. Digital image has been wide used in people's lives it's a lots of integral and your safety is a lot of important. Currently, the protection of photography has become a very important issue within the field of computer [2]. Secure connections via insecure channels, such as the Web, have been a revenue goal associated with research since the start of the digital age. With the continued development of symmetrical and asymmetric key coding, digital communication across the network is created in a secure and personal way [3]. Cryptography may be a kind of image security technique; it offers the secure method of transmittal and storing the image on the net. Security is the main concern of any system to maintain the integrity, confidentiality, and authenticity of the image. Although cryptography is the efficient method, it also faces the problem of security if data with gray levels are more numerous [4].

S. Agrawal (✉) · M. Patel · A. Sinhal
Department of Computer Science and Engineering, Geetanjali Institute of Technical Studies (GITS), Udaipur, India
e-mail: speedins214@gmail.com

M. Patel
e-mail: mayank.patel@gits.ac.in

A. Sinhal
e-mail: hodcse@gits.ac.in

© Springer Nature Singapore Pte Ltd. 2021
S. D. Purohit et al. (eds.), *Proceedings of International Conference on Communication and Computational Technologies*, Algorithms for Intelligent Systems,
https://doi.org/10.1007/978-981-15-5077-5_25

2 Image Encryption

Nowadays, we tend to hear the word cryptography in virtually every place. Cryptography is that technique of encrypting a message or data exceedingly, and there are various methods that exclusively approved parties can access and folks unauthorized cannot do so. In other words, encryption is a method or mechanism that allows you to hide your message or data so that only authorized people can access it. The two-key encryption scheme includes two uses: public and personal keys. Using the public key, the plain text is converted into the encrypted text, and this process is called encryption. The encryption process is done at the sender end. At the receiving end, get the original plaintext from the encrypted text using the private key and this method is called decryption. In order to preserve the secrecy of the non-public key, the key of the final public is discovered to the general public. The general public secret is used for authentication to form certain the message comes from the required sender. The general public key coding system additionally guarantees confidentiality. Solely the recipient's personal key will decipher the encrypted text or the sender's image. The communication is often established from the messages in an exceedingly secure approach, and as a result, the information of the general public keys is not enough to decipher the encrypted text [5]. There are primarily two sorts of encoding formula technique that are used: symmetric encoding and asymmetric encoding. Several symmetrical and asymmetric coding techniques are used to code and decode functions. Symmetric works during a common key, whereas uneven works with a public key and a private key [6, 7]. In such an encryption using asymmetric RSA algorithm, use the public key and for decryption using another key that differs from the public key which is called private key. Comfortably totally different keys are used for cryptography and decipherment. The RSA algorithmic rule is additionally referred to as asymmetric cryptography [8].

3 Grayscale Image and Color Image

In digital photography, computer-generated images, and color measurement, the grayscale or grayscale image is an image in which the value of every component may be a single sample representing simply the number of sunshine, that is, it transmits data from density [9]. Grayscale images, which are monochromatic black, white, or gray, are exclusively gray [10]. Variation varies from black in the lowest intensity to white in strongest intensity. The grayscale image contains only black and white tones. It does not have any color information like chrominance. A grayscale image is typically a 8-pixel image and each pixel has 256 shades of gray. The color image is usually 24 bits with 8 bits of red, 8 bits of green, and 8 bits of blue information [11, 12]. Combining these three basic colors can create 16,777,216 color groups per pixel.

Grayscale image is one-dimensional image (channel). Grayscale image is derived from the RGB image and incorporates a price of intensity [13]. It's the intensity value, exploitation the below equation to cipher the grey scale Image [14].

$$Y = 0.299\ \mathrm{R} + 0.587\ \mathrm{G} + 0.114\ \mathrm{B} \tag{1}$$

The RGB color model is an associate mixture color model during which red, green, and blue light area unit supplementary in different ways that to provide a large variety of colors. The different color comes from the initials of the three basic colors red, green and blue, reason behind this due to these three sq. colors of primary color.

The main objective of the RGB color model is to watch, represent, and visualize pictures in electronic systems, like televisions and computers, though it has been employed in recent photography. Before the electronic age, the RGB color model already had a strong theory that supported human perception of colors.

Digital color images consist of pixels and sets of basic colors represented in a series of symbols. The channel during this context is the grayscale image of same size because the color image consist from just one of those basic colors. For example, the quality photographic camera image can have a red, green, and blue channel. The grayscale image contains one channel.

As we see in both above images (Figs. 1 and 2) which shown Fig. 1 grayscale image we can't differentiate in every gem stone because there is no specific color difference among them. But in Fig. 2 color image, we can easily separate among them easily. This concludes that color image encryption is better than grayscale image.

Fig. 1 Gray image

Fig. 2 Color image

4 Proposed Asymmetric RSA Algorithm for Color Image Encryption

RSA is one among the most algorithms that care for the overall secret writing key and also the secret decipherment key. RSA is the most used encryption technology to date [9].

The task of coloring an image involves distribution RGB values to a picture that varies on solely the brightness, hue, and saturation price. Since completely different colors will have identical physical property worth however vary in hue or saturation, the matter of coloring grayscale pictures does not solve the matter properly. Owing to these ambiguities, human interactions typically play a crucial role within the staining method.

Here the grayscale image is portrayed by solely the luminousness values which will be matched between the two pictures. Due to this reason, we cannot use the grayscale image encryption process. Once the image is converted in grayscale, it is impossible to get it back in color. Because while converting it in grayscale, we lose hue and saturation of image. Due to the presence of luminance values in grayscale image, we cannot able to differentiate among the object of the same character and size in an image. So, in the proposed algorithm, we encrypt the image in color format and decrypt it back in color format keeping the hue, saturation, and luminance in entire image encryption process. We can easily understand from the below image that object is easily classified in color image rather grayscale image.

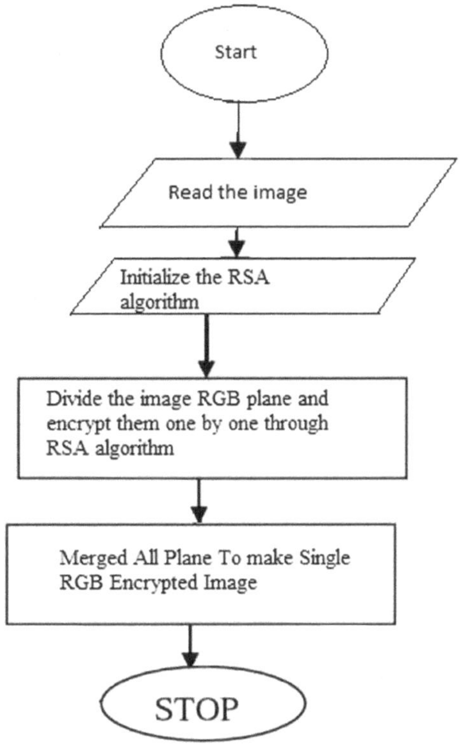

Fig. 3 Flowchart of proposed image encryption technique

The proposed color image encryption using asymmetric RSA algorithm design flowchart is shown in Fig. 3.

4.1 Simulation and Result

In Fig. 4 shown, the original color in Fig. 4a, b shown the encrypted image of the color image which is totally different from the original image and also we get the decrypted image (Fig. 4d) which is same as the original image from the encrypted image (Fig. 4c).

In Fig. 5a shown the original color of gem stone in Fig. 5a, b shown the encrypted image of the color image(gem stone) which is totally differ then the original image and also we get the decrypted image (Fig. 5d) which is same as the original image (gem stone).

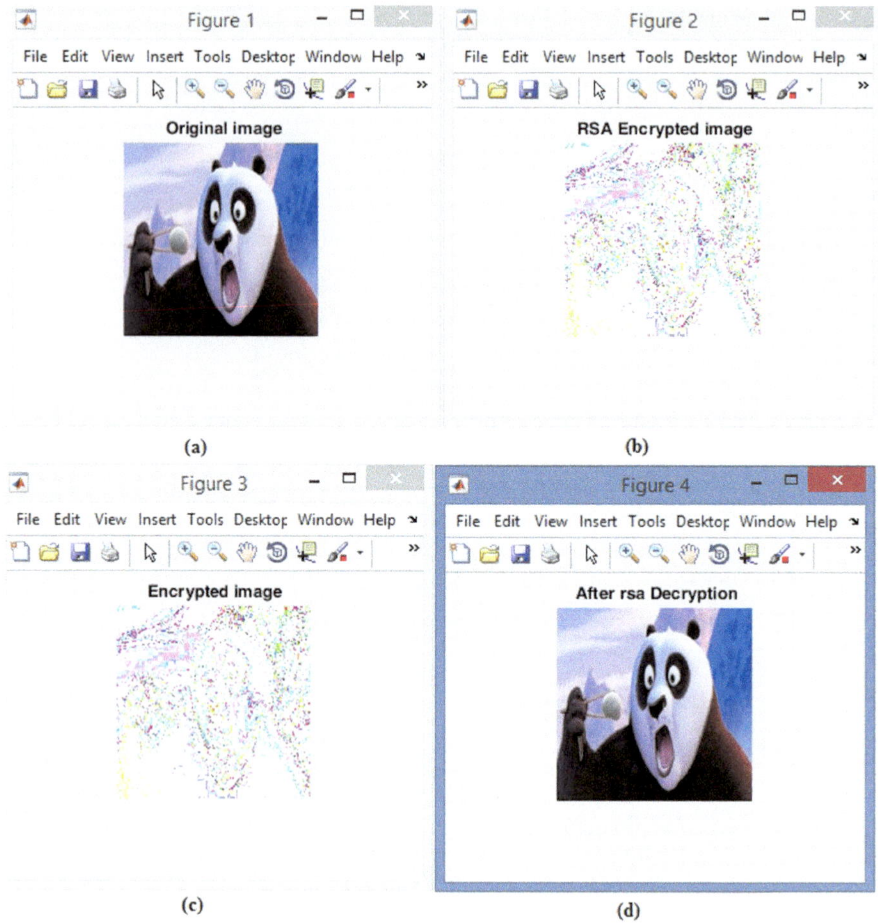

Fig. 4 Color image encryption and decryption using asymmetric RSA algorithm

5 Conclusion

The asymmetric encryption algorithm of RSA makes encryption more secure and receiver has not too worried to give each sender a different key to secure the communication. And one more advantage of RSA algorithm is that RSA algorithm is difficult to crack because it involves factorization of prime numbers that are difficult to factorize.

In this work, we are going to process the color image security using the RSA algorithm. The grayscale image encryption has many drawbacks. Here, the grayscale image is drawn solely by the physical property values which will correspond between the two digital pictures. Thanks to this reason we tend to cannot use the grayscale

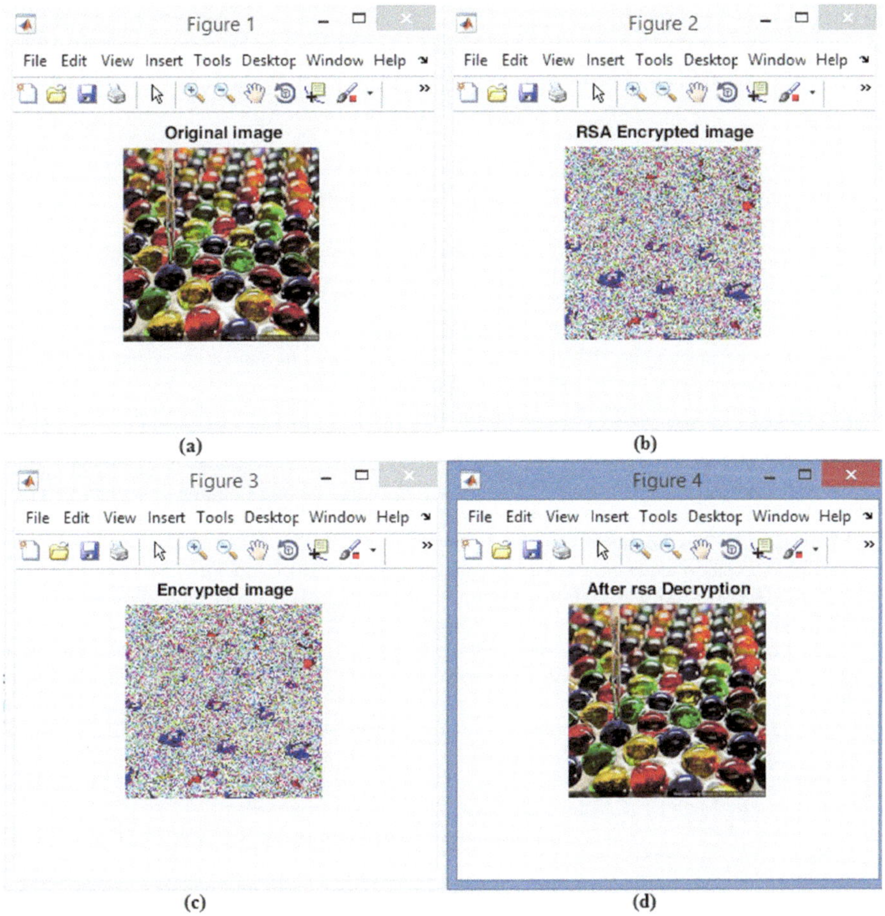

Fig. 5 Color image (gem stone) encryption and decryption using asymmetric RSA algorithm

image cryptography method. Once the image is converted in grayscale, it is impossible to get it back in color because while converting it in grayscale, we lose hue and saturation of image. To solve this issue, we encrypt color image and get decrypted image also in color.

By doing image encryption in color, we save the hue, saturation, and luminance of original image. Image that get in color image encryption makes object classification easy and precise. This algorithm has many benefits like it makes encryption strong, it provides sender and receiver different keys, and it makes object classification and recognition better.

References

1. Kumar BS, Raj VR, Nair A (2017) Comparative study on AES and RSA algorithm for medical images. In: IEEE international conference on communication and signal processing, pp 501–504
2. Li T, Wang Y (2010) Study on image encryption algorithm based on Arnold transformation and chaotic system. In: IEEE international conference on intelligent system design and engineering application, pp 449–451
3. Mare SF, Vladutiu M, Prodan L (2011) Secret data communication system using Steganography, AES and RSA. In: IEEE 17th international symposium for design and technology in electronic packaging (SIITME), pp 339–344
4. Panda PK, Chattopadhyay S (2017) A hybrid security algorithm for RSA cryptosystem. In: IEEE international conference on advanced computing and communication systems (ICACCS-2017), Coimbatore, India, 06–07 Jan 2017, pp 1–6
5. Mishra S, Minni R, Sultania K, Vincent DR (2013) An algorithm to enhance security in RSA In: 4th ICCCNT 2013, 4, Tiruchengode, India, July 4–6 2013, pp 1–4
6. Rathod C, Advani N, Gonsai AM (2019) Comparative study of various cryptographic algorithms used for text, image, and video. Emerging trends in expert applications and security, advances in intelligent systems and computing. Springer, Singapore, pp 393–399
7. Mukherjee S, Chakrabarti S, Sinha S, Mukhopadhyay T (2017) A meticulous implementation of RSA Algorithm using MATLAB for image encryption, IEEE, pp 1–6
8. Yen JC, Guo JI (2000) An efficient hierarchical chaotic image encryption algorithm and its VLSI realization. IEEE Proc Vis Image Process, 167–175
9. Chaturvedi R, Sharma A, Dwivedi U, Kumar S, Praveen A (2016) Security enhanced image watermarking using mid-band DCT coefficient in YCbCr space. Int J Control Theory Appl 9(23):277–284
10. Gupta P, Kumar S (2014) A comparative analysis of SHA and MD5 algorithm. Int J Comput Sci Inf Technol 5(3):4492–4495
11. Kumar A et al (2019) An enhanced quantum key distribution protocol for security authentication. J Discret Math Sci Cryptogr 22(4):499–507
12. Kumar A et al (2019) An improved quantum key distribution protocol for verification. J Discret Math Sci Cryptogr 22(4):491–498
13. Gupta S et al. (2019) Tier application in multi-cloud databases to improve security and service availability. Handbook of research on cloud computing and big data applications in IoT. IGI Global, 82–93
14. Dwivedy P, Ray A, Potnis A, Soofi S, Bhade U (2017) Comparative study of AES, RSA, genetic, affine transform with XOR operation, and watermarking for image encryption. In: IEEE proceeding international conference on recent innovations is signal processing and embedded systems (RISE-2017), pp 274–278

Chapter 26
Social Media Analytics and Mutual Fund Recommendation

Samkit Shah and Harshal Trivedi

1 Introduction

Social media data has great influence on fintech market, and exploration of social media data can give some valuable financial insights. This paper illustrates exploration of social media data and gated the user segmentation on different user profiles with the help of clustering algorithms. Furthermore, we have created unbiased mutual fund recommender with the help of Long-Short Term Memory (LSTM) and rated mutual funds by stock holdings of specific mutual funds.

2 Analysis of Methodologies Used

2.1 Analysis and Methodologies Used in Social Media Analytics

Social media analytics represents area of computing adapted from user profiles. The aim of author is provide terminologies for social media analytics and review software tools and how to utilize them to scrape, cleanse and analyse the user profile data of social media [1]. Analysing social media, for instance, Twitter feeds for sentiment analysis and scrapping data from various Websites and logs have become a major research and business activity due to the availability of Web-based application programming interfaces (APIs) provided by Twitter, Facebook, News services and

S. Shah (✉)
Government Engineering College, Sec-28, Gandhinagar, Gujarat, India
e-mail: shahsamkit408@gmail.com

H. Trivedi
Softvan Technologies, Ahmedabad, Gujarat, India

© Springer Nature Singapore Pte Ltd. 2021

S. D. Purohit et al. (eds.), *Proceedings of International Conference on Communication and Computational Technologies*, Algorithms for Intelligent Systems,
https://doi.org/10.1007/978-981-15-5077-5_26

many social media giants [1]. All of these utilizations led to an 'explosion' of data services, software tools for scraping and social media analytics platforms. Research area in these fields is undergoing numerous changes and evolution due to commercial activities and the potential for using social media data for computational research [1]. There are a number of approaches within the literature which deal with social media data collection, processing and computing techniques.

Although this literature appears to be no formal techniques of these different approaches and has not stated the use of these techniques for financial domain, this paper classifies the literature into the topics proposed by Philip and Bogdan [1] and augments different mining and scrapping techniques which provides data in the form of either structured or unstructured.

Author has proposed different terminologies used for data collection, processing and computing like natural language processing for textual data mining, news analytics for unstructured news data coming from various sources, scrapping for unstructured Web data, sentiment analysis on textual data and textual mining for predictive analytics.

Author has illustrated different sources of data like social media platforms, news Websites and programmable interface data (APIs) and types of data like historic, real-time and raw data available in HTML, XML, JSON or CSV files. Author has described strategies and techniques to collect data from programmable interfaces and store it as structured form.

Another illustration where Yigitcan [2] stated methods to predict stock market daily returns and trading volume by training facebook textual data. In this paper, Author has proposed measure of investor sentiment and sentiment for stock markets that is based on data coming from millions of people. Furthermore, Author has used Facebook's Gross National Happiness (GNH) [2] measure to predict investor sentiment, and by training autoregressive models, author found correlation between GNH and stock market activities. GNH purely depends on textual analysis of emotion words posted by millions of users on Facebook and according to correlation GNH is important factors when stocks are rising as sentiment of textual data is positive.

There is a wide acceptance of these techniques, whereby these are seen as great data collection methods from various sources and APIs, but correct and efficient usage of these methods in different domains are clearly not stated in research outcomes, and here, our aim is to use these methods to gain financial insights from user profile and social media data.

2.2 Analysis and Methodologies Used in Investments and Finance

As financial social media sites and analytical outcomes have grown, analysts, traders, technologists and academics have begun to make sense of the data and activities on exploring social data has seen big rise in several years.

Financial analytics mainly focused on three core things like equity sentiment analysis, breaking news discovery, macroeconomic trend analysis and different companies have made their analytical tools to predict several sentiment scores which can improve financial insights [3].

Yang and Liu [3] have proposed methods that recommend stocks from S&P 500 using buy and hold strategy and select top 20% stocks dynamically. Authors have taken various features and stock indicators and trained the model with machine learning methods like linear regression, ridge regression, stepwise regression, random forest and generalized boosted regression and choose the model with the lowest Mean Square Error in each period to rank stocks. Furthermore, they have tested stocks dynamically and perform different portfolio allocation statistical methods such as equally weighted, mean variance and minimum-variance. Authors have suggested different methodologies to decide the final prediction such as Rolling Window Based Data Separation in which training cycle ranges from 16 quarter to 40 quarter. Furthermore, they have processed data by evaluating various financial ratios like revenue growth, cash ratios, earning per share (EPS), return on equity (ROE), return on asset (ROA), gross profit margins, net profit margins, operating margin, working capital ratio, debt-to-equity ratio and quick ratio. They have proposed statistical measures like transaction cost and mean variance measures.

Batres-Estrada [4] has illustrated multivariate time series analysis using deep learning. Author proposed deep learning methods to forecast financial outcomes. He has trained neural network (DBN) coupled with multilayer perceptron (MLP) which is used to choose stocks to form portfolios. Author has performed testing on S&P 500 and used various statistical measure to find accurate results. All illustrated results have shown that neural network model outcome betters from all other methods and gives support due to reliability and efficient outcomes [4].

Chen et al. [5] performed LSTM-based approach for stock prediction. They trained the model and make prediction on Chinese stock market data for 30 days sequences with ten model features and three-day learning rate labelling. They have trained the network based on LSTM and taken data from the yahoo finance and used technical analysis parameters and various economic measures as feature for the models. Author has illustrated various sequence patterns found accuracy on closed prices as model features [5].

Coyne et al. [6] have suggested natural language processing models which successfully filter out important twits from StockTwit platforms and find correlation of sentiment and stock prices through the user twits. Authors have used $[Stock_name] to filter out the correct stocks from the StockTwits and used financial data from Quandl and made sentimental analysis tool which can predict bullish or bearish sentiment for several trading session for specific stocks. They have created three models such as linear regression model, sentimental analyzer and stock forecaster smart user model. In linear regression model, they found average vector for each stock by Term Frequency Inverse Term Frequency method and then aggregate it with percent change information per day which can predict target percentage change per day. In sentimental analyzer, they have trained model by labelling the twits manually into bullish or bearish and then train classifier which can predict the sentiment of incoming tweets. In the smart user forecaster model, they have combined sentiment classifier with smart users who have successfully predicted for the n times with defined accuracy and trained classifier to predict outcomes.

Culkin and Das [7] have approached deep learning method to predict price of options. They have trained artificial neural network on several features such as Stock Price, Strike Price, Maturity, Dividend Rate, Risk Free Rate, Volatility and Call price and tested it over 60,000 outcomes and used RMSE as evaluation parameter.

Chatzis [8] and Leon [9] examined the stock market and global financial crisis using deep learning methods and clustering, respectively. Carlos trained model using agglomerative clustering, which describes correlation factor of world equity markets based on cross section similarity (CSE). Author has used dendograms for hierarchical classification outcome which is generated by agglomerative clustering. In Dendogram, clusters appear by successive merge and Euclidian distance. Chatzis has trained data of different countries stock market, bond market and currency using deep neural networks, classification and regression tree method, extreme gradient boosting and random forest algorithms and predict the global market crisis event with the help of trained model. Nayak [10] has used supervised learning approach for historic data of BSE stock market and trained model which are combination of sentiment analyser, news analyzer, volume variation and up/down data for stocks and predict the up or down movement for specific stocks. Munjal et al. [11–13] simulated some natural phenomenon [11] for social networking, Twitter sentiment analysis [12] and new model based on Ostwald Repening [13].

Here, all of the methods in financial and stock market domains are accepted worldwide and have some similarities such as LSTM for stock prediction, sentiment analysis, regression-based predictions and financial crisis evaluation, but all of these proposed methods still have some edge of research where user-based portfolio recommendation and mutual fund analytics have taken as core area. Here, our primary research is to make unbiased user-based mutual fund recommendation by exploring social media data.

Looking at Research Gap, our paper proposed unbiased user-based mutual funds as recommendation according to specific clusters which are generated by exploration of social media profiles by user estimated income to spend, and our main aim here is to recommend mutual funds to users by appropriate clusters.

3 Proposed Flow

See Fig. 1.

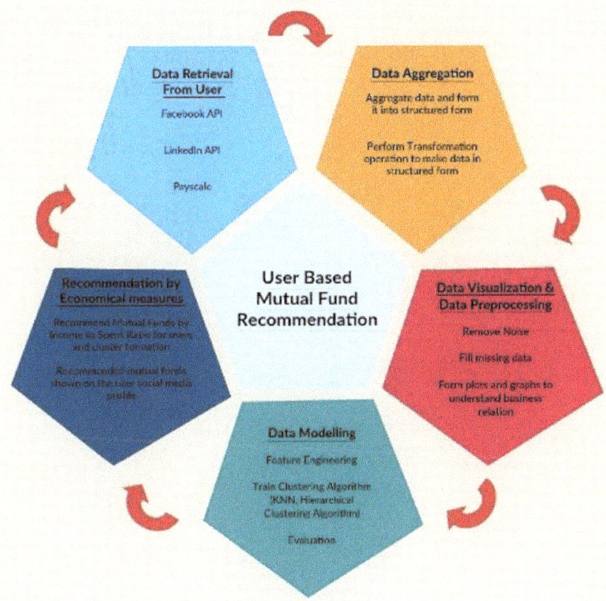

Fig. 1 User-based mutual fund recommendation cycle

3.1 Data Retrieval

Vast amount of social media data is available and accessible via APIs and various tools. We have used different data retrieval methods and collected social media profile data from Facebook's Graph API and LinkedIn API with help of access tokens. We have used Facebook Graph API for retrieving user profile and collected professional social network data from LinkedIn.

API in unstructured forms like JSON and XML which can be used as data sources for processing and analytical insights. One can retrieve batch updates, user profiles and real-time updates via graph API.

3.2 Data Aggregation

Data aggregation is required to transform unstructured JSON and XML data into structured CSV file. Furthermore, it requires some join operation to form single relational dataset for final modelling. Estimated income is generated according to average salary from Payscale Website according to work profile, work experience and location, and spent income is generated by different economical measures such as location, relationship status and work profile (Fig. 2).

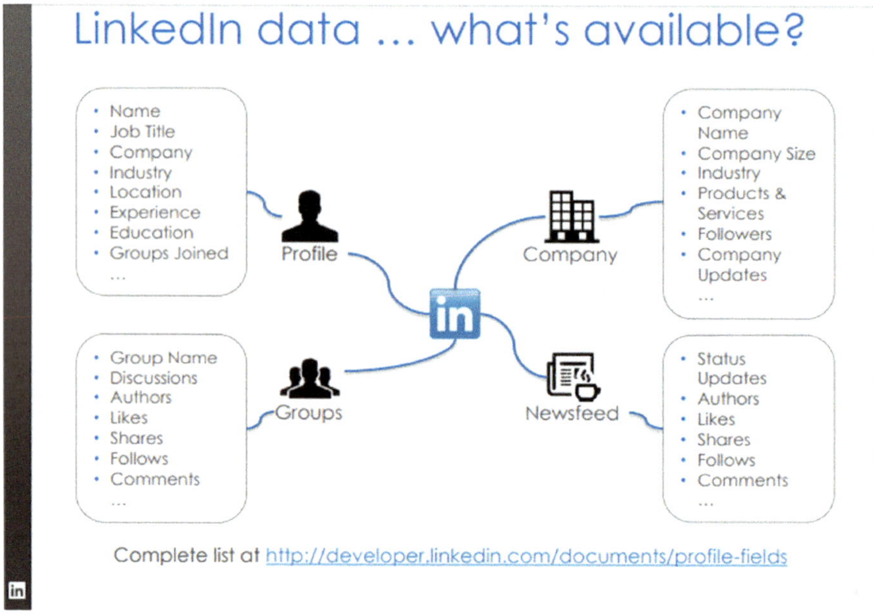

Fig. 2 Types of data available from LinkedIn API [14]

3.3 *Data Preprocessing and Data Visualizations*

Data preprocessing activities include missing data handling, remove noise and unwanted data and normalize the data for the smooth computation. Data visualization activities performed on the data to find visual insight of relations and business analysis.

3.4 *Data Modelling*

In data modelling phase, we have applied various clustering algorithms on estimated earnings and estimated spent. We have decided number of clusters according to various methods like elbow method and dendograms.

3.5 *Recommendation by Economical Measures*

Our recommendation engine provides mutual funds by cluster formation and risk analysis for time horizon.

4 Proposed Algorithms

We have applied various unsupervised algorithms such as K-means clustering and hierarchical clustering on estimated income and estimated spent. The main aim of both approaches is to find groups according to user income to spent details in the data and the number of groups represented by the variable K which are decided as per various methods. The algorithm works iteratively and finds the nearest objects and forms a cluster by smallest Euclidean distance with centroid.

4.1 K-Means Algorithm

Optimal numbers of clusters are decided by elbow method which decides clusters based on minimal within-cluster sum of square (WSS). The optimal number of clusters is selected by following approach:

Compute clustering algorithm to find number K.
Calculate within-cluster sum of square WSS.
Find number of clusters by minimal WSS.

In K-means algorithm, different data points are placed into clusters by feature similarities. K-means algorithm trained based on smallest Euclidean distance between centroid and data point. K-means algorithm is iterative algorithm which performs calculation of centroid every time when new data point is added in cluster. Collection of feature similarities represented as clusters, and centroid of a cluster represented as qualitative measures (Figs. 3, 4 and 5).

4.2 Hierarchical Clustering

Hierarchical clustering finds each specific observation as separate cluster. Then, it identifies the two closest clusters by Euclidean distance and merges the two most similar clusters. It converges when all of the clusters are merged accordingly. The hierarchical relationship of clusters is represented as dendograms.

There are two approaches for hierarchical clustering

1. Bottom-Up Approach (Agglomerative clustering)
2. Top-Down Approach (Divisive Clustering)

Both of these methods have used different linkage methods such as min distance, max distance, inter cluster distance and group average.

Figures 6 and 7 illustrated dendogram for our proposed solution.

In Dendogram, users are stated on the X-axis, and on the Y-axis, Euclidean distance is given. At the first step, there are users at the bottom, various vertical lines up to

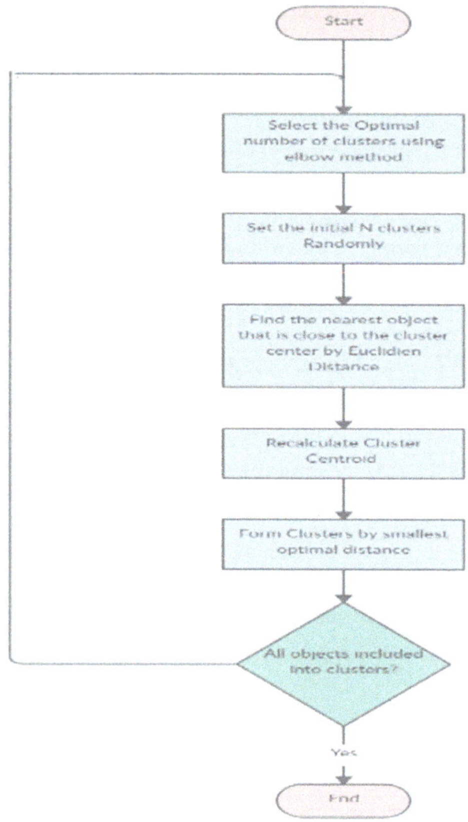

Fig. 3 Flowchart of *K*-Means algorithm with elbow method

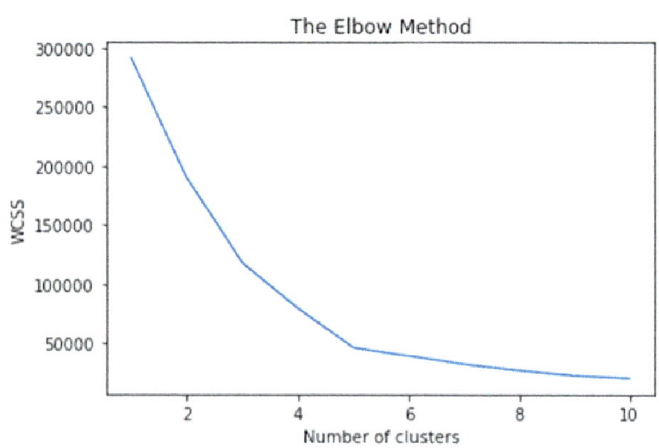

Fig. 4 Elbow method plot

1. Initialize **cluster centroids** $\mu_1, \mu_2, \ldots, \mu_k \in \mathbb{R}^n$ randomly.
2. Repeat until convergence: {

For every i, set
$$c^{(i)} := \arg\min_j ||x^{(i)} - \mu_j||^2.$$

For each j, set
$$\mu_j := \frac{\sum_{i=1}^{m} 1\{c^{(i)} = j\} x^{(i)}}{\sum_{i=1}^{m} 1\{c^{(i)} = j\}}.$$

}

Fig. 5 K-mean algorithm [15]

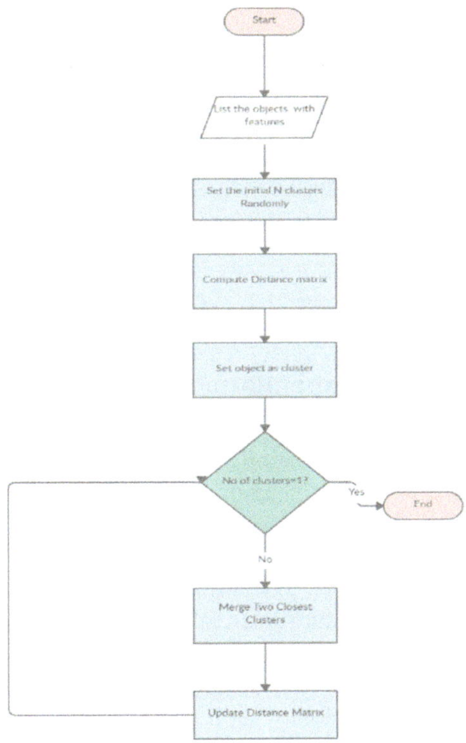

Fig. 6 Flowchart of hierarchical algorithm

horizontal lines are formed, and height of those horizontal lines tell us the distance which this users are merged cluster.

- Here, horizontal lines are cluster merges, while vertical lines are for clusters were part of merge forming that new cluster
- Heights of the horizontal lines give distance which needed to form the new cluster.

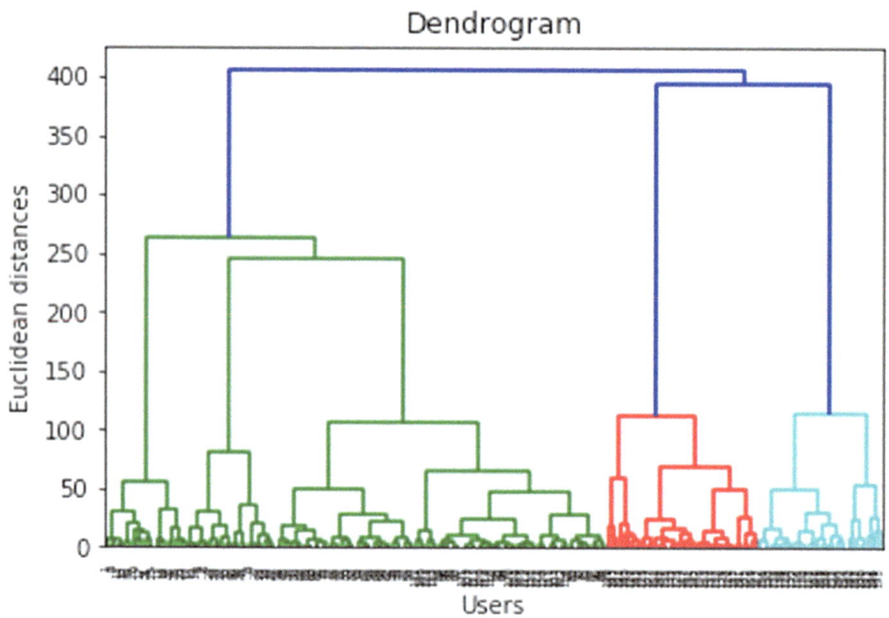

Fig. 7 Dendogram

5 Recommendation Process

Our recommendation system gives recommendation of mutual funds according to market future potential, crude prices and currency potential along with mutual funds top holdings analysis and sectorial analysis for specific mutual funds which are predicted by Long-Short Term Memory (LSTM). We have taken various historic data of Sensex, Nifty, BSE Midcap, BSE Smallcap, Crude Oil and trained LSTM model to predict the future potential prices. Additionally, we have evaluated positive potential by training LSTM on top ten holdings of specific mutual funds. Furthermore, we have evaluated all outcomes and recommendations with some economical measures which are as following (Fig. 8).

1. Fundamental Analysis
   Analysts view mutual funds by their performance and outlook which provided by the asset management companies.
2. Rating of mutual funds and Risk Ratings
   Various ratings are decided according to future potential of mutual fund which is decided by future stock price prediction rather than previous year returns and many other factors like cash ratios and beta of mutual funds.
3. Sentimental analysis of stock markets

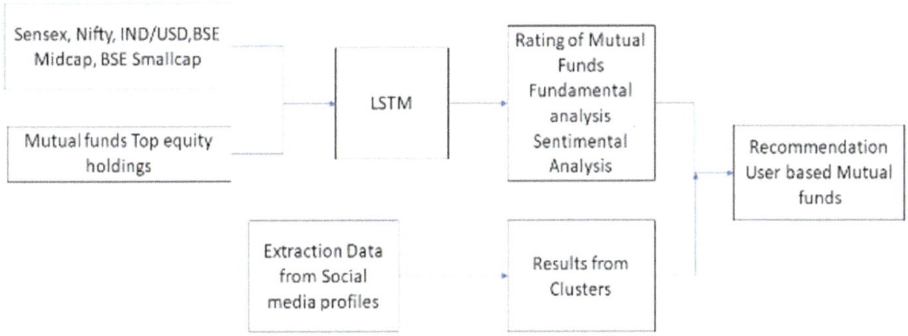

Fig. 8 Recommendation process

Sentiment analysis on different stock market news, company-specific news, sectorial news by natural language processing methods where we have classified news outcomes in 1 (Positive) and 0 (Negative). We have trained LSTM model on textual data after performing data analysis and feature engineering.

6 Results

As the first step of evaluation, user segmentation produced by clustering algorithms in which users are segmented according to their social media profiles, estimated income and spent (Fig. 9).

As the second step of evaluation, we recommend different mutual funds according to clusters by their cluster category. We have filtered out various mutual funds by their risk rating, fundamental analysis and potential future outcome which predicted by LSTM [16].

Fig. 9 Cluster results

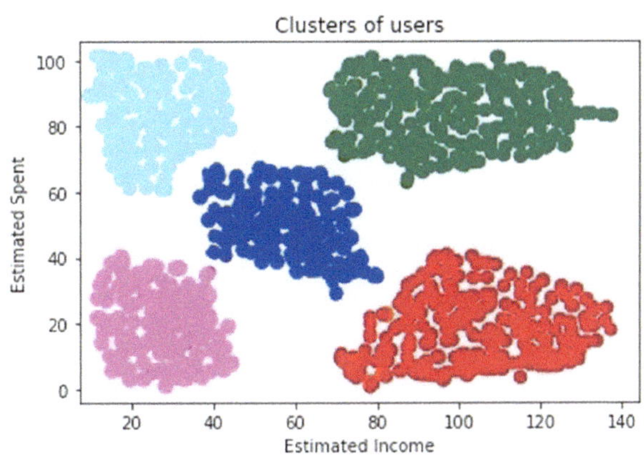

Table 1 Model results

Stock/indices name	Train (RMSE)	Test (RMSE)
Sensex	0.01	0.01
Nifty	0.03	0.04
RIL	0.01	0.01
HDFC Bank	0.01	0.01
Wipro	0.02	0.01
TCS	0.03	0.04
SBI	0.01	0.02
ICICI Bank	0.02	0.032

Results of stock predictions are as following, and we rate different mutual funds according to future potential of their holdings instead of their past performance (Table 1).

6.1 *Original Price Versus Model Price Results*

See Figs. 10, 11, 12, 13 and 14; Table 2.

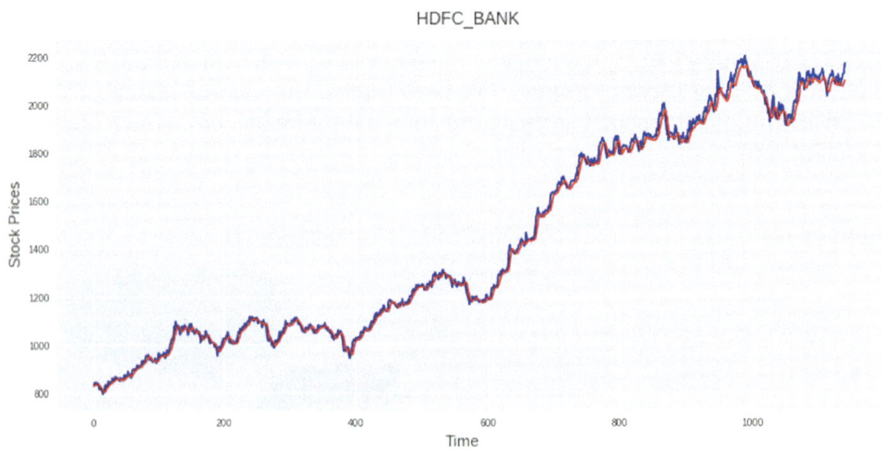

Fig. 10 HDFC Bank

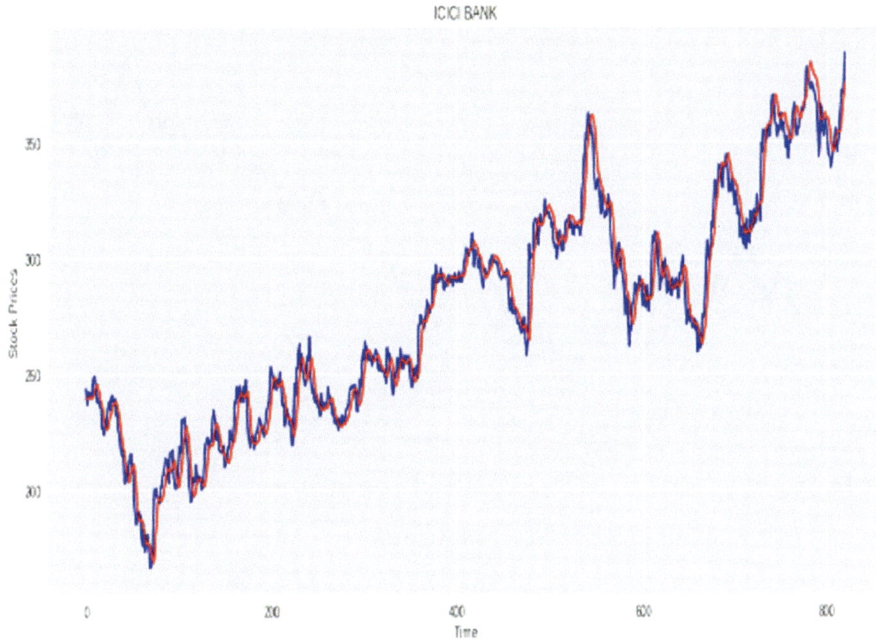

Fig. 11 ICICI Bank

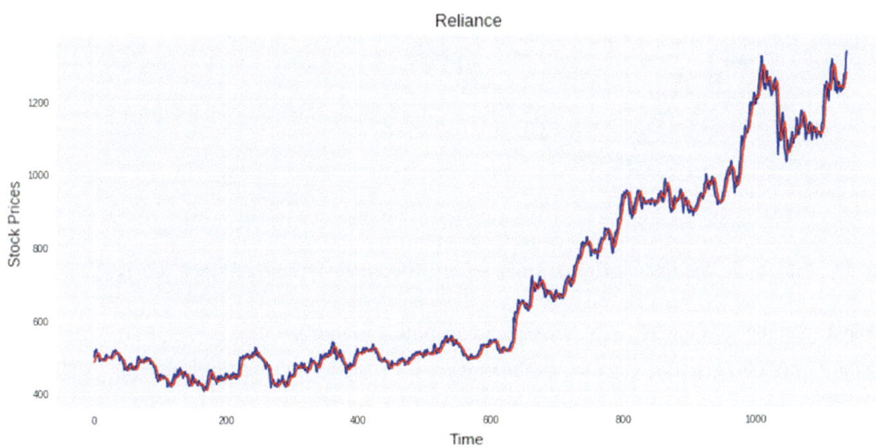

Fig. 12 Reliance industries

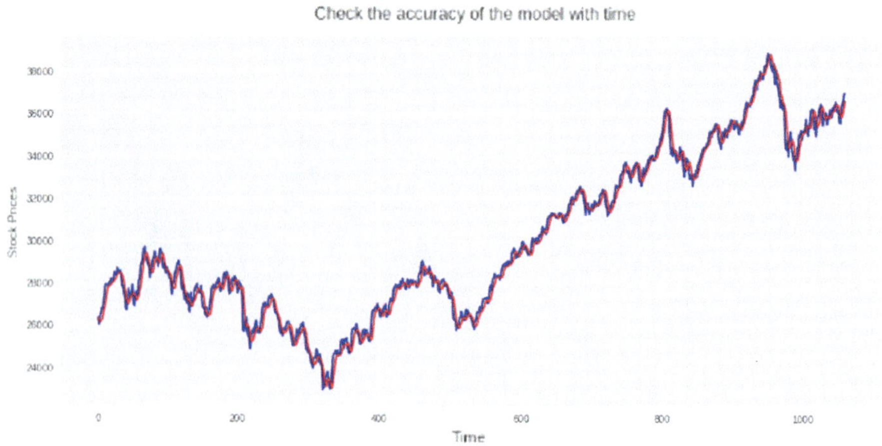

Fig. 13 Sensex

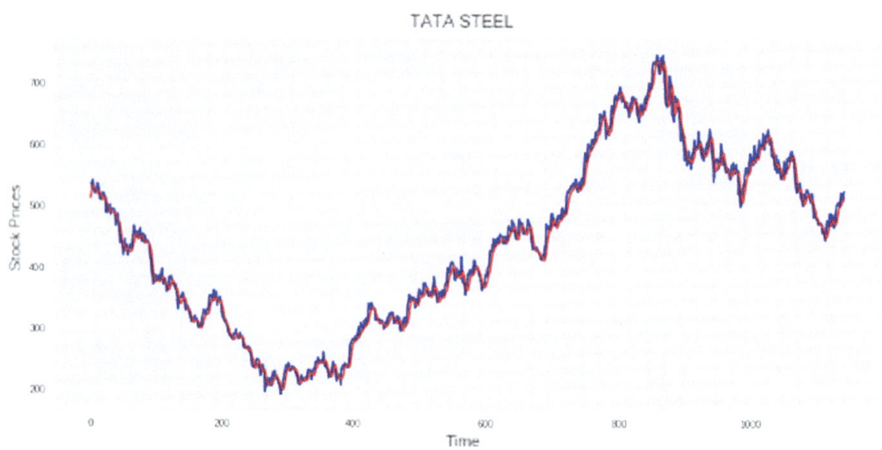

Fig. 14 Tata steel. Red line—original stock prices. Blue line—predicted stock prices

Table 2 Number of mutual funds recommend according to profile

Categories of clusters (By savings)	Low	Moderate low	Moderate	Moderate high	High
Low	2	1	1	1	0
Moderate low	1	2	1	1	0
Moderate	0	1	2	1	0
Moderate high	0	1	1	2	1
High	0	1	1	1	2

Appendix: List of Mutual Fund by Category

Low Risk Funds with Decent and Good Ratings

- Indiabulls Liquid Fund
- Reliance Liquid Fund
- DSP Blackrock Liquidity Fund
- ICICI Prudential Liquid Fund
- Indiabulls Short-Term Fund
- Axis Liquid Fund
- Tata Money Market Fund
- UTI Liquid Cash Plan
- Edelweiss Liquid Fund
- JM Liquid Fund
- HDFC Liquid Fund
- Aditya Birla Sun Life Money Manager
- TATA Liquid fund
- IDFC Cash Fund
- HDFC Money Market fund.

Moderate Low Risk Funds

- HDFC Short-Term Debt Fund
- Reliance Low Duration Fund
- ICICI Prudential Money Market Fund
- ICICI Prudential Savings Plan Directly
- Kotak Banking & PSU Debt Fund
- Relaince Short-Term Fund
- SBI Magnum Constant Maturity Fund
- L&T Short-Term Bond Fund
- BOI AXA Ultra Short Duration Fund
- Relaince Arbitrage fund
- HDFC Floating Rate Debt Fund
- Kotak Equity Arbitrage Fund
- Edelweiss Arbitrage Fund.

Moderate Risk Funds

- Franklin India Low Duration Fund
- Franklin India Ultra Short Bond Fund

- Franklin India STIP
- ICICI Prudential Balanced Advantage
- ICICI Prudential All Seasons Bond Fund
- Aditya Birla Sun Life Banking & PSU Debt
- Franklin India Credit Risk Fund
- SBI Magnum Constant Maturity Fund
- Axis Strategic Bond Fund
- SBI Magnum Medium Duration Fund
- Reliance Gilt Securities Fund
- HDFC Medium Term Debt Fund
- HDFC Corporate Bond Fund
- Franklin India Income Opportunities Fund.

Moderate High Funds

- HDFC Small Cap Funds
- Mirae Asset Emerging Bluechip Fund
- SBI Bluechip Fund
- Motilal Oswal Multicap 35 Fund
- Franklin India Prima Fund
- DSP Blackrock Tax Saver Fund
- HDFC Balanced Funds
- IDFC Tax Advantage (ELSS)
- ICICI Prudential Equity & Debt Fund
- Reliance Equity Hybrid Fund
- Aditya Birla Sun Life Pure Value Fund
- ICICI Prudential Nifty Next 50 Index Fund.

High Risk Funds

- L&T Emerging Businesses Fund
- TATA Digital India Fund L&T MidCap Fund
- UTI Transportation & Logistic Funds
- DSP BlackRock Natural Resources Fund
- Franklin Build India Fund
- L&T Infrastructure Fund
- SBI Technologies Opportunities Fund.

References

1. Bogdan B, Treleaven PC (2015) Social media analytics: a survey of techniques, tools and platforms. AI & Soc 30:89–116. 10.1007/s00146-014-0549-4
2. Karabulut Y (2013) Can Facebook predict stock market activity? Goethe University Frankfurt This Draft, 21 Oct 21
3. Yang H, Liu XY, Wu Q (2018) A practical machine learning approach for dynamic stock recommendation. In: 2018 17th IEEE international conference on trust, security and privacy in computing and communications/12th IEEE international conference on big data science and engineering (TrustCom/BigDataSE). IEEE, pp 1693–1697
4. Batres-Estrada G (2015) Deep Learning for Multivariate Financial Time Series 4 June 2015
5. Chen K, Zhou Y, Dai F (2015) A LSTM-based method for stock returns prediction: a case study of China stock market. In: 2015 IEEE international conference on big data (Big Data). IEEE, pp 2823–2824
6. Coyne S, Madiraju P, Coelho J (2017) Forecasting stock prices using social media analysis. In: 2017 IEEE 15th international conference on dependable, autonomic and secure computing, 15th international conference on pervasive intelligence and computing, 3rd international conference on big data intelligence and computing and cyber science and technology congress (DASC/PiCom/DataCom/CyberSciTech). IEEE, pp 1031–1038
7. Culkin R, Das SR (2017) Machine learning in finance: the case of deep learning for option pricing. J Investment Manag 15(4):92–100
8. Chatzis SP, Siakoulis V, Petropoulos A, Stavroulakis E, Vlachogiannakis N (2018) Forecasting stock market crisis events using deep and statistical machine learning techniques. Expert Syst Appl 112:353–371
9. León C, Kim GY, Martínez C, Lee D (2017) Equity markets' clustering and the global financial crisis. Quant Fin 17(12):1905–1922
10. Nayak A, Pai MM, Pai RM (2016) Prediction models for Indian stock market. Procedia Comput Sci 89:441–449
11. Munjal P, Kumar S, Kumar L, Banati A (2017) Opinion dynamics through natural phenomenon of grain growth and population migration. In: Hybrid intelligence for social networks. Springer, Cham, pp 161–175
12. Munjal P, Narula M, Kumar S, Banati H (2018) Twitter sentiments based suggestive framework to predict trends. J Stat Manag Syst 21(4):685–693
13. Munjal P, Kumar L, Kumar S, Banati H (2019) Evidence of Ostwald Ripening in opinion driven dynamics of mutually competitive social networks. Phys Stat Mech Appl
14. www.valueresearchonline.com
15. http://stanford.edu/~cpiech/cs221/handouts/kmeans.html
16. Diwaker C et al (2018) Prediction of software reliability using bio inspired soft computing techniques. J Med Syst 42(5):93

Chapter 27
Hyperspectral Endmember Extraction Algorithm Using Spearman's Rank Correlation

Dharambhai Shah and Tanish Zaveri

1 Introduction

Imaging spectroscopy, also known as Hyperspectral imaging, captures tens or hundreds of wavelengths at continuous narrow bandwidth in contrast to multispectral imaging which captures only a few wavelengths at discrete bands. This capacity of hyperspectral sensors is crucial to elaborate on the observations of the earth in remote sensing. Many remote sensing applications like vegetation [16], water quality [11], mineral mapping [13], soil study [10] are greatly influenced by this hyperspectral imaging technology. A hyperspectral image is mainly considered as a data cube as shown in Fig. 1. Hyperspectral image is three dimension (3D) image with two spatial dimensions and one spectral dimension. These two types of information (spatial and spectral) in the hyperspectral data cube makes identification and interpretation of ground-based objects easy. But due to limited spatial resolution, microscopic mixtures of materials on the ground and multiple scattering, mixed pixels widely exist in the remotely sensed hyperspectral images. Mixed pixels are pixels which contains more than one material. Due to these mixed pixels, the accuracy of identification and classification of ground objects from remote sensors is negatively affected. The decomposition of these mixed pixels into pure materials along-with their fractions is called as Spectral Unmixing (SU).

There are two types of spectral unmixing [3]. The first one is based on Linear Mixing Model (LMM) which assumes linear combinations of pure materials in the mixing. The second one is Non-Linear Mixing Model (NLMM) which is assuming non-linear combinations. Most of the researchers assume the problem as LMM as it is

D. Shah (✉) · T. Zaveri
Institute of Technology, Nirma University, Ahmedabad, Gujarat, India
e-mail: shahdharam7@gmail.com

T. Zaveri
e-mail: ztanish@gmail.com

© Springer Nature Singapore Pte Ltd. 2021

S. D. Purohit et al. (eds.), *Proceedings of International Conference on Communication and Computational Technologies*, Algorithms for Intelligent Systems,
https://doi.org/10.1007/978-981-15-5077-5_27

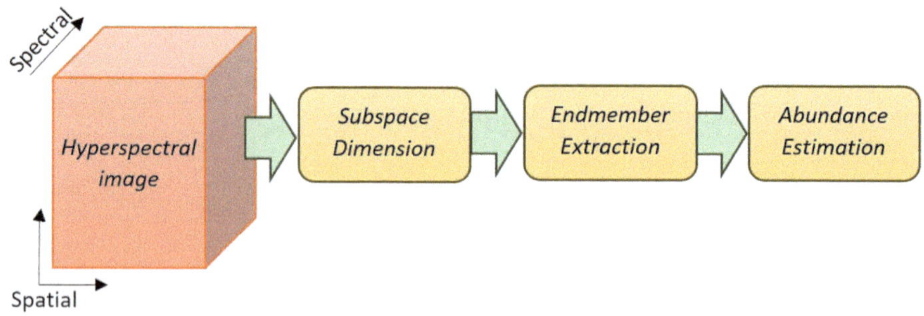

Fig. 1 Spectral Unmixing

a simple and acceptable approximation to the real-world scenario. Blind hyperspectral unmixing can be considered as three sub-problems [1] as shown in Fig. 1. The first sub-problem, Hyperspectral Subspace Dimension, estimates the number of pure materials in the image. The second sub-problem, Endmember Extraction, extracts pure endmember signature from the image. The final one, Abundance Estimation, estimates fractions of each material in all pixels. The focus of this paper is on the second block.

In the last decade, many researchers have attempted endmember extraction sub-problem by assuming LMM with convexity based and orthogonal projection-based approach. Many well-known algorithms like N-point Finder (NFINDR) [18], Simplex Growing Algorithm (SGA) [5], Successive Volume Maximization (SVMAX) [4], Alternating Volume Maximization (AVMAX) [4] N-point Finder (NFINDR) [18], Simplex Growing Algorithm (SGA) [5], Successive Volume Maximization (SVMAX) [4], Alternating Volume Maximization (AVMAX) [4] are developed using convexity based criteria because purity of endmembers. Pixel Purity Index (PPI) [7] algorithm explores the convexity using orthogonal projection score. Independent Component Analysis (ICA) [17] assumes the pure materials as an independent component for endmember extraction. The concept of orthogonal projection is further explored by Automatic Target Generation Procedure (ATGP) [14].

A novel algorithm which explores the convexity criteria using Spearman's Rank Correlation is proposed. Spearman's Rank Correlation Endmember Extraction (SRCEE) algorithm calculates correlation coefficient proposed by spearman [15] to find highly non-correlated bands. The proposed algorithm applies the concept of convexity into these highly non-correlated bands.

The rest of the paper is organized as follows. Section 2 formulates the SU problem mathematically with convex geometry background. Section 3 proposes a novel algorithm, Spearman's Rank Correlation Endmember Extraction (SRCEE). Section 4 includes the definition of evaluation parameters along with simulation results. Finally, the paper ends with concluding remark.

2 Problem Statement and Background Concepts

2.1 Problem Statment

Let $\mathbf{y}$ be a mixed pixel vector of size $L \times 1$ defined as,

$$\mathbf{y} = \mathbf{M}\mathbf{a} + \mathbf{n} \tag{1}$$

where, $\mathbf{M}$ is the matrix of pure spectral signature with size $L \times Q$. The fractional abundance vector $\mathbf{a} = [a_1, a_2, \ldots, a_Q]^T$ is of size $Q \times 1$. $\mathbf{n}$ is the additive perturbation (errors due to modelling and noise) vector of size $L \times 1$. L is number of bands in hyperspectral image. Q is number of pure materials in the data. There are various algorithms [2, 6] which can find values of Q. In HU jargon, the fractional abundance vector is limited with the following two constraints.

1. Abundance Non-negativity Constraint (ANC)

$$a_i \geq 0, \qquad i = 1, 2, \ldots, Q \tag{2}$$

2. Abundance Sum-to-one Constraint (ASC)

$$\sum_{i=1}^{Q} a_i = 1 \tag{3}$$

2.2 Background Concepts

In mathematics, the concepts of convex sets in the Euclidean space is covered in the convex geometry. In concepts of convex geometry, the intersection of all affine sets containing set $\mathbf{S}$ is defined as the affine hull for set $\mathbf{S}$. All affine combinations of elements of $\mathbf{S}$ make the affine hull aff($\mathbf{S}$) as follow,

$$\text{aff}(\mathbf{S}) = \left\{ \sum_{i=1}^{Q} a_i s_i \;\middle|\; s_i \in \mathbf{S},\ \sum_{i=1}^{Q} a_i = 1, \forall_i : a_i \in \mathbb{R} \right\} \tag{4}$$

It has been observed from (4) and (1) that both equations are giving weight to some values and creating mixed values. But the difference between both is that weighting value in (1) is non-negative as per (2). While in (4), it takes values from $\mathbb{R}$ which can be positive or negative. This non-negativity constraint can be added in the affine hull if we consider positive weight values. Instead of using affine hull, mathematicians generally uses convex combinations which considers only positive real numbers ($\mathbb{R}_+$) for abundances. Convex hull of $\mathbf{S}$ is defined as,

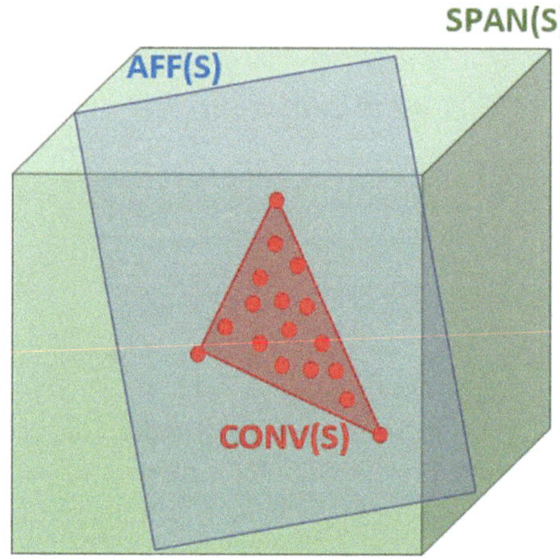

Fig. 2 Affine hull and Convex hull

$$\text{conv}(\mathbf{S}) = \left\{ \sum_{i=1}^{Q} a_i s_i \;\middle|\; s_i \in \mathbf{S},\ \sum_{i=1}^{Q} a_i = 1, \forall_i : a_i \in \mathbb{R}_+ \right\} \tag{5}$$

The above concept of convex geometry in the linear mixing model (1) can be easily explained using Fig. 2 of affine hull and convex hull for $Q = 3$. Convex hull considers only positive wights so that it is triangle while the affine hull is considering all real values so that it is plane. The span of the set is the whole cube which is not considering (2) and (3). It is clear from Fig. 2 that convex combination is best suited to mixed pixel vector ($\mathbf{y}$) than affine combination. This geometrical view is exploited by many endmember extraction algorithms.

3 Proposed Algorithm

The proposed algorithm is implemented in four steps. The first step normalizes each band of the data to make all bands uniform. The second step selects two bands based on minimum Spearman's Rank Correlation coefficient. The third step finds convex set points. Final step removes extra points if exists. The flow chart of the proposed algorithm is shown in Fig. 3. Original hyperspectral image of size $M \times N \times L$ is converted into 2-dimensional matrix ($\mathbf{Y}$) of size $L \times P$ before processing.

$$\mathbf{Y} = [\mathbf{b_1}, \mathbf{b_2}, \ldots, \mathbf{b_L}]^T = \begin{bmatrix} b_1^1 & b_1^2 & b_1^3 & \ldots & b_1^P \\ b_2^1 & b_2^2 & b_2^3 & \ldots & b_2^P \\ \vdots & \vdots & \vdots & \ddots & \vdots \\ b_L^1 & b_L^2 & b_L^3 & \ldots & b_L^P \end{bmatrix} \tag{6}$$

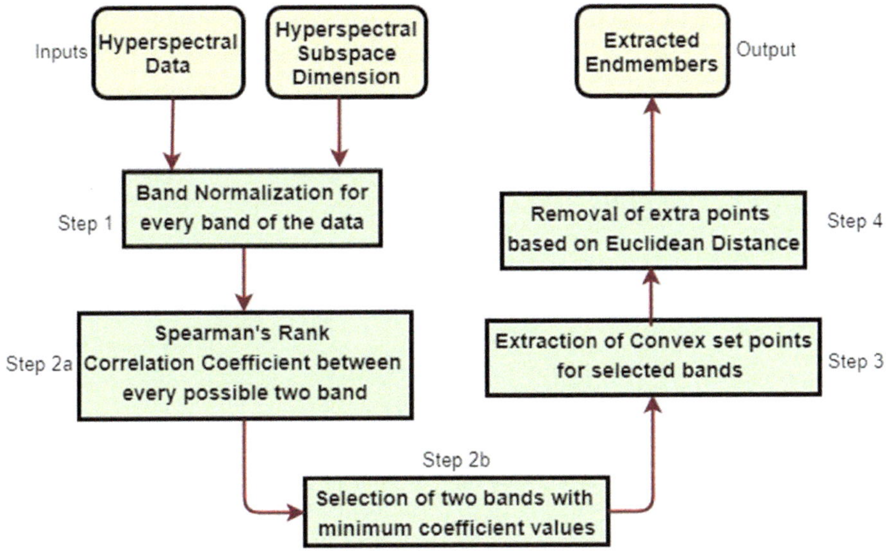

Fig. 3 Flowchart of proposed algorithm

3.1 Band Normalization

Each band of hyperspectral data has a wide dynamic range. To make all bands uniform, all bands are normalized as follows.

$$\bar{b}^j = \frac{b^j - \mu_i}{\sigma_i}, \forall j \in \mathbf{b}_i \tag{7}$$

Equation (7) converts each pixel b^j of ith band ($\mathbf{b}_i$) to $\bar{b}^j$ in the hyperspectral image $\mathbf{Y}$. In this equation, μ_i and σ_i represents mean and standard deviation for band ($\mathbf{b_i}$). Each pixel b^j is replaced by $\bar{b}^j$ in the hyperspectral data. Each new band is represented as $\bar{\mathbf{b}}_i$ instead of $\mathbf{b}_i$.

3.2 Band Selection Based on Spearman's Rank Correlation Coefficient

In hyperspectral image $\mathbf{Y}$, bands are considered as variables. Spearman's Rank Correlation coefficient ρ between two bands $\mathbf{b}_i$ and $\mathbf{b}_j$ is calculated as,

$$\rho_{\mathbf{ij}} = \frac{\text{Var}(\mathbf{b}_i, \mathbf{b}_j)}{\sigma_i \sigma_j} \tag{8}$$

where, Var shows variance between two bands. The above equation is computed repeatedly for all possible combinations of bands. Two bands ($\mathbf{b}_m$ and $\mathbf{b}_n$) are selected which is having minimum Spearman's Rank Correlation coefficient ρ as per Eq. (8).

3.3 Convex Set Points

Convex set points in 2-dimensional euclidean space are boundary points of dataset. The optimum convex set can be found using following optimization problem.

$$\begin{aligned} &\underset{i \in T}{\text{minimize}} \ \text{boundary}(S, i) \\ &\text{subject to } N_S \geq Q \end{aligned} \tag{9}$$

where, boundary(S, i) is basically finding convex set points as per (5) with shrinking factor i. Shrinking factor is ranging from zero to one. zero value of shrinking factor is for null shrink while one is for highest shrink. In optimization problem, $T = \{0, 0.001, 0.002, \ldots, 1\}$. N_S is the number of convex set points from *boundary* function.

3.4 Removal of Extra Points

The number of points (N_S) extracted by the algorithm should be equal to Q. The reason for putting constraint in (9) is that the number of materials extracted by the algorithm should not be less than the desired one. Due to this constraint, there are two possibilities ($N_S = Q$ and $N_S > Q$) in Eq. (9).

1. $N_S > Q$: This is the case where the number of materials (N_S) extracted from the algorithm is greater than the subspace dimension (Q). In this case, extra $(N_S - Q)$ points are removed using Euclidean Distance. The points which are at close Euclidean Distance from its neighbour are removed.
2. $N_S = Q$: This is the perfect scenario where the number of extracted materials from the (9) is the same as the subspace dimension. In this case, no points are removed.

The algorithm returns matrix $\bar{\mathbf{M}}$ which has Q pure materials of L-dimension.

4 Simulation Results

This section includes evaluation parameters used for comparison and discussion on simulation results with a synthetic and real dataset.

4.1 Evaluation Parameters

The output of extracted algorithm is $\bar{\mathbf{M}} = [\bar{\mathbf{m}}_1, \bar{\mathbf{m}}_2, \ldots, \bar{\mathbf{m}}_Q]$. Each endmember in mixing matrix is represented as $\bar{\mathbf{m}}_i = [\bar{m}_i^1, \bar{m}_i^2, \ldots \bar{m}_i^L]^T$ while Ground Truth (GT) endmember is $\mathbf{m}_i = [m_i^1, m_i^2, \ldots m_i^L]^T$. Endmember extraction algorithms are compared based on following three standard evaluation parameters.

1. Spectral Angle Error (SAE) : SAE between two vectors $\mathbf{m_i}$ and $\bar{\mathbf{m}}_i$ is defined as,

$$\theta_i = \cos^{-1}\left(\frac{\mathbf{m_i} \cdot \bar{\mathbf{m}}_i}{|\mathbf{m_i}||\bar{\mathbf{m}}_i|}\right) \tag{10}$$

   SAE verctor $\boldsymbol{\theta} = [\theta_1, \theta_2, \ldots \theta_Q]^T$ of size $Q \times 1$ is defined for all Q endmembers.
2. Spectral Information Divergence (SID) : SID is an information theory-based measurement parameter to test spectral variability between two spectra. Probability vectors $\mathbf{p} = [p^1, p^2, \ldots p^L]^T$ and $\mathbf{q} = [q^1, q^2, \ldots q^L]^T$ are defined for L-dimensional two spectra $\mathbf{m}$ and $\bar{\mathbf{m}}$ respectively. Each component of probability vectors is defined as,

$$p^i = \frac{m^i}{\sum_{i=1}^{L} m^i} \quad , \quad q^i = \frac{\bar{m}^i}{\sum_{i=1}^{L} \bar{m}^i} \tag{11}$$

   SID value between two spectra $\mathbf{m_i}$ and $\bar{\mathbf{m}}_i$ is defined as,

$$\phi_i = \sum_{j=1}^{L} p^j \log\left(\frac{p^j}{q^j}\right) + \sum_{j=1}^{L} q^j \log\left(\frac{q^j}{p^j}\right) \tag{12}$$

   The first term in (12) is the relative entropy of $\mathbf{m_i}$ with respect to $\bar{\mathbf{m}}_i$. Q-dimensional SID vector $\boldsymbol{\phi} = [\phi_1, \phi_2, \ldots \phi_Q]^T$ is SID error vector defined for all Q endmembers.
3. Normalized Cross Correlation (NXC) : NXC between two spectra $\mathbf{m_i}$ and $\bar{\mathbf{m}}_i$ of size $L \times 1$ is defined as,

$$\nu = \frac{1}{(L-1)} \times \sqrt{\frac{\sum_{i=1}^{L} (m^i - a_m)(\bar{m}^i - a_{\bar{m}})}{s_m \times s_{\bar{m}}}} \tag{13}$$

   Here, a_m and $a_{\bar{m}}$ are average or mean values of spectra $\mathbf{m}$ and $\bar{\mathbf{m}}$ respectively. s_m and $s_{\bar{m}}$ are standard deviation of spectra $\mathbf{m}$ and $\bar{\mathbf{m}}$ respectively. Normalized Cross Correlation vector is represented as ($\boldsymbol{\nu} = [\nu_1, \nu_2, \ldots \nu_Q]^T$).

Parameters (θ, ϕ, ν) are defined between two vectors. But to evaluate for all vectors, Root Mean Square (RMS) of all vectors is used. Root Mean Square (RMS) value of SAE vector defined in (14) is RMSSAE [12]. Similarly, RMSSID [12] and RMSNXC are defined in (15) and (16) respectively.

$$\text{RMSSAE} : \epsilon_{\theta} = \left(\frac{1}{Q} E[\| \boldsymbol{\theta} \|_2^2]\right)^{1/2} \tag{14}$$

$$\text{RMSSID} : \epsilon_{\phi} = \left(\frac{1}{Q} E[\| \boldsymbol{\phi} \|_2^2]\right)^{1/2} \tag{15}$$

$$\text{RMSNXC} : \epsilon_{\nu} = \left(\frac{1}{Q} E[\| \boldsymbol{\nu} \|_2^2]\right)^{1/2} \tag{16}$$

where $E[.]$ denotes the expectation operator.

4.2 Discussion

4.2.1 Simulation with Synthetic Dataset

Hyperspectral Imagery Synthesis toolbox (HIST) package [9] is used to generate different images. Five different distributions (Rational Gaussian (Fig. 4a), Spheric Gaussian (Fig. 4b), Matern Gaussian (Fig. 4c), Exponential Gaussian (Fig. 4d), Legendre (Fig. 4e) are used to test proposed algorithm on variety of data. All five images are of size $128 \times 128 \times 431$. Five endmembers (vinyl-plastic-gds372, fiberglass-gds374, brick-gds350, sheet metal-gds352, asphalt-gds367) as shown in Fig. 5 are selected from the USGS spectral library [8] as GT endmembers in this five images.

Endmembers extracted from all algorithms are the same as GT endmembers as shown in Fig. 5. No comparison is made for the synthetic dataset as all algorithms output is the same as GT. This simulation with synthetic dataset just verifies that the proposed algorithm is working as good as other algorithms.

4.2.2 Simulation with Real Dataset

All algorithms are evaluated on two real datasets (Urban and Jasper) in this simulation. The detail description of both the datasets is in [19]. Brief description of two real datasets is as follow.

(a) Rational (b) Spheric (c) Matern (d) Exponential (e) Legendre

Fig. 4 Synthetic dataset

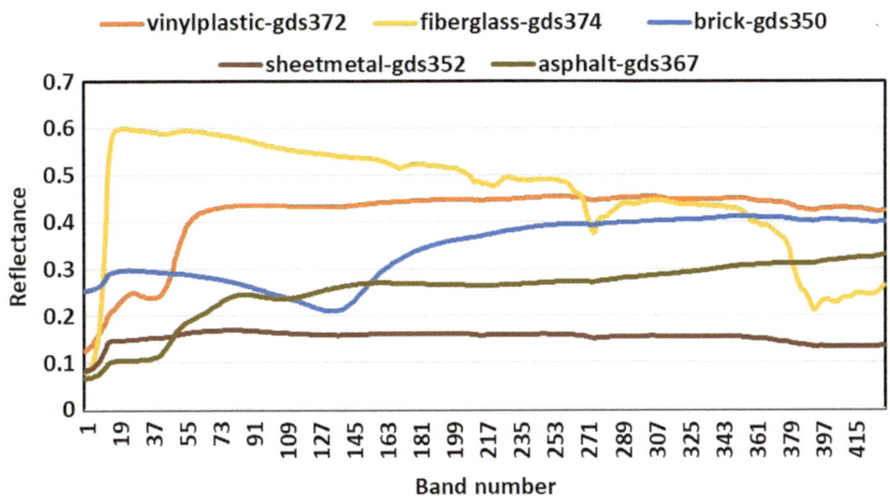

Fig. 5 Spectral signature of five USGS materials

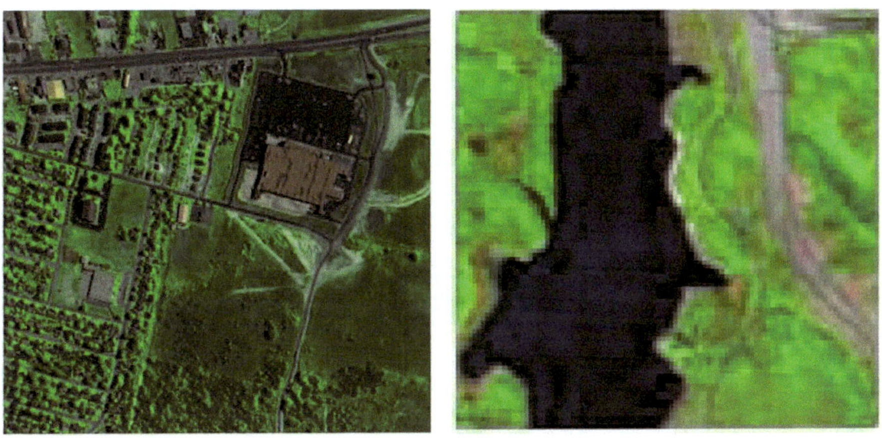

(a) Urban (b) Jasper

Fig. 6 Real dataset

- **Urban** image is popular image in the hyperspectral unmixing study. After removal of low Signal to Noise Ratio (SNR) bands, the image of size 307 × 307 × 162 is processed [19]. The image shown in Fig. 6a consists of six endmembers defined in Table 1.
- **Jasper** image captured by Airborne Visible/InfraRed Imaging Spectrometer (AVIRIS) sensor is collected over the Jasper bridge. The image of size 95 × 95 × 156 is taken after removal of noisy and water absorption bands [19]. The image shown in Fig. 6b has four endmembers defined in Table 2.

Table 1 Urban GT

Notation	Endmember
U_1	Asphalt
U_2	Grass
U_3	Tree
U_4	Roof
U_5	Metal
U_6	Dirt

Table 2 Jasper GT

Notation	Endmember
J_1	Road
J_2	Soil
J_3	Water
J_4	Tree

Table 3 Urban dataset results

Parameter	Algorithm	U_1	U_2	U_3	U_4	U_5	U_6	RMS
SAE	NFINDR	3.490	6.213	18.065	33.606	9.829	3.325	16.402
	SGA	3.490	2.706	8.934	39.911	22.254	3.419	19.145
	SVMAX	3.490	2.706	8.934	30.342	9.829	3.261	13.707
	AVMAX	3.490	6.213	40.521	23.886	8.683	3.261	19.788
	PPI	3.490	2.706	8.934	26.311	8.683	3.179	12.091
	ICA	36.259	6.007	3.754	37.110	6.927	19.955	23.051
	ATGP	3.490	2.706	8.934	39.911	22.254	3.419	19.145
	Proposed	5.098	6.213	18.046	3.223	9.441	3.179	**9.128**
SID	NFINDR	0.235	0.894	8.572	25.281	1.729	0.198	10.928
	SGA	0.235	0.182	2.660	42.053	10.119	0.204	17.692
	SVMAX	0.235	0.182	2.660	19.601	1.729	0.186	8.107
	AVMAX	0.235	0.894	40.699	10.452	1.312	0.186	17.167
	PPI	0.235	0.182	2.660	13.332	1.312	0.187	5.578
	ICA	29.103	1.280	0.758	30.120	1.003	7.778	17.406
	ATGP	0.235	0.182	2.660	42.053	10.119	0.204	17.692
	Proposed	0.528	0.894	10.710	0.199	1.595	0.187	**4.442**
NXC	NFINDR	0.950	0.988	0.908	-0.294	0.713	0.974	0.841
	SGA	0.950	0.996	0.987	-0.340	0.570	0.968	0.841
	SVMAX	0.950	0.996	0.987	-0.510	0.713	0.970	0.874
	AVMAX	0.950	0.988	0.316	-0.811	0.797	0.970	0.838
	PPI	0.950	0.996	0.987	-0.840	0.797	0.976	0.928
	ICA	0.150	0.992	0.997	0.105	0.831	0.689	0.728
	ATGP	0.950	0.996	0.987	-0.340	0.570	0.968	0.841
	Proposed	0.974	0.988	0.956	0.974	0.792	0.976	**0.946**

Table 4 Jasper dataset results

Parameter	Algorithm	J_1	J_2	J_3	J_4	RMS
SAE	NFINDR	8.930	51.299	6.654	6.102	26.424
	SGA	8.930	51.299	7.652	6.102	26.491
	SVMAX	53.801	8.330	3.397	17.510	28.645
	AVMAX	53.801	4.520	3.397	17.510	28.430
	PPI	53.801	8.330	9.651	4.165	27.724
	ICA	8.167	51.279	8.322	39.180	32.789
	ATGP	53.801	4.520	3.397	37.124	32.805
	Proposed	51.670	3.380	5.251	6.051	**26.198**
SID	NFINDR	3.290	49.747	1.163	1.067	24.940
	SGA	3.290	49.747	1.451	1.067	24.944
	SVMAX	66.373	6.466	0.221	7.604	33.560
	AVMAX	66.373	2.621	0.221	7.604	33.429
	PPI	66.373	6.466	2.401	0.238	33.365
	ICA	3.095	53.596	1.366	26.631	29.972
	ATGP	66.373	2.621	0.221	23.515	35.232
	Proposed	43.631	5.056	0.584	0.829	**21.967**
NXC	NFINDR	0.985	-0.312	0.968	0.928	0.847
	SGA	0.985	-0.312	0.948	0.928	0.841
	SVMAX	-0.307	0.995	0.987	0.766	0.813
	AVMAX	-0.307	0.995	0.987	0.766	0.813
	PPI	-0.307	0.995	0.909	0.919	0.830
	ICA	-0.399	0.992	0.916	-0.629	0.771
	ATGP	-0.307	0.995	0.987	-0.457	0.753
	Proposed	-0.316	0.987	0.965	0.956	**0.854**

The proposed algorithm is compared with prevailing algorithms (NFINDR, SGA, SVMAX, AVMAX, PPI, ICA, ATGP). The performance of all algorithms on Urban and Jasper dataset is shown in Tables 3 and 4 respectively. SAE, SID and NXC values for all endmembers ($U_1 - U_6$, $J_1 - J_4$) are calculated and tabulated for all algorithms in both tables. SAE and SID values are measured in degree. In the last columns of both the table, RMS values of all algorithms for all parameters are shown. It has been observed from both tables that the proposed algorithm is giving the least SAE and SID errors. It is also observed that GT endmembers and extracted endmembers for proposed algorithm are highly correlated as NXC value for the proposed one is highest. Comparative graphs (Figs. 7 and 8) shows that proposed algorithm outperforms.

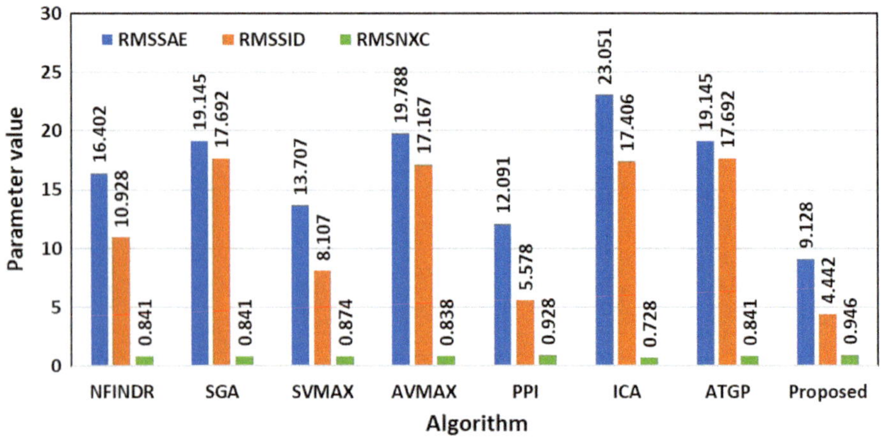

Fig. 7 Comparison of algorithms on urban dataset

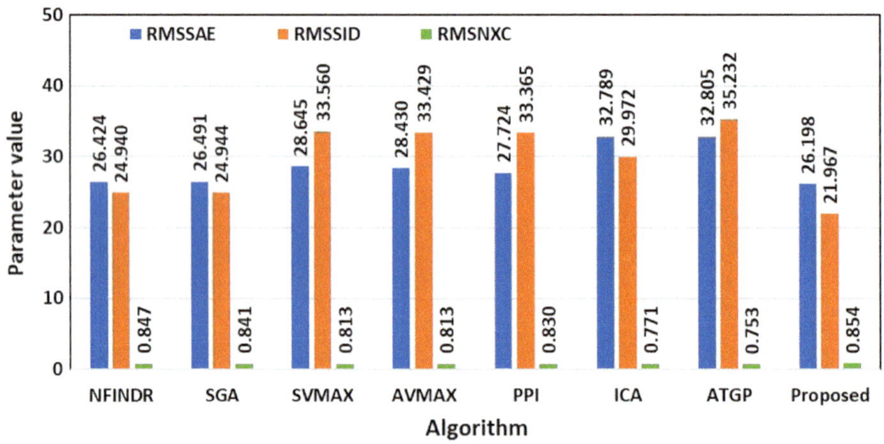

Fig. 8 Comparison of algorithms on jasper dataset

5 Conclusion

Endmember extraction from the hyperspectral image is addressed in this paper. A novel endmember extraction algorithm which explores the convexity with the Spearman's rank correlation of each band. The proposed algorithm is compared with other convexity-based approaches and orthogonal projection-based approaches. The proposed algorithm is evaluated on real and synthetic dataset. Simulation results demonstrate that the proposed algorithm outperforms other prevailing algorithms. The endmembers extracted from the proposed algorithm is highly correlated with the GT. The effectiveness of the proposed algorithm in endmember extraction is extremely useful for many applications of hyperspectral remote sensing.

Acknowledgements This work has been carried out under the grant received from Vishveshvarya Ph.D. scheme by Government of India. The authors of this paper are also thankful to the management of Nirma University for providing the necessary infrastructure and support.

References

1. Bhatt JS (2014) Novel approaches for spectral unmixing of hyperspectral data. Ph.D. thesis, Dhirubhai Ambani institute of information and communication technology
2. Bioucas-Dias JM, Nascimento JM (2008) Hyperspectral subspace identification. IEEE Trans Geosci Remote Sens 46(8):2435–2445
3. Bioucas-Dias JM, Plaza A, Dobigeon N, Parente M, Du Q, Gader P, Chanussot J (2012) Hyperspectral unmixing overview: geometrical, statistical, and sparse regression-based approaches. IEEE J Sel Top Appl Earth Obs Remote Sens 5(2):354–379
4. Chan T, Ma W, Ambikapathi A, Chi C (2011) A simplex volume maximization framework for hyperspectral endmember extraction. IEEE Trans Geosci Rem Sens 49(11):4177–4193. https://doi.org/10.1109/TGRS.2011.2141672 Nov
5. Chang CI, Wu CC, Liu W, Ouyang Y (2006) A new growing method for simplex-based endmember extraction algorithm. IEEE Trans Geosci Rem Sens 44(10):2804–2819. https://doi.org/10.1109/TGRS.2006.881803 Oct
6. Chang CI, Du Q (2004) Estimation of number of spectrally distinct signal sources in hyperspectral imagery. IEEE Trans Geosci Rem Sens 42(3):608–619
7. Chang CI, Plaza A (2006) A fast iterative algorithm for implementation of pixel purity index. IEEE Geosci Rem Sens Lett 3(1):63–67. https://doi.org/10.1109/LGRS.2005.856701 Jan
8. Clark RN, Swayze GA, Wise R, Livo KE, Hoefen T, Kokaly RF, Sutley SJ (2007) Usgs digital spectral library splib06a. US geological survey, digital data series 231:2007
9. Computational Intelligence Group (2010) University of the Basque Country / Euskal Herriko Unibertsitatea (UPV / EHU), S.: Hyperspectral imagery synthesis (eias) toolbox
10. Ghosh G, Kumar S, Saha SK (2012) Hyperspectral satellite data in mapping salt-affected soils using linear spectral unmixing analysis. J Indian Soc Rem Sens 40(1):129–136. https://doi.org/10.1007/s12524-011-0143-x Mar
11. Mishra R, Shah D, Zaveri T, Ramakrishnan R, Shah P (2017) Separation of sewage water based on water quality parameters for south karnataka coastal region. vol 2017-October. Asian Assoc Rem Sens
12. Nascimento JM, Dias JM (2005) Vertex component analysis: a fast algorithm to unmix hyperspectral data. IEEE Trans Geosci Rem Sens 43(4):898–910
13. Oskouei MM, Babakan S (2016) Detection of alteration minerals using hyperion data analysis in lahroud. J Indian Soc Rem Sens 44(5):713–721. https://doi.org/10.1007/s12524-016-0549-6
14. Ren H, Chang CI (2003) Automatic spectral target recognition in hyperspectral imagery. IEEE Trans Aerosp Electron Syst 39(4):1232–1249
15. Schober P, Boer C, Schwarte LA (2018) Correlation coefficients: appropriate use and interpretation. Anesthesia Analgesia 126(5):1763–1768
16. Wang M, Niu X, Yang Q, Chen S, Yang G, Wang F (2018) Inversion of vegetation components based on the spectral mixture analysis using hyperion data. J Indian Soc Rem Sens 46(1):1–8. https://doi.org/10.1007/s12524-017-0661-2 Jan
17. Xia W, Liu X, Wang B, Zhang L (2011) Independent component analysis for blind unmixing of hyperspectral imagery with additional constraints. IEEE Trans Geosci Rem Sens 49(6):2165–2179. https://doi.org/10.1109/TGRS.2010.2101609 June

18. Xiong W, Chang C, Wu C, Kalpakis K, Chen HM (2011) Fast algorithms to implement n-findr for hyperspectral endmember extraction. IEEE J Sel Top Appl Earth Obs Rem Sens 4(3):545–564. https://doi.org/10.1109/JSTARS.2011.2119466 Sept
19. Zhu F (2017) Hyperspectral unmixing: ground truth labeling, datasets, benchmark performances and survey. arXiv preprint arXiv:1708.05125

Chapter 28
Optimization Algorithms for Antenna Design, Array Thinning, and Radiation Pattern Synthesis

Madhuri Sahal, Vivekanand Tiwari, and Tejpal

1 Introduction

With the advent of technology, the last few decades have witnessed a tremendous increase in the use of wireless communication devices for different applications as Bluetooth, Wi-Fi/WiMax, etc. Consequently there is a growing need to be always connected. The antennas are a key component of all wireless communication systems.

Traditional approach for design of antennas is to start with an initial antenna design and a set of specifications. Then modify the design through application of key concepts for achieving the desired design objectives. The antenna designer then fine-tunes the performance of the initial design by means of experience-driven parametric optimization study. The parametric optimization technique does not always lead to optimum designs as atmost only one or two parameters can be dealt with simultaneously. Often an antenna design involves a large number of interrelated antenna design parameters that must be dealt with simultaneously to achieve the optimum design. Moreover, the parametric optimization cannot deal with far-field properties such as gain or radiation efficiency. For fine-tuning the antenna performance, it becomes relevant to apply some form of numerical optimization technique to automate the fine-tuning of antenna parameters to achieve the desired antenna performance.

2 Classifications of Optimization Algorithms Used for Antenna Design

In the recent past, a lot of research has been done to apply numerical optimization techniques to automate the design optimization of antennas. The optimization

M. Sahal (✉) · V. Tiwari · Tejpal
Manipal University Jaipur, Jaipur, India
e-mail: madhuri.sahal@jaipur.manipal.edu

© Springer Nature Singapore Pte Ltd. 2021
S. D. Purohit et al. (eds.), *Proceedings of International Conference on Communication and Computational Technologies*, Algorithms for Intelligent Systems,
https://doi.org/10.1007/978-981-15-5077-5_28

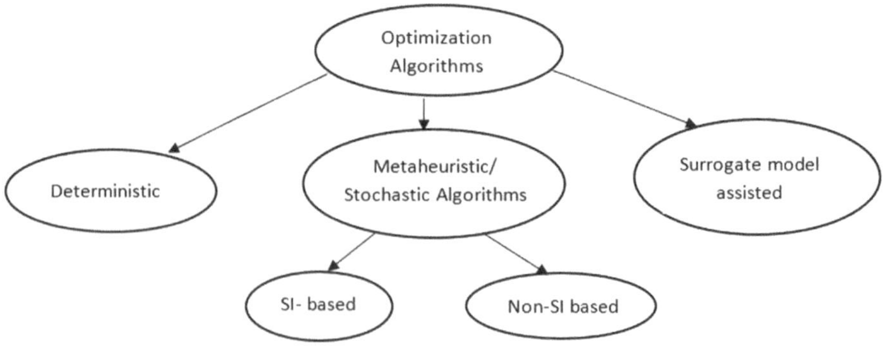

Fig. 1 Classification of optimization algorithms for antenna design

algorithms used for antenna design can be broadly classified as deterministic and stochastic. The deterministic algorithms like Nelder–Mead (NM) [1] have a possibility of getting trapped into local maxima/minima. Stochastic algorithms are better at achieving global optima, due to randomness involved in the generation of solutions at each iteration. This leads to a better ability to explore the search space. Stochastic algorithms can be segregated as swarm intelligence-based (SI-based) and non-swarm intelligence-based (non-SI-based). Performance of stochastic algorithms requires careful tuning of algorithm-specific parameters and their performance is highly dependent on their ability to explore the entire solution space for searching solutions. Surrogate model-assisted algorithms use a surrogate model to predict the function values instead of the computationally expensive EM simulation solver using GP machine learning, SVM, ANN, or any other AI-based approach and use an evolutionary algorithm to obtain the final optimum solution. Surrogate model-assisted algorithms though complex involve less number of cost function computations and hence are advantageous in terms of computation time and cost (Fig. 1).

The performance of an algorithm depends on several factors—the application for which it is to be used, tuning of algorithm-specific parameters, number of function calculations/computation time, ability to explore the entire search space, etc. The choice of algorithm depends on application—an algorithm may perform well for a given application, but it may be outperformed by some other algorithm in another application. Certain algorithms are very sensitive to tuning of algorithm-specific parameters; hence, careful parameter tuning is desired. A good optimization algorithm must have the ability to explore/search the entire solution space in order to achieve global optima and not get trapped in local optima. Simultaneously, it must also be computationally efficient and converge to global optima fast.

Optimization algorithms have been used for design optimization of antennas [1–13], for array thinning [14, 15] and pattern synthesis [14–22] of antenna arrays.

3 Applications of Optimization Techniques in Antenna Design

3.1 Performance Enhancement of Individual Antenna Design

Several algorithms and their modifications/hybridizations have been explored for achieving different antenna design objectives such as single [1]/dual-band impedance matching [2, 3], wideband operation [4, 5], compact design configuration [6, 9, 11], broadband circular polarization [7], antenna gain/directivity optimization [1, 8], and antenna arrays for different problems as sidelobe reduction [14–17, 20, 22], radiation pattern synthesis [18, 19], and array thinning [15].

The authors have hybridized very simple optimization (VSO) and Nelder–mead (NM) algorithm [1]. They have demonstrated its application to minimize the antenna gain to *Q*-factor based on current distribution for different antenna designs, leading to an increase in radiation efficiency. The robustness of the technique is demonstrated by applying the VSO-NM to three different antenna designs at three different frequency bands. The first being a UWB meshed circular disc monopole, the second is a microstrip patch antenna array in submillimetre frequency range, and the third is a nanostrip patch antenna operating in optical spectrum band. The fast convergence of the algorithm is a result of global deterministic search optimization which is followed by local NM optimization to improve the fitness.

Surrogate model-assisted techniques trade-off simplicity for convergence speed are therefore popular for design optimization of individual antennas. Surrogate models have been implemented using adaptive response correction [9] for maximizing front-to-back ratio of a DRA, feature-based optimization [10] for design optimization of UWB antennas, trust-region-based gradient search [11] for achieving compact configuration and gain stability in desired frequency range. For designs involving multi-objective optimization as compact CP antenna as in [12], the use of surrogate model-assisted algorithms would be more time-efficient. The authors have the surrogate model in conjunction with hypervolume-based PoI and PoF [13] for design of corner truncated patch antenna for GPS application. The use of estimation of local minima (ELM) approach [8] has been applied to design a 12-element yagi-uda antenna for maximum gain, where the cost function has several local minima.

Dual-band/Wideband antennas Non-SI-based metaheuristic algorithms like GA, DE, and IWO have been extensively used for the design optimization of antennas. Design of dual-band antennas has been carried out by minimizing in-band reflections in each of the two bands [2, 3]. Li [2] has demonstrated the use of GA to design of dual-band antenna using pixelization of antenna periphery. The design parameters of modified T-shape antenna with slotted DGS have been optimized for dual-band operation in X and Ku band applications using differential evolution algorithm (DE) [3].

Besides, SI-based metaheuristic algorithms like particle swarm optimization (PSO), ant colony optimization (ACO), moth flame optimization (MFO), artificial bee colony (ABC), social spider optimization (SSO), etc., are popular to achieve automated design optimization of antennas. The use of hybrid DE-ACO to design of wideband E-shaped patch antenna has been demonstrated [4]. The authors have demonstrated the use of social spider optimization (SSO) algorithm to optimize the FSS parameters for improving impedance matching and bandwidth of a UWB microstrip ring monopole with FSS [5].

Dual/broadband antennas involve the minimization of in-band reflections that are represented by the return loss coefficient in the desired frequency range. The cost function [2] to be minimized is

$$F = \sum_{i=1}^{N_{\text{freq}}} S_{11}(i) - S_{11,\text{th}}(i) \tag{1}$$

where $S_{11,\text{th}}$ (i) is the desired minimum return loss at ith frequency, and N_{freq} is the total number of frequencies under consideration.

Compact antennas Size reduction of a rectangular patch is achieved through pixelization of the antenna [6] and using GA to define the best configuration of metallized pixels. The cost function is the sum of the return loss coefficient at all desired frequencies (Eq. 1).

Broadband circular polarized (CP) antennas A hybrid GA-direct search algorithm has been used [7] to overcome the slow convergence of GA in late generations for the design of a broadband circular polarized (CP) stacked patch design. Hybridization of algorithms allows to combine the good points of both the algorithms under consideration. The fitness function [7] for design of broadband CP antennas is

$$F = \sum_{i=1}^{N_{\text{freq}}} A_i + \begin{cases} 0, & \text{if}|\Gamma_i| < \Gamma\text{max} \\ 2|\Gamma_i, & \text{if}|\Gamma_i| > \Gamma\text{max} \end{cases} \tag{2}$$

Antenna gain/directivity enhancement Montaser have used VSO-NM technique [1] to minimize the gain to Q-factor ratio to improve the gain and radiation efficiency of antennas by optimizing the surface current distribution for three different antennas in SHF, submillimeter, and optical frequency band.

3.2 Performance Enhancement of Antenna Arrays

Metaheuristic and surrogate model-assisted algorithms have found wide application in design of antenna arrays for problems such as array thinning, sidelobe reduction, and array pattern synthesis and/or null control.

Some metaheuristic techniques used in antenna array synthesis are hybrid DE-ACO algorithm [4] for linear antenna array pattern synthesis, invasive weed optimization (IWO) for SLL reduction of linear antenna arrays and thinning of non-uniform planar antenna arrays [14], artificial bee colony (ABC) algorithm for synthesis of linear antenna arrays [16], moth flame optimization (MFO) for uniform linear antenna array synthesis and non-uniform circular antenna array for reduced SLL [17], improved fruit fly optimization algorithm (FOA) for linear and planar array synthesis, null control and array thinning [15]. Hybridizations of evolutionary techniques, for instance, hybrid GA-PSO optimization algorithm [18] have also been implemented for array pattern synthesis, to allow for combining the best features of the individual algorithms.

Surrogate model-based evolutionary algorithms have found to expedite the optimization process by using the surrogate model, often based on an artificial intelligence technique to replace the computationally complex EM solver for reducing the number of cost function computations. Different approaches have been used to build the surrogate model like Gaussian process machine learning [21] for achieving different objectives as diverse as maximizing coupling while reducing crosstalk, optimizing antenna gain, maximizing radiation efficiency, self-adaptive DE(SaDE) for synthesis of uniformly and non-uniformly excited and thinned elliptical cylindrical antenna array (ECAA) with reduced sidelobe level and better directivity [22], ELM for the synthesis of linear broadside array with reduced SLL [8].

Antenna pattern synthesis Certain applications require the design of antennas with a specific radiation pattern as for satellite antennas, which require isoflux radiation pattern to account for the earth's curvature. This requires the design of antennas with a specific radiation pattern. The authors have used [19] PSO to optimize the inter-element spacing and excitation amplitudes of a non-uniform concentric circular antenna for synthesis of isoflux radiation pattern.

Sidelobe reduction Often it is required to direct the energy of an antenna into the desired direction simultaneously minimizing radiation into the sidelobes. Such applications can be viewed as a sidelobe reduction problem. The cost function for sidelobe reduction [20] is of the form:

$$F = \frac{|\mathrm{AF}(\theta_{\mathrm{msl}},\ I,\ d)|}{|\mathrm{AF}(\theta_0,\ I,\ d)|} + |\mathrm{BWFN}_{\mathrm{des}} - \mathrm{BWFN}(I,\ d)| \tag{3}$$

where θ_{msl} is the angle of maximum sidelobe level and θ_0 is the maximum radiation angle.

Sidelobe suppression Certain applications, like mobile communication, typically require point-to-point communication, simultaneously suppressing the signal from interferes. This would require the use of sidelobe suppression in certain specific angular regions.

The fitness function for sidelobe suppression [15] is of the form:

$$F = \sum_{i=1}^{K} \frac{1}{\theta_{\mathrm{ui}} - \theta_{\mathrm{li}}} \int_{\theta_{\mathrm{li}}}^{\theta_{\mathrm{ui}}} |\mathrm{AF}(\theta)| \mathrm{d}\theta \tag{4}$$

where K = number of regions for SLL suppression, $\mathrm{AF}(\theta)$ is the array factor of linear/planar array, and $[\theta_{\mathrm{li}}, \theta_{\mathrm{ui}}]$ region of sidelobe suppression.

Array thinning Array thinning involves antenna radiation pattern synthesis with reduced number of antenna elements by leaving certain antenna elements unexcited in the array, and the inter-element spacing and excitation amplitudes of other elements are optimized to synthesize the required radiation pattern [15].

3.3 Design of Metasurfaces for Antenna Performance Enhancement

Metasurfaces are artificially engineered surfaces used to control the wavefronts and polarization of EM waves by redistributing the phase [23]. Metaheuristic optimization techniques can be used to design metasurfaces for enhancing antenna performance. Recently, particle swarm optimization has been used to optimize the design parameters of unit cell of metasurface to achieve a 14–29 dB SLL reduction and gain enhancement of 1.2–2.3 dB of a pyramidal horn antenna [24].

4 Conclusion

The application of deterministic algorithms for antenna design is limited by the possibility of the algorithm being trapped in local optima. Metaheuristic optimization techniques have successfully been used to the design of antenna and antenna arrays, but the performance of these algorithms is dependent on the tuning of algorithm-specific parameters [25]. These parameters must be selected so as to ensure that these can explore the entire solution space and avoid being trapped in local optima. Surrogate model-assisted optimization techniques are computationally fast but are relatively complex as compared to metaheuristic algorithms. Surrrogate model-assisted algorithms are suited to the design of antennas and antenna array radiation pattern synthesis and array thinning problems [26]. Though there is no single algorithm that is suited to all antenna design problems, but as the antenna structure gets complex, example for structures having complex design or multi-layered dielectric geometry, use of surrogate model-assisted algorithms would be more time-efficient.

References

1. Montaser AM, Mahmoud KR (2017) Antenna current optimization for optimal antenna design in different frequency bands using VSO-NM algorithm. Int J RF Microw Comput-Aided Eng, e21104
2. Li Y (2010) Simulation-based evolutionary method in antenna design optimization. Math Comput Model 51:944–955
3. Nitika, Khanna R, Kaur J (2017) Optimization of modified T shape microstrip patch antenna using differential algorithm for X and ku band applications. Microw Opt Technol Lett 60: 219–229
4. Chang L, Liao C, Lin WB, Chen LL, Zheng X (2012) A hybrid method based on differential evolution and continuous ant colony optimization and its application on wideband antenna design. Prog Electromagn Res 122:105–118
5. Oliveira PS, Assuncao AGD, Souza EAM, Peixeiro C (2016) A fast and accurate technique for FSS and antenna designs based on the social spider optimization algorithm. Microw Opt Technol Lett 58(8):1912–1917
6. Lamsalli M, Hamichi AE, Boussouis M, Touhami NA, Elhamadi T (2016) Genetic algorithm optimization for microstrip patch antenna miniaturization. Prog Electromagn Res Lett 60:113–120
7. Zentner R, Sipus Z, Bartolic J (2001) Optimization synthesis of broadband circularly polarized microstrip antennas by hybrid genetic algorithm. Microw Opt Technol Lett 31(3):197–201
8. Kolundzija BM, Olcan DI (2006) Multiminima heuristic methods for antenna optimization. IEEE Trans Antennas Propag 54(5):1405–1415
9. Koziel S, Ogurtsov S (2014) Design optimisation of antennas using electromagnetic simulations and adaptive response correction technique. IET Microw Antennas Propag 8(3):180–185
10. Koziel S, Bekasiewicz A (2017) Expedited simulation-driven design optimization of UWB antennas by means of response features. Int J RF Microw Comput-Aided Eng, e21102
11. Koziel S, Cheng QS, Li S (2017) Optimization-driven antenna design framework with multiple performance constraints. Int J RF Microw Comput-Aided Eng, e21208
12. Bekasiewicz A, Koziel, S, Cheng, QS (2018) Analysis of circular polarization antenna design trade-offs using low-cost EM-driven multiobjective optimization. Int J RF Microw Comput-Aided Eng, e21483
13. Singh P, Rossi M, Couckuyt I, Deschrijver D, Rogier H, Dhaene T (2017) Constrained multi-objective antenna design optimization using surrogates. Int J Numer Model, e2248
14. Karimkashi S, Kishk AA (2010) Invasive weed optimization and its features in electromagnetics. IEEE Trans Antennas Propag 58(4):1269–1278
15. Darvish A, Ebrahimzadeh A (2018) Improved fruit-fly optimization algorithm and its applications in antenna arrays synthesis. IEEE Trans Antennas Propag 66(4):1756–1766
16. Zaman MA, Gaffar M, Alam MM, Mamun SA, Matin MA (2011) Synthesis of antenna arrays using artificial bee colony optimization algorithm. Int J Microw Opt Technol 6(4):234–241
17. Das A, Mandal D, Ghoshal SP, Kar R (2018) Moth flame optimization based design of linear and circular antenna array for side lobe reduction. Int J Numer Model
18. Liang Z, Ouyang J, Yang F (2018) A hybrid GA-PSO optimization algorithm for conformal antenna array pattern synthesis. J Electromagn Waves Appl
19. Reyna A, Panduro MA, del Rio-Bocio C (2011) On the design of concentric ring array for isoflux radiation in MEO satellites based on PSO. Prog Electromagn Res M 20:243–255

20. Panduro MA, Mendez AL, Dominguez R, Romero G (2006) Design of non-uniform circular antenna arrays for side lobe reduction using the method of genetic algorithms. Int J Electron Commun (AEÜ) 60:713–717
21. Liu B, Aliakbarian H, Ma Z, Vandenbosch GAE, Gielen G, Excell P (2014) An efficient method for antenna design optimization based on evolutionary computation and machine learning techniques. IEEE Trans Antennas Propag 62(1)
22. Lema GG, Tesfamariam GT, Mohammed MI (2016) A novel elliptical-cylindrical antenna array for radar applications. IEEE Trans Antennas Propag 64(5):1681–1688
23. Holloway CL, Kuester EF, Gordon JA, O'Hara J, Booth J, Smith DR (2012) An overview of the theory and applications of metasurfaces: the two-dimensional equivalents of metamaterials. IEEE Antennas Propag Mag 54(2):10–35
24. Moradi A, Mohajeri F (2018) Side lobe level reduction and gain enhancement of a pyramidal horn antenna in the presence of metasurfaces. IET Microwaves Antennas Propag 12(3):295–301
25. Singh V et al. (2019) Source redundancy management and host intrusion detection in wireless sensor networks. Recent Patents Comput Sci 12(1)
26. Bhushan S, Saroliya A, Singh V (2013) Implementation and evaluation of wireless mesh networks on MANET routing protocols. Int J Adv Res Comput Commun Eng 2(6)

Chapter 29
Implementation of a Simple Plug-and-Play Module for Vehicle to Vehicle Communication Over 434 Megahertz Band

Anand Sabne, Abruti Panda, Rohan Shevade, Divyansh Kasture, and Dr. Vaishali Ingale

1 Introduction

There has been a dramatic rise in road accidents worldwide. The number of annual traffic deaths has reached 1.35 million, according to the global status report on road safety 2018, launched by WHO in December 2018. Traffic casualties are now the leading killer of people between the ages of 5 and 29 years [1]. As a matter of fact, road traffic accidents are ranked tenth globally, according to the number of deaths [2]. Concept of vehicular communication has evolved as a potential solution to this problem.

Vehicle to Vehicle (V2V) communication is defined as wireless transmission of data like speed, acceleration, Global Positioning System (GPS) coordinates, etc. between vehicles. This assists the driver in taking decisions and hence avoid road accidents. The 802.11p standard [3] for wireless access in vehicular environment defines a standard for vehicular communication over 5.9 gigahertz band. However, technology capable of using the 5.9 gigahertz band is still considered as off the shelf

A. Sabne (✉) · A. Panda · R. Shevade · D. Kasture · Dr. V. Ingale
Department of Electronics and Telecommunication, College of Engineering, Pune, Maharastra, India
e-mail: sabneanand@gmail.com

A. Panda
e-mail: panda.abruti@gmail.com

R. Shevade
e-mail: rohanshevade2012@gmail.com

D. Kasture
e-mail: kasturedp15.extc@coep.ac.in

Dr. V. Ingale
e-mail: vvi.extc@coep.ac.in

© Springer Nature Singapore Pte Ltd. 2021
S. D. Purohit et al. (eds.), *Proceedings of International Conference on Communication and Computational Technologies*, Algorithms for Intelligent Systems,
https://doi.org/10.1007/978-981-15-5077-5_29

and is expensive to build upon and hence has not been deployed yet on a large scale. At this point, simple and low-cost plug-and-play modules as the one proposed here come to use and can serve the purpose until the standard technology builds up. After that, the same system can be used by replacing the 434 megahertz radio transceivers by 5.9 gigahertz transceivers and some protocol modifications.

2 Methodology

The proposed system can be broken down into two modules:

1. Module establishing communication among vehicles
2. Module displaying information using a User Interface for driver assistance.

2.1 System Implementation Details

Refer Fig. 1. The master controller based on ATmega2560 is used to handle the communication between the vehicles. GPS modules like Ublox NEO-6M or SIM28 are interfaced with the master controller in order to get the GPS coordinates of the moving vehicles. RF transmitter and receiver working on 434 megahertz with ASK (Amplitude Shift Keying) modulation scheme are connected to the master controller. Also, both the transmitter and receiver modules have a provision for antenna and thus, various types of antennas may be used to control the directionality and range of the transmission. Monopole antennas of appropriate length are used in this system at both transmitter and receiver. A secondary controller, ATmega328P-PU is used to drive the TFT screen and is connected to the master controller by UART (Universal

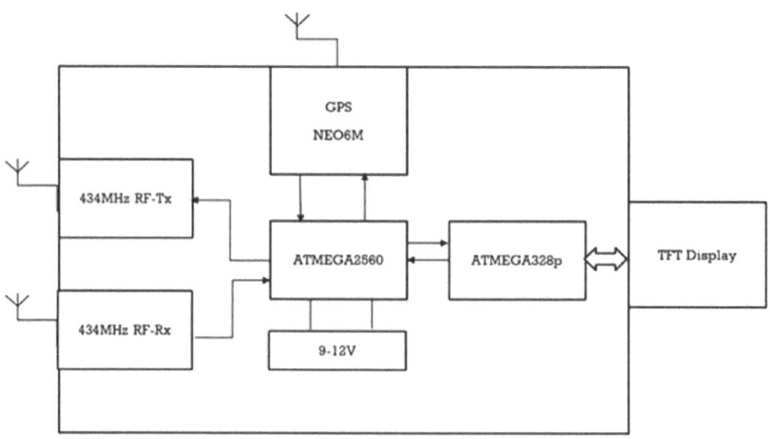

Fig. 1 Block diagram of the implemented system

Asynchronous Receiver Transmitter) interface. The entire system is powered up by a 5 V supply available in the car.

2.2 Working of the Proposed System

All the vehicles are synchronized by the time obtained using the GPS module. The multiple access technique described in Sect. 2.4 is used to decide the time slot of the transmission of GPS coordinates by a vehicle. At a particular time, only one vehicle is configured to transmit while all other vehicles are configured as receivers. The transmitting vehicle's transmitter interfaced to the master controller broadcasts a frame which consists of its ID and GPS coordinates. The receiving vehicles' receiver interfaced with the master controller receives the frame and forwards it to the secondary controller with the TFT display using UART protocol. The secondary controller processes the frame to separate out the GPS coordinates and vehicle ID fields. It then does appropriate computations to plot the user interface on the TFT display as described in Sect. 2.5. The maximum number of vehicles that can be displayed on the user interface at a time is software-defined, with its maximum value limited by RAM availability in the controllers involved, which is usually more than that required in a practical scenario. Hence, although the system is tested using three plug-and-play modules, the system is designed in such a way that it can be put to use in a practical scenario.

2.3 Protocol Used

The proposed system is a peer-to-peer broadcast type network. The application layer frame is constructed by concatenating data such as GPS coordinates and vehicle ID along with separating characters. This is further passed to the data link layer which encodes it using CRC codes. This encoded frame is passed to ASK modulator which generates the physical layer by sending appropriate symbols over the RF channel, in its designated time slot. This system can accommodate a network layer that can allow communication from source to the destination instead of broadcasting but is not required in the application and hence is not included. Thus, the system is developed with its own protocol and does not use any standard networking protocol. Also, the data here is not encrypted as the data that is being broadcast, is related to the GPS coordinates and vehicle ID and does not include any critical information. As the system operates at short distances, for an eavesdropper, receiving the data from long distances for gathering vehicle data is difficult, and receiving it from shorter distances is pointless as GPS coordinates would be already known in that case. Moreover, as the system is processing data directly without invoking any complex services or internet, the possibility of hacking into the system is negligible. However, the system is sensitive to jamming.

2.4 Multiple Access Technique Used: Time Division Multiple Access (Time Slicing)

The vehicles are given unique IDs and the ID is used to decide the time slot in which the vehicle must transmit its location. The vehicle whose ID matches the time slot transmits its GPS coordinates on the configured transmitter pin while all other vehicles are configured as receivers which receive the GPS coordinates. In this way, the algorithm is designed in such a way that in a given time slot, only one vehicle can transmit, whereas all others must listen to and thereby, receive the coordinates of the transmitting vehicle. Each vehicle must wait until its time slot arrives before transmitting its GPS coordinates as the time slots are coded to be in round robin manner.

2.5 User Interface for Driver Assistance

The user interface will display a birds' eye view of the road traffic surrounding the vehicle up to a distance of 100 m (this value can be changed in the software to cover a smaller or a larger radius, the maximum value of which is limited by the transmitter RF power supply voltage and the type of antennas used). As soon as the GPS coordinates of the transmitting vehicle have been received by a vehicle, its supporting microcontroller based on Atmega328P-PU which controls the TFT screen will compute the relative distance of the transmitting vehicle with respect to itself and plot the point by mapping it appropriately on a circular area of 100 m radius (blue circle shown in Fig. 2) on the TFT display. The program on the controller board based on ATmega328P-PU is designed to receive the GPS coordinates from the master controller using standard UART protocol along with the ID of the vehicle. It then computes the relative distance and maps the GPS coordinates (latitude, longitude)

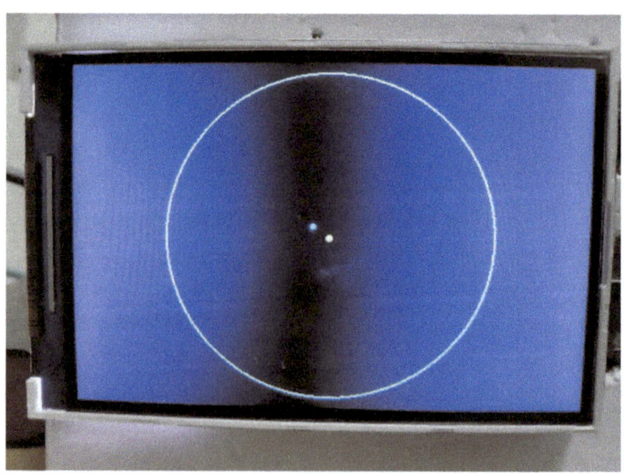

Fig. 2 User interface

to rectangular (x, y) coordinates to be plotted on the TFT screen. In Fig. 2, yellow dot shows position of self and blue dot shows position of nearby vehicle (only one nearby vehicle is shown for the sake of demonstration).

3 Results

The following parameters have been evaluated on the hardware system developed. The testing has been done on three such units developed and mounted one on each vehicle.

- *RF Range* :

- 100–150 m (line of sight)
- Non-line of sight detection at a road intersection with building at corner with path separation 100 m
- Conditions:

1. Omnidirectional monopole antenna at both transmitter and receiver
2. *Transmitter voltage*: 5 V
3. *Transmitted RF power*: 25 mW
4. *Receiver used with sensitivity*: −105 dBm.

The RF range can be controlled by controlling the power supply voltage to the transmitter from 3 to 12 V. Use of properly tuned antennas or other types of antennas such as a ground plane antenna along with power control can allow for a larger coverage area beyond 150 m radius.

- *Bit rate*: 2000 bps, software-defined, limited by hardware to 50 kbps at 434 MHz
- *Power consumption*: less than 400 mA at 5 V, i.e. 2 W
- *Number of nodes*: No limit to the number of nodes, algorithm can be appropriately modified.
- *Error rate*: RF processing is equipped with CRC error detection, and signal discard rate is around 10%.
- *Collision possibility*: GPS requires around 10 min for time to get synchronized with global time. RF collisions occur during this interval as time is not synchronized and time slots overlap. Multiple transmissions cause signal jamming. After settling, collision possibility is very low.
- *Nearby vehicle information update time*: Depends on the limit set on total number of vehicles connected at a time. Time slots are divided accordingly. Current system with three units gives an update time of 3 s.

4 Related Work

The domain of connected mobility has been in the limelight since the past few years and hence significant work in various subdomains has been done. Vehicular Ad hoc Networks (VANETs) have been developed for this purpose. Channel allocation using time slicing has been discussed by the author of [4] with work related to 802.11p protocol. Work done by [5, 6] is similar to that proposed here. The work presented here includes a user interface that shows position of nearby vehicles with respect to the user, which is additional to that published in [5, 6]. Significant work has also been done on 802.11p standard and algorithms related to it [7, 8]. This system can be upgraded to the 802.11p standard by replacing the wireless transceiver hardware with those compliant with the standard and by making appropriate modifications in the protocol.

5 Discussions

Many different methods, protocols, and algorithms for vehicle to vehicle communication have been developed over recent years. Standards such as 802.11p for vehicle to vehicle communication have also been developed. Moreover, with the advent of new technology and advancement in wireless communication systems, work and research beyond the existing standards and methods is also being done.

The work proposed here describes a simple wireless communication system for vehicle to vehicle communication over the 434 MHz band. A custom protocol is developed for the said application along with time slicing based multiple access technique. Much of the related work is based on the 5.9 GHz communication band using 802.11p protocol. The results obtained from the proposed system quantify the performance of the system in parameters related to vehicular communication systems and show that this system can serve the purpose of vehicle to vehicle communication and improvement can be done in certain features or implementation methods. The performance cannot be directly compared with a system following the standard as the underlying technology itself is different. Also, as there is no universal way of implementing the standard, the performance varies from system to system.

Vehicle to infrastructure (V2I) communication systems are also being developed to enhance the vehicle to vehicle network [9]. The current system can be easily reconfigured to incorporate V2I communication capabilities.

The concept of vehicle to vehicle communication further leads to the concept of vehicle trains [10]. Vehicle trains have huge potential to change the transportation industry owing to its advantages. Existing vehicle to vehicle communication system needs to be modified to incorporate the functionality of vehicle trains.

The system developed has several advantages. As communication is of broadcast nature, it does not involve a long procedure for authentication before pairing takes place. This is very significant in the case of vehicular communication as the latency required is very less (of the order of few milliseconds) for the system to be effective. Also, there is no central server involved which makes the system secure and independent. Also, the system is developed as a plug-and-play module which makes it very simple to use. The system can be used in old vehicles as retrofit modules that do not have significant driver assistance systems built in them.

Moreover, in latest automobiles, the system can be easily developed as the processor, the GPS and the display (infotainment screen) are already present. Only RF transmitter and receiver need to be interfaced, which can be done through the USB port available in the car. A software application can be developed with similar functionality in the vehicle's infotainment system (on existing platforms) to execute the logic in place of microcontrollers as in the proposed system. This will make it very easy to incorporate a vehicle to vehicle communication system in existing cars with minimum external components and cost.

6 Conclusion

A plug-and-play vehicle to vehicle communication system is developed (refer Fig. 3) and communication is established over 434 MHz RF link. Nearby vehicle's location relative to the position of user is plotted on a TFT screen, with timely updates. Thus, an overview of nearby vehicles that may or may not be in the field of vision is obtained, which assists the driver in taking decisions and enhances overall safety. All vehicles using the system are able to communicate on a peer-to-peer broadcast

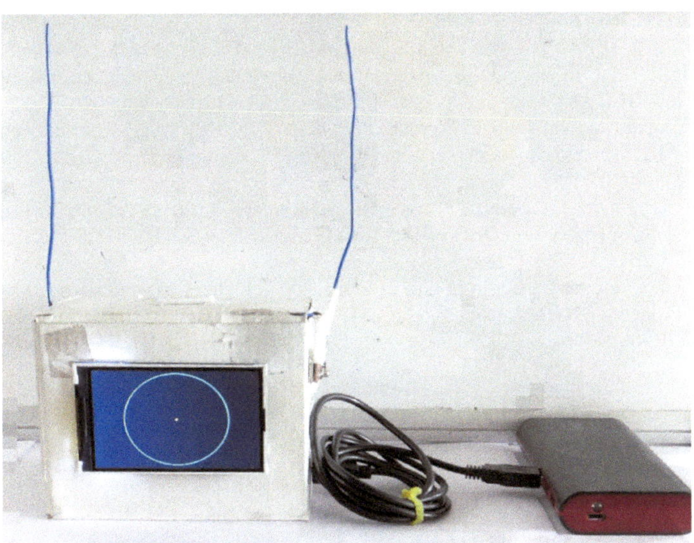

Fig. 3 Plug-and-play system powered up using a power bank

network, in real-time without prior pairing of devices. There is no dependency on any central infrastructure such as base stations or the internet for communication and hence this system is truly independent. This system is developed as a simple plug-and-play module for a vehicle and has no connections and interfaces to the vehicles' internal electronic system. Hence, the vehicle cannot be hacked via this system. The system can be used in old vehicles as retrofit modules that do not have significant driver assistance systems built in them. It can also be used easily in latest automobiles as a plug-and-play module or with a software application and minimum additional hardware.

References

1. Global status report on road safety 2018, World Health Organization (2018) vol 1. https://www.who.int/violence_injury_prevention/road_safety_status/2018/en
2. Road Traffic Accidents Increase Dramatically Worldwide, PRB (2006) vol 1. https://www.prb.org/roadtrafficaccidentsincreasedramaticallyworldwide/
3. Gupta R, Agarwal S (2011) A survey on wireless access in vehicular environment. Int J Adv Commun Eng 3(2):1–6
4. Bilstrup K, Uhlemann E, Ström E et al (2009) J Wirel Com Netw 2009:902414. https://doi.org/10.1155/2009/902414
5. Ghatwai NG, Harpale VK, Kale M (2016) Vehicle to vehicle communication for crash avoidance system. In: 2016 international conference on computing communication control and automation (ICCUBEA), Pune, pp 1–3. https://doi.org/10.1109/iccubea.2016.7860118
6. Anadu D, Mushagalusa C, Alsbou N, Abuabed ASA (2018) Internet of things: vehicle collision detection and avoidance in a VANET environment. In: 2018 IEEE international instrumentation and measurement technology conference (I2MTC), Houston, TX, pp 1–6. https://doi.org/10.1109/I2MTC.2018.8409861
7. Lv F et al. (2016) An empirical study on urban IEEE 802.11p vehicle-to-vehicle communication. In: 2016 13th annual IEEE international conference on sensing, communication, and networking (SECON), London, pp 1–9. https://doi.org/10.1109/sahcn.2016.7732969
8. Han C, Dianati M, Tafazolli R, Kernchen R, Shen X (2012) Analytical study of the IEEE 802.11p MAC sublayer in vehicular networks. IEEE Trans Intell Transp Syst 13(2):873–886. https://doi.org/10.1109/TITS.2012.2183366
9. Milanes V, Villagra J, Godoy J, Simo J, Perez J, Onieva E (2012) An intelligent V2I-based traffic management system. IEEE Trans Intell Transp Syst 13(1):49–58. https://doi.org/10.1109/TITS.2011.2178839
10. Bergenhem C, Hedin E, Skarin D (2012) Vehicle-to-vehicle communication for a platooning system, Procedia Soc Behav Sci 48: 1222–1233, ISSN 1877-0428, https://doi.org/10.1016/j.sbspro.2012.06.1098

Chapter 30
Performance Analysis of Hello Flood Attack in WSN

Sourabh Banga, Himanshu Arora, Smita Sankhla, Gopesh Sharma, and Bhavesh Jain

1 Introduction

Wireless sensor networks (WSN), a large-scale computer part, are presently wont to monitor the time period environmental state of affairs. However, this sensing element works underneath the utmost power limit and is meant for applying. coming up with a brand new wireless sensing element node may be a terribly tough task and involves reviewing the assorted parameters needed by the target application, together with scope, kind of antenna, elements, memory, storage, power, proposed technology, security, communication technology, size, programming interface and application [1]. In order to satisfy the large type of applications that this technology has to be compelled to support, several domains among the WSN domain want analysis and active work. There's little question that security is one in all the essential components of the network style that has to be addressed within the initial place [2]. A wireless sensor network (WSN) comprises cheap and low-power systems referred to as nodes. These nodes are wirelessly connected to observe/management environmental impacts wherever they're enforced. In WSNs, some contract nodes are referred to as supply nodes that notice the event within the surroundings, as an example activity,

S. Banga (✉) · H. Arora (✉) · S. Sankhla · G. Sharma · B. Jain
Arya College of Engineering and Research Centre, Jaipur, Rajasthan, India
e-mail: Sourabh.banga07@gmail.com

H. Arora
e-mail: dr.himanshuarora17@gmail.com

S. Sankhla
e-mail: smita.sankhla@gmail.com

G. Sharma
e-mail: gopeshsharma92@gmail.com

B. Jain
e-mail: bhaveshjain678@gmail.com

© Springer Nature Singapore Pte Ltd. 2021

S. D. Purohit et al. (eds.), *Proceedings of International Conference on Communication and Computational Technologies*, Algorithms for Intelligent Systems,
https://doi.org/10.1007/978-981-15-5077-5_30

pressure, no signal, etc. While there's another special sort of contract that collects info regarding these events from all nodes. These nodes are referred to as sinks or base stations (BS). Due to its uncontrolled, uncontrolled, and adverse environmental conditions, the WSN is exposed to several DoS attacks. During a DoS attack, the attacker attempts to prevent the information generated by the provisioning nodes from reaching the destination or destination nodes. The most common denial-of-service attacks are Flood, Sink Hole, Wormhole, INA, Black Hole, Rushing, and so on. Some WSN routing protocols need that nodes to be declared for adjacent nodes through a "Hello" message. This message permits incoming contracts within the neighborhood, anyone who can hear this message assumes that the transmitter is in the radio band. This can be easy thanks to put together a network device. However, there's a haul with this easy mechanism, as a result of this ends up in hello flood attacks [3].

2 Related Work

Singh et al. described a network of wireless sensors consists of the many necessary components, like sensors, base stations, and users. A device will live several non-electrical quantities, like pressure, temperature, sound, etc. and transmit this info using interior receiver. The protection of this transmitted data is extremely necessary as a result will contain important information. Since wireless device networks have several civil and military applications, the protection of wireless device networks is changing into a key concern. A Sybil attack could be an essential attack that will have an effect on the routing protocols, the honest allocation of resources, the aggregation of information, and also the configuration of detection of misconduct within the network [4]. Shakhov proposed to develop effective methods to counter the threat. We tend to discuss first of all the practicability of premature fatigue in battery performance. Next, we tend to propose a model to analyze energy consumption below the attack. And the countermeasures against the attack are mentioned [5]. Mezrag proposes a secure replacement protocol supported by the well-recognized LEACH routing protocol Hybrid Cryptography, based primarily on topics for secure electronic communication in WSN clusters (HCBS). As an approach to multi-constraint criteria, HCBS relies on a combination of Elliptic Curves-supported encryption technology to exchange keys that use circular keys for encryption and hacking [6]. Networks are vulnerable to various attacks [7] and routing efficiency is one of the major issues [8]. Some approaches to detect and prevent attacks are discussed in [9].

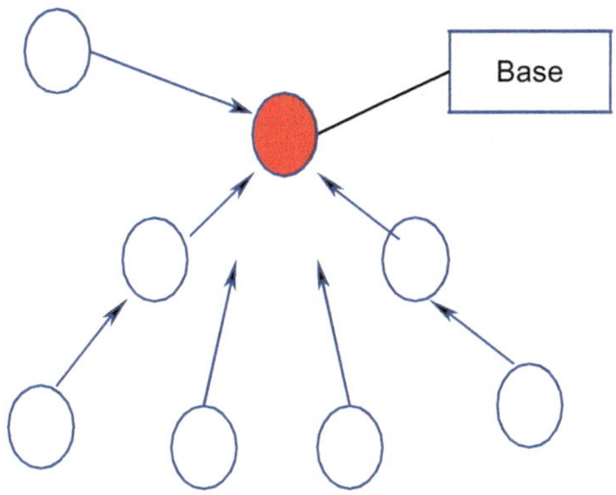

Fig. 1 Hello flood attack

3 Hello Flood Attack

WSN has a routing protocol that requires nodes to send greetings messages to notify their neighbors. When the attacker plays the upper limit message, make sure that all nodes in the network are surrounded by the attacker [10]. Many protocols that use salutation packages assume naively that receiving this packet implies that the transmitter is among the radio vary and thus is close. The opponent will use a high-energy transmitter to transform an oversized space of nodes into a basic cognitive process, which are the neighbors of that sent node. If the opponent sends a lie on a higher route to the valley Station, these nodes could send them to the offensive node even if many of them are out of the radio [11, 12] (Fig. 1).

3.1 Performance Parameters

1. Throughput Packet
2. Delivery Ratio
3. Dropped Packets
4. Received Packets.

4 Result and Analysis

- In this Section, the experimented and simulated work results are presented and then through graph does an analysis of the result obtained.
- Flood impact and flood attack are analyzed. The effect of a hello flood attack is analyzed with the difference in the number of nodes and also its effect on the packet delivery relationship, packet received, average performance (kbps), the total packet loss is analyzed. The contract number is 10, 30, 50, 70, 90, and 110.

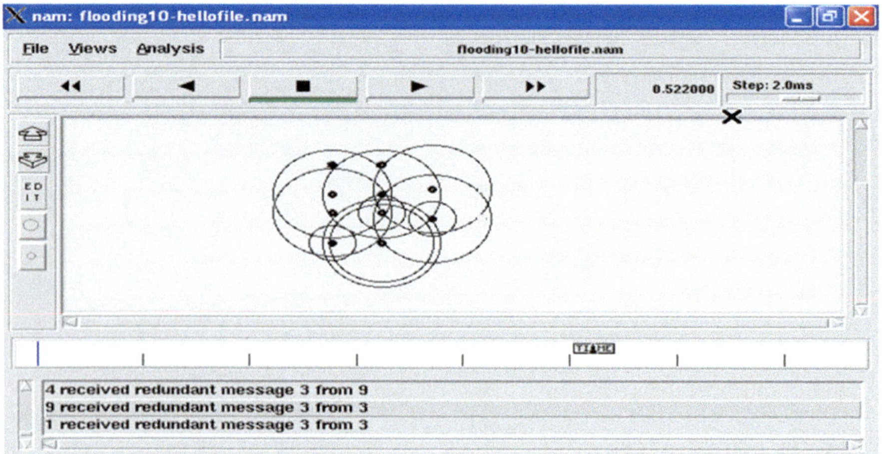

Fig. 2 Output of flooding10-hellofile

He analyzed the effect of flood and flood attack on the variable number of nodes and analyzed the overall performance of the network when the flood occurred the attack.

4.1 Simulation Result

All Simulation results are shown below as follows (Fig. 2).

The impact of a flood-welcoming attack on the wireless sensor network appears when there are 10 points. It also provides packet delivery rate, total packet loss, packet received and average performance per kbps (Fig. 3).

The effect of a flood-welcoming attack on the wireless sensor network appears when there are 30 nodes. It also provides packet delivery rate, total packet loss, packet received, and average performance per kbps (Fig. 4).

The effect of a flood-welcoming attack on the wireless sensor network appears when there are 50 nodes. It also provides packet delivery rate, total packet loss, packet received, and average performance per kbps (Fig. 5).

The impact of a flood attack on the wireless sensor network appears when there are 70 knots. It also provides packet delivery rate, total packet loss, packet received, and average performance per kbps.

Figure 6 shows the output of the Hello-Flooding90.tcl file. The impact of a flood-welcoming attack on a wireless sensor network appears when there are 90 knots. It also provides packet delivery rate, total packet loss, packet received, and average performance per kbps.

Figure 7 shows the output of the Hello-Flooding110.tcl file. The effect of the flood attack on the wireless sensor network appears when there are 110 nodes. It

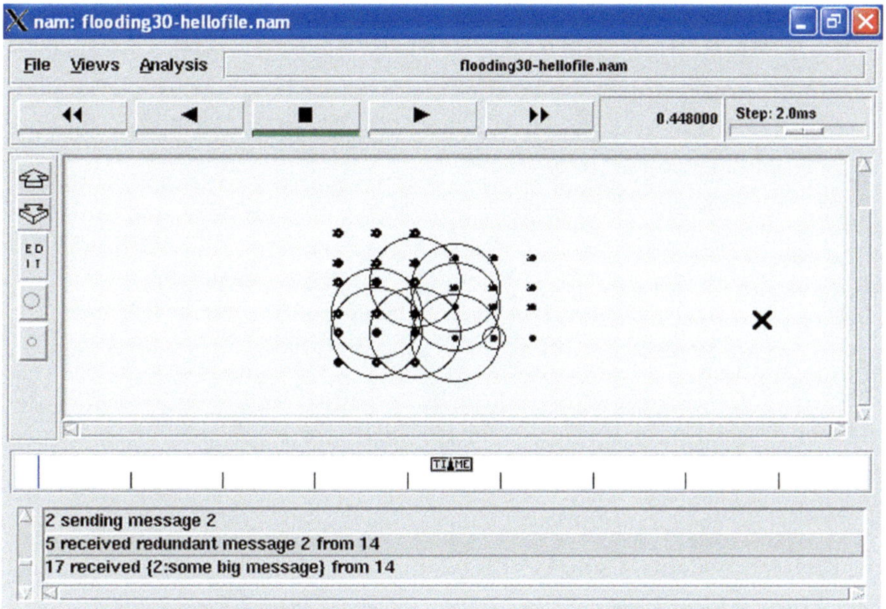

Fig. 3 Output of flooding30-hellofile

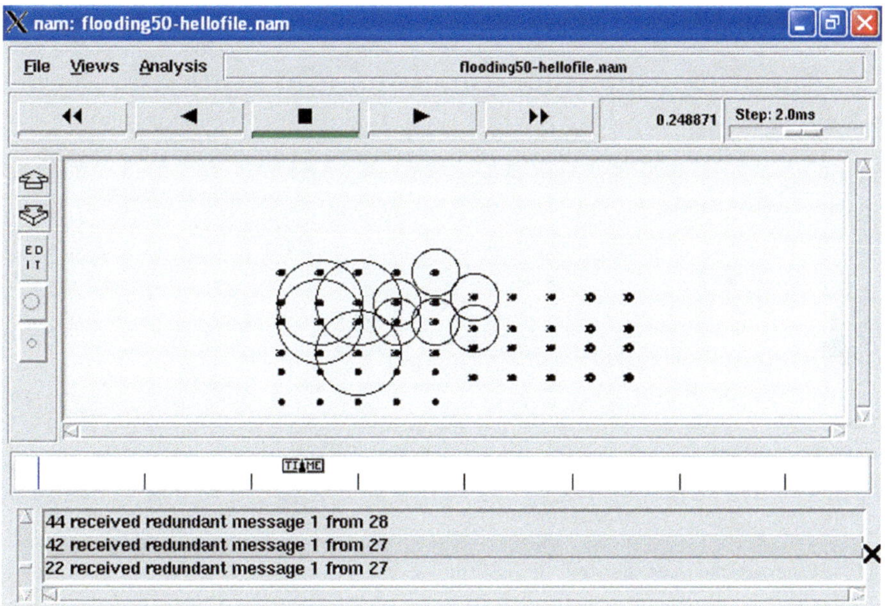

Fig. 4 Output of flooding50-hellofile

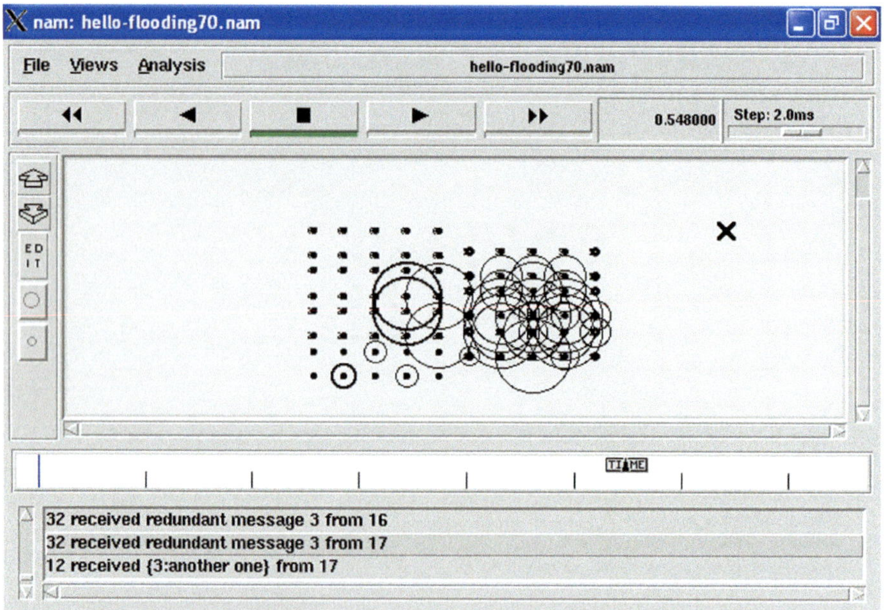

Fig. 5 Output of hello-flooding70

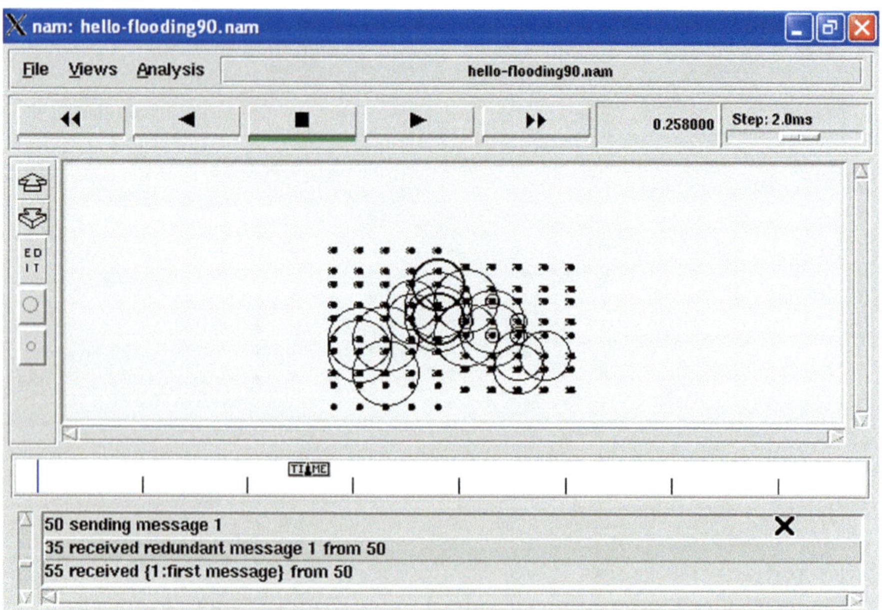

Fig. 6 Output of hello-flooding90

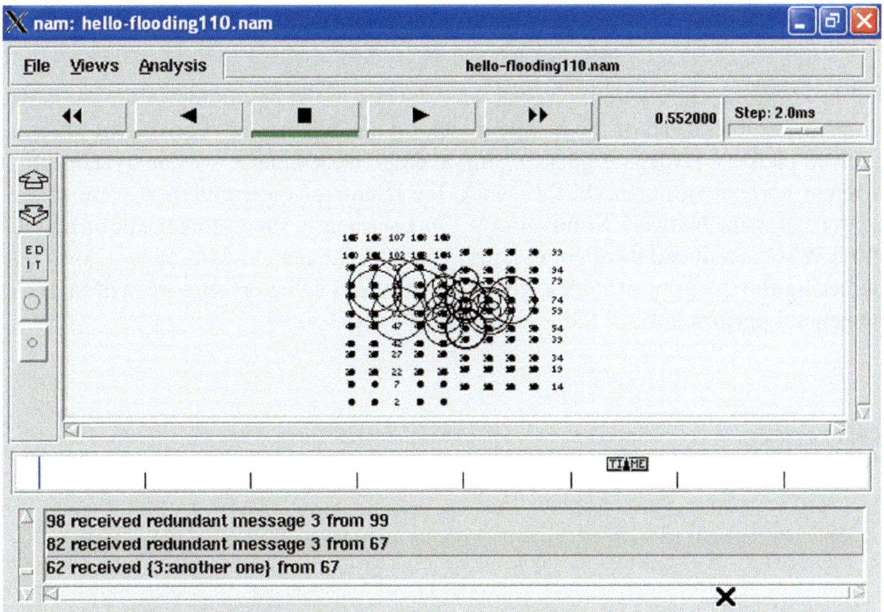

Fig. 7 Output of hello-flooding110

also provides packet delivery rate, total packet loss, packet received, and average performance per kbps.

Table 1 presented all simulation results at a glance. In this table, packet delivery, packets received, total dropped packets, average performance in kbps are provided with variable number of nodes. This table describes general work at a glance. Describe what happens when the flood occurs in the wireless sensor network.

Table 1 Simulation result at a glance

Number of nodes	Packet delivery ratio (PDR)	Received packets	Total dropped packet	Average throughput (in kbps)
10	47.5806	118	130	548.29
30	28.3837	216	545	2123.96
50	28.7356	250	620	1739.73
70	27.5544	658	1730	583.33
90	24.7117	750	2285	1117.3
110	25.1055	1130	3371	1865.51

5 Conclusion

In this work, the clearly defined goal was reached at the beginning of the message. In this paper, the effect of the hi-tide attack on the delivery relationship of packets, received packets, projected packets, and average performance was analyzed with a different number of nodes. In this work, the results of the simulation were implemented using the Network Simulator (NS2). This clearly shows that the performance of the WSN is reduced throughout a flood attack. The result additionally shows that increasing the quantity of nodes within the wireless detector network will increase the general performance of the wireless sensor network.

References

1. Potdar V, Sharif A, Chang E (2009) Wireless sensor networks: a survey. In: IEEE international conference on advanced information networking and applications workshops, pp 636–641
2. Sarigiannidis P, Karapistoli E, Economides AA (2016) Analysing indirect Sybil attacks in randomly deployed wireless sensor networks. In: IEEE 27th annual IEEE international symposium on personal, indoor and mobile radio communications—(PIMRC): mobile and wireless networks, pp 1–6
3. Saghar K, Kendalla D, Bouridane A (2015) RAEED: a solution for hello flood attack. In: IEEE proceedings of 2015 12th international Bhurban conference on applied sciences and technology (IBCAST), pp 248–253
4. Singh S, Saini HS (2018) Security approaches for data aggregation in wireless sensor networks against Sybil attack. In: IEEE 2nd international conference on inventive communication and computational technologies (ICICCT 2018), pp 190–193
5. Shakhov V (2016) On a new type of attack in wireless sensor networks: depletion of battery. In: IEEE information and communication technologies, pp 491–494
6. Mezrag F, Bitam S, Mellouk A (2017) Secure routing in cluster-based wireless sensor networks. IEEE, pp 1–6
7. Kumar S, Goyal M, Goyal D, Poonia RC (December, 2017) Routing protocols and security issues in MANET. In: 2017 international conference on infocom technologies and unmanned systems (trends and future directions) (ICTUS), pp 818–824. IEEE
8. Sagtani VK, Kumar S (2014) Modern approach to enhance routing recitation in MANET. Int J Emerg Technol Adv Eng 4(7):265–270
9. Goyal M, Poonia SK, Goyal D (2017) Attacks finding and prevention techniques in MANET: a survey. Adv Wireless Mob Commun 10(5):1185–1195
10. Gaware A, Dhonde SB (2016) A survey on attacks in wireless sensor network. In: IEEE international conference on computing for sustainable global development, pp 536–539
11. Magotra S, Kumar K (2014) Detection of HELLO flood attack on LEACH protocol. In: IEEE international advance computing conference (IACC), pp 193–198
12. Singh V et al (2019) Source redundancy management and host intrusion detection in wireless sensor networks. Recent Pat Comput Sci 12(1)

Chapter 31
Performance Enhancement of Satellite Communication Subsystem Using 4 × 4 MIMO

Gopesh Sharma, Himanshu Arora, Sourabh Banga, and Vijay Kumar Sharma

1 Introduction

A number of challenges are faced by the wireless system engineers to provide higher data rate and higher channel capacity within available spectrum. These require new technologies that improve channel efficiency as well as energy efficiency. MIMO technology [1] is one of the technologies which promises to provide these capabilities. This paper has proven the possibility of MIMO system over satellite communication. Nowadays, satellite communication is very important wireless communication for a country and its people. Satellite communication has the important applications like military, navigation, radio and broadcast, global telephone, and global positioning system. A satellite communication subsystem provides channel for data transmission, the data may be audio, video, or any other signal format, to users; thus, it generates revenue for satellite designer. So, one should think about to increase channel capacity without increasing the power consumption. To improve the channel capacity, many successive satellites are introduced with large capacity and lower cost, but the number of users are increasing day by day which needs more channel bandwidth and

G. Sharma (✉) · H. Arora · S. Banga · V. K. Sharma
Department of Electronics and Communication Engineering, Arya College of Engineering and Research Centre, Jaipur, Rajasthan, India
e-mail: gopeshsharma92@gmail.com

H. Arora
e-mail: dr.himanshuarora17@gmail.com

S. Banga
e-mail: sourabh.banga07@gmail.com

V. K. Sharma
e-mail: vijaykumarshrm2@gmail.com

Department of Computer Science and Engineering, Arya College of Engineering and Research Centre, Jaipur, Rajasthan, India

© Springer Nature Singapore Pte Ltd. 2021
S. D. Purohit et al. (eds.), *Proceedings of International Conference on Communication and Computational Technologies*, Algorithms for Intelligent Systems,
https://doi.org/10.1007/978-981-15-5077-5_31

higher data transmission rate. So, transponders are used with communication satellite to fulfill user demand. Early transponders of 250 MHz or 500 MHz were used to provide bandwidth at 6 GHz uplink and 4 GHz downlink. Further, satellites use the transponder which provides more bandwidth at 14/11 GHz. For more improvement in channel capacity, bandwidth, and signal-to-noise ratio, there may be combination of MIMO with satellite communication [2]. The satellite designer may inherit the properties of MIMO technology in the design of transponder. Thus, it may provide more bandwidth within limited resources. In satellite MIMO communication subsystem, the available bandwidth can be used very efficiently without any more additional power requirement and with high signal-to-noise ratio, because the MIMO supports very important feature that is diversity [2]. Diversity means "link reliability." There are multiple paths between earth station and the satellite transponder uplink antennas. By using MIMO, the system will able to choose most efficient (bandwidth, power) path to transmit/receive the data due to link reliability. Thus, the path having highest SNR will be chosen. So, there will be maximum utilization of available bandwidth at minimum power consumption. Basically, MIMO is based on beam forming, spatial multiplexing, and diversity coding.

2 Satellite Link Without MIMO

A classical model of satellite link without MIMO contains a downlink transmitter, AWGN channel (without MIMO).

This model contains three main parts.

- Satellite Downlink Transmitter.
- Ground Station Downlink Receiver.

2.1 Satellite Downlink Transmitter

- Random Integer Generator—Creates a random data stream.
- QPSK modulator, modulates the input signal using the quaternary phase shift keying method.
- Memory less Nonlinearity (High Power Amplifier)—Model of a traveling-wave tube amplifier (TWTA) using the Saleh model.
- Gain (Tx. Dish Antenna Gain)—Gain of the transmitter parabolic dish antenna on the satellite.

2.2 Ground Station Downlink Receiver

- Receiver Thermal Noise (satellite Receiver System Temp)—Adds white Gaussian noise that represents the effective system temperature of the receiver.
- Gain (Rx. Dish Antenna Gain)—Gain of the receiver parabolic dish antenna at the ground station.
- Phase Noise—Introduces random phase perturbations that result from 1/*f* or phase flicker noise.
- I/Q Imbalance Compensator—Estimates and removes I/Q imbalance from the signal by a blind adaptive algorithm.
- Phase/Frequency Offset (Doppler and Phase Compensation)—Rotates the signal to represent correction of phase and Doppler error on the link. This block is a static block that simply corrects using the same values as the Phase/Frequency Offset block.
- QPSK demodulator, demodulates the input signal using the quaternary phase shift keying method.

Figure 1 is a part of satellite downlink receiver without MIMO, which contains a QPSK demodulator to get the original signal. A block named with SNR estimation is used to estimate the SNR of received signal. A SNR.mat block is used to store the estimated values of received signal's SNR.

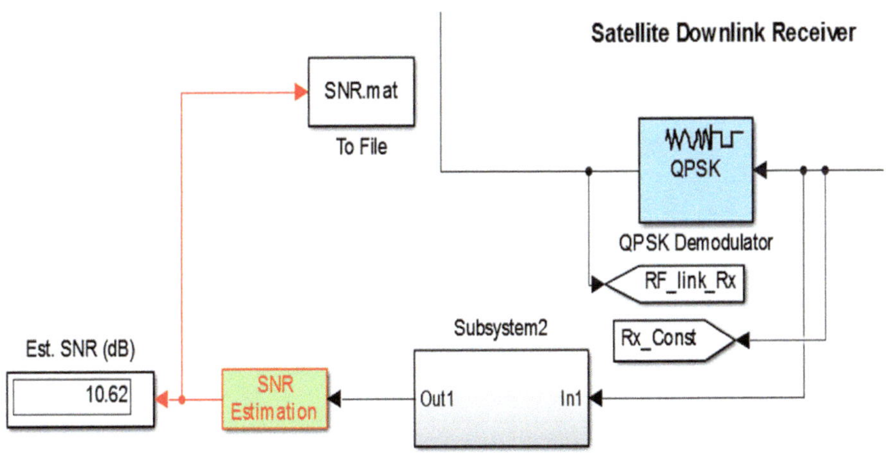

Fig. 1 Zoomed view of satellite downlink receiver without MIMO

3 Satellite Link with MIMO (Proposed Model)

This model is a model of satellite communication subsystem having the features of new technology named MIMO. A transponder is responsible to provide communication facilities to the users. This model is modeled based on adaptive MIMO technology [3]. And orthogonal space time block coding (OSTBC) is used as diversity coding. And, the signal-to-noise ratio is calculated. And a frame error rate calculation block is also added in the proposed model to calculate the total number of transmitted frames, upcoming errors, and frame error rate (FER). SNR estimation is used to estimate the SNR of receiving signal. A block is showing the calculated SNR in dB. SNR.mat block is used to store the value of calculated SNR each time to show the appropriate graph (Fig. 2).

This model has three basic blocks as given

- Satellite Downlink Transmitter
- Downlink path
- Ground Station Satellite Downlink Receiver.

Further Fig. 3 shows the calculated SNR of a satellite link which is fully featured

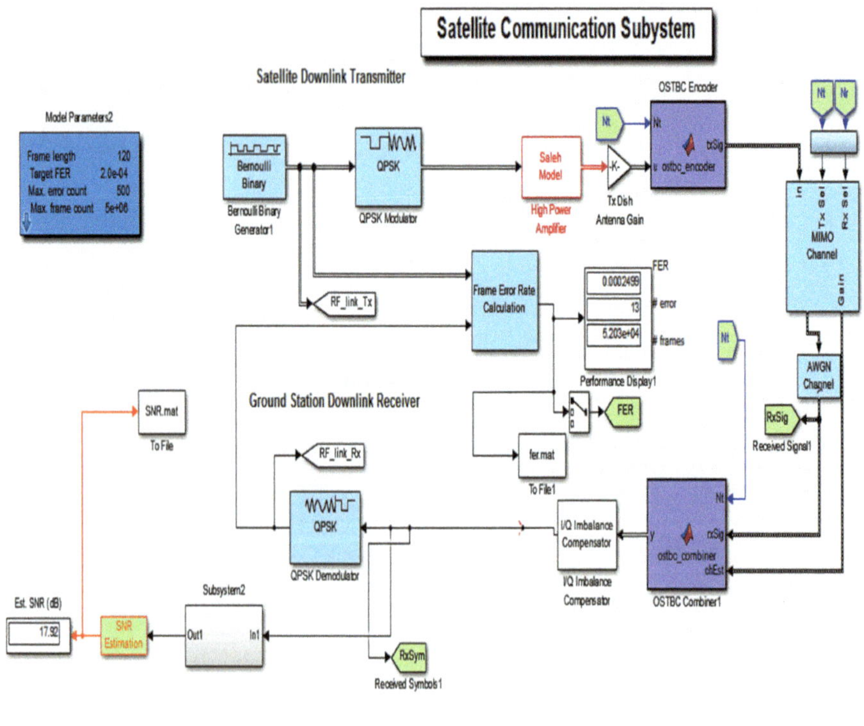

Fig. 2 Proposed MIMO satellite communication subsystem

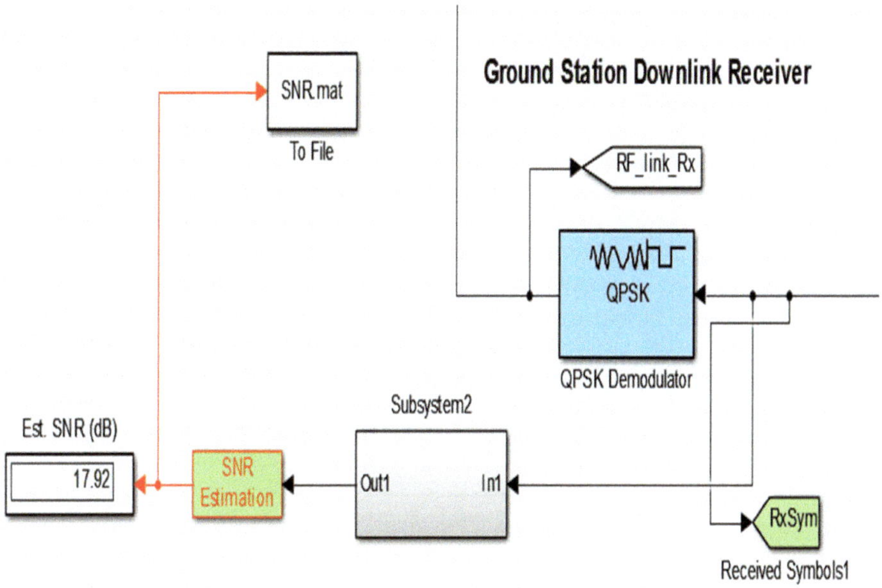

Fig. 3 Zoomed view of satellite downlink receiver with MIMO

with MIMO and shows a downlink receiver part of satellite link with the use of MIMO. The QPSK demodulator is use to demodulate the received signal. SNR estimation is used to estimate the SNR of receiving signal.

4 Simulation and Results

This section is showing the simulated results of the proposed satellite link with MIMO. Figure 5 is showing the estimated values of SNR with respect to time. It shows that the SNR is approximately 18 dB. At time $t = 1$ s, the value of SNR is slightly improving just because of low FER. At time $t = 3$, once again the same thing is happening. Figures 4 and 5 show the basic relation between the SNR and FER. Both parameters are inversely proportional to each other. At time $t = 1$ s (approximately), the FER (Fig. 6) is decreasing, and at the same time, the SNR (Fig. 4) is increasing. Figure 5 is showing that the FER is decreasing with respect to time. The model is set to get target FER which is approximately 2×10^{-4}. At time $t = 1$ s, it is showing the graph which is decreasing, and Fig. 4 shows that the SNR is increasing.

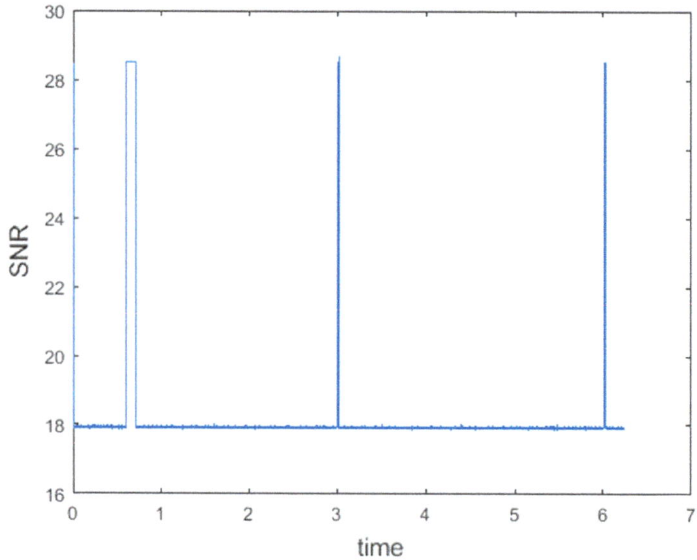

Fig. 4 SNR (dB) versus time

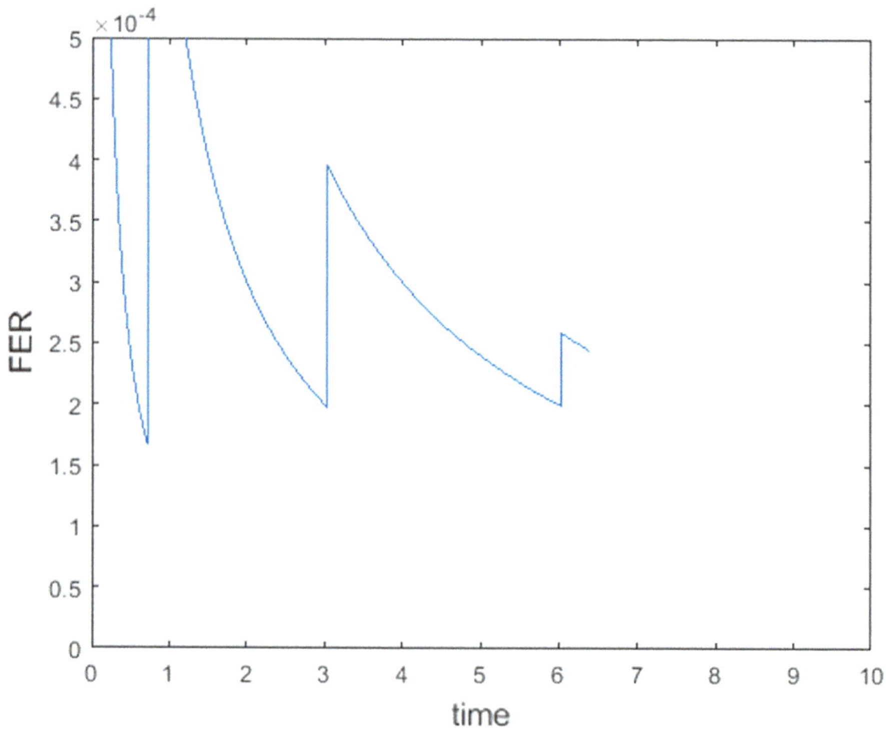

Fig. 5 FER versus time

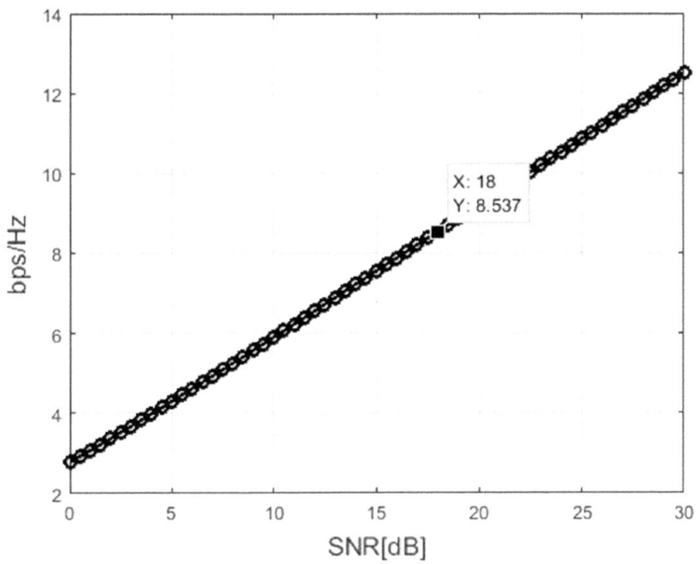

Fig. 6 Capacity versus SNR (with MIMO)

Table 1 Comparison of SNR

System	SNR in dB
Satellite link without MIMO	11
Satellite link with MIMO	17.92

4.1 Comparison of SNR

Table is showing the comparison between the satellite link with MIMO and without MIMO. It is showing that the satellite link without MIMO has the value of SNR 11 dB, and a satellite link with MIMO has the value of SNR 17.92 dB. The satellite link with MIMO has the more value of SNR as compared to satellite link with MIMO (Table 1).

4.2 Comparison of Capacity

Figure 6 shows the comparison in capacity, the satellite link with MIMO has the capacity 8.537 bps/Hz, and Fig. 7 shows the satellite link without MIMO has the capacity 6.22 bps/Hz.

The above graph is showing the value of capacity with respect to SNR, when MIMO is not in use.

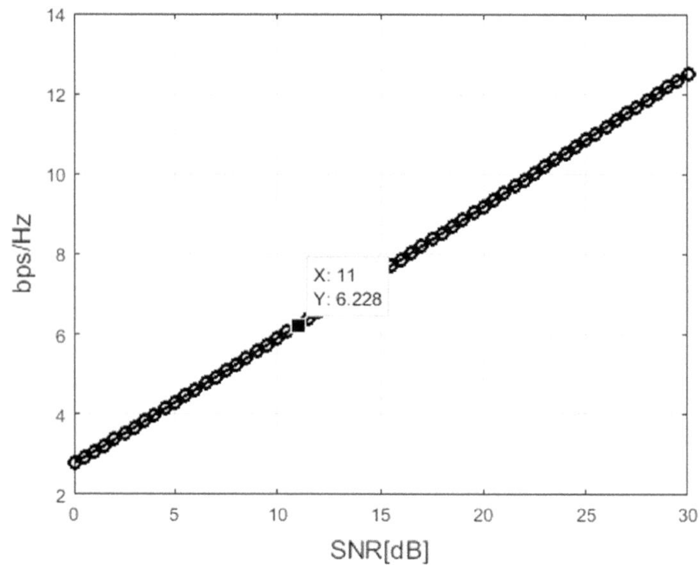

Fig. 7 Capacity versus SNR (without MIMO)

Table 2 Comparison of capacity

System	Capacity in bps/Hz
Satellite link without MIMO	6.228
Satellite link with MIMO	8.537

4.3 Comparison of Capacity

See Table 2.

5 Conclusion

The whole dissertation describes the implementation of MIMO technology in satellite communication subsystem and is shown by using MIMO technology that the SNR is improving, and thus, the channel capacity is improving. MIMO's features like spatial multiplexing and diversity [4] are responsible for increment in data rate and decrement in fading, etc., within available limited resources (energy efficient). So, it has been proven that the channel capacity of a satellite communication subsystem can be improved by using the MIMO technology.

6 Future Scope

The proposed model with MIMO can further modify with the use of massive MIMO to improve overall performance of satellite communication subsystem.

References

1. Sharma G, Tharani L (April, 2016) Application of MIMO in telemetry tracking command and monitoring to control a satellite. In: Proceedings of IEEE international conference on computing, communication and automation (ICCCA). Galgotias University, Noida, pp 29–30
2. Sharma G, Tharani L (9–11 September, 2016) Application of MIMO in satellite communication subsystem. In: Proceedings of Taylor and Francis Group CRC press International Conference on Communication and Computing System (ICCCS), Dronacharya College of Engineering Gurgaon Haryana, pp 27–28
3. Ngo HQ, Larsson EG, Marzetta TL (April, 2013) Energy and spectral efficiency of very large multiuser MIMO systems. IEEE Trans Commun 61:1436–1449
4. Chong Z, Jorswieck EA (May, 2011) Energy efficiency in random opportunistic beamforming. In: Proceedings of the 73rd vehicular technology conference (VTC-Spring'11). Budapest, Hungary. Piscataway, NJ, USA: IEEE, pp 15–18

Chapter 32
Design of Compact Dual-Band Antenna for Wi-Max and WLAN Applications

Megha Shringi, Rajveer Singh, and M. L. Meena

1 Introduction

Microstrip antennas are widely used in wireless communication applications due to their small size, lightweight, low cost, and easy integration with different feeds. They can be easily integrated with circuits and allow multi-frequency operation to be achieved. The designing of an antenna is based on its geometry, dimensions, and characteristics of the dielectric substrate [1]. On the basis of these aspects, radiation pattern, directivity, resonant frequencies, return loss, and other parameters show major variations.

The compact design of proposed antenna gives it edge over other antennas at same frequency bands [2–8]. The geometry of the antenna is one of the important parameters as stated above; hence, many antennas are designed keeping in concern the different geometries. The fork shape antenna [9], Gemini shape [10], and various other complicated structures make it difficult to fabricate such antennas. The proposed antenna has simple geometry and shape which gives ease in manufacturing and integration.

M. Shringi (✉) · R. Singh
Rajasthan College of Engineering for Women, Jaipur, Rajasthan, India
e-mail: meghashringi6@gmail.com

R. Singh
e-mail: rajv.singh.2007@gmail.com

M. L. Meena
Department of Electronics Engineering, Rajasthan Technical University, Kota, Rajasthan 324010, India
e-mail: madan.meena.ece@gmail.com

© Springer Nature Singapore Pte Ltd. 2021

S. D. Purohit et al. (eds.), *Proceedings of International Conference on Communication and Computational Technologies*, Algorithms for Intelligent Systems,
https://doi.org/10.1007/978-981-15-5077-5_32

2 Antenna Design

The geometrical layout of the proposed microstrip patch antenna is depicted in Fig. 1 with the dimensions as given in Table 1. Designed antenna consists of three layers which are patch, substrate, and partial ground. In this antenna, FR-4 (lossy) dielectric substrate is used. The dimension of the substrate is 23 × 22 × 1.5 mm^3. Copper annealed is used as the patch and ground material. It is the shape of the antenna that makes it possible to resonate at different frequencies.

Proposed design consists of modified structure of conventional rectangular microstrip patch antenna with symmetrical stepped slots at left and right sides of

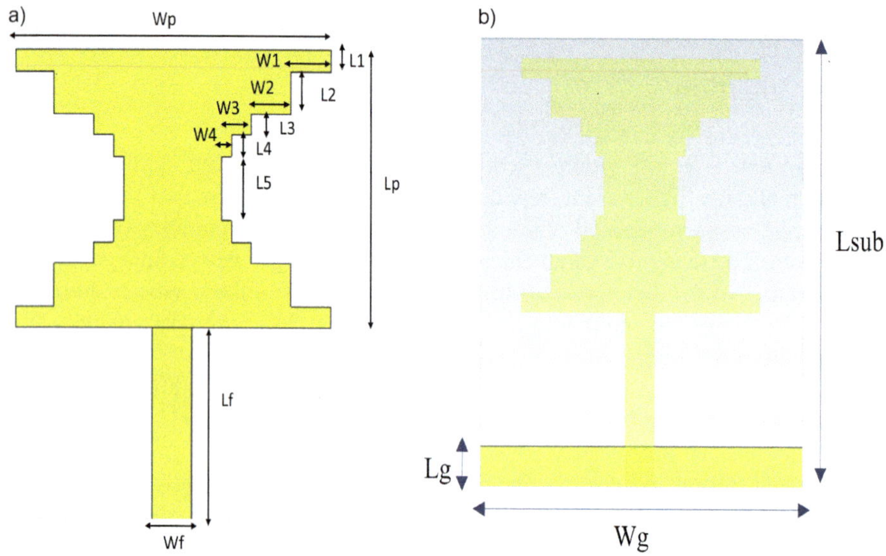

Fig. 1 Geometry of the proposed antenna **a** front view **b** back view

Table 1 Design parameters of proposed antenna

Parameter	Value (mm)	Parameter	Value (mm)
Lg	2.15	*L*1	1
Wg	22	*W*1	2
*L*sub	23	*L*2	2
*W*sub	22	*W*2	2
*T*sub	1.5	*L*3	1
Lf	9	*W*3	1
Wf	2	*L*4	1
Lp	13	*W*4	0.5
Wp	16	*L*5	3

the radiating patch above feed line. The radiating patch has a length *L* and width *W* as shown in Fig. 1a. The *Wf* and *Lf* are the width and length of the feed line, which connect the patch with SMA connector. On the other side of the substrate, a conducting partial ground plane is placed.

The dimensions of the partial ground plane are width *Wg* and length *Lg* (see Fig. 1b). The width of the microstrip feed line is fixed at 2 mm. The antenna is printed on a $23 \times 22 \times 1.5\ \text{mm}^3$ and a FR-4 epoxy substrate of thickness 1.5 mm, permittivity 4.3, and loss tangent $\tan\delta = 0.002$. The design parameters to reach the proposed antenna are optimized [11] as given in Table 1.

3 Results and Discussions

This section presents the performance results of the proposed antenna at WLAN and Wi-Max communication system frequencies at 2.6 and 5.1 GHz. The performance results such as return loss, VSWR, surface current, and radiation pattern at the desired frequencies are analyzed.

3.1 Parametric Analysis

This section presents the parametric analysis of various design parameters on the antenna performance. It has been noticed in the simulation that the operating bandwidth and resonant frequencies of the antenna is dependent on the width of the feed line (*Wf*) and the length (*Lg*) of partial ground. So, these parameters have to be optimized for optimum return loss for proposed antenna.

3.1.1 Effect of Width of the Microstrip Feed Line (*Wf*)

Figure 2 shows the variations in the values of S11 of the antenna with the variations in width of the feed line. Them simulated values are quite symmetrical at the two resonant frequencies when *Wf* =2 mm and the curve deviates from the symmetry for other values of *Wf*. It can be seen that the resonance frequency shifts from 2.6 to 5.1 GHz as *Wf* changes. Also, it can be noted that the value of S11 is high when *Wf* = 1.25 mm at the 5.1 GHz resonant frequency. This shows that the impedance of the planar antenna matches with that of 50 Ω SMA connector.

3.1.2 Effect of Partial Ground Plane Length (*Lg*)

Figure 3 shows the variations in the values of S11 of the antenna with the variations in length of the ground plane. The simulated S11 is quite symmetrical at the two

Fig. 2 Effect of width of feed line on return loss

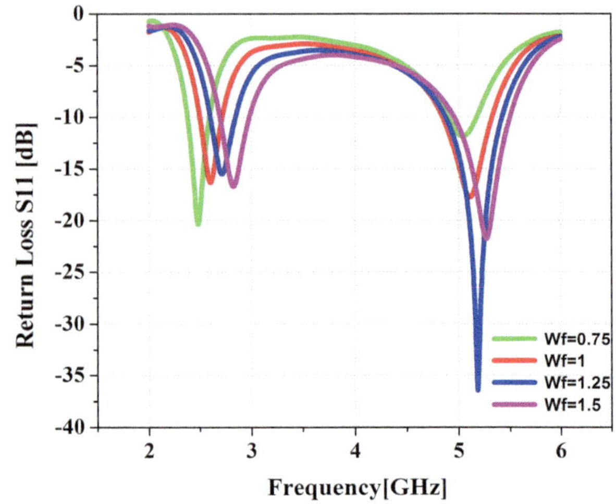

Fig. 3 Effect of length of partial ground on return loss

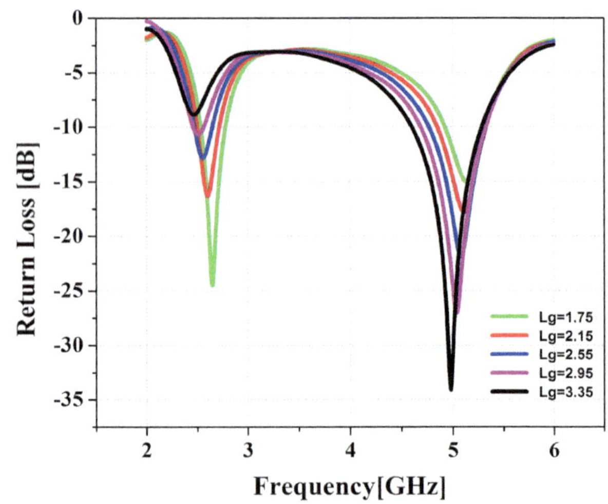

resonant frequencies when $Lg = 2.15$ mm and the curve deviates from the symmetry for other values of Lg. There is shift in the frequency and value of S11 is greatly affected. For instance, it can be noted that the value of S11 is low when $Lg = 2.55$ mm (2.6 GHz) and $Lg = 1.75$ mm (5.1 GHz). This shows that the impedance of the planar antenna matches with that of 50 Ω SMA connector.

Figure 4 shows the return loss of the proposed antenna for three operating frequency bands. From the parametric analysis, the optimum reflection coefficient (S11) for the first radiator at 2.6 GHz (2.49–2.71 GHz) is −16.33 dB, and the bandwidth is about 219 MHz ($f1$–$f2$). While the S11 for a second radiator is −17.78 dB at frequency of 5.1 GHz (4.84–5.32 GHz), the simulated bandwidth for second radiator is at 484 MHz ($f1$–$f2$).

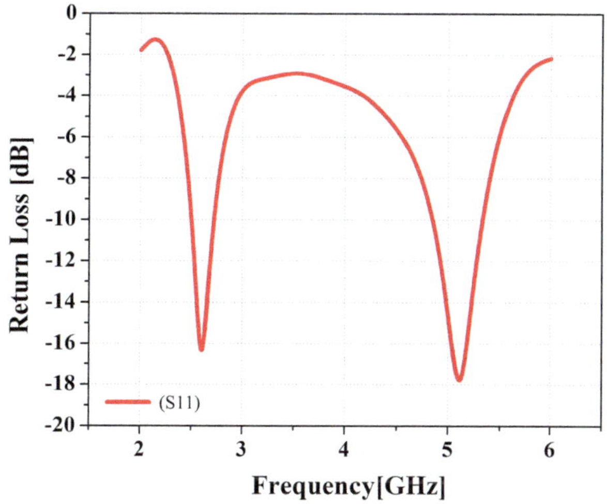

Fig. 4 Reflection coefficient S11 (dB)

3.2 *Voltage Standing Wave Ratio (VSWR)*

It can be seen from VSWR graph (Fig. 5) that this parameter is less than 1.5 for both the frequencies and hence shows good impedance matching.

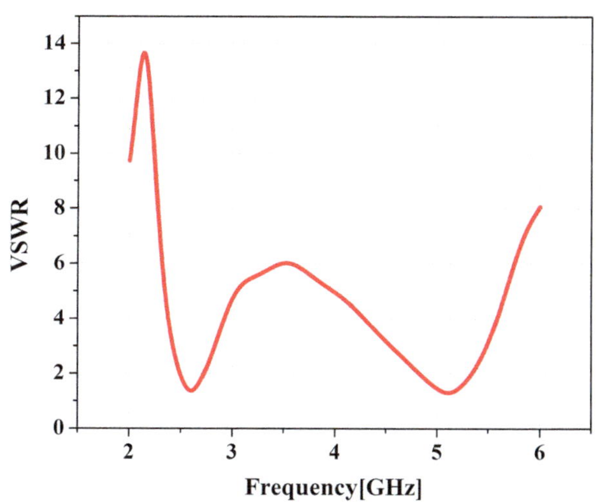

Fig. 5 Voltage standing wave ratio

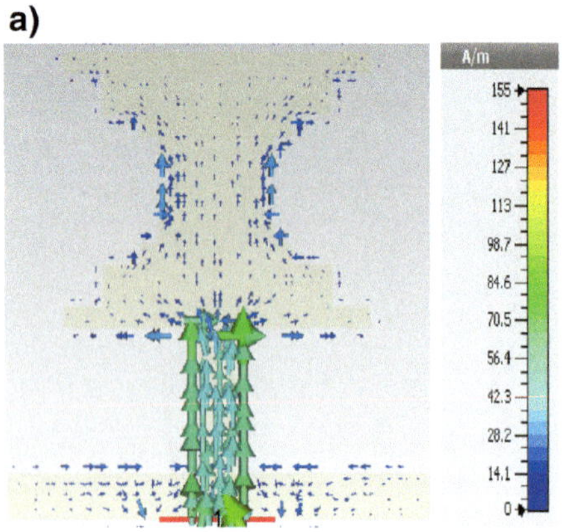

Fig. 6 Current density at **a** 2.6 GHz **b** 5.1 GHz

3.3 Surface Current Density

It is found that by symmetrically incorporating the slots at right and left edges of the radiator patch, additional resonance frequency at 5.1 GHz is excited. Figures 6a, b demonstrate the current distributions at 2.6 GHz and 5.1 GHz, respectively [12]. As we can see, the currents at 5.1 GHz mostly concentrate on the slots at left and right sides of patch which clearly indicates that the slots generate the second resonant frequency at 5.1 GHz together, and the operating frequency of microstrip patch is 2.6 GHz which is the first resonance frequency.

3.4 Radiation Field Patterns

Nearly, omni-directional patterns are achieved at central frequencies of 2.6 GHz and 5.1 GHz, respectively, as shown in Fig. 7a, b.

4 Measured Results

The antenna design is optimized, and the measured or experimental result has been obtained using vector network analyzer (Fig. 8a). Antenna prototype has been fabricated (Fig. 8b). The simulated and experimental result shows reasonable agreement. The slight variations in results are due to fabrication tolerance.

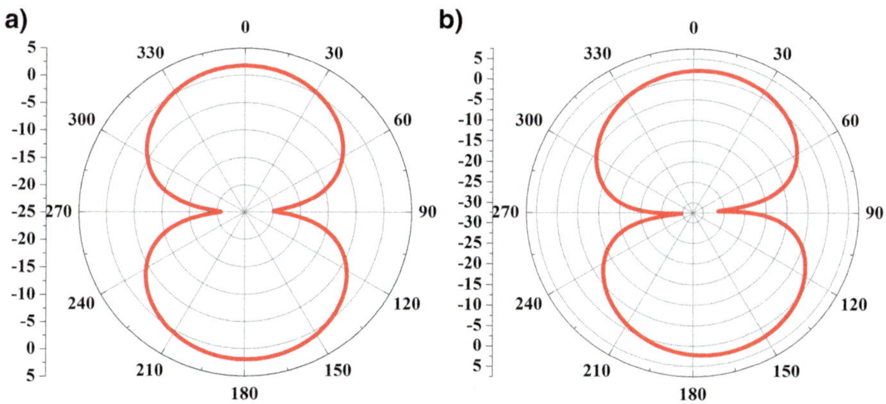

Fig. 7 Radiation pattern at **a** 2.6 GHz **b** 5.1 GHz

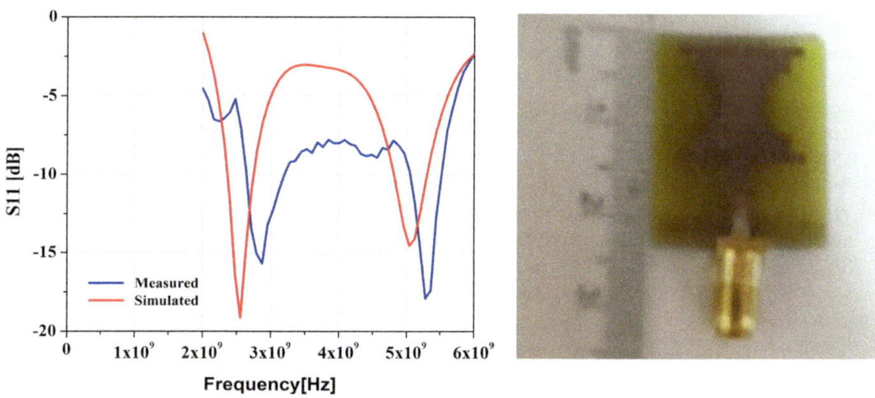

(a) (b)

Fig. 8 **a** Comparison between measured and simulated S11 **b** fabricated antenna

5 Conclusion

A simplified design of a compact printed dual-band monopole antenna capable of wireless applications in 2.6 and 5.1 GHz range has been presented. The simulated results show that the antenna exhibits a very low return loss (with VSWR < 1.5) at both the resonating frequencies 2.6 and 5.1 GHz. Also, the compact size and radiation elevation field patterns of the antenna show very low side lobes, thus making it suitable and economical for the wireless applications.

References

1. Alam MJ, Faruque MRI, Hossain MI, Islam MT, Abdullah S (2017) A combined double H-shaped microstrip patch antenna for X-band operation. In: International Conference on Electrical, Computer and Communication Engineering (ECCE). Bangladesh, pp 514–517
2. Boutejdar A, Halim BI, El Hani S, Bellarbi L, Afyf A (2018) Design of multiband microstrip antenna using stepped cut method for WLAN/Wi-Max and C/Ku-band applications. Int J Electr Commun Eng 12(6)
3. Thaher RH, Jamel ZS (2018) New design of dual-band microstrip antenna for Wi-Max and WLAN applications. In: 1st international scientific conference of engineering sciences—3rd scientific conference of engineering science (ISCES), pp 131–134
4. Sahithi S, Kumar SA, Shanmuganantham T (2017) A U-slot microstrip patch antenna for Wi-Max applications. In: Proceedings of 2017 IEEE international conference on circuits and systems (ICCS), pp 109–111
5. Tang X, Jin S (2015) CPW-fed dual-band antenna for WLAN/Wi-Max applications. In: IEEE sponsored 2nd international conference on electronics and communication system (ICECS), pp 1525–1529
6. Parbat RS, Mahamine SD, Bodake SH, Aher MP (2016) Dual polarized triple band microstrip antenna for GSM/Wi-Max/WLAN applications. In: International conference on automatic control and dynamic optimization techniques (ICACDOT), pp 1137–1141. International institute of information technology (I^2IT), Pune
7. Boutejdar A, Salamin MA, Bennani SD, El Hani S (2017) Design of compact monopole antenna using double U-DMS resonators for WLAN, LTE, and Wi-Max applications. TELKOMNIKA 15(4):1693–1700
8. Kumari A, Badhai RK (2017) A dual-band high-gain base-station antenna for WLAN and Wi-Max applications. In: International conference on innovations in information, embedded and communication systems (ICIIECS)
9. Shi F, Jiang T, Li Y (2017) A fork-like dual-band antenna with an inverted U-shaped parasitic element for WLAN and WIMAX applications. In: Progress in electromagnetic research symposium—fall (PIERS—FALL). Singapore, pp 124–127
10. Djengomemgoto G, Altunok R, Karabacak C, Imeci T, Durak T (2017) Dual-band gemini-shaped microstrip patch antenna for C-band and X-band applications. In: International applied computational electromagnetics society symposium (ACES). IEEE, Italy
11. Balanis C (2005) Antenna theory analysis and design, 3rd edn. Wiley, Inc.publications
12. Singh V, et al. (2019) Source redundancy management and host intrusion detection in wireless sensor networks. Recent Pat Comput Sci 12(1)

Chapter 33
Quantum Hilbert Image Scrambling and SVD-Based Watermarking Scheme for Rightful Ownership

Naveen Kumar and Vijay Kumar Sharma

1 Introduction

The wider availability of Internet, computer system, and its related hardware equipment, such as printers and scanners makes the date exchange simple tasks. However, making the digital data accessible there is some chance to make the salable copies of the copyrighted content without any kind of authorization of the owner of contents. The digital watermarking schemes are used to protect the copyright of multimedia file or data [1, 2].

An efficient watermarking technique must have the following properties.

(1) Imperceptibility: it states that the visual difference between the original and watermarked file must be unnoticeable to the observer, or in other word, we can say that watermark should not be visible with the media being confined or protected.
(2) Trustworthiness: an acceptable watermarking system must be guaranteed that it is unfeasible to generate fake watermarks and should present trustworthy proof to defend the correct or rightful ownership.
(3) Robustness: for the given watermarked file or document, an unconstitutional or unauthorized person should not be capable to demolish the watermark, that is the watermark should be robust against intentional and common attacks.

Steganography and watermarking are some similar techniques but have different usefulness according to the need for data protection. Steganography is used for

N. Kumar (✉) · V. K. Sharma
Rajasthan Institute of Engineering and Technology, Jaipur, Rajasthan, India
e-mail: halonaveenkumar@gmail.com

V. K. Sharma
e-mail: vijaymayankmudgal2008@gmail.com

V. K. Sharma
Manipal University Jaipur, Jaipur, Rajasthan, India

© Springer Nature Singapore Pte Ltd. 2021

S. D. Purohit et al. (eds.), *Proceedings of International Conference on Communication and Computational Technologies*, Algorithms for Intelligent Systems,
https://doi.org/10.1007/978-981-15-5077-5_33

hidden communication of important documents. It is an art of hiding secret information whereas the watermarking is used to introduce the trustworthiness, copyright protection of information.

Watermarking is the process of embedding a message on the multimedia data [3, 4]. Placing a watermark in an image or other media file serves to identify the artist or author of the work, i.e., it is used for copyright protection. It is not so much an attempt to hide a message as it is to tag a document for later identification. A watermark can be either visible or invisible.

2 Related Work

A lot of work have been done in watermarking. Discrete wavelet transform (DWT) and SVD-based watermarking scheme are proposed by [5]. In [6], fragile watermark-based technique using self-fractional Fourier function (SFFF) is proposed. A Hilbert quantum scrambling-based technique is proposed in [7]. It is a progressively generation of scrambling image by using sequence of Hilbert map shapes. SVD- and DWT-based robust watermarking is proposed in [8], and here the technique is robust against some powerful attacks. The concept of watermarking scheme is quite similar to steganography. An alpha blending is used to mix up the secret image (watermark in watermarking) into cover image. The alpha blending-based steganography using graph wavelet is proposed in [9]. Arnold transform-based watermarking scheme is presented in [10], and Arnold transform is used to increase the confidentiality by scrambling the image and this scrambled image is hidden inside the cover image and got the watermarked image. Arnold scrambling increases visual imperceptibility of image. Various image segmentation [14] and watermarking [15] schemes are proposed by researchers with traditional and non-traditional approaches [11–13].

3 The Proposed Watermarking Scheme

In this paper, we proposed the singular value decomposition (SVD) method for implanting the watermark which is mainly used in almost all image processing applications such as image watermarking and image compression etc.

The watermark is embedded into the host media according to the category of embedding (i.e., spatial domain, transform domain) [8]. The strength of watermark depends on the scaling factor or embedding strength factor (alpha) used during the embedding process. In current work, alpha is used to control the strength of embedding the watermark. Alpha is used to mix up two image signals jointly and generate the new image. Larger the value of alpha (α) results in more alteration or distortion of the host media quality and which also results in the stronger robustness. On the other hand, lesser α results in better host image quality and weaker robustness.

The proposed method consists of two processes—the watermark embedding and retrieving process. In watermark embedding process, first we scrambled the watermark image using quantum Hilbert image scrambling and get scrambled watermark image. This quantum Hilbert image scrambling introduces more security and robustness in the watermarking. Then, allying the SVD-based watermark hiding process is followed by alpha blending method. The alpha here denotes the strength of embedding. The watermark retrieving process is the just overturn of the embedding process. So, the reverse steps are applied to extract the watermark. The proposed method introduces the good visible quality inside the watermarked image and also more confidentiality is introduced in the watermark using the quantum scrambling method.

3.1 Scrambling-Based Quantum Hilbert Image Scrambling

Hilbert image scrambling is done with the help of scrambling quantum circuits [7]. In Hilbert scrambling, the quantum circuits are recursive and layered progressively to generate the scrambled image. Using the Hilbert curve, the image can be scrambled efficiently. In the year 1890, G. Peano presented the family of curves which can pass throughout all position in space (Fig. 1).

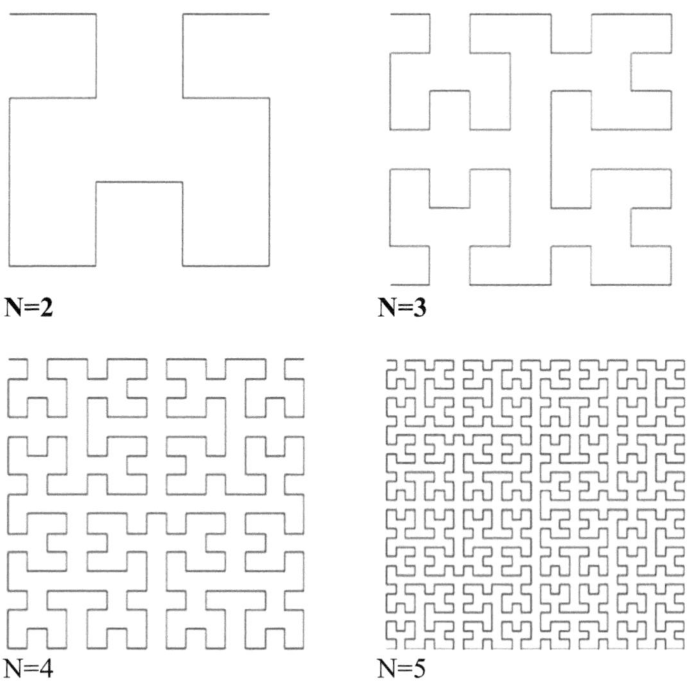

Fig. 1. Hilbert curve image

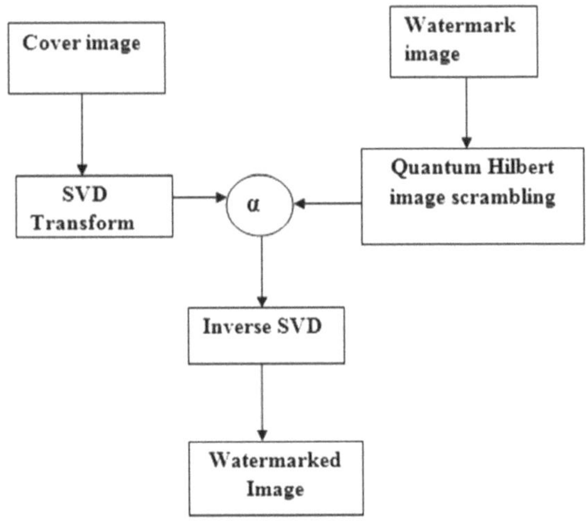

Fig. 2. Block diagram encoding watermark image

The algorithmic steps of embedding process are shown in Fig. 2.

Step 1. Take cover image as input

Step 2. Apply SVD transform at host image, and this can be done by applying following

$$X = USV^{\mathrm{H}} \quad (1)$$

where U and V are unitary matrices and S is a real diagonal matrix that has the singular values (SVs) of X.

Step 3. Take a watermark image as input

Step 4. Scramble watermark using quantum Hilbert image scrambling

Step 5. Perform alpha blending operation and generate an image singular value acquired that is added to the every pixel value of quantum Hilbert image scrambled watermark (w) to get watermarked image.

$$S' = S + \alpha \mathrm{w} \quad (2)$$

Here, S' is the updated S matrix which includes the watermark information, α represents the strengthening factor for watermark, and w is scrambled version of watermark image.

Step 6. Apply inverse SVD at S' image and generate watermarked image, by modifying singular value of generated image after blending.

$$US'V^{\mathrm{H}} = \mathrm{SW} \quad (3)$$

Here, U and V^{H} are the matrixes as obtained in Eq. (4.1) and SW is the watermarked image

The watermark extraction process is just the reverse process of the above embedding process.

4 Experimental Results and Analysis

The performance of the proposed method is evaluated by implementing it. By taking lenna.tiff as the host image or cover image, pepper.tiff as the watermark image, the consequent experimental results are shown in Fig. 3.

The performance of the projected method is measured in terms of some of image quality measurements like peak signal-to-noise ratio (PSNR), mean square error (MSE), and normalized cross-correlation (NCC) of cover image and the watermarked image.

Average difference (AD), maximum difference (MD), and structure contents(SC) [9, 10].

The experimental results of the proposed techniques are tested on the different host image and the watermark image at the varying values of alpha. Table 1 shows the result obtained after performing the proposed technique in terms of image quality measurements by taking different host image and same watermark image (Logo.jpg image). Here, we derive results of proposed technique but they are not compared to the existing technique, because of not availability of the results on all parameters. So, we compared our technique with the other existing techniques on PSNR and NCC.

The image of logo is taken from the existing technique [8] of Shandong University which is as shown in Fig. 4.

Figure 5 shows the obtained results after performing the proposed technique.

Bar graph has been plotted for measurement of the corresponding results shown in Table 1, and it can be clearly seen from the graph that the proposed technique gives the excellent results than the existing technique (Tables 2 and 3; Figs. 6 and 7).

Performance of the proposed method is measured on different picture quality measurements as shown in the table. Our techniques provide excellent results than the previously proposed technique or existing technique [8].

5 Conclusion

The proposed SVD and Hilbert scrambling-based watermarking technique introduced more security and robustness. Watermark is invisible inside the cover image because of the use of Hilbert scrambling which introduced excellent visual quality in watermarked image. It is clearly observed from the results.

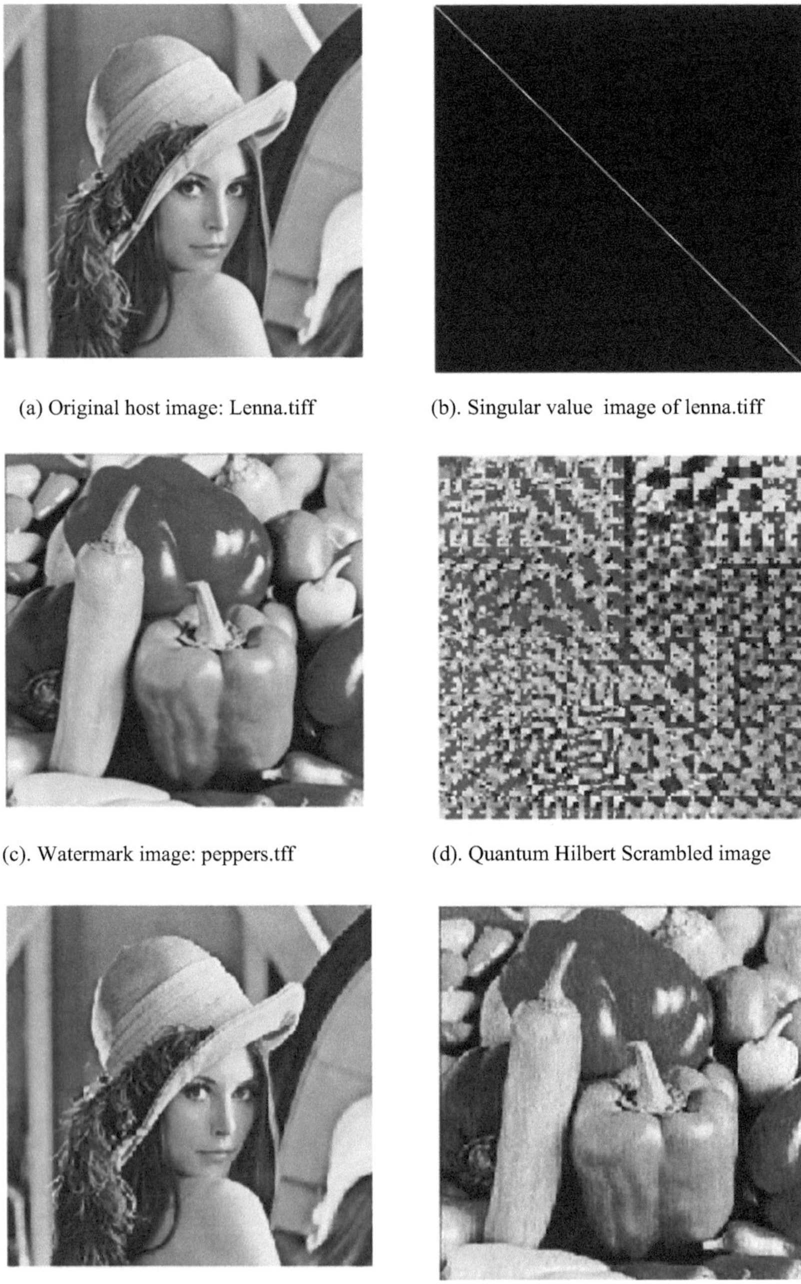

(a) Original host image: Lenna.tiff

(b). Singular value image of lenna.tiff

(c). Watermark image: peppers.tff

(d). Quantum Hilbert Scrambled image

(e). Watermarked image

(f). Extracted image

Fig. 3. Encoding and decoding of cover image (lenna.tiff) and watermark image (pepper.tiff)

Table 1 Comparison of proposed watermark technique with the existing technique in Psnr

Cover image	Proposed technique	Existing technique
Lenna	81.0542	49.07
Peppers	81.1740	48.28
Barbara	80.6473	49.99
Boat	77.6532	49.09
Airplane	77.4074	49.13
Baboon	77.6061	49.06
Average PSNR =	475.5422/6 = 79.25703	294.62/6 = 49.10333

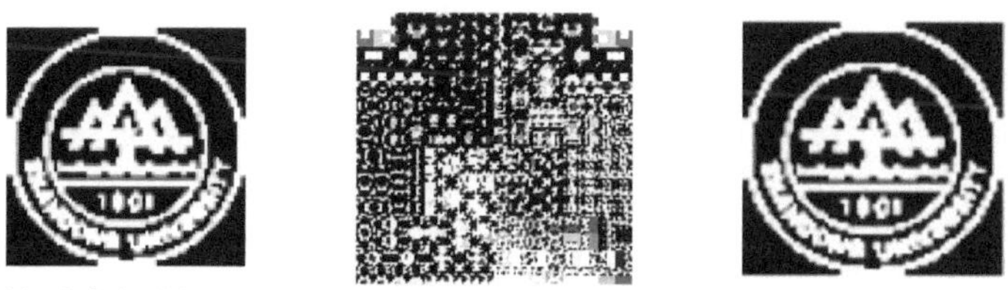

(a). Original logo (b). scrambled logo (c). Extracted logo

Fig. 4. Original logo image of Shandong University, scrambled logo and extracted logo after extraction process

Fig. 5. Original cover image and their corresponding watermarked image and extracted watermark

Table 2 Comparison of proposed watermark technique with the existing technique in Ncc

Cover image	Proposed technique	Existing technique
Lenna	1.0000	0.5014
Peppers	1.0000	–
Barbara	1.0000	0.4883
Boat	1.0000	–
Airplane	1.0000	–
Baboon	1.0000	0.4910

Table 3 Comparison of the proposed technique with the existing on AD, SC, and MD

Cover image	AD	SC	MD
Lenna	−0.0019	0.9999	0.1624
Peppers	−0.0020	0.9999	0.1760
Barbara	−0.0019	0.9999	0.1539
Boat	−0.0019	0.9999	0.1940
Airplane	−0.0020	1.0000	0.2075
Baboon	−0.0020	0.9999	0.2269

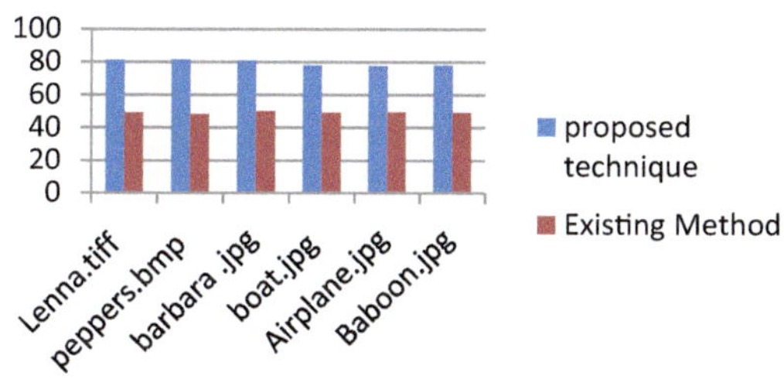

Fig. 6. Bar graph for PSNR values obtained in proposed work and previous work

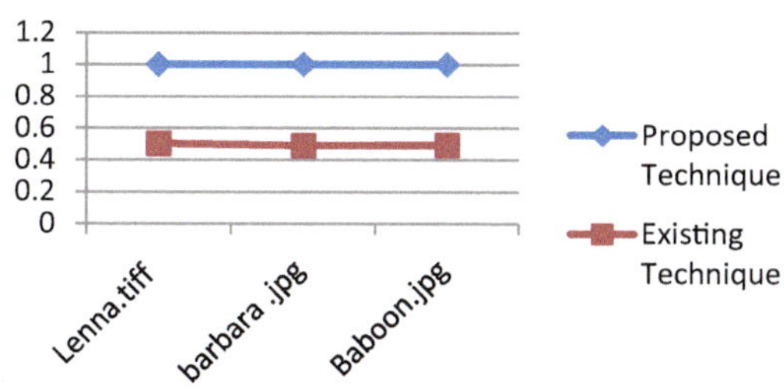

Fig. 7. Bar graph plotted for NCC values obtained in proposed work versus previous work

References

1. Fazli S, Moeini M (2016) A robust image watermarking method based on DWT, DCT, and SVD using a new technique for correction of main geometric attacks. Opt Int J Light Electron Opt 127:964–972
2. Singh D, Singh SK (2017) DWT-SVD and DCT based robust and blind watermarking scheme for copyright protection. Mult Tools Appl 76:13001–13024
3. Mishra A, Agarwal C, Sharma A, Bedi P (2014) Optimized gray-scale image watermarking using DWT-SVD and firefly algorithm. Expert Syst Appl 41:7858–7867
4. T. Kumar Tewari, V. Saxena (2018) Collusion attack resistant SVD based secured audio watermarking scheme with chaotic encryption. In: 2018 11th international conference on contemporary computing (IC3), Noida, pp 1–5
5. P Shah, T Meenpal, A Sharma, V Gupta, A Kotecha (2015) Notice of removal: a DWT-SVD based digital watermarking technique for copyright protection. In: 2015 international conference on electrical, electronics, signals, communication and optimization (EESCO), Visakhapatnam, pp 1–5
6. Sharma JB, Sharma KK, Sahula (2013) Digital image dual watermarking using self-fractional fourier functions, bivariate empirical mode decomposition and error correcting code. J Optics 42(3)214–227
7. Jiang N, Wang L, Wu W-Y (2014) Quantum Hilbert image scrambling. Int J Theor Phys 53(7):2463–2484
8. Zhang H, Wang C, Zhou X (2017) A robust image watermarking scheme based on SVD in the spatial domain. Fut Int 9(3):45
9. Sharma VK, Srivastava DK, Mathur P (2018) Efficient image steganography using graph signal processing. IET Image Proc 12(6):1065–1071
10. Khalili Mehdi, Asatryan David (2013) Colour spaces effects on improved discrete wavelet transform-based digital image watermarking using Arnold transform map. IET Signal Proc 7:177–187
11. Gupta R, Kumar S, Yadav P, Shrivastava S (2018) Identification of age, gender, & race SMT (scare, marks, tattoos) from unconstrained facial images using statistical techniques. In: 2018 international conference on smart computing and electronic enterprise (ICSCEE). IEEE, pp 1–8
12. Gupta R, Yadav P, Kumar S (2017) Race identification from facial images using statistical techniques. J Stat Manage Syst 20(4):723–730
13. Yadav P, Gupta R, Kumar S (2019) Video image retrieval method using dither-based block truncation code with hybrid features of color and shape. Engineering vibration, communication and information processing. Springer, Singapore, pp 339–348
14. Sharma A, Chaturvedi R, Dwivedi UK, Kumar S, Reddy S (2018) Firefly algorithm based effective gray scale image segmentation using multilevel thresholding and Entropy function. Int J Pure Appl Math 118(5):437–443
15. Chaturvedi R, Sharma A, Dwivedi U, Kumar S, Praveen A (2016) Security enhanced image watermarking using mid-band DCT coefficient in YCbCr space. Int J Control Theo Appl 9(23):277–284

Chapter 34
Compensating Effect of Environmental Noise Sources in FSO Networks

Arashid Ahmad Bhat, Afreen Aijaz, Mir Andleeb, Murtaza Waheed and Maroof Bhat

1 Introduction

Light has an extensive budding to transfer huge data and high optical frequencies facilitate to use large bandwidth in a coherent manner. During recent years, FSO has been growing interest for most users. It is a technology which employs visible to near infrared (NIR) light that passes via atmosphere to deliver the information. FSO communication is attracting attention as it resolves the last-mile bottleneck in local area networks (LANS). FSO communication is based on the line of sight technology and has relatively low pass implementation. It provides a non-licensed spectrum less power consumption charges more immune and secures then any microwave and RF technologies. FSO provides LAN to LAN connectivity easily this can turn out in breaking up and absorption of visible and NIR beams, and thus lowering the FSO link performance [1].

2 Contribution and Related Works

A large no. of theoretical work has been carried out measuring the atmospheric attenuation due to external factors light snow, rain, fog, haze, etc. on FSO performance various model were proposed like Kim cruise model based on lamberts law that provides the various attenuation values of different external factors on FSO link [2, 3]. However, no such model provides the practical improvement of these attenuations on system performance Table 1 shows the various attenuation values of various external conditions over a specific range.

A. A. Bhat (✉) · A. Aijaz · M. Andleeb · M. Waheed · M. Bhat
Department of Electronics and Communication Engineering, Baba Ghulam Shah Badshah University, Rajauri, Jammu and Kashmir, India
e-mail: arshidbhat@bgsbu.ac.in

© Springer Nature Singapore Pte Ltd. 2021

S. D. Purohit et al. (eds.), *Proceedings of International Conference on Communication and Computational Technologies*, Algorithms for Intelligent Systems,
https://doi.org/10.1007/978-981-15-5077-5_34

Table 1 Attenuation of various environmental factors in FSO systems

S. No.	Title	Attenuation (db/km)
01	Heavy fog	125
02	Moderate fog	42.2
03	Light fog	20
04	Rain	9.2
05	Moderate rain	5.8
06	Dry snow	128
07	Wet snow	110

3 Environmental Noise in FSO

Environmental Noise plays a vital role in limiting the FSO performance hence acts as a serious performance bottleneck in FSO implementation. The various External Noise factors in FSO are as follows.

3.1 Fog Attenuation

Fog consists of water droplets, smoke, ice or the combination of all in the air near the surface of air. Its presence in the air scatters the light at a given attenuation value which reduces the system range. Fog is found to be either in different sizes as heavy fog, moderate fog, and light fog that affects the range at their corresponding values proposed equation. The attenuation due to fog at a distance R can be determined by Beer-Lambert law.

$$\mathrm{Att}_{\mathrm{fog}} = \mathrm{e}^{-\propto \mathrm{fog} R} \quad (1)$$

where $\propto$ fog is the attenuation coefficient.

3.1.1 Experimental Setup

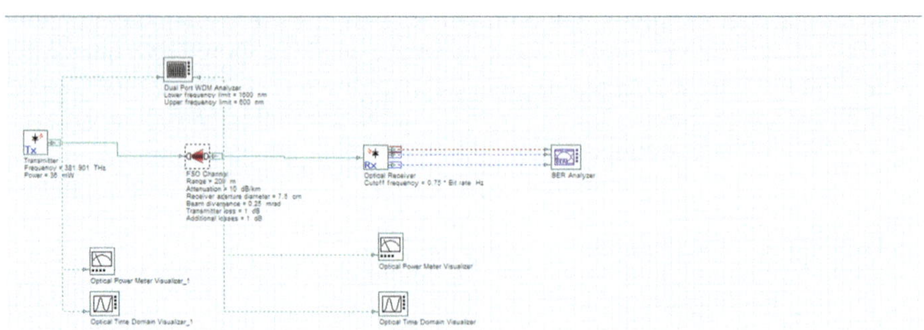

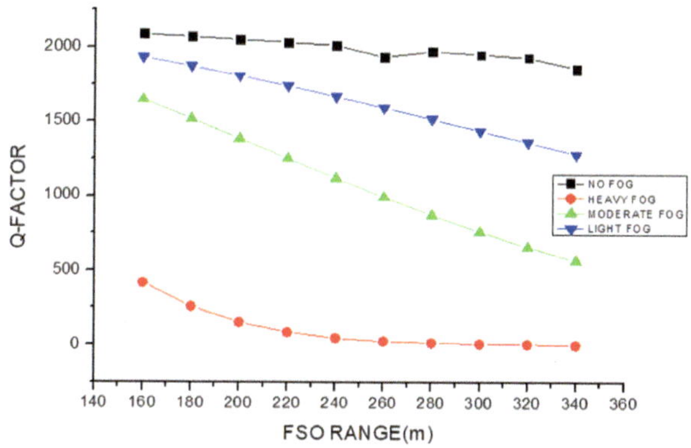

Based on all the theoretical and practical measurements the fog attenuation was implemented graphically and pictorially as Eye the FSO performance was evaluated under heavy, Moderate, and light foggy conditions, and mutual comparison of FSO performance was drawn and plotted.

(1) Under Heavy Fog conditions, the FSO performance is reduced the most followed by moderate and light fog
(2) Faso performance shows a marginal drop in comparison to no-fog conditions
(3) FSO range is severally limited by heavy fog.

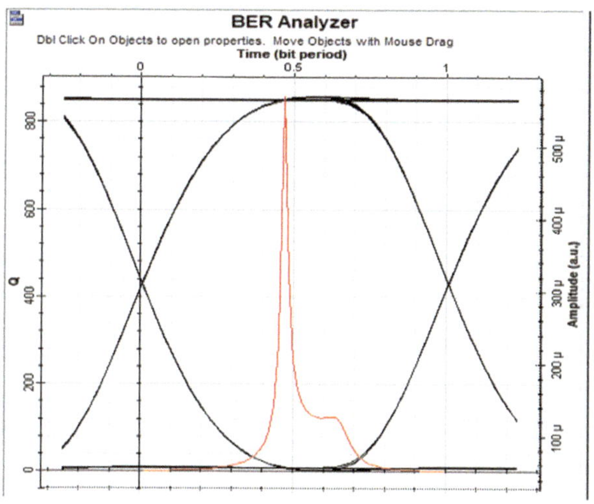

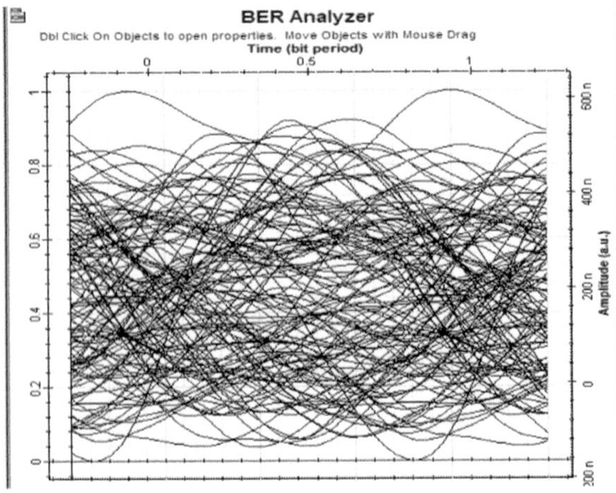

3.2 *Snow Attenuation*

Snow affects the FSO, as it causes the light to scatter and causes the laser beam power to attenuate [4]. Snow attenuation is of two types' dry snow and wet snow. The dry snow has less density and can be drifted by wind whereas wet snow has more density and is partially melted. The attenuation due to both dry snow and wet snow can be calculated by

$$\text{Att}_{\text{snow}} = b_1 S^{b_2} \tag{2}$$

Here S is snowfall rate in mm/h and b_1 and b_2 are functions of wavelength

3.3 *Rain Attenuation*

Rain causes attenuation in electromagnetic waves by absorbing and scattering phenomenon. Attenuation due to rain causes path loss degrades the system performance. Rain attenuation in this paper is modeled by data collected from different sources. The proposed equation for the same is given as.

$$\text{Att}_{\text{rain}} = k_1 R^{K_2} \tag{3}$$

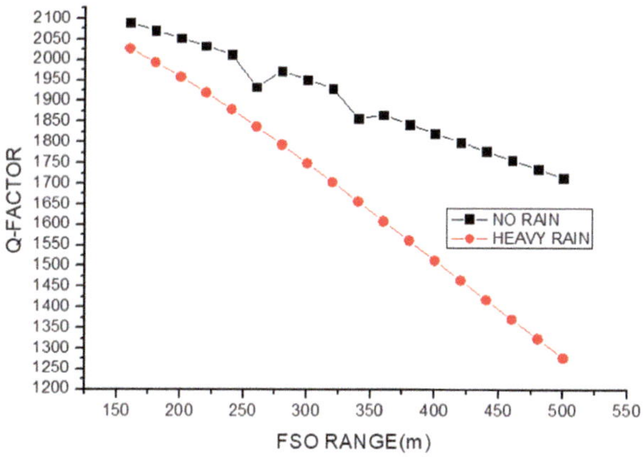

The FSO performance was measured under RAINY conditions and plotted. The inferences from the plot are as follows:

(1) Rain reduces the FSO Performance considerably
(2) FSO range is limited by Rain.

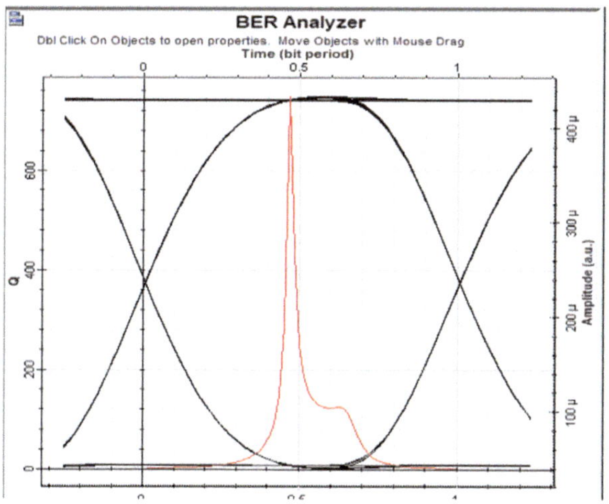

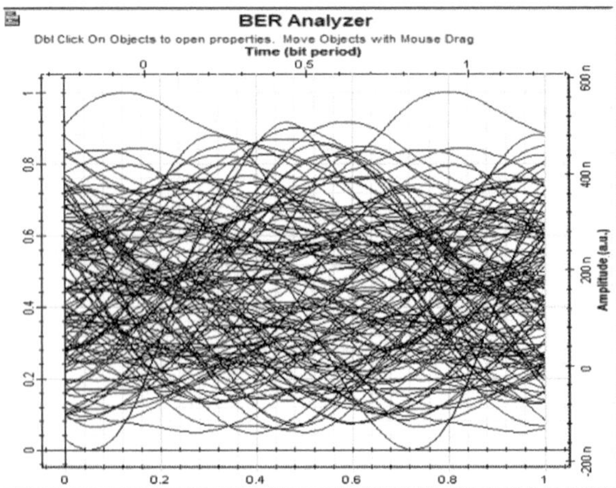

4 Minimum Transmission Range

From the past few years, significant efforts have been made to develop a channel model for predicting the weather effects on FSO transmission. By the Kim cruise [5]

$$\mathrm{Pr} = p_{\mathrm{t}} \frac{A_{\mathrm{r}}}{(\emptyset R)^2} \mathrm{Att} \tag{4}$$

In the above equation p_{t} is transmitted power, $\emptyset$ is beam divergence, R is link length, A_{R} is receiver aperture effective area, at is atmospheric attenuation which has different values for different atmospheric weather effects like rain, snow, and fog. For transreceiver, the minimum achievable transmission range depends on the predominating nature condition on FSO links. The range can be obtained as [6]

$$R_{\mathrm{fog}} = \frac{2w_{\mathrm{o}} \frac{(Q_{\mathrm{fog}})}{2\emptyset} \sqrt{\frac{P_{\mathrm{t}} A_{\mathrm{r}}}{P_{\mathrm{r}}}}}{Q_{\mathrm{fog}}} \tag{5}$$

$$R_{\mathrm{rain}} = \frac{2w_{\mathrm{o}} \frac{(\ln z_1)}{2\emptyset} \sqrt{\frac{P_{\mathrm{t}} A_{\mathrm{r}}}{P_{\mathrm{r}}}}}{\ln z_1} \tag{6}$$

$$R_{\mathrm{snow}} = \frac{2w_{\mathrm{o}} \frac{(\ln z_2)}{2\emptyset} \sqrt{\frac{P_{\mathrm{t}} A_{\mathrm{r}}}{P_{\mathrm{r}}}}}{\ln z_2} \tag{7}$$

where the values of $z_1 = 10^{k_1 R^{K_2}}$ and $z_2 = 10^{b_1 S^{b_2}}$, w_{o} denotes the principal real-valued branch of the Lambert (w) function. Based on this theory, different experiments are done on the various values of receiver aperture area and beam divergence.

It has been seen that on increasing and decreasing the value of receiver aperture area and beam divergence respectively the overall optimum performance on the signal is increased. After implementing the loss equations incase of FSO systems, it was observed that FSO performance is a function of Receiver Aperture Diameter hence in order to Improve FSO performance Receiver A The system was run under wet snow conditions with various Receiver Aperture Diameters and the inferences from the plot are as follows:

(1) The FSO performance increases with Increase in Receiver Aperture Diameter
(2) The FSO performance showed considerable improvement with respect to original design when run at aperture of 25 cm
(3) The FSO performance decreases with increase in FSO range
(4) The maximum supported FSO range got increased largely by running the setup at an aperture of 25 cm.

After implementing the loss equations incase of FSO systems, it was observed that FSO performance is a function of Beam Divergence hence in order to Improve FSO performance Beam Divergence can be a limiting factor. The system was run under wet snow conditions with various Beam Divergences and the inferences from the plot are as follows:

(1) The FSO performance Increases by Decreasing the Beam Divergence
(2) The FSO performance showed a considerable improvement with respect to original design when run at a beam divergence of 0.01 mrad
(3) FSO performance decreases exponentially with increase in FSO range
(4) The maximum supported FSO range got increased considerably by Limiting the Beam Divergence to 0.01.

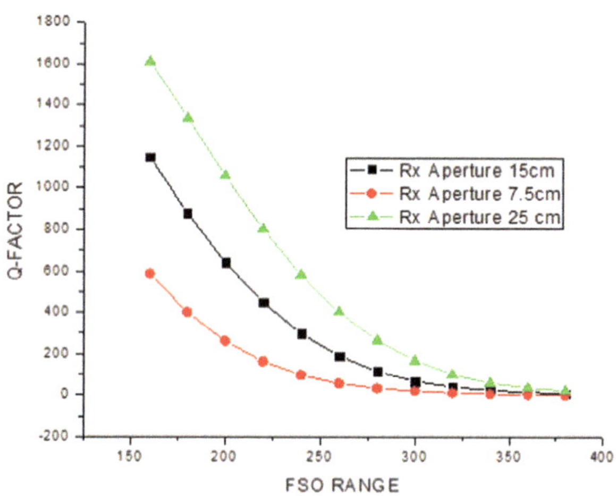

5 Conclusion

In this project environmental noise impact on the FSO performance was evaluated to compensate signal loss due to external factors such as fog, snow, rain, haze, scintillation was evaluated and methodologies were proposed to minimize the effect of such factors on FSO performance. The proposed methodologies showed considerable improvements in the signal quality under the effect of external factors.

References

1. Bhat AA (2018) Design and performance analysis of FSO (free space optical) networks. Int J Adv Eng Res Deve 5
2. Singh V et al (2019) Delay tolerant networks architecture, protocols, and its application in vehicular Ad-Hoc networks. In: Hidden link prediction in stochastic social networks. IGI Global, pp 135–161
3. Singh V, Reena D (2017) Routing optimization by minimizing one-way delay in delay tolerant enabled vehicular ad-hoc networks using ferry selection approach. In: 2017 international conference on infocom technologies and unmanned systems (trends and future directions) (ICTUS). IEEE, 2017
4. Singhal P (2015) Basic concept of free space optics communication (FSO). In: IEEE ICCSP 2015 conference
5. Jain D, Mehra R (2017) Performance analysis of free space optical communication system for S, C and L band. In: International conference on computer, communications and electronics, 2017
6. Chaudhary S, Amphawan A, Nisar K (2014) Realization of free space optics with OFDM under atmospheric turbulence. Optik Int J Light Electron Optics 125(18):5196–5198

Chapter 35
Whale Optimization Algorithm-Based Design of PID Controller for Controlling Two-Area Load Frequency with Nonlinearities

Jitendra Bikaneria, Kapil Parkh, Tilkesh Sharma, and Sakshi Saxsena

1 Introduction

In interconnected power system, several power systems are connected together to share the generation to meet out load demand. These areas are connected together by tie lines. Hence, it is elementary to have a constant frequency for the stable operation of such a system. This requires a regular observation of frequency and tie-line power flows by employing automatic generation control (AGC) for the individual area. It computes the total change required in power as per the demand variation, and it is called as area control error (ACE). An interconnected power system is constructed in many areas. It is well known that for stable operation tie-line power transfer and constant frequency are desired. In each area, the AGC tracks the system frequency and tie-line power flows, which computes the net variation in generation needed according to the variation in demand to adjust the set point of the generators within the area to maintain the change in frequency and tie-line power at low price [1]. In an automatic generation control scheme, the ACE is taken as the controlled output and it is forced to be set zero by control schemes to have both frequency and tie-line power error zero. The PID controller and its several variants are reported in the literature like integral (I) [2], proportional integral (PI), integral derivative (ID), integral double derivative (IDD) [3], proportional–integral–derivative (PID) [4] and PID with derivative filter (PIDN) [5]. They have been used, and their performances were compared for an AGC system in the established literature. In [6], several artificial intelligence techniques were discussed for finding out the controller's parameters. Gozde et al. [7] demonstrated that artificial bee colony (ABC) optimized controller gives a better

J. Bikaneria (✉) · T. Sharma
Lecturer, Government Polytechnic College, Rajsamand, Rajasthan, India
e-mail: jitendra.bikaneria@gmail.com

K. Parkh · S. Saxsena
Electrical Engineering, Shrinathji Institute of Technology and Engineering, Nathdwara, Rajasthan Technical University Kota, Kota, India

© Springer Nature Singapore Pte Ltd. 2021

S. D. Purohit et al. (eds.), *Proceedings of International Conference on Communication and Computational Technologies*, Algorithms for Intelligent Systems,
https://doi.org/10.1007/978-981-15-5077-5_35

performance than PSO-based controller for interconnected reheat thermal power system. Panda et al. [8] have utilized a method of lines algorithm to optimize the PID controller parameters and observed its performance over ABC, PSO and differential evolution (DE) approaches in automatic voltage regulator system. Mohanty et al. [9] presented DE-based AGC for two-area power system considering nonlinearity. Sahu et al. [10] employed firefly algorithm (FA) to optimize the parameter of the classical Controller for AGC of combined cycle gas turbine power plant. Das et al. [11] proposed a bat algorithm (BAT) optimized PD-PID cascade controller in AGC. A grey wolf optimizer (GWO)-based regulator design for AGC is explained in [12]. In [13], hybrid firefly algorithm–pattern search (hFA-PS) technique was presented for AGC of multi-area power system with the consideration of generation rate constraint (GRC). Recently, Mirjalili et al. [14] proposed a nature-inspired meta-heuristic optimization algorithm called whale optimization algorithm (WOA).

In this paper, the WOA is used to optimize parameters of PID controller for AGC of two-area interconnected power system including non-linearity. Initially, PID controller is considered for each area and the WOA is utilized to find the optimal controller parameters by reducing the time-domain objective function. The performance of the proposed controller is its excellent damping characteristic to system oscillations over a broad range of loading conditions and vast change in system parameters. A comparative analysis is also carried out with other controller based on (i) hybrid firefly algorithm–pattern search, (ii) bacteria foraging optimization algorithm, (iii) genetic algorithm and (iv) Ziegler–Nichols-based PID controller for similar interconnected power system to indicate the excellence of the proposed WOA algorithm.

2 System Modelling

The two-area power system is considered in this paper, which consists of the two non-reheat thermal power plants as shown in Fig. 1. Rating of each area is 2000 MW with a nominal load of 1000 MW. For design and analysis point of view, this system is mostly used in the literature [13]. In Fig. 1, ΔP_{G1} and ΔP_{G2} are the governor output command in p.u.; ΔF_1 and ΔF_2 are the system frequency deviations in Hz; T_{12} is the synchronizing coefficient; K_{P1} and K_{P2} are the gains of power system; T_{P1} and T_{P2} are the time constants of power system in seconds; ΔP_{tie} is the incremental change in tie-line power in p.u.; ΔP_{D1} and ΔP_{D2} are changed in load demand; ΔP_{T1} and ΔP_{T2} are the change in turbine output power; T_{T1} and T_{T2} are the turbine time constants in seconds; ΔP_{V1} and ΔP_{V2} are the change in governor valve position in p.u.; T_{G1} and T_{G2} are the speed governor time constants in seconds; $R_1$and R_2 are the governor speed regulation parameter in p.u.; u_1 and u_2 are the controller outputs; B_1 and B_2 are the frequency bias parameters; and ACE_1 and ACE_2 are area control errors.

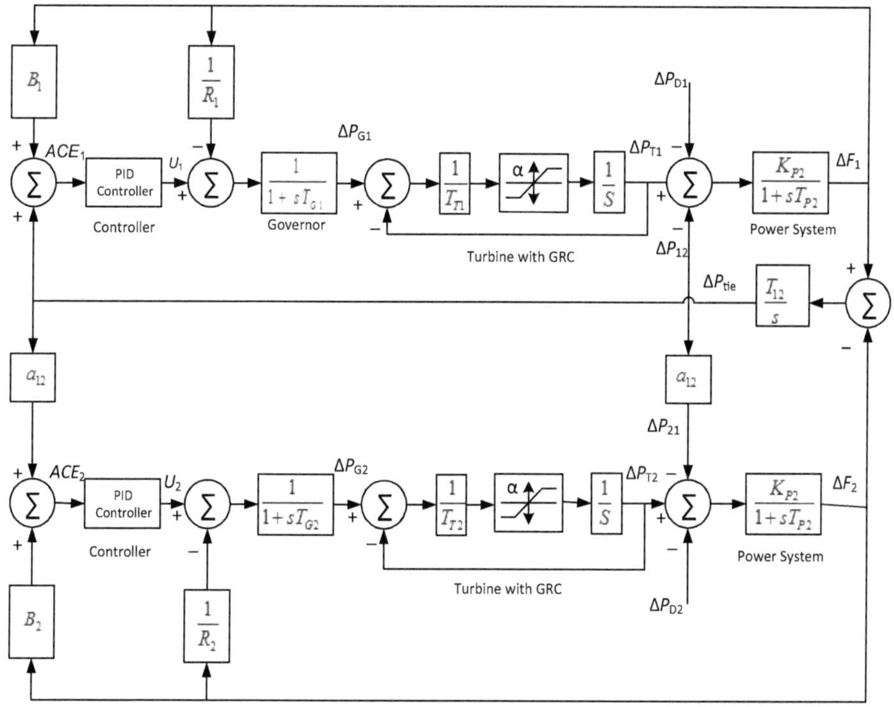

Fig. 1 Two-area non-reheat thermal power plant with GRC

In steam power plants, power generation will vary at a specific maximum rate referred to as generation rate constraints (GRC) and its value for a thermal power plant are 2.5% per min. ACEs are input to controllers, as in Fig. 1, and given by Eqs. (1) and (2),

$$\text{ACE}_1 = B_1 \Delta F_1 + \Delta P_{\text{tie}} \tag{1}$$

$$\text{ACE}_2 = B_2 \Delta F_2 + a_{12} \Delta P_{\text{tie}} \tag{2}$$

3 Whale Optimization Algorithm (WOA)

Mirjalili and Lewis proposed a meta-heuristic nature-inspired population-based whale optimization algorithm in 2016 [14]. The mathematical model of WOA has been divided into three sections:

1. Encircling prey
2. Bubble net hunting method
3. Search the prey.

Random sets of solutions are initially taken to start WOA. All search agents improve their positions iteratively till the best solution obtained. For exploration and exploitation, parameter c is lowered from 0 to 2. If $|\overrightarrow{E}| > 1$, random search is chosen and when $|\overrightarrow{E}| < 1$, the best solution is selected to improve search agent's position. WOA can shift between spiral and circular movement depending on Q (random number) value. At last, it stops when termination criteria are fulfilled.

4 Objective Function

Controller's design relies on an explicit objective function which meets system's requirements and constraints. Here, the objective functions are the ITAE of frequency variation of both areas and tie-line power and are given by,

$$\text{ITAE} = J = \int_{0}^{t_{\text{sim}}} (|\Delta F_1| + |\Delta F_2| + |\Delta P_{\text{tie}}|)t\,\mathrm{d}t \tag{3}$$

where, ΔF_1 and ΔF_2—variation in system frequencies, ΔP_{tie}—tie-line power variation and t_{sim}—simulation time range. Based on this performance index 'J', the design problem can be formulated as the following optimization problem, Minimize J,

Subjected to

$$\begin{aligned} K_P^{\min} &\le K_P \le K_P^{\max} \\ K_i^{\min} &\le K_i \le K_i^{\max} \\ K_d^{\min} &\le K_d \le K_d^{\max} \end{aligned} \tag{4}$$

5 Implementation of WOA in LFC Problem

The proposed algorithm is explained in the following steps:

I. WOA's input parameter initialization like size of population, the control variable number according to the controller structure, total number of generation and upper and lower limit of each search agent.

II. At the time of starting, search agents are produced randomly between upper and lower limits in the search space, i.e. parameters of the controllers (K_{p}, K_{i}, K_{d}).
III. Fitness function using Eq. (3) for each search agent is computed for getting an initial best agent.
IV. Thereafter, updating c, E, G, l, s and Q for every search agent. Here, s, l and Q are random numbers.
V. Go to Step 6 if ($Q < 0.5$) else Step 8.
VI. If $|\overrightarrow{E}| \geq 1$, using Eq. (5) for updating position of the current search agent.

$$\overrightarrow{Y}(t+1) = \overrightarrow{Y}_{\mathrm{rand}} - \overrightarrow{E} * \overrightarrow{F} \tag{5}$$

where $\overrightarrow{Y}_{\mathrm{rand}}$—random whales, which are selected from the current population.
VII. If $|\overrightarrow{E}| < 1$, the position of current search agent is updated by Eqs. (6) and (7).

$$\overrightarrow{Y}(t+1) = \overrightarrow{Y^{*}}(t) - \overrightarrow{E}\,.\,\overrightarrow{F} \tag{6}$$

$$\overrightarrow{F} = \left|\overrightarrow{G} \cdot \overrightarrow{Y^{*}}(t) - \overrightarrow{Y}(t)\right| \tag{7}$$

where t-present iteration, $\overrightarrow{E}$ and $\overrightarrow{G}$—coefficient vectors, $\overrightarrow{Y^{*}}$—obtained the best solution's position vector, $\overrightarrow{Y}$—position vector.
VIII. Using Eq. (8), updating position of current search agent

$$\overrightarrow{Y}(t+1) = \overrightarrow{F'}.\mathrm{e}^{dl}.\cos(2\pi l) + \overrightarrow{Y^{*}} \tag{8}$$

where logarithmic spiral shape is shown by constant d and l is a random number [−1,1].
IX. Modifying search agent (PID controller parameters) if it goes beyond the search space.
X. Compute the fitness of each search agent (controller parameters).
XI. Improve the best search agent (controller parameter) if there is a better solution.
XII. If the termination criterion is reached, step 13 is taken, else repeat from step 3.
XIII. Print the optimal solution.

6 Result and Discussion

The proposed algorithm is developed in MATLAB™. Simulations are conducted on an Intel, Core i-3 CPU of 2.53 GHz and 2 GB RAM computer in MATLAB environment. In this work, two GRC values are taken as 0.025 and 0.05. The results of the proposed WOA technique are shown in Table 1, and a convergence curve of WOA corresponding to the best result is shown in Fig. 2.

The upper and lower limits of PID controller parameters are selected as 1.0 and −1.0, respectively [13]. Tables 2 and 3 represent the comparative performance of WOA with other recently published conventional and heuristic algorithms [13]

The following cases are considered for studying system's dynamic performance for step load increase (SLI) with different values of GRC at different locations.

Table 1 Parameters of controller

Controller variables	PID controller parameters based on hFA-PS		PID controller parameters based on WOA	
	GRC = ±0.05	GRC = ±0.025	GRC = ±0.05	GRC = ±0.025
K_p	0.3834	0.1898	1	−0.2848
K_i	0.6127	0.3164	0.7616	0.3363
K_d	0.4021	0.4528	0.4295	0.0912

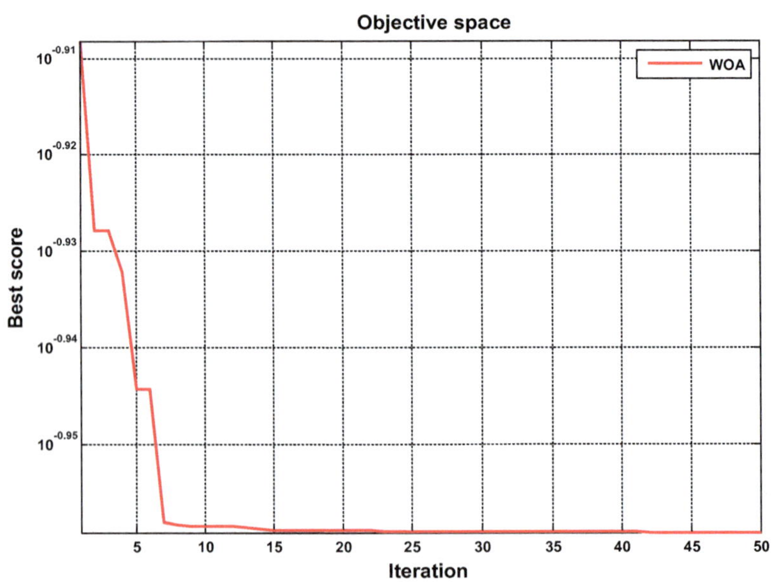

Fig. 2 Convergence curve of WOA

Table 2 Performance indices and settling time with GRC = ± 0.05

Techniques	GRC = ±0.05				
	Settling time Ts (s)				Performance indices
	ΔF_1	ΔF_2	ΔP_{tie}		ITAE
Proposed WOA	3	3.1	3.7		0.1098
hFA—PS [14]	2.8	4.5	4		0.2782
FA [14]	3.1	4.9	4.3		0.324
BFOA [15]	4.7	6.4	5.1		0.4788
GA [15]	6.9	8	5.7		0.5513
ZN [15]	8.1	9.2	6.7		0.604

Table 3 Performance indices and settling time with GRC = ±0.025

Techniques	GRC = ± 0.025			
	Settling time Ts (s)			Performance indices
	ΔF_1	ΔF_2	ΔP_{tie}	ITAE
Proposed WOA	4.7	5	6.1	0.4255
hFA—PS [14]	6.9	5.2	7.5	0.7405
FA [14]	7.8	6.3	7.9	0.8023
BFOA [15]	9	7.9	8.3	1.5078
GA [15]	11.1	11.2	11	2.4668
ZN [15]	15.3	14.1	15.3	3.4972

Case A SLI in area 1 with limit of saturation (GRC = ± 0.05): At t = 0 s 5% step load increase (SLI) is applied. It is clear from Figs. 3, 4, 5, 6 and 7 that system

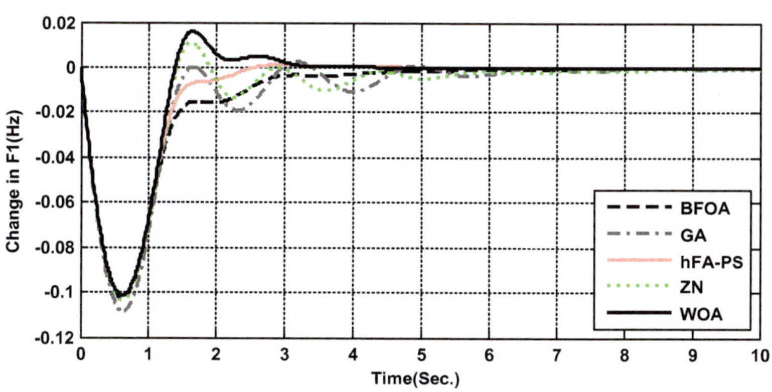

Fig. 3 Change in frequency of area 1 for 5% SLI in area-1, with GRC = ±0.05

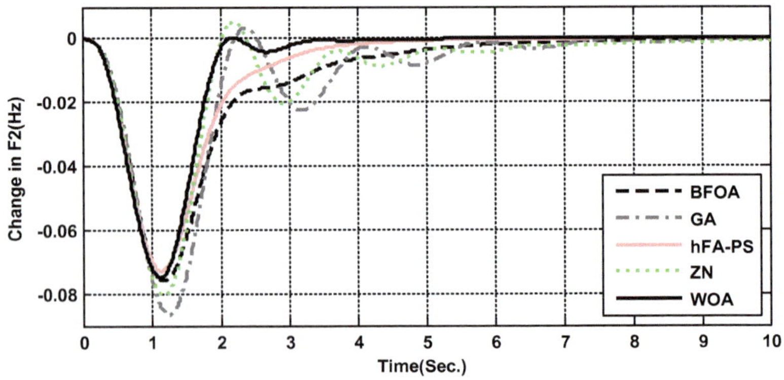

Fig. 4 Change in frequency of area-2 for 5% SLI in area-1, with GRC = ±0.05

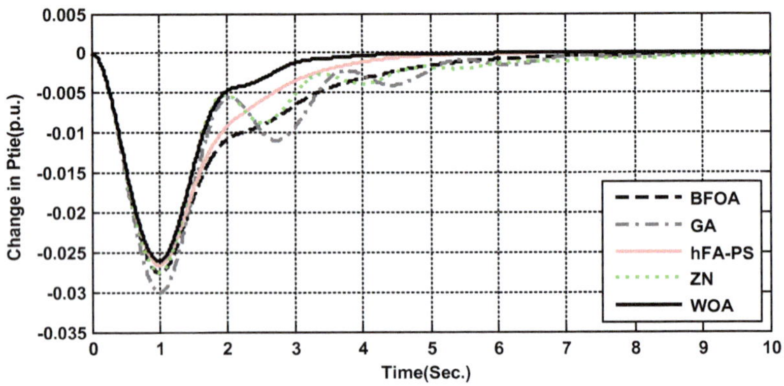

Fig. 5 Change in tie-line power for 5% SLI in area-1, with GRC = ±0.05

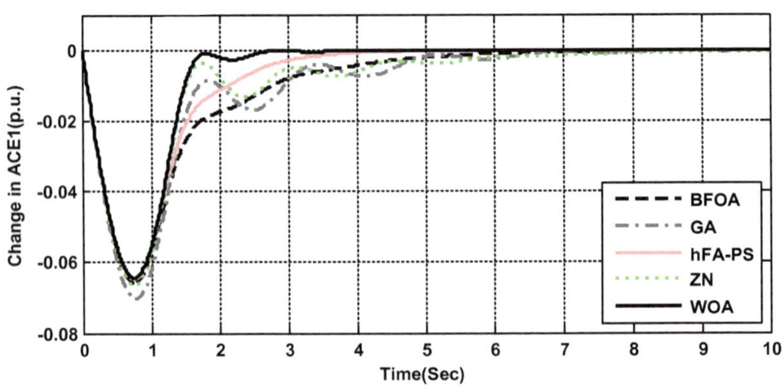

Fig. 6 Change in ACE1 for 5% SLI in area 1, with GRC = ± 0.05

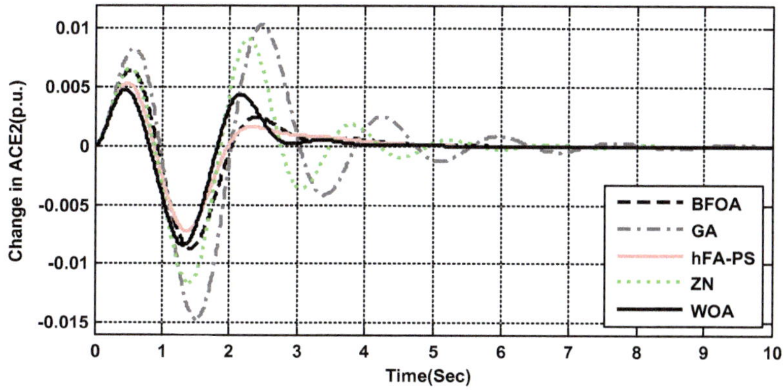

Fig. 7 Change in ACE2 for 5% SLI in area 1, with GRC = ± 0.05

dynamic responses of the proposed PID controller based on WOA perform better than PID controller based on other algorithms, viz. ZN, GA, BFOA, hFA-PS.

Case B SLI in area 2 with limit of saturation (GRC = ± 0.05): At t = 0 s 5% SLI is applied. Figures 8, 9, 10, 11 and 12 show system's dynamic responses, which are superior with proposed controller.

Case C SLI in area 1 and area 2 at the same time with limit of saturation (GRC = ±0.05): A 5% SLI in area 1 and 2% SLI in area 2 are considered. The proposed controller is robust in nature and performs well subjected to different location and size of disturbance changes as shown in Figs. 13, 14 and 15.

Case D Limit of saturation effect (GRC = ±0.025): At t = 0 s, a 5% SLI in area 1 is applied. Figures 16, 17 and 18 show system dynamic responses from which it is seen that the given controller gives a better dynamic response having minimum value of ITAE as well as settling times.

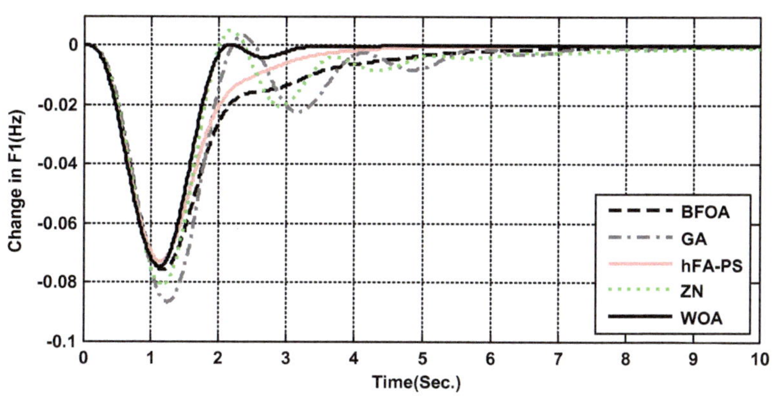

Fig. 8 Change in frequency of area-1 for 5% SLI in area 2, with GRC = ±0.05

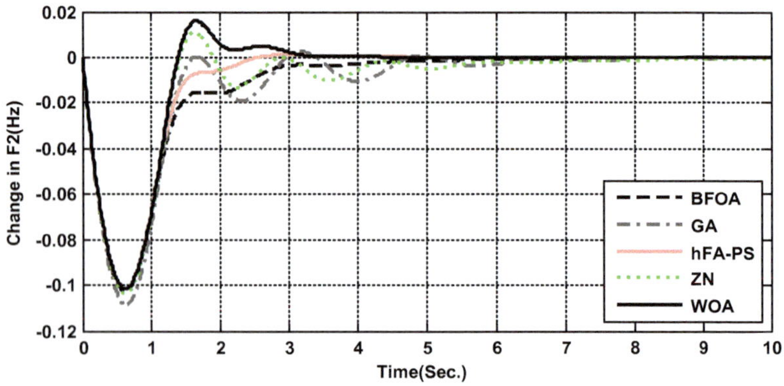

Fig. 9. Change in frequency of area-2 for 5% SLI in area 2, with GRC = ±0.05

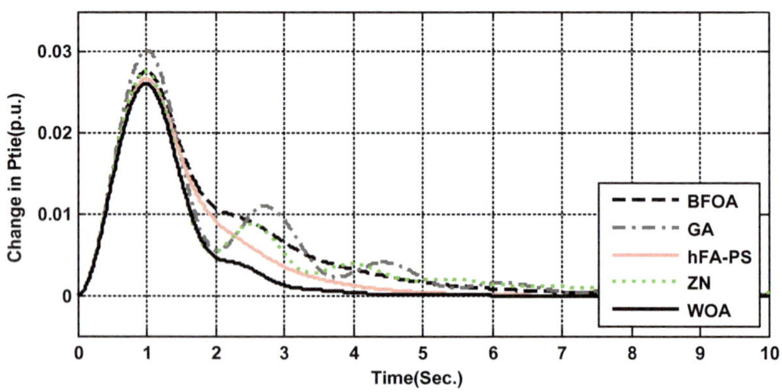

Fig. 10 Change in tie-line power for 5% SLI in area 2, with GRC = ±0.05

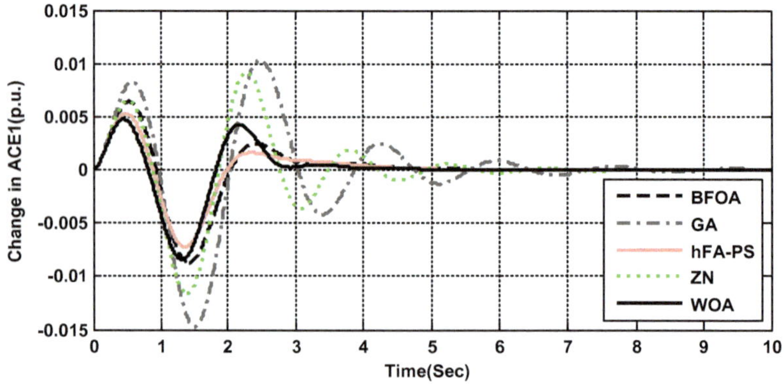

Fig. 11 Change in ACE1 for 5% SLI in area 2, with GRC = ± 0.05

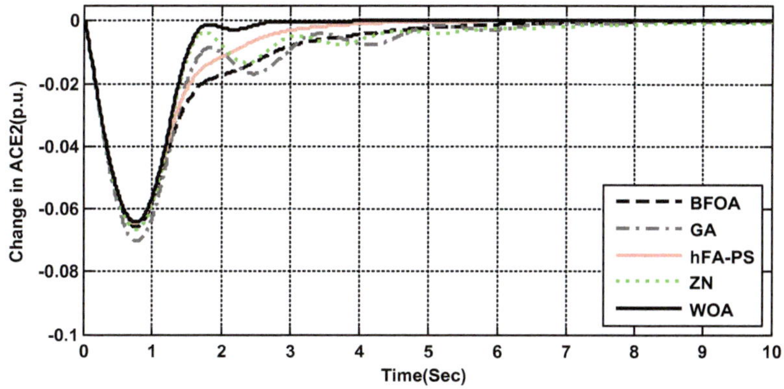

Fig. 12 Change in ACE2 for 5% SLI in area 2, with GRC = ±0.05

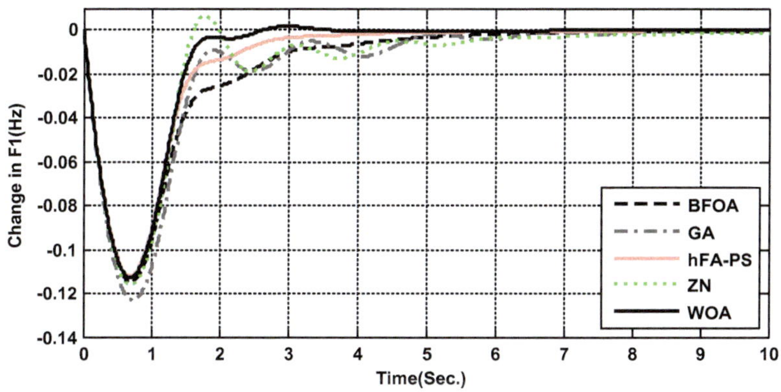

Fig. 13 Frequency deviation of area 1 for 5% SLI in area 1 and 2% SLI in area 2, with GRC = ±0.05

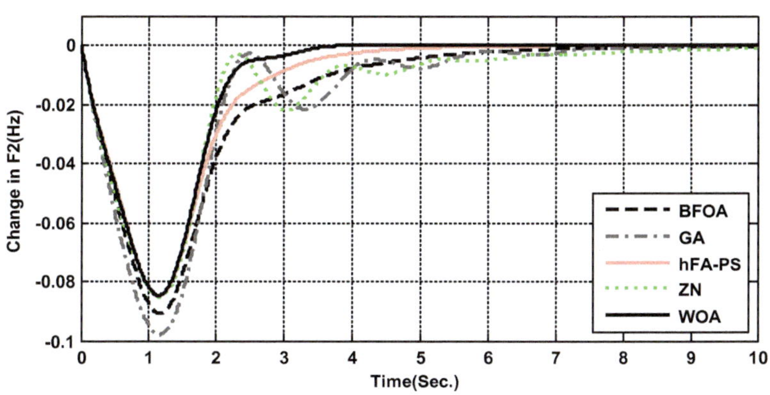

Fig. 14 Change in frequency of area 2 for 5% SLI in area 1 and 2% SLI in area 2, with GRC = ±0.05

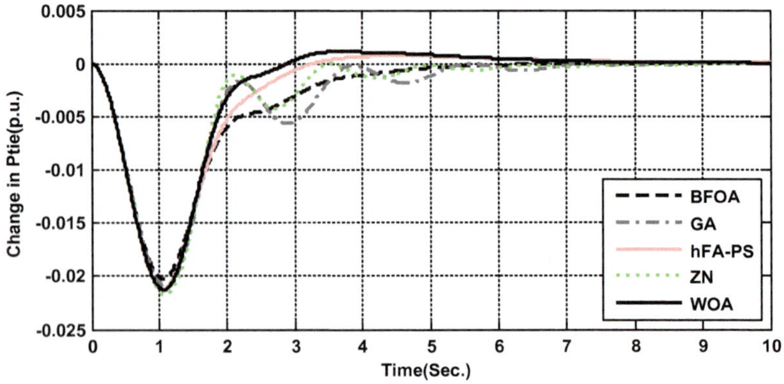

Fig. 15 Change in tie-line power for 5% SLI in area 1 and 2% SLI in area 2, with GRC = ±0.05

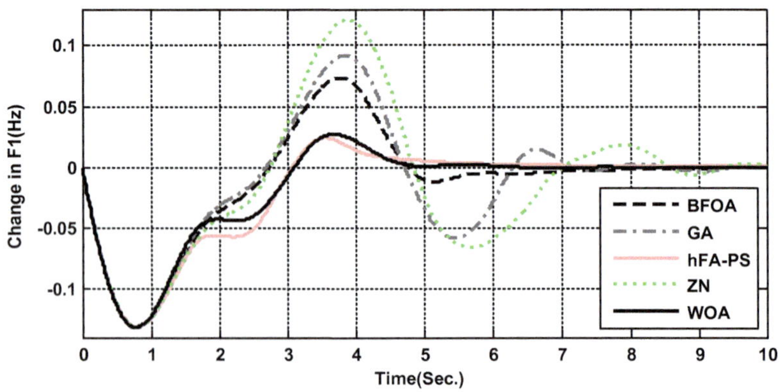

Fig. 16 Frequency deviation of area 1 for 5% SLI in area 1, with GRC = ±0.025

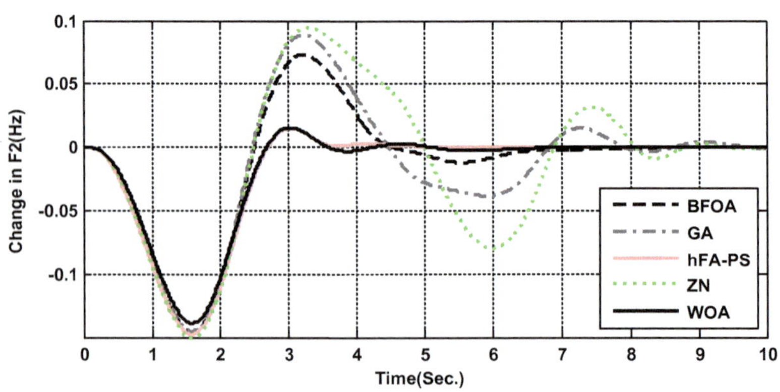

Fig. 17 Change in frequency of area 2 for 5% SLI in area 1, with GRC = ±0.025

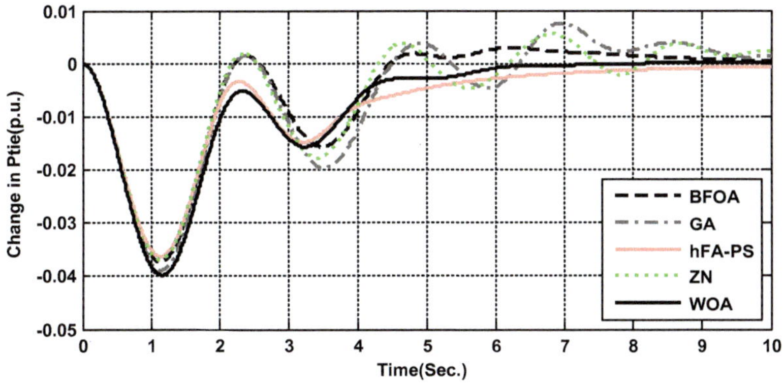

Fig. 18 Change in tie-line power for 5% SLI in area 1, with GRC = ±0.025

Sensitivity analysis: Power system's sensitivity is analysed to check the robustness of the system for changing operating conditions and system parameters. These operating conditions and parameters of systems are taken one by one, viz. operating load condition, speed governor time constant, power of tie line and turbine time constant. Their values are changed from nominal values (Appendix A) within + 50 to −50%. By changing their operating load conditions, which affects the power system parameters K_{PS} and T_{PS}. Table 4 gives parameters under varying operating load conditions (GRC 0.05 and 0.025). Tables 5 and 6 show the sensitivity analysis of the system for a 5% SLI in area-1 under the above-varying conditions with GRCs of 0.05 and 0.025. It can be concluded from Tables 5 and 6 that settling time and ITAE values vary within satisfactory ranges and close equal to the respective values obtained with nominal system parameter. It can be observed from Tables 5 and 6 that the proposed controller gives a better dynamic performance than those of the latest hFA-PS optimized PID controller for all the cases. Figures 19, 20, 21, 22 and

Table 4 Parameters of PID controller for variation in system parameters

Change in system parameters	% Change	GRC = ±0.05			GRC = ±0.025		
		K_p	K_i	K_d	K_p	K_i	K_d
Nominal loading condition	0	1	0.7616	0.4295	−0.2848	0.3363	0.0912
	50	0.7857	0.7674	0.3491	−0.27215	0.32909	0.099833
	−50	0.6515	0.7333	0.3328	−0.30306	0.32833	0.075162
T_G	50	0.9873	0.7338	0.5339	−0.1714	0.3214	0.1999
	−50	0.9961	0.7952	0.3239	−0.3249	0.3508	0.0529
T_T	50	0.6633	0.7055	0.4351	−0.3125	0.2988	0.1339
	−50	0.6541	0.7637	0.3136	−0.3804	0.3512	0.0352
T_{12}	50	0.4693	0.7374	0.3382	−0.2199	0.3691	0.0821
	−50	1	0.7787	0.3788	−0.2977	0.3116	0.1135

Table 5 Sensitivity analysis with GRC = ±0.05

Change in system parameters	% Change	Proposed WOA technique				hFA-PS technique			
		ST (s)			ITAE	ST (s)			ITAE
		ΔF_1	ΔF_2	ΔP_{tie}		ΔF_1	ΔF_2	ΔP_{tie}	
Nominal loading condition	0	3	3.1	3.7	0.1098	2.8	4.5	4	0.2782
	50	2.73	3.2	3.6	0.1089	2.8	4.1	3.9	0.2799
	−50	2.78	3.2	3.8	0.1139	2.8	4.3	4	0.2713
T_G	50	2.7	3	3.8	0.1144	2.8	4.6	4.4	0.2705
	−50	3.1	3.2	3.2	0.107	3.3	4.9	4.2	0.2931
T_T	50	3.3	3.3	4.2	0.1239	3.7	5	4.4	0.2669
	−50	2.3	2.8	3.2	0.1072	2.7	4.1	3.6	0.2739
T_{12}	50	2.9	2.7	3.7	0.1111	3.1	3.8	4.1	0.2714
	−50	3.8	4.1	4.3	0.1708	5.4	5.2	4.6	0.2877

Table 6 Sensitivity analysis with GRC = ±0.025

Change in system parameters	% change	Proposed WOA technique				hFA-PS technique			
		ST (s)			ITAE	ST (s)			ITAE
		ΔF_1	ΔF_2	ΔP_{tie}		ΔF_1	ΔF_2	ΔP_{tie}	
Nominal loading condition	0	4.7	5	6.1	0.4255	6.9	5.2	7.5	0.7405
	50	4.7	5.8	6.2	0.44728	7.2	3.8	7.7	0.6537
	−50	4.6	6	6.1	0.4478	6.3	5.3	7.3	0.8878
T_G	50	5.7	4.6	7.1	0.4654	6.6	5.3	7.5	0.8695
	−50	4.7	4.9	6	0.3949	9.5	8.3	7.7	0.9415
T_T	50	6.2	5.2	7.5	0.508	7.4	5.3	8.1	0.7681
	−50	4.9	4.2	4.6	0.3805	7.3	4.3	7.2	0.6825
T_{12}	50	5.1	5	6.5	0.3935	7	6.1	9.2	0.9098
	−50	6.1	3.4	6.2	0.4903	10.5	7.6	7	1.194

23 show the dynamic response of the system under varying loading conditions and system time constants. Effect on system performance is negligible due to varying operating load conditions and time constants of system as shown in Figs. 19, 20, 21, 22 and 23. Therefore, it can be easily observed that the proposed control strategy gives robust control with widely changing load conditions and system parameters.

7 Conclusion

The whale optimization algorithm (WOA) has been used to optimize PID controller's parameters for load frequency control of two-area non-reheat thermal power system

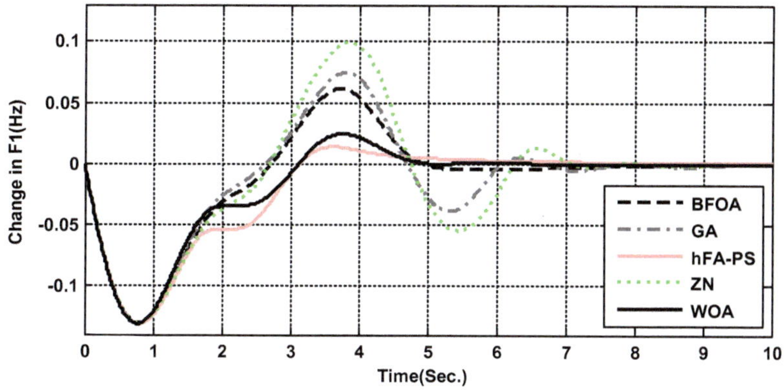

Fig. 19 Change in frequency of area 1 due to decrease in T_G by 50% with

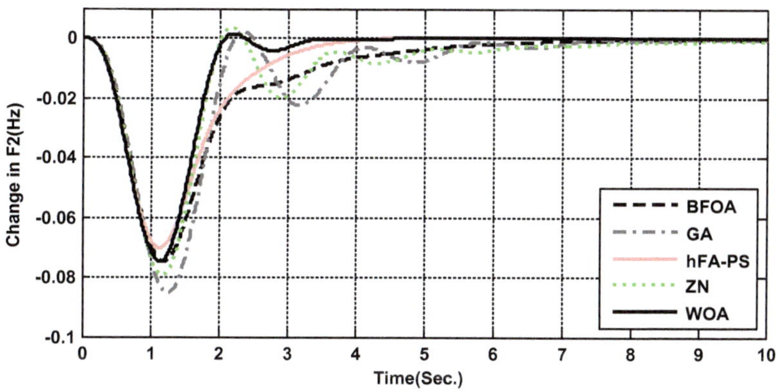

Fig. 20 Change in frequency of area 2 due to increase in nominal loading by 50%, with GRC = ±0.05

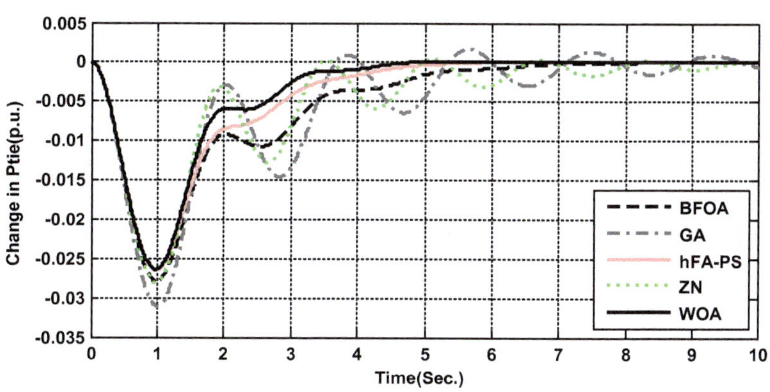

Fig. 21 Change in tie-line power due to increase in T_T by 50%, with GRC = ±0.05

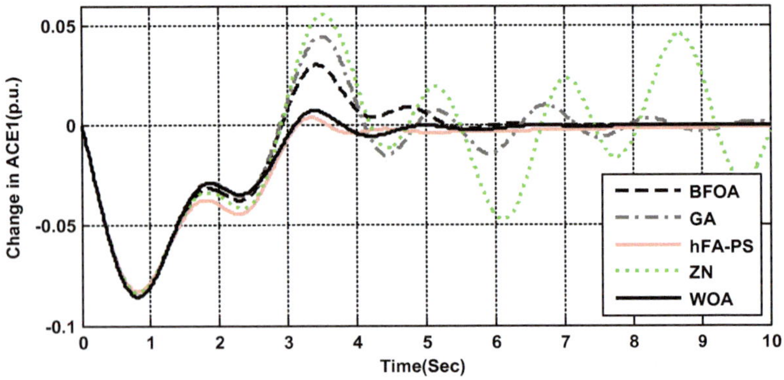

Fig. 22 Change in ACE1 due to increase in T_{12} by 50%, with GRC = ±0.025

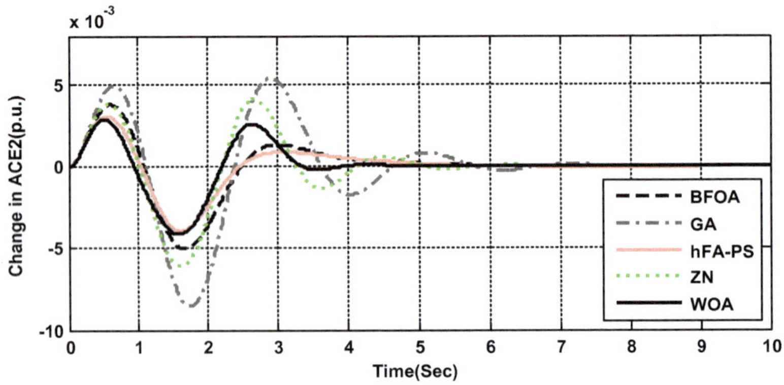

Fig. 23 Change in ACE2 due to increase in T_{12} by 50%, with GRC = ±0.05

including nonlinearities. The given approach enhances system dynamic performance by improving its settling times and overshoots. All simulations validate the excellence of the proposed algorithm by comparing the results with recently reported approaches (ZN, GA, BFOA and hFA-PS-based PID controller) for the same interconnected power system. Moreover, power system's sensitivity is analysed for varying operating conditions and parameters of system under nominal values (+50 to −50%) and the proposed algorithm has shown effective and robust performance.

Appendix

Nominal parameters of two-area non-reheat thermal–thermal power system [13]

Parameter	Value	Parameter	Value
F	60 Hz	$T_{G1} = T_{G2}$	0.08 S
$B_1 = B_2$	0.425 p.u. MW/Hz	$T_{T1} = T_{T2}$	0.3 S
$R_1 = R_2$	2.4 Hz/p.u.	$K_{P1} = K_{P2}$	120 Hz/p.u.
a_{12}	-1	T_{12}	0.545 p.u.
$T_{p1} = T_{p2}$	20 S	$P_{r1} = P_{r2}$	2000 MW
P_L	1000 MW	GRC	$\pm 0.05\% \pm 0.025\%$

References

1. Kundur P, Balu NJ, Lauby MG (1994) Power system stability and control. McGraw-Hill, New York
2. Nanda J, Kaul B (1978) Automatic generation control of an interconnected power system. Proc Inst Electr Eng 125:385–390
3. Saikia LC, Nanda J, Mishra S (2011) Performance comparison of several classical controllers in AGC for multi-area interconnected thermal system. Int J Electr Power Energy Syst 33:394–401
4. Gaing ZL (2004) A particle swarm optimization approach for optimum design of PID Controller in AVR system. IEEE Trans Energy Convers 19:384–391
5. Panda S, Yegireddy NK (2013) Automatic generation control of multi-area power system using multi-objective non-dominated sorting genetic algorithm-II. Int J Electr Power Energy Syst 53:54–63
6. Pandey SK, Mohanty SR, Kishor N (2013) A literature survey on load–frequency control for conventional and distribution generation power systems. Renew Sustain Energy Rev 25:318–334
7. Gozde H, Taplamacioglu MC, Kocaarslan I (2012) Comparative performance analysis of artificial bee colony algorithm in automatic generation control for interconnected reheat thermal power system. Int J Electr Power Energy Syst 42:167–178
8. Panda S, Sahu B, Mohanty P (2012) Design and performance analysis of PID controller for an automatic voltage regulator system using simplified particle swarm optimization. J Frankl Inst 349:2609–2625
9. Mohanty B, Panda S, Hota P (2014) Differential evolution algorithm based automatic generation control for interconnected power systems with non-linearity. Alex Eng J 53:537–552
10. Saikia LC, Sahu SK (2013) Automatic generation control of a combined cycle gas turbine plant with classical controllers using Firefly algorithm. Int J Electr Power Energy Syst 53:27–33
11. Dash P, Saikia LC, Sinha N (2015) Automatic generation control of multi area thermal system using Bat algorithm optimized PD–PID cascade controller. Int J Electr Power Energy Syst 68:364–372
12. Gupta E, Saxena A (2016) Grey wolf optimizer based regulator design for automatic generation control of interconnected power system. Cogent Eng 3:1151612
13. Sahu RK, Panda S, Paddhan S (2015) A hybrid firefly algorithm and pattern search technique for automatic generation control of multiarea power systems. Electr Power Energy Syst 64(2015):9–23
14. Mirjalili S, Lewis A (2016) The whale optimization algorithm. Adv In Eng Softw 95(2016):51–67
15. Ali ES, Abd-Elazim SM (2013) BFOA based design of PID Controller for two area load frequency control with nonlinearities Elect. Power Energy Syst 51(2013):224–231
16. Reddy PDP, Reddy VCV, Monohar TG (2017). Whale optimization algorithm for distribution system. Renewables 4:3

Chapter 36
Predicting Chemical Reactions with Artificial Intelligence and Machine Learning

Akshat Mathur, Shivani Dave, and Sushma Dave

1 Introduction

1.1 A Subsection Sample

Engineering reactions can be a complicated task demanding much experience and effort from the chemist. As in all fields of science, it is virtually impossible for a single person to keep track of all the developments taking place, even in this seemingly narrow field of specialization. Clearly, it would seem beneficial to have a computer help with which the chemist digests the available information, evaluate, correlate, and extrapolate it to provide ideas for new experiments for advancement of knowledge in this field and for solving important synthetic problems. Artificial intelligence and machine learning are fast emerging as very useful and reliable tool for reducing this effort and time duration by studying and predicting reactions.

It is difficult to predict that what will be the outcome when the chemicals are mixed because a huge number of variables are involved. To have a device which does it will eventually speed up the growth of useful new materials, especially drugs used in medical sciences [1]. The novel approach involves treating chemical reactions as translation problems by rephrasing elements involved in predictions as letters and words rather than atoms and molecules. This changes the way of analyzing the chemical reactions where prediction will be done by looking at how the words are being translated from one form to another which is mostly solved by artificial intelligence.

The fundamental ability of AI to quickly deduce and identify logical paths has made it an immensely useful tool for identifying and directing the course of chemical

A. Mathur · S. Dave · S. Dave (✉)
Jodhpur Institute of Engineering and Technology, Jodhpur, Rajasthan, India
e-mail: sushmadave26@gmail.com

MBM Engineering College Jodhpur, Jodhpur, Rajasthan, India

© Springer Nature Singapore Pte Ltd. 2021

S. D. Purohit et al. (eds.), *Proceedings of International Conference on Communication and Computational Technologies*, Algorithms for Intelligent Systems,
https://doi.org/10.1007/978-981-15-5077-5_36

research along with guiding the researchers with tasks such as reaction engineering. It also serves as a useful tool in predicting and/or identifying the reaction path. This is a very useful ability for applications such as drug discovery in medical sciences.

2 Predicting Chemical Reactions with AI

While predicting a chemical reaction, the following are the few key considerations which are primarily taken into account by the chemist.

- How to synthesize a specific compound?
- What conditions of catalyst, solvent, reactants, temperature, pressure, and other parameters are required for the reaction to be most regioselective (chemo, stereo selective) and to produce the highest possible yield of products?
- What will be the rate, yield, and the time duration of the given reaction?
- What major and minor product(s) are to be expected from the reaction conducted under certain specified conditions.
- How to find a reaction similar to this one in either a commercial or open-access database in order to get a reference of the procedure for the synthesis of a specified compound.
- What types of reactions are stored in the database and whether there are enough examples of the type of reaction of interest?

Nevertheless, the initial approaches handling the information about organic reactions which were described during the creation of software products for computer-assisted synthesis were not left in vain instead they proved to be a base for new generation software systems which are capable of dealing with big data covering the information on an enormous number of organic reactions. Software products designed for retrosynthetic analysis are, in particular, ARChem (route designer) (developed by SimBioSys, Canada), ICSYNTH (developed by Info Chem GmbH, Germany), and CHEMATICA [2].

For application of artificial intelligence in chemical engineering, retrosynthetic analysis technique is employed which is an approach toward problem-solving in organic reaction synthesis. Retrosynthesis is a technique to plan the synthesis of organic molecules but is a challenge to solve. Although the search depth rarely exceeds more than 15–20 steps, the search tree explodes combinatorically if solutions are explored exhaustively [1]. The branching factor for a complex retrosynthetic problem can even go up to tens of thousands.

Retrosynthetic analysis focuses on standardization and simplification of reactants which are then treated individually in engineering the reaction. The principal goal of retrosynthetic analysis is structural simplification of participating organic compounds.

Corey and Wipke have formalized the origin in their efforts to introduce computer assistance to synthesis with logic and heuristics applied to synthetic analysis (LHASA). [3, 4] Corey's approach to codifying retrosynthesis involved the explicit identification of molecular structures which lend themselves to disconnection or, rather, can be produced by known reactions in the forward direction. More or less every approach to programmed retrosynthesis involves the use of reaction templates submolecular patterns that encode changes in atom connectivity. Recursively applying retrosynthetic templates to a target molecule produces a candidate synthesis tree a successful forward synthesis.

Marwin et al. [1] show that the Monte Carlo tree search (MCTS) technique along with the deep neural network policies and an in-scope filter can be used to effectively perform chemical synthesis planning with advantage of providing full retrosynthetic pathways within time duration of a few seconds. In comparison to earlier approaches, this approach is purely based on data-driven guess, and initial setup can be done within few days without any expert encoding or curation, which may be plagued by human bias, and is equally applicable to both in-house and discipline-scale datasets. They demonstrate that their 3 N-MCTS approach had the best performance characteristics in terms of speed and number of problems solved as compared to other established search methods.

However, a number of challenges, such as that of investigating natural product synthesis, are still beyond the capabilities of their method and pose a significant challenge. Furthermore, the unavailability of substantial training data in this area remains a challenge for machine learning approaches.

Templates are locally defined pattern-matching rules, inherently naive to what is present in the rest of the molecule. Therefore, for a planned retrosynthetic disconnection, it is very general to be unviable in the forward direction. Once a synthetic pathway has been projected, it is critical to evaluate each step in the forward direction to recognize these challenges. When approaching reaction with such a manner, the possible routes for a reaction to successfully conclude can be more than one. In such a case, AI is a very useful tool which can identify the optimum path for the reaction to progress.

The following two types of approaches are usually employed for predicting organic reactions:

- Template-based Approach:
  Existing machine learning models for product prediction are mostly built on the templates of a set of common reactions. In these approaches, different templates are specified for a given reaction and the final product is selected from multiple candidates which originate from these templates. For example, Wei et al. [5] learn to select from among 16 pre-specified templates with fingerprints of reactants and reagents. It develops on a narrow range of chemical reaction types but is among the first implementations that demonstrate the potential of neural models for analyzing chemical reactions with more recent work that has demonstrated the power of neural methods on a broader set of reactions. Among these, Segler and Waller [1] and Coley et al. [3] have used a data-driven methodology to obtain a large set of templates, and then employ a neural model to rank the candidates. In the case of Segler and Waller [1], molecules are represented based on their Morgan fingerprints, while Coley et al. [3] have represented reactions which feature the

atoms and bonds in the reaction center. However, the template-based architecture limits both of these methods in scaling up to larger datasets of various kinds of reactions.

- Template-free Approach:
  For a template-free approach, Kayala et al. [6] present an innovative model somewhat similar to the Weisfeiler-Lehman network (WLN). However, there are several key differences between their model and the WLN. First, their model operates at the mechanistic level and identifies elementary mechanistic steps. Since most reactions consist of many mechanistic steps, their approach requires multiple intermediate predictions to predict the entire reaction. In comparison to this, WLN approach operates at the graph level, i.e., it predicts transformations from reactants to products in a single step. The second key difference is that mechanistic descriptions of reactions are not available in existing reaction databases. Therefore, Kayala et al. created their own training set based on a mechanistic-level template-driven expert system. In contrast to this, the WLN model learned directly from real-world experimental data as it was freely and widely available for its consumption. Third, Kayala et al. used chosen feed-forward neural networks for their model where atoms and graphs are represented by molecular fingerprints and additional hand-crafted features, whereas theWLN approach predicts from graph neural networks to encode graph structures.

Connor W Coley et al. show a similar AI model to predict organic chemical reactions in [2]. Their technique predicts the result of a chemical reaction in a twofold way by applying overgeneralized forward response layouts to a pool of reactants to create a set of possible products and by assessing which potential product is the major product as a multiway classification problem utilizing artificial intelligence and machine learning.

In the first approach, the authors apply a library of forward engineered layouts to define which products could possibly form depending on the reactants. Identification was through advance amalgamation investigation, but since numerous particular templates can theoretically match with a given set of reactants and create hundreds or thousands of products, the authors note that the basic count is not a valuable tool to identify real-world results. So, if a specific path has a feasible possibility of synthesis but proceeds at a relatively slower pace, it can be excluded from the path to be considered. However, if a reaction path continues at a rate which is relatively significant, it can be further evaluated. In this manner, through the first step, the authors note that the chemist does not has to think through all possible reaction paths when assessing the reasonability of a forward response step.

In the second stage, the authors propose that every applicant response is scored exclusively by the AI model. This model evaluates the probability of reactivity, similar to a response rate, in disconnection from contending responses. The scores from all possible paths are looked at in a softmax system layer (i.e., an exponential initiation work that maps a rundown of numbers to a rundown of probabilities that aggregate to one) to create probabilities depicting which item is anticipated to be the most abundant [7–11].

A key segment of this methodology is this "two-step formalization." By creating possible items in the first step, in effect, existing response databases are increased with negative response models which expand the amount of data at the disposal of the learning algorithm. This also bypasses the constraint of just having high-yielding response information.

Therefore, by identifying that one reaction which has progressed through in the real world, the AI model proposed by the authors can spot the reaction in its own database and therefore learn about predicting chemical reactions more thoroughly [12–17].

They proposed a novel template-free approach for chemical reaction prediction. Instead of generating candidate products by reaction templates, they first predict a small set of atoms/bonds in reaction center and then produced candidate products by enumerating all possible bond configuration changes within that particular set. Compared to the template-based approach, their model ran 140 times faster, allowing them to scale to relatively much larger reaction databases. The reaction center identifier and candidate ranking model both were built from Weisfeiler-Lehman network and its variants that learn compact representation of graphs and reactions.

The reaction predictor system utilizes neural networks to predict the products of multistep reactions by considering the reaction mechanisms and comparing the feasibility of the formal electron transfer schemes in the elementary steps. An advanced version of the reaction predictor system is applicable not only to reactions that follow a polar mechanism, but also to reactions occurring by pericyclic and radical mechanisms. An alternative method was also proposed for combining the reaction predictor system based on machine learning techniques with quantum chemical calculations. In this model, the transition states and activation barriers were to be calculated for the most favorable reaction steps predicted by the system. The calculation results are fed back into the reaction predictor and are used for additional training of neural networks. This increases the reliability of future predictions [17–20].

Another method of neural network prediction of reaction products was proposed by Aspuru-Guzik's research group. The strategy was tested on a database comprising of only 17 types of organic reactions. However, the model had quite an ability to be extended to a very wide range of reactions. The authors used special neural network fingerprints derived from chemical structures by means of a convolutional neural network operating on the deep learning principle as the basis for the model's predictions.

The authors concluded that through expanded atom and bond-level featurization of the reaction core, the model can be further extended to achieve even higher predictive performance. Through the use of reaction examples with more complete information obtained from commercial sources, a context-dependent model could also be trained to understand reactivity in a subtler manner [2]. Such a model would have a greatly enhanced predictive precision as the context-based approach would reduce the number of desired or possible products and would be helpful in incorporating other subtle factors into the predictive model for improved precision.

Prediction is a very important task associated with new drug development and estimation of drug toxicity and pharmacokinetic properties. The development of quantitative structure-retention relationships (QSSR) have been a great tool empowering reaction predictions [21, 22].

3 Conclusion

The use of artificial intelligence and machine learning can reduce the time duration and efforts required in the decision-making process for certain chemical reactions. The progress of computer analysis of chemical reactions and its application in chemistry would not only generate tools for predicting characteristics and conditions of the reactions, but would also compete with humans in the ability to choose the optimal pathway a reaction to conclude. Combining both the human and the machine approaches to reaction prediction would finally bring us a mechanism and methodology which will be able to provide a chemist with the information that he/she is interested in and make decisions as to which compounds are to be synthesized and how this is to be done in a relatively efficient manner. The experimental equipment used in chemistry already has well-developed electronic interfaces which allow computer control of processes. The reaction prediction models could also be integrated with them and may be enabled to control the process as well to maximize the probability of achieving the desired/predicted product. It is helpful in advancing critical researches, such as drug synthesis, quickly and efficiently without having the chemist to invest a great majority of their time and energy on the reaction's decision-making process. Furthermore, retrosynthesis, which is used extensively in these models, is a familiar technique which can be easily empowered using AI and ML. A large combination of pathways for a reaction to progress can be easily depicted by using retrosynthetic analysis and be studied in a matter of seconds using AI models which places tremendous capabilities in a chemist's hands and facilitate chemical synthesis greatly.

References

1. Marwin S, Mike P, Mark PW (2017) Towards "AlphaChem": chemical synthesis planning with tree search and deep neural network policies. arXiv.org > cs >. arXiv:1702.00020
2. Szymkuć S, Gajewska EP, Klucznik T, Molga K, Dittwald P, Startek M, Bajczyk M, Grzybowski BA (2016) Computer-assisted synthetic planning: the end of the beginning. Angew Chem Int Ed Engl 55(20):5904–5937
3. Corey EJ, Wipke WT (1969) Computer-Assisted Design of Complex Organic Syntheses. Science 166:178–1921
4. Corey EJ (1967) General methods for the construction of complex molecules. Pure Appl Chem 14:19–38. https://doi.org/10.1016/B978-0-08-020741-4
5. Wei JN, Duvenaud D, Aspuru-Guzik A (2016) Neural networks for the prediction of organic chemistry reactions. ACS Cent Sci 2(10):725–732

6. Kayala MA, Azencott CA, Chen JH, Baldi P (2011) Learning to predict chemical reactions. J Chem Inf Model 51(9):2209–2222
7. Jorgensen WL, Laird ER, Gushurst AJ, Fleischer JM, Gothe SA, Helson HE, Paderes GD, Sinclair S (1990) CAMEO: a program for the logical prediction of the products of organic reactions. Pure Appl Chem 62:1921–1932
8. Gasteiger J, Hutchings MG, Christoph B, Gann L, Hiller C, Löw P, Marsili M, Saller H, Yuki K (1987) Organic synthesis, reactions and mechanisms. Springer, Berlin, pp 19–73
9. Connor WC, Regina B, Tommi SJ, William HG, Klavs FJ (2017) Prediction of organic reaction outcomes using machine learning. ACS Cent Sci 3(5):434–443
10. Chen JH, Baldi P (2009) No electron left behind: a rule-based expert system to predict chemical reactions and reaction mechanisms. J Chem Inf Model 49(9):2034–2043
11. Christ CD, Zentgraf M, Kriegl JM (2012) Mining electronic laboratory notebooks: analysis, retrosynthesis, and reaction based enumeration. J Chem Inf Model 52(7):1745–1756
12. Todd MH (2005) Computer-aided organic synthesis. Chem Soc Rev 34(3):247–266
13. Warr WA (2014) A short review of chemical reaction database systems, computer-aided synthesis design, reaction prediction and synthetic feasibility. Mol Inf 33(6–7):469–476
14. Wei JN, Duvenaud D, Aspuru-Guzik A (2016) Neural networks for the prediction of organic chemistry reactions. ACS Cent Sci 2(10):725–732
15. Dragone V, Sans V, Henson AB, Granda JM, Cronin L (2017) An autonomous organic reactions reaction search engine for chemical reactivity. Nat Commun 8:15733
16. Gómez-Bombarelli R et al (2018) Automatic chemical design using a data-driven continuous representation of molecules. ACS Cent Sci 4:268–276
17. Kayala MA, Baldi P (2012) ReactionPredictor: prediction of complex chemical reactions at the mechanistic level using machine learning. J Chem Inf Model 52:2526
18. Sadowski P, Fooshee D, Subrahmanya N, Baldi P (2016) Synergies between quantum mechanics and machine learning in reaction prediction. J Chem Inf Model 56:2125
19. Wei JN, Duvenaud D, Aspuru-Guzik A (2016) ACS Centr Sci 2725
20. Duvenaud DK, Maclaurin D, Aguilera-Iparraguirre J, Gomez-Bombarell R, Hirzel T, Aspuru-Guzik A, Adams RP (2015) Paper presented at the advances in neural information processing systems. arXiv:1509.09292v2[cs.LG]3Nov
21. Halberstam NM, Baskin II, Palyulin VA, Zefirov Mendeleev NS Commun (2002) 185
22. Panwar D, Tomar P, Singh V (2018) Hybridization of Cuckoo-ACO algorithm for test case prioritization. J Stat Manage Syst 21(4):539–546

Chapter 37
Text Line Segmentation for Medieval Devnagari Manuscript

Nikita Mehta and Jyotika Doshi

1 Introduction

OCR is a technique to extract optical component of a digital image in terms of alphabets, numeric, and special symbols. The character recognition is achieved with different stages like preprocessing, segmentation, and feature extraction. Preprocessing will produce a clean, noise-free, and normalized image for further processing [1].

After preprocessing, the next stage of recognition is segmentation. It is an important step in OCR procedure which will affect the accuracy of the character recognition. Segmentation divides the digital image to sub-components. It will extract each line from the document and then each word and character, respectively, from the lines. Segmentation of handwritten documents is much difficult than segmentation of printed documents. And in addition to that, this paper addresses the segmentation of medieval Devnagari manuscripts, which is a big challenge by itself.

Medieval documents are aged from hundreds to thousand years, degraded due to dust, fungus, insects, humidity, etc. Further, these documents are written on bark, palm leaf, cloth, etc. It had space constraints, and writers tried to accommodate as much content as possible on a single page. This increases the segmentation complexities further.

N. Mehta (✉)
School of Doctoral Research and Innovation, GLS University, Ahmedabad, India
e-mail: nikipshah@gmail.com

J. Doshi
GLS(SRP)ICT, GLS University, Ahmedabad, India

© Springer Nature Singapore Pte Ltd. 2021

S. D. Purohit et al. (eds.), *Proceedings of International Conference on Communication and Computational Technologies*, Algorithms for Intelligent Systems,
https://doi.org/10.1007/978-981-15-5077-5_37

1.1 A Brief Introduction of Medieval Manuscripts

Any handwritten composition which is at least seventy-five years old is considered as a manuscript. Lithographs and printed volumes are not manuscripts. These scripts are written on many different materials like paper, bark, cloth, metal, palm leaf, etc. They have significant scientific, historical, or esthetic value. Manuscripts are found in many different languages (like Hindi, Gujarati, Sanskrit, etc.) and scripts (like Brahmi, Oriya, Kharoshti, Sharda, Granth, Devnagari, etc.). Among these, Devnagari was one of the most popular scripts. In manuscripts, Devnagari script was used to write many languages like Hindi, Sanskrit, Pali, Prakrit, Gujarati, etc. [2] (Fig. 1).

India possesses an estimate of five million manuscripts, probably the largest collection in the world. These cover a variety of themes, textures, esthetics, scripts, languages, calligraphies, illuminations, and illustrations. They are on various subjects like religious, cultural, economic, scientific researches, astrology, and stories. Together, they constitute the "memory" of India's history, heritage, and thoughts [2].

As manuscripts are having scientific, social, and historical importance, it is necessary to conserve and preserve them, and this is the most crucial issue for manuscript libraries. These manuscripts are aging for hundreds to thousands years, so they are damaged due to ill age, insects, humidity, fungus, weather, etc., and human negligence also leads them to distortion.

There are several organizations which work for manuscript and even Indian government has taken initiative to collect, conserve, and preserve this ancient treasure as a project called National Mission for Manuscripts (Namami).

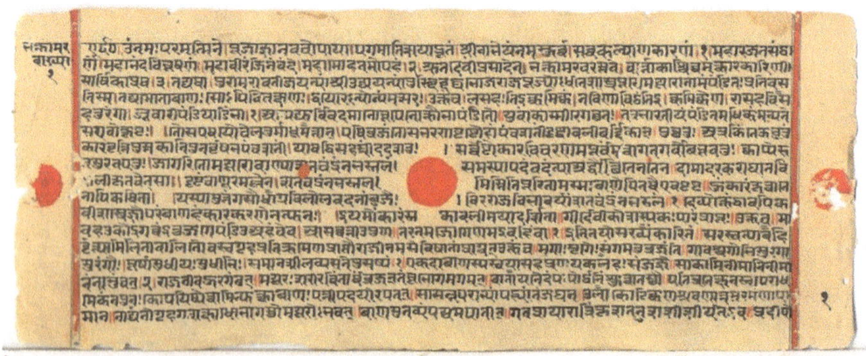

Fig. 1 Sample medieval Devnagari script

1.2 How Medieval Devnagari Manuscripts Written?

Medieval manuscripts are written on material which is not easily available. For example, if a manuscript is written on palm leaf, cutting down palm leaves to make them writable for a document was a time-consuming and lengthy process in that non-technically equipped era.

- Writers used to utilize the resources maximum.
- No spacing other than margins were left, which means there is no spacing between two words.
- Characters were joined in up-down way to save spaces.

1.3 Problems During Line Segmentation of Medieval Devnagari Manuscript

Freestyle handwritten text is difficult to segment, and hence, it is medieval handwritten scripts where the task becomes more challenging due to following reasons:

1. As it is aged, even after noise removal, image is noisy at some context. So, it is difficult to extract lines properly.
2. Characters are not of equal sizes in a single line.
3. Many decorative marks are used in a manuscript.
4. Characters are joined in up-down way, so line segmentation is difficult.
5. There is uneven spacing between lines
6. Modifiers in one line may touch the upper or lower line characters which will make line segmentation more difficult.
7. A single character occurring in different places of document may vary in sizes.

2 Literature Review

There is a significant work reported for Indian scripts, but it is much laser than the work done for scripts used for English, Japanese, Chinese, etc., languages [3]. As per our knowledge and observation, no work has been reported for character recognition of medieval Devnagari script, though there are a few researchers who have worked for some other historical scripts.

In broad way, segmentation methods can be categorized in three categories [4]:

1. Projection-based method: Projection profile is sum of all pixels along with row or column.
2. Hough transform-based method: It considers any image to compose of straight lines. It creates an angle, offset plane in which the local maxima are assumed to correlate with text lines [5].

3. Smearing methods: A short white part is filled with black pixels to make it a large body of black pixels. This large body is considered as data part, and accordingly, it is segmented.

There are several other methods which, in one way or another, are created by modifying and extending above three.

Singh et al. [6] have proposed segmentation technique for ancient Brahmi script. They used horizontal projection profile for line segmentation.

Pal et al. [7] have used projection-based method for segmentation of unconstrained Bangla text. For line segmentation, they divides document in vertical stripes. Then, horizontal histogram for all stripes is created, and the relationship of minimal values is used for segmentation.

Bounding box technique is majorly used for line segmentation in machine-printed documents, but it is not suitable for freestyle handwritten documents as a text line could be curvilinear with insignificant line gaps [8]. Connected component labeling can also give a false result if a line touch another by a single handwritten stroke [8].

Jindal et al. proposed segmentation technique for old Gurumukhi documents [9]. They divided document into different text blocks, and each text block contains upper, middle, and lower zones. And then after connecting, component labeling is applied to the blocks.

Louloudis et al. [10] used Hough transform method for segmentation in handwritten document. They do this with three steps: In first step, an average character height is calculated and average character width is assumed, secondly, Hough transform is applied, and the last step deals with the identified false alarms in the segments.

Nicolaou et al. [5] proposed a novel approach by shredding the text lines through its surface. In this approach, image is blurred, and then, it traces the white most and black most paths through right to left and left to right.

3 Proposed Segmentation Technique

This paper presents the line segmentation for medieval Devnagari manuscripts, though technique proposed here is script independent and can be used for any other script. This technique is based on projection profile with few improvements for connected lines and upper and lower modifiers (Fig. 2).

1. Scanned historical image is converted to a binary image (contains only black and white pixels) with a lot of preprocessing done on it. Preprocessing steps are
   (a) Convert image into gray scale
   (b) Apply a noise removal technique (threshold) on it.
   (c) After threshold application, image contains black pixels—which are data values and white pixels—which are background values (Fig. 3).

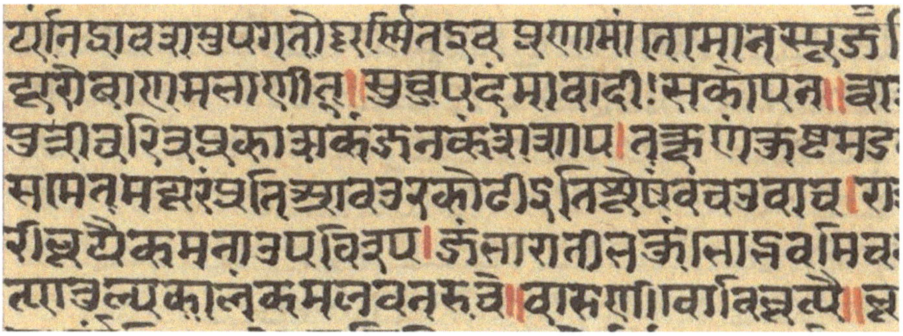

Fig. 2 Sample document image

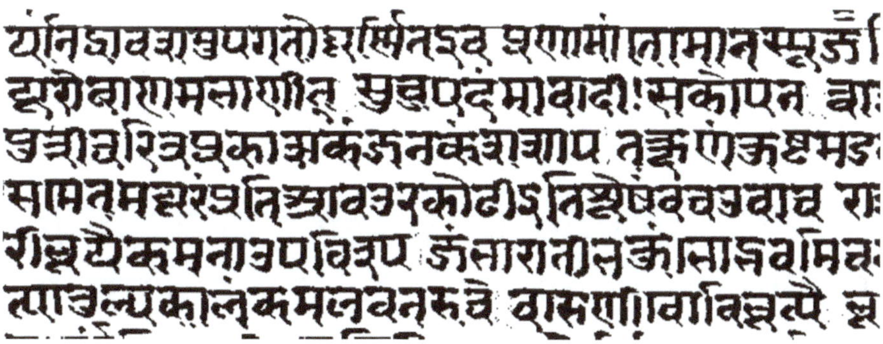

Fig. 3 Image after preprocessing

2. Black and white image is then mapped to a binary array, where all 1 s represent data pixels and 0 s represent background pixels.
3. A horizontal histogram for ON pixels is created.

$$f(y) = \sum_{1 \le x \le N} f(x, y)$$

where $f(y)$ is sum of all pixels in the row y.

All picks and valleys are identified in the histogram,

4. Ideally, center point of the valley is considered as segmentation point. But in case of Devnagari script, this segmentation will lose upper and lower modifiers (Fig. 4). So, here an improvised approach is used. Valley portion is attached in both segmented part (Fig 5).
5. Lines are segmented from segmentation point (Fig. 6).

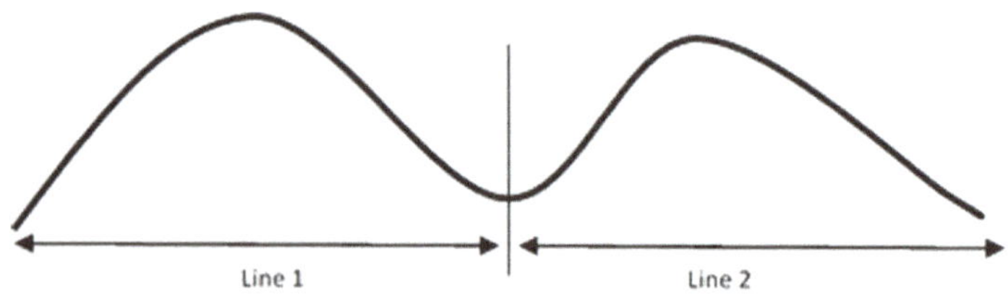

Fig. 4 Projection-based line segmentation

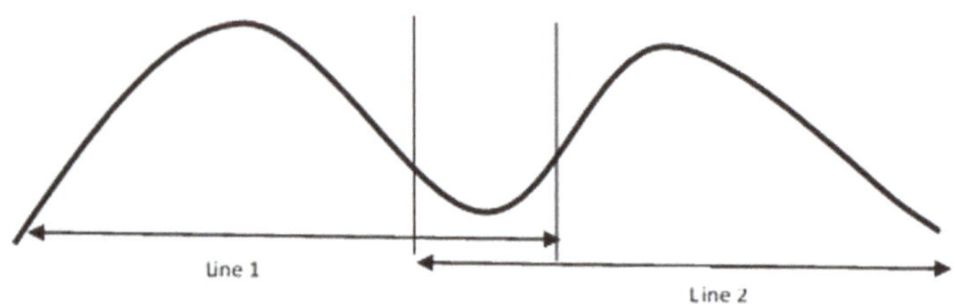

Fig. 5 Modified line segmentation

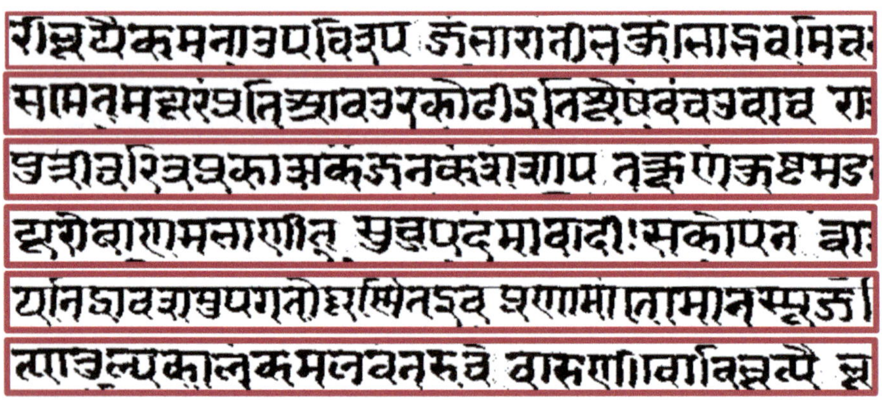

Fig. 6 Segmented lines

4 Result of Proposed Technique

For testing purpose, the proposed algorithm is implemented on Windows platform using Visual Studio 2013. Almost 60 manuscripts, of a single author, with total 993 lines were considered as input. Among 993 lines, 952 lines were segmented correctly using the proposed technique which is 95.87% accuracy (Table 1).

As shown in Fig. 1, each document contains a rounded mark at the center of the document and middle of right and left margins. This mark affects the histogram value and may produce incorrect result (Fig. 7). This problem was solved using an

Table 1 Comparison with existing techniques

Authors	For languages	Total lines	Correct segmented lines	Accuracy (%)
Garg et al. [11]	Handwritten Hindi	200	189	94.5
Palakollu et al. [12]	Handwritten Hindi	500	468	93.6
Proposed method	Handwritten medieval Devnagari	993	952	95.9

Fig. 7 Incorrect segmentation due to middle mark

Fig. 8 Overlapped lines

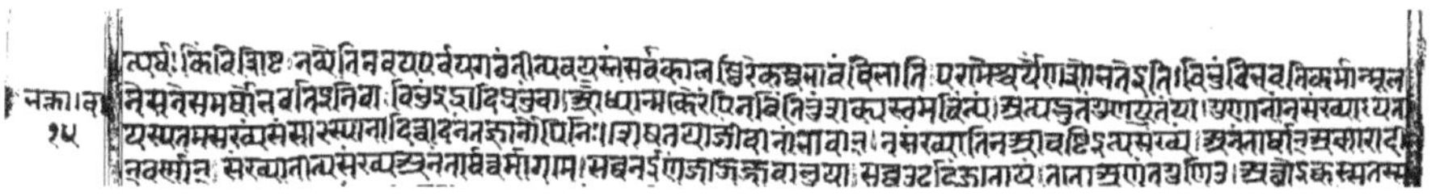

Fig. 9 Skewed lines

algorithm applied on every page to remove the middle mark.

Another major problem with these documents is that because of aging, many pages stick to each other and leave impression of one page's content on another page. These overlapped lines causes noisy image after preprocessing and results in incorrect segmentation (Fig. 8).

As mentioned earlier, proposed technique is not suitable for skewed image. So, if any document contains skewed lines, it will not be segmented properly (Fig. 9).

5 Conclusion

The line segmentation technique proposed in this paper is a projection-based method. This works very well with medieval scripts which are written with significant line gaps. But, it has some drawbacks, like if text is skewed or lines are overlapped, this method may give false results.

6 Future Expansion

Some better techniques for noise removal needed to be identified to solve the problem of overlapped lines because modifications in segmentation technique may segment the overlapped lines, but it will not result in correct character identification.

References

1. Chaudhuri A, Mandaviya K, Badelia P, Ghosh SK (2017) Optical character recognition systems. In: Optical character recognition systems for different languages with soft computing. Studies in fuzziness and soft computing, vol 352. Springer, Cham
2. http://www.namami.org/whatismenuscript.htm
3. Jain AA, Arolkar HA (September, 2018) A survey of gujarati handwritten character. Int J Res Appl Sci Eng Technol (IJRASET) 461–465
4. Sharma A, Chaturvedi R, Dwivedi UK, Kumar S, Reddy S (2018) Firefly algorithm based effective gray scale image segmentation using multilevel thresholding and entropy function. Int J Pure Appl Math 118(5):437–443
5. Nicolaou A, Gatos B (July, 2009) Handwritten text line segmentation by shredding text into its lines. In: 2009 10th international conference on document analysis and recognition, pp 626–630. IEEE
6. Singh AP, Kushwaha AK (2019) Analysis of segmentation methods for Brahmi script. DESIDOC J Libr Inf Technol 39(2):109–116
7. Pal U, Datta S (August, 2003) Segmentation of Bangla unconstrained handwritten text. Null, p 1128, IEEE
8. Li Y, Zheng Y, Doermann D, Jaeger S (2008) Script-independent text line segmentation in freestyle handwritten documents. IEEE Trans Pattern Anal Mach Intell 30(8):1313–1329
9. Jindal S, Lehal GS (December, 2012) Line segmentation of handwritten Gurmukhi manuscripts. In: Proceeding of the workshop on document analysis and recognition, pp 74–78. ACM
10. Louloudis G, Gatos B, Pratikakis I, Halatsis C (2009) Text line and word segmentation of handwritten documents. Pattern Recogn 42(12):3169–3183
11. Garg NK, Kaur L, Jindal MK. (April, 2010) A new method for line segmentation of handwritten Hindi text. In: 2010 seventh international conference on information technology: new generations, pp 392–397. IEEE
12. Palakollu S, Dhir R, Rani R (2012) Handwritten Hindi text segmentation techniques for lines and characters. In: Proceedings of the world congress on engineering and computer science, vol 1, pp 24–26

Chapter 38
Audit Fraud Data Prediction Using Machine Learning Algorithms

Ankita Sharma, Amit Sinhal, Manish Tiwari, and Mayank Patel

1 Introduction

Fraud is intentional or purposely false committed crimes for the benefit of an individual or group of person. Individuals are able to commit fraud because they see an opportunity or due to pressure or greed or due to rationalization. Auditing practices are conducted or accountable for detection of a committed fraud. It is the process of verification of on-site financial detail of any firms to check whether their financial details are verified with different principle and standard accounting laws. It is challenging task to spot firms in detecting frauds, spotting errors, and exposing employee of being guilty of aiding an illegal transaction. Audits performed by external groups over private companies can be exceedingly supportive in eliminating unfairness when it narrows down to company's finances. Audits actually seek for what is called "material error" in any statement on any distinct object. Machine learning is an active area of research, and researchers are working on solving fraud prediction issues using machine learning techniques [1]. So, designing a machine learning problems using different machine learning classifiers helps in building a prediction model that would predict whether the firm has done any fraud or not.

In this research work, after preprocessing of data, various machine learning algorithms are trained to check the performance of classification. The performance and quality measure of the best classifier are being compared with the state-of-the-art classifiers models like SMO, etc., using various performance metrics like accuracy, FP rate, MCC, area under the curve, sensitivity, etc.

This paper is organized into various sections as follows: Sect. 2 briefly describes the classification technique that is used in the proposed framework. Section 3 discusses the data, its features, and experimental setup. Section 4 summarizes the

A. Sharma (✉) · A. Sinhal · M. Tiwari · M. Patel
Department of Computer Science, Geetanjali Institute of Technical Studies (GITS), Udaipur, India
e-mail: smile.ankita55@gmail.com

© Springer Nature Singapore Pte Ltd. 2021

S. D. Purohit et al. (eds.), *Proceedings of International Conference on Communication and Computational Technologies*, Algorithms for Intelligent Systems,
https://doi.org/10.1007/978-981-15-5077-5_38

result of the experiments performed and performance comparison. And at last in Sect. 5, conclusion and future scope have been discussed.

2 Materials and Methods

This research paper's main objective is to apply machine learning application to develop a computing machine which detects fraud firm based on data classification.

2.1 Proposed Framework

The complete prediction model is presented in Fig. 1. The proposed framework outcome helps to predict the fraudulent audit firm of any nominated firm and to understand and analyze the audit risk analysis and the workflow of the company by in-depth interview with the audit expertise, and to proffer a decision-making foundation for risk assessment of firms in the time of audit planning.

2.2 Machine Learning Classifiers

2.2.1 Logistic Function

In logistic model, work on dependent and independent variable, the relationship is calculated between the definite dependent variable and one or more independent variables by evaluating and calculating probabilities using logistic function approach.

2.2.2 Bayesian Network (BN)

An acyclic graph theorem and probabilities-based model is presented in and work as a graphical model which shows a number of features and their various conditional dependencies through a specified acyclic graph [2].

2.2.3 Random Forest

Random forest algorithm is generally applied on classification algorithm. As the name suggests, this algorithm creates the forest with a number of trees. Decision tree concept is more to the rule-based system. Given the training dataset with targets and features, the decision tree algorithm will come up with some set of rules. The same set rules can be used to perform the prediction on the test dataset [3].

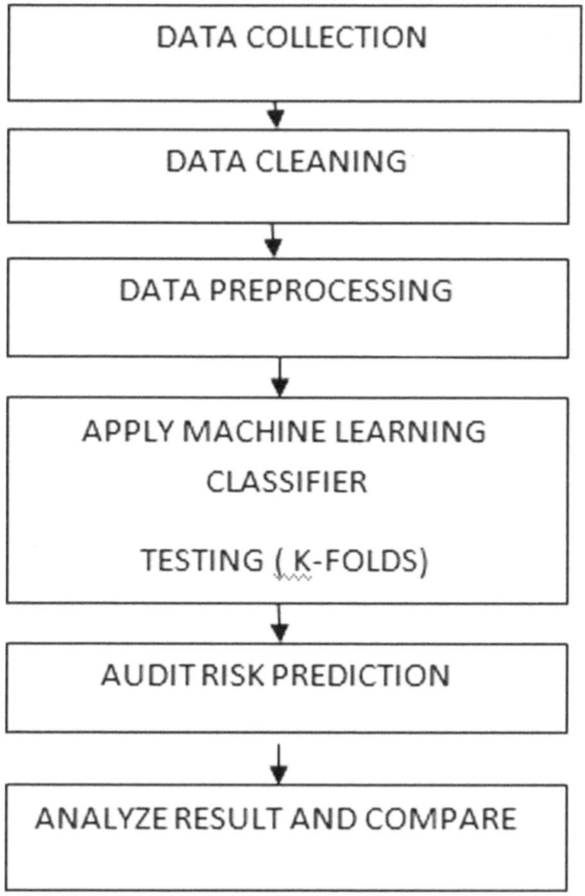

Fig. 1 Prediction model

2.2.4 AdaBoost

Boosting is a process to make strong classifiers from number of weak classifiers. It used a mechanism of correcting error from previous created model means first model is created using training dataset and calculate error of this model and create second model which solve errors of first model and rectifies it. In this way, a number of models are created until training dataset is not predicted perfectly. In mostly binary classification problem, AdaBoost algorithm is used. It boosts the power of decision tree algorithm in binary classification problem [4].

2.2.5 Naive Bayes

The Naive Bayes classifier technique is based on the so-called Bayesian theorem and is particularly suited when the dimensionality of the inputs is high. Despite its simplicity, Naive Bayes can often outperform more sophisticated classification methods. Naive Bayes is among one of the most simple and powerful algorithms for classification based on Bayes' theorem with an assumption of independence among predictors.

2.2.6 J48

The C4.5 algorithm for building decision trees is implemented in Weka as a classifier called J48. Classifiers, like filters, are organized in a hierarchy: J48 has the full name weka.classifiers.trees.J48. The additional features of J48 are accounting for missing values, decision trees pruning, continuous attribute value ranges, derivation of rules, etc. In the WEKA data mining tool, J48 is an open-source Java implementation of the C4.5 algorithm [5–7].

3 Experimental Investigation

This section discusses the dataset and experimental setup.

3.1 Dataset

The dataset was collected from 777 different firms from 6 different sectors.

3.2 Experimental Setting

The "Weka" is a tool for model building, feature selection, and class balancing techniques; these all are machine learning features. Machine learning classifier applies the principle of k-fold cross-validation technique First dataset is divided into k same size subsets; best classifier is used as base classifier to train 9 subset and test it on the basis of last folded subset [8]. To measure the quality and performance of designed framework, various parameters like accuracy, error rate, F-measure, false positive rate, Mathew's correlation coefficient (MCC), and area under curve (AUC) are used [9].

Table 1 Confusion matrix

	True reference	
Predicted condition	Condition positive	Condition negative
Fraud	True positive *X*	False positive *Z*
No fraud	False negative *Q*	True negative *Y*

Table 2 Performance metric formula

Performance metric	Formula
Sensitivity	$X/(X + Z)$
Specificity	$Y/(Q + Y)$
Accuracy	$(X + Y)/(X + Z + Q + Y)$
MCC	$(X * Y) - (Q * Z)/\text{SQRT}((X + Q) + (X + Z) + (Y + Q) + (Y + Z))$

4 Results and Discussion

This section discusses parameter evaluation metrics to measure the performance of various machine learning algorithms. The results of 10-fold cross-validation method are presented graphically and discussed much in detail.

4.1 Performance Evaluation

The performance of the proposed framework is evaluated on the basis of confusion matrix as shown in Table 1. The various evaluation metrics calculated from the Table 1 are presented in Table 2.

4.2 Experimental Results

This section explains results and discusses the performance of various machine learning classifiers. The results are also presented graphically, and the reason different performances of classifiers are also discussed (Figs. 2 and 3)..

Performance of the proposed prediction model has been evaluated and analyzed with 7 different performance metric parameters, and the results are summarized in Table 3. It can be observed that Bayes net has the highest performance for the prediction of suspicious firm. Hence, they are highlighted in Table 3.

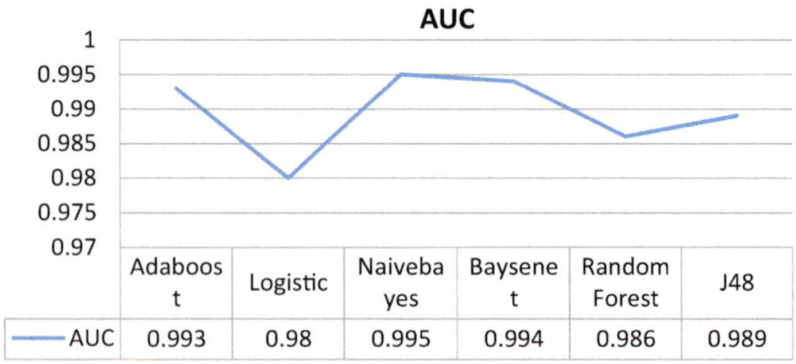

Fig. 2 Comparison of AUC of different ML classifiers

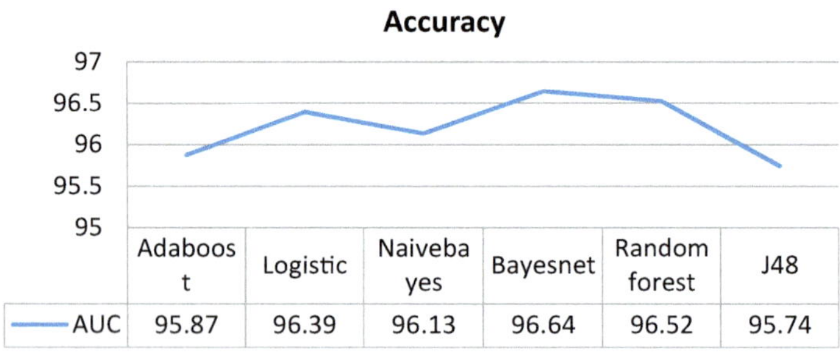

Fig. 3 Comparison of accuracy of different ML classifiers

Table 3 Result analysis

Classifier	Accuracy	Error	TP rate	FP rate	F-measure	MCC	AUC
AdaBoost	95.87	4.13	0.95	0.054	0.95	0.914	0.99
Random forest	96.52	3.48	0.96	0.042	0.96	0.927	0.98
Logistic	96.39	3.61	0.96	0.043	0.96	0.924	0.98
J48	95.74	4.26	0.95	0.052	0.95	0.911	0.98
Naïve Bayes	96.13	3.87	0.96	0.052	0.94	0.919	0.99
Bayes net	**96.64**	3.36	0.96	0.045	0.96	0.93	0.99

Byesnet classifier given highest accuracy which is shown in bold in Table 3

5 Conclusion

The objective of this research paper is to use machine learning techniques to find and detect the fraudulent firm in the time of audit planning. To find out low risk firm at the time of audit planning, auditor will be fully prepared with a complete Audit Field Work decision support kit to check list low risk firms. Auditor can easily find out high risk firm at the starting phase of audit. To complete this work, collect dataset, preprocess on it, apply machine learning classifiers with 10 folds, and conclude result. Result is higher than previous result.

In future, hybrid, and new machine learning techniques can be applied to increase accuracy of result.

References

1. Meenatkshi R, Sivaranjani K (July, 2016) A comparative study on fraud detection in financial statement using data mining technique. Int J Comput Sci Mob Comput 5(7)
2. Rosario A, Liburd HB, Kogan A, Appelbaum D, The Research Assistance of Jamie Freiman (2016) Research ideas for artificial intelligence in auditing: the formalization of audit and workforce supplementation. J Emerg Technol Account 13(2). https://doi.org/10.2308/jeta-10511
3. Chen S (2016) Detection of fraudulent financial statements using the hybrid data mining approach. Springer Plus
4. Tangod K, Kulkarni G (July, 2015) Detection of financial statement fraud using data mining technique and performance analysis. Int J Adv Res Comput Commun Eng 4(7)
5. Panwar D, Tomar P, Singh V (2018) Hybridization of cuckoo-ACO algorithm for test case prioritization. J Stat Manag Syst 21(4):539–546
6. Apparao G, Singh A, Rao GS, Lalitha Bhavani B, Eswar K, Rajani D (2009) Financial statement fraud detection by data mining. Int J Adv Netw Appl 1(3):159–163
7. Ravisankar P, Ravi V, Raghava Rao G, Bose I (2011) Detection of financial statement fraud and feature selection using data mining techniques. P. Ravisankar Et Al. Decision Support Systems 491–500/Elsevier
8. Perols J (2011) Financial statement fraud detection: an analysis of statistical and machine learning algorithms. J Pract Theory 30(2)
9. Tiwari A, Hooda N (June, 2018) Machine learning framework for audit fraud data prediction. Res Gate Publ 7(6)

Chapter 39
Parkinson's Disease Prediction Using a Stacked Generalization Ensemble of kNN and Its Variants with mRMR Feature Selection

Vinod J. Kadam, Shivajirao M. Jadhav, Mahesh R. Shirsath, and Atharv A. Kurdukar

1 Introduction

Parkinson's disease (PD) is a kind of movement disorder and chronic, long-term, progressive nervous system disorder (or simply brain disorder) that harms certain nerve cells or neurons (cells become impaired and/or die) in the human brain (including a region called the substantia nigra) that produces a vital brain chemical called dopamine (a chemical messenger or neurotransmitter). Various medical research papers reported that the dysphonic measurements are vital reference index in the diagnosis of PD (90% of PD patients show such signs and symptom) [1, 2]. Many researchers proposed various feature selection and classification approaches to construct PD diagnosis systems using dysphonia features. Important feature selection and extraction approaches used by researchers to extract relevant features from PD dataset in order to increase accuracy of base classifier are: correlation-based feature selection [3], fuzzy-based nonlinear transformation [4], fuzzy entropy measures [5], feature weighting method [6], genetic programming [7], mRMR criterion [8], principal component analysis [4, 9], etc. Important classification algorithms available in the literature to construct PD diagnosis systems using dysphonia features are ANN [10, 11], decision tree learning [10], DMNeural analysis [10], ensemble classifier

V. J. Kadam (✉) · S. M. Jadhav · M. R. Shirsath · A. A. Kurdukar
Department of Information Technology, Dr. Babasaheb Ambedkar Technological University, Lonere, India
e-mail: vjkadam@dbatu.ac.in

S. M. Jadhav
e-mail: smjadhav@dbatu.ac.in

M. R. Shirsath
e-mail: maheshshirsath106@gmail.com

A. A. Kurdukar
e-mail: 3atharvkurdukar@gmail.com

© Springer Nature Singapore Pte Ltd. 2021
S. D. Purohit et al. (eds.), *Proceedings of International Conference on Communication and Computational Technologies*, Algorithms for Intelligent Systems,
https://doi.org/10.1007/978-981-15-5077-5_39

with RF algorithm [3], fuzzy k-nearest neighbor [9], improved EM algorithm [7], kernel tricks [12], k-nearest neighbor [6], mRVM models [13], mixture models based on the Dirichlet process (DP) [14], regression analysis [10], SVM-based approaches [4, 8, 12], similarity classifier using similarity measure [5], stacked autoencoders [1, 15], softmax regression [1, 15–17], etc. In this study, we proposed the minimum redundancy maximum relevance approach [18] for relevant and non-redundant feature selection and a stacked generalization ensemble classifier [1] of kNN algorithm (based on different distance functions) and its variants (fuzzy kNN and weighted kNN) to construct Parkinson's disease diagnosis system based upon vocal representations. We proposed logistic regression classifier as the combiner to make a final collective prediction.

2 Proposed Approach

The proposed model includes two steps: feature ranking using the mRMR approach and classification using ensemble classifier. mRMR is a filter-based feature selection method which uses both the maximum relevance criteria and minimum redundancy criteria for useful feature ranking. In the second step, we constructed a novel ensemble classifier using five component classifiers: fuzzy kNN, weighted kNN, and three different kNNs with Euclidean, Mahalanobis, and correlation distance functions. The stacked generalization approach was used to combine these 5 kNN classification models via a logistic regression classifier. Figure 1 shows the proposed approach.

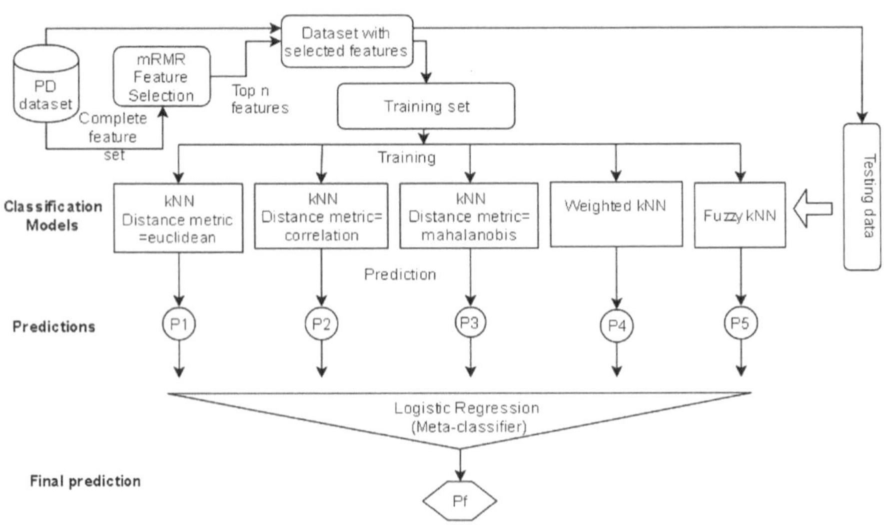

Fig. 1 Proposed model: stacked generalization ensemble of kNN and its variants with mRMR feature selection

2.1 Minimum Redundancy Maximum Relevance Feature Selection

Several feature selection methods have been proposed by various researchers to increase classification accuracy (or to improve the generalization performance) and reduce dimensionality. Different studies have reported that highly relevant and non-redundant features significantly enhance the accuracy of the base classification system [16]. In our study, we used minimum redundancy maximum relevance (mRMR) method [18] for feature selection. This method of feature selection is a filtering algorithm based on mutual information. The mutual information is defined as

$$I(X, Y) = \sum_{i,j} P(X_i, Y_j) log \frac{P(X_i, Y_j)}{P(X_i)P(Y_j)} \tag{1}$$

For two discrete features (X and Y), the joint probability distributions $P(X_i, Y_j)$ and $P(X_i)$, $P(Y_j)$ are the marginals. The mRMR algorithm chooses attributes or variables that have a high correlation with respect to the output (maximum relevance) and a low correlation between themselves (minimum redundancy). Therefore, the following two requirements have to be satisfied. The first requirement is maximum relevance: MaxMR (selecting the features with the highest relevance to the target class h)

$$\text{maxMR, MR} = \frac{1}{|S|} \sum_{i \epsilon S} I(h, i) \tag{2}$$

The second requirement is minimum redundancy: minMV

$$\min \text{MV, MV} = \frac{1}{|S|^2} \sum_{i, j \epsilon S} I(i, j) \tag{3}$$

where h is the class variable. S is the set of features. There are two ways to combine the above two requirements: mutual information difference (MID) (additive combination) max(MR-MV) and mutual information quotient (MIQ) (multiplicative combination) max(MR/MV) [18].

2.2 Stacked Generalization Ensemble of kNN and Its Variants

k-nearest neighbor classifier is a nonparametric approach widely used for classification tasks. It is one of the well-known instance-based learnings or lazy learning approaches. The main advantages of kNN include ease to interpret the output, low calculation time and better predictive power. To find the class label of the given test sample, kNN computes its K-nearest neighbors and determines the class label by majority voting. Steps of kNN algorithm are as follows

1. Load the training data, input the parameter k
2. Input the test data sample
3. Iterate from 1 to N (N = total samples in training data)
   Compute the distance between test data sample and each training data sample
4. Sort the computed distances in ascending order.
5. Select top k rows from the sorted array and obtain the most common class of these rows
6. Return the class label.

Here, we proposed a novel ensemble classifier of fuzzy kNN, weighted kNN, and three different kNN algorithms with Euclidean, Mahalanobis, and correlation distance functions. Weighed kNN is based on the concept that such training samples, which are especially near to the given test sample, should get the greater weight than such neighbors that are far away from the test sample [19]. Fuzzy kNN computes a fuzzy degree of membership of each instance to the classes of the problem [20]. The stacked generalization approach was used to combine 5 kNN classification models via a logistic regression classifier. The k-nearest neighbor algorithm uses the whole training dataset as the model. In the training phase, all five models were loaded with the training dataset. All five sub-models individually made a prediction on every training data sample. Using these predictions on the training dataset, we created a new stacked dataset. The (meta-classifier) logistic regression was then trained on the stacked dataset. In the testing phase, the test sample was provided to each sub-model, and all five sub-models separately made a prediction on the given test sample. These predictions were given as an input sample to meta-classifier. The meta-classifier provided the final class label of the test sample.

3 Experimentation and Results

3.1 Dataset

Parkinson's disease dataset of biomedical voice measurements adopted in this study was donated and generated by Dr. Little et al. (also referred to as Oxford Parkinson's Disease Detection Dataset in many studies). This dataset contains 197 instances (48 normal and 147 PD cases) and 22 attributes. The dataset is composed of biomedical voice measurements from 23 PD patients and eight healthy individuals [21, 22].

Table 1 Feature rankings obtained by mRMR feature selection algorithm

Rank	Feature	Rank	Feature	Rank	Feature
1	Spread1	9	D2	16	MDVP:Shimmer
2	MDVP:Fo (Hz)	10	MDVP:Jitter (Abs)	17	MDVP:PPQ
3	Spread2	11	MDVP:APQ	18	Shimmer:DDA
4	MDVP:Shimmer (dB)	12	Shimmer:APQ5	19	Jitter:DDP
5	MDVP:Flo (Hz)	13	HNR	20	MDVP:Fhi (Hz)
6	RPDE	14	Shimmer:APQ3	21	MDVP:Jitter (%)
7	PPE	15	MDVP:RAP	22	NHR
8	DFA				

3.2 Experimentation

Both feature selection and ensemble classifier modules ware implemented within the MATLAB 2016b environment. For evaluation of the proposed classifier, the tenfold cross-validation scheme was adopted. Initially, mRMR feature selection approach was employed for the selection of the relevant and non-redundant features. Here, we employed the mutual information difference (MID) scheme. Feature rankings obtained by mRMR feature selection algorithms are given in Table 1.

After feature selection, we constructed an ensemble classifier of fuzzy kNN (parameter $m = 2$), weighted kNN (Euclidean distance), and three different kNNs with Euclidean, Mahalanobis, and correlation distance functions. To aggregate these trained KNN classifiers and to get a combined prediction, we applied an upper-layer logistic regression classifier. Figure 2 shows the relation between the number of features and the classification accuracy of the proposed ensemble classifier on UCI Oxford PD detection dataset. For each case, we tried various values of k in the range of 1–10, and the best result was reported. We obtained a promising classification accuracy 98.46% (sensitivity 99.32% and specificity 95.83%) using eight top-ranked features and $k = 2$. The performance of the proposed ensemble classifier using eight top-ranked features is further illustrated in Fig. 3.

We also compared our proposed approach with different methods present in the literature (Table 2). The statistical analysis of Table 2 shows that the proposed method is also comparable with these approaches.

4 Conclusion

In this study, we investigated a novel hybrid diagnosis method based on minimum redundancy maximum relevance (mRMR) feature selection algorithm and stacked generalization ensemble classifier of fuzzy kNN, weighted kNN, and three different kNN algorithms with Euclidean, Mahalanobis, and correlation distance functions for

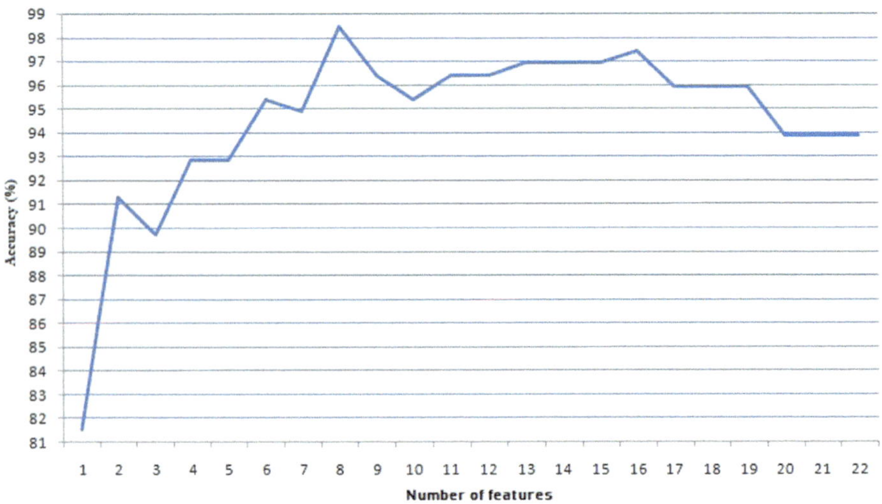

Fig. 2 True classification accuracy (using tenfold CV) with the increase in the number of features selected using mRMR algorithm (for each case, we tried the values of *k* in the range of 1–10, and best result was reported)

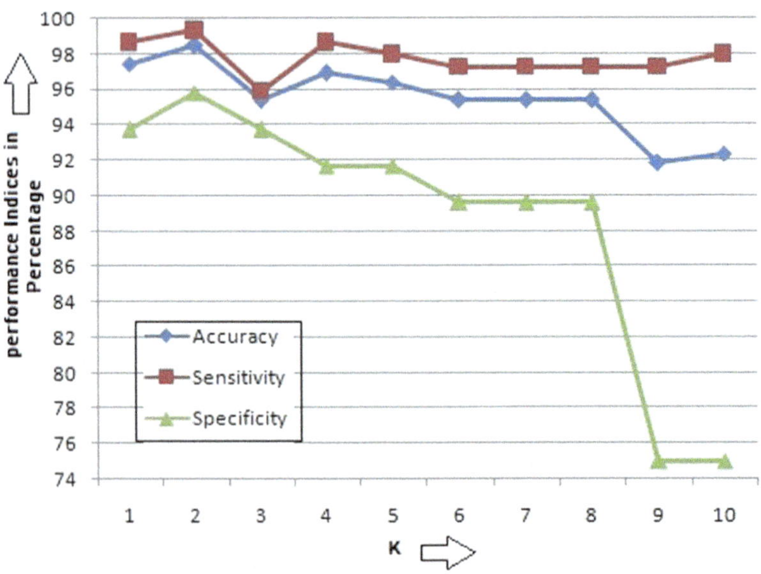

Fig. 3 Classification performance indices (accuracy, sensitivity, and specificity) versus value of *K* (performance of the proposed classifier using eight top-ranked features)

Table 2 Comparison with other approaches in the literature on Oxford PD detection dataset

Method (Ref.)	Accuracy (%)
Stacked sparse autoencoders and softmax regression [1]	92.19 (tenfold CV)
Feature ensemble using stacked autoencoders [1]	93.84 (tenfold CV)
CFS + ensemble with rotation forest [3]	87.10 (tenfold CV)
Fuzzy-based transformation + support vector machine [4]	93.47 (hold out)
Fuzzy entropy measures + similarity classifier [5]	85.03 (hold out)
Support vector machine [12]	91.40 (bootstrap)
Artificial neural network [10]	92.90 (hold out)
Parallel neural network [11]	91.20 (hold out)
Fuzzy *c*-means clustering-based feature weighting + kNN [6]	97.93 (hold out)
Genetic programming + expectation maximization algorithm [7]	93.10 (tenfold CV)
Nonlinear models using Dirichlet process mixtures [14]	87.70 (fivefold CV)
Deep neural network [15]	93.79 (tenfold CV)
mRMR+ support vector machine [8]	92.75 (bootstrap)
PCA-fuzzy *k*-nearest neighbor [9]	96.07 (tenfold CV)
Multiclass relevance vector machines [13]	89.47 (tenfold CV)
mRMR + ensemble of kNN and its variants (this study)	**98.46** (tenfold CV)

Bold emphasis indicates Best performance

addressing the Parkinson's disease problem. To evaluate the performance of the proposed method, the UCI Oxford PD detection dataset was applied. The mRMR feature selection algorithm assisted us to decrease over-fitting, increase the true classification accuracy, sensitivity and specificity by electing the best vocal representations and by eliminating less relevant features. The prediction results obtained by the proposed approach were very promising. In addition, the proposed method outperforms many methods available in the literature. Experiment outcomes also indicate that the proposed ensemble classifier can be used as a computer-aided decision-making tool for early diagnosis of Parkinson's disease.

References

1. Kadam VJ, Jadhav SM (2019) Feature ensemble learning based on sparse autoencoders for diagnosis of Parkinsons disease. In: Iyer B, Nalbalwar S, Pathak N (eds) Computing, communication and signal processing, vol 810. Advances in intelligent systems and computing. Springer, Singapore
2. Kadam VJ, Yadav SS, Jadhav SM (2020) Soft-margin SVM incorporating feature selection using improved elitist GA for arrhythmia classification. In: Abraham A, Cherukuri A, Melin P, Gandhi N (eds) Intelligent systems design and applications. ISDA 2018. Advances in intelligent systems and computing, vol 941. Springer, Cham

3. Ozcift A, Gulten A (2011) Classifier ensemble construction with rotation forest to improve medical diagnosis performance of machine learning algorithms. Comput Meth Program Biomed 104:443–451
4. Li DC, Liu CW, Hu SC (2011) A fuzzy-based data transformation for feature extraction to increase classification performance with small medical data sets. Artif Intell Med. 52(1):45–52. https://doi.org/10.1016/j.artmed.2011.02.001 (Epub 13 Apr 2011)
5. Pasi L (2011) Feature selection using fuzzy entropy measures with similarity classifier. Exp Syst Appl 38(4):4600–4607. https://doi.org/10.1016/j.eswa.2010.09.133
6. Polat Kemal (2012) Classification of Parkinson's disease using feature weighting method on the basis of fuzzy C-means clustering. Int J Syst Sci 43(4):597–609 (Taylor & Francis)
7. Guo P-F, Bhattacharya P, Kharma N (2010) Advances in detecting Parkinson's disease. In: Zhang D, Sonka M (eds) Medical biometrics. ICMB 2010, vol 6165. Lecture notes in computer science. Springer, Berlin
8. Sakar CO, Kursun O (2010) Telediagnosis of Parkinsons disease using measurements of dysphonia. J Med Syst 34:591. https://doi.org/10.1007/s10916-009-9272-y
9. Chen H (2013) An efficient diagnosis system for detection of Parkinson's disease using fuzzy k-nearest neighbor approach. Exp Syst Appl 40(1):263–271
10. Das R (2010) A comparison of multiple classification methods for diagnosis of Parkinson disease. Exp Syst Appl 37(2010):1568–1572
11. Åström F, Koker R (2011) A parallel neural network approach to prediction of Parkinson's disease. Exp Syst Appl 38(10):12470–12474
12. Little MA, McSharry PE, Hunter EJ, Spielman J, Ramig LO (2009) Suitability of dysphonia measurements for telemonitoring of Parkinson's disease. IEEE Trans Biomed Eng 56:1015. https://doi.org/10.1109/TBME.2008.2005954
13. Psorakis Damoulas T, Girolami MA (2010) Multiclass relevance vector machines: sparsity and accuracy. IEEE Trans Neural Netw 21(10):1588–1598. https://doi.org/10.1109/TNN.2010.2064787
14. Shahbaba Babak, Neal Radford (2009) Nonlinear models using Dirichlet process mixtures. J Mach Learn Res 10:1829–1850
15. Caliskan A, Badem H, Basturk A, Yuksel ME (2017) Diagnosis of the parkinson disease by using deep neural network classifier. Istanbul Univ J Electr Electron Eng 17(2):3311–3318
16. Kadam VJ, Jadhav SM, Yadav SS (2019) Bagging based ensemble of support vector machines with improved elitist GA-SVM features selection for cardiac arrhythmia classification. Int J Hybrid Intell Syst Pre-press(Pre-press):1–9. https://doi.org/10.3233/HIS-190276
17. Kadam VJ, Jadhav SM, Vijayakumar K (2019) Breast cancer diagnosis using feature ensemble learning based on stacked sparse autoencoders and softmax regression. J Med Syst 43:263. https://doi.org/10.1007/s10916-019-1397-z
18. Peng H, Long F, Ding C (2005) Feature selection based on mutual information criteria of max-dependency, max-relevance, and min-redundancy. IEEE Trans Pattern Anal Mach Intell 27(8):1226–1238
19. Dudani SA (1976) The distance-weighted k-nearest-neighbor rule. IEEE Trans Syst Man Cybernet SMC-6(4):325–327
20. Keller JM, Gray MR, Givens JA (1985) A fuzzy K-nearest neighbor algorithm. IEEE Trans Syst Man Cybernet SMC-15(4):580–585
21. Little MA, McSharry PE, Roberts SJ, Costello DAE, Moroz IM (2007) Exploiting nonlinear recurrence and fractal scaling properties for voice disorder detection. BioMed Eng 6:23
22. Dua D, Graff C (2019) UCI machine learning repository. University of California, School of Information and Computer Science, Irvine, CA. http://archive.ics.uci.edu/ml

Chapter 40
Multi-classification of UNSW-NB15 Dataset for Network Anomaly Detection System

Aditi Roy and Khundrakpam Johnson Singh

1 Introduction

Because of the expanded utilization of networked computers, computer network security is drawing in expanding consideration and network intrusions have turned into a critical danger as of late. Firewall frameworks cannot distinguish present-day attack scenarios and incapable of investigating network packets in depth. In view of these facts, IDSs are intended to accomplish high insurance for the cyber security framework [1].

An intrusion detection system (IDS) is kind of security software that screens a network traffic or systems for any harmful activity or infringement to the security policy. Kinds of IDS vary in scope from single computers to enormous networks. The IDS can be divided as host-based intrusion detection systems (HIDS) and network intrusion detection systems (NIDS). HIDS is a system that screens significant files of operating system, while in a case of a NIDS it scrutinizes incoming network traffic.

2 Dataset Description

The existing benchmark datasets especially DARPA (Department of Defense Advanced Research Projects Agency) [2], KDD99 (Knowledge Discovery and Data Mining, 1999) [3], and NSL-KDD [4] have several drawbacks, such as lack of modern low footprint attacks, absence of false positives, training and testing sets having different distribution, etc. That is why UNSW-NB15 dataset is chosen in our research which [5] is a modern KDD-99 alternative.

A. Roy (✉) · K. J. Singh
Department of Computer Science and Engineering, National Institute of Technology, Manipur, Imphal, Manipur, India
e-mail: aditiroy500@gmail.com

© Springer Nature Singapore Pte Ltd. 2021

S. D. Purohit et al. (eds.), *Proceedings of International Conference on Communication and Computational Technologies*, Algorithms for Intelligent Systems,
https://doi.org/10.1007/978-981-15-5077-5_40

There were in total 49 features in the full dataset excluding the feature "id", shown in Table 1. The dataset records have been parted into training and testing sets, shown in Table 2. Features numbered from 46 to 50 are not included in the training and testing datasets. The above-mentioned distribution of training dataset has been used in this research work with cross-validation, and test dataset was kept separate.

3 Related Work

There exists a lot of research on feature selection and classification of attacks. Some of them are mentioned below.

Janarthanan and Zargari [6] proposed a subset of attributes in their study, and the discoveries are contrasted with the past researches in connection to attribute selection in the KDD'99 dataset. A couple of techniques for attribute selection were employed using UNSW-NB15 dataset. The recommended attributes by these techniques were analyzed, and a couple of machine learning algorithms were applied on the UNSW-NB15 dataset.

Kumar et al. [7] proposed the feature selection technique by ranking them utilizing the different feature selection techniques. In this paper, they had utilized just 1 classifier to compare the performances of feature selection methods which is biased to other classifiers. In our research 2 classifiers were used as performance predictors.

Mogal et al. [8] suggested a network intrusion detection system (NIDS) utilizing central points (CP) of attribute values and for preprocessing used apriori algorithm of ARM. By eliminating the irrelevant features, the central point provides an opportunity to choose the highest ranked features. These algorithms are executed on the KDDCUP99 and UNSW-NB15 datasets. To separate among normal and attack records, Naïve Bayes and logistic regression are utilized. The proposed system was able to advance the detection accuracy and decrease the false alarm rate by reducing the processing time where larger processing time was the major drawback of this proposal.

Ben Amor et al. [9] proposed an approach using KDDCUP99 dataset for reasonable analysis of Naive Bayes against decision tree. They demonstrated that decision tree provides somewhat preferable outcomes over Naive Bayes.

Breiman [10] proposed an ensemble approach, random forest, which was first applied to intrusion detection by Zhang and Zulkernine [11] to accomplish anomaly and misuse detection based on network traffic.

Hu et al. [12] suggested a procedure for intrusion detection dependent on the AdaBoost algorithm. AdaBoost algorithm joins the weak classifiers (such as decision stumps) into a strong classifier for continuous features as well as for the categorical features.

Mukkamala et al. [13] suggested support vector machine (SVM) and neural networks (NN) for anomaly detection.

Sammany et al. [14] proposed a neural network-based technique with two hidden layers of neuron. They had assumed that an attack will noticeably affect system

Table 1 Features of UNSW dataset

Number	Feature	Number	Feature	Number	Feature	Number	Feature	Number	Feature
1	id	11	sttl	21	Swin	31	response_body_len	41	ct_src_ltm
2	dur	12	dttl	22	Stcpb	32	ct_srv_src	42	ct_srv_dst
3	Proto	13	sload	23	Dtcpb	33	ct_state_ttl	43	is_sm_ips_ports
4	Service	14	Dload	24	Dwin	34	ct_dst_ltm	44	attack_cat
5	State	15	sloss	25	Tcprtt	35	ct_src_dport_ltm	45	Label
6	spkts	16	dloss	26	Synack	36	ct_dst_sport_ltm	46	Srcip
7	Dpkts	17	sinpkt	27	Ackdat	37	ct_dst_src_ltm	47	Sport
8	sbytes	18	dinpkt	28	Smean	38	is_ftp_login	48	Dstip
9	dbytes	19	sjit	29	Dmean	39	ct_ftp_cmd	49	Stime
10	rate	20	Djit	30	trans_depth	40	ct_flw_http_mthd	50	Ltime

Table 2 A part of UNSW-NB15 dataset distribution

Category	Training set	Testing set	Description
Normal	56,000	37,000	Normal exchange information
Analysis	2000	677	It includes various assaults of port sweep, spam and html documents infiltrations
Backdoor	1746	583	A procedure where a security technique of a system is passed round furtively to get to a computer or its data
DoS	12,264	4089	A pernicious endeavor making a server or a system asset inaccessible to clients, more often than not by briefly hindering or blocking the host services associated with the Internet
Exploits	33,393	11,132	The invader aware of a vulnerability inside a software piece or an operating system and use that information by abusing the loophole
Fuzzers	18,184	6062	Endeavoring to make a program or network blocked by supplying it the haphazardly created data
Generic	40,000	18,871	A procedure that builds a dictionary of input/output pairs for example from past plaintext/cipher text for a fixed key with given block size. This attack comes in action while an input or output in the dictionary is used again
Reconnaissance	10,491	3496	It involves all the strikes that can reproduce assaults that accumulate information
Shellcode	1133	378	An attack wherein the attacker enters a small bit of code beginning from a shell to govern the compromised machine
Worms	130	44	A malicious code or program which has the ability to copy and spread itself on different computers
Total records	175,341	82,332	

performance or behavior. This assumption does not take into account some attack types such as spy attacks that adjust their style to seem alike normal behaviors.

Duque and Bin Omar suggested an ideal for intrusion detection system with k-means on NSL-KDD dataset [15]. Their work has issues such as false positive rate is high and detection rate is low.

Nawir et al. [16] suggested an ideal for multi-classification on UNSW-NB15 dataset using online averaged one-dependence estimator (AODE) algorithm.

Belouch et al. [17] introduced a classifier with 2 stages dependent on REPTree algorithm and subset of protocols for NIDS. Here, the efficiency of detection is substantially lower on TCP protocol and very high in UDP, i.e., 85.13% and 98.13%, respectively.

4 Methodologies

The methodology that adopted in this paper is shown in Fig. 1.

4.1 Preprocessing and Feature Selection

For preprocessing, the following steps were taken:

Encoding the dataset (making it all numerical except the feature "attack_cat") using the anaconda platform for python 2.7 on windows 8.

Removal of the features named "id". Attribute id is nothing but the number of rows in the dataset as our aim is multi-classification of attacks, so this attribute is unnecessary here.

In order to select the best suitable features, an experiment was performed using both feature ranking as well as feature subset selection methods (individually) on the dataset to select the efficient attribute selection technique. In our research, 2 predicting classifiers, which are REPTree and Naïve Bayes, have been used. The feature ranking techniques which are used along with ranker search method are info-gain, chi-square, gain-ratio, symmetric-uncertainty, significant, OneR attribute evaluator. In total 8 combination of 2 feature subset selection techniques (consistency and CFS subset evaluator) and 4 search methods (best first, greedy step-wise, genetic, particle swarm optimization) are used here. Notably this experiment is limited to filter-based feature selection because wrapper approaches are inefficient in terms of time, particularly in high-dimensional data applications, filter approaches are considered.

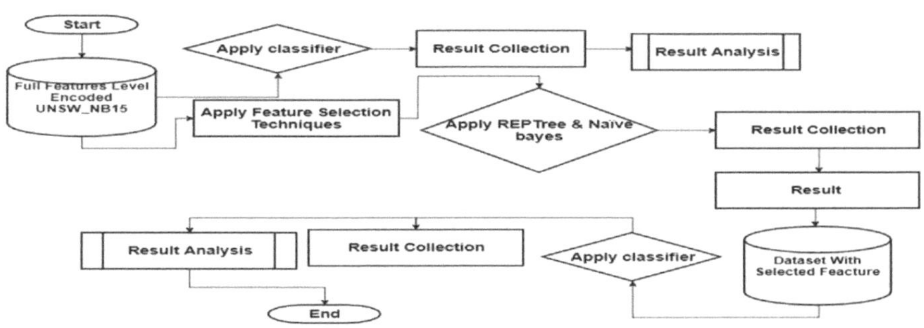

Fig. 1 Methodological workflow adopted in this paper

4.2 *Classifiers*

In this research, various machine learning algorithms have been used with 10-fold cross-validation.

4.2.1 J48 Classifier (Java Version of the C4.5 Algorithm in WEKA Data Mining Tool)

C4.5 depends on the ID3 algorithm created by Quinlan [18], with extra features to address issues that ID3 was unfit to deal. By and by, the J48 is a decision tree classifier algorithm which makes a decision tree dependent on the feature values of the accessible training data.

4.2.2 Naive Bayes Classifier

The Naive Bayes is a classification algorithm, in view of Bayesian theorem. The working of the procedure is simple to catch for input values that are categorical or binary.

Bayes' theorem gives a way that we can compute the likelihood of a hypothesis given our earlier knowledge. Bayes' theorem is expressed in Eq. (1):

$$P(v|a) = \frac{P(a|v) \times P(v)}{P(v)} \tag{1}$$

where $P(v|a)$ is the probability of hypothesis t given the data a.

$P(a|v)$ is the probability of data a given that the hypothesis v was true.

$P(v)$ is the probability of hypothesis v being true (regardless of the data).

$P(a)$ is the probability of the data (regardless of the hypothesis).

In Naive Bayes, it is assumed that attributes are conditionally independent of each other so instead of calculating the values of each attribute $P(a1, a2, a3|v)$, given the target value it is calculated as $P(a1|v) \times P(a2|v)$, etc.

4.2.3 Bagging Classifiers

Bagging [19] is a "bootstrap" [20] ensemble strategy that makes individuals for its ensemble by training every classifier on an arbitrary reallocation of the training set. Bagging is a general strategy that can be utilized to decrease the variance for the algorithms having high variance.

4.2.4 MLP Classifier (Multi-layer Perceptron)

In WEKA, the MLP classifier implements a multi-layer perceptron with one hidden layer utilizing BFGS algorithm to minimize the given loss function plus a quadratic penalty. In this case, all attributes are standardized, including the target. There are a few parameters. The edge parameter is utilized to decide the penalty on the weights' size. Here, what should be the number of hidden units that can be specified. Huge numbers take long time for training. At long last, it is conceivable to utilize conjugate gradient descent as opposed to BFGS updates, which might be quicker for cases with numerous parameters. For improving speed, an approximate version of the logistic function is utilized as the default activation function for the hidden layer, yet other activation functions can be determined. In the output layer, the sigmoid function is utilized for classification. On the off chance that the approximate sigmoid is specified for the hidden layers, it is likewise utilized for the output layer.

4.2.5 Random Forest Classifier

Random forest adds extra haphazardness to the model, while developing the trees. The random forest algorithm changes the methodology of choosing the most significant element for ideal split point with the goal that the learning algorithm is constrained to an irregular example of attributes of which to look. The quantity of features that can be looked at each split point must be indicated as a parameter to the algorithm.

4.2.6 MultiBoostAB Classifier

MultiBoosting is an expansion to the exceptionally popular algorithm called "AdaBoost" for making decision panels [21]. MultiBoosting can be thought of as consolidating AdaBoost with wagging. It can outfit both AdaBoost's high bias and variance decrease with wagging's superior variance decrease. Utilizing C4.5 as the base learning algorithm, MultiBoosting is exhibited to deliver decision panels with smaller error than either wagging or AdaBoost fundamentally more regular than the converse. It has other advantages over AdaBoost of fitting parallel execution.

4.2.7 IBK (Instance-Based Learning with k = No. of Neighbors)

It is a basic instance-based learner that utilizes the class of the closest k training samples for the class of the test samples. It produces classification predictions utilizing just explicit samples. These algorithms do not keep up a lot of abstractions got from explicit instances. This algorithm modified the nearest neighbor classifier, which has huge storage prerequisites [22].

4.2.8 Online AODE (A1DEUpdateable as in WEKA Tool)

Averaged one-dependence estimators (AODE) is a probabilistic classification learning strategy. It was created to address the feature-independence issue of the popular Naive Bayes classifier. It much of the time grows significantly more exact classifiers than Naive Bayes at the expense of a modest increase in the amount of computation. Since online AODE (unlike Naïve Bayes) does not perform model selection and does not utilize tunable parameters, it results in low variance [23].

4.2.9 REPTree Classifier

This classifier is a quick decision tree learning algorithm and depends on the principle of computing the information gain with entropy and minimizing the error arising from variance [24]. REPTree applies regression tree logic and produces numerous trees in altered iterations. Then, it picks best one from all generated trees. This algorithm tree uses variance and information gain to construct the regression/decision tree and reduced error pruning with back fitting method for pruning; for numeric attributes, it sorts the values once at the beginning. For missing values, this algorithm splits the corresponding instances into pieces.

4.2.10 SMO (Sequential Minimal Optimization)

It is a java version of algorithm proposed by John Platt's for training a support vector classifier. It replaces all missing values globally and does nominal to binary attribute transformation. It also normalizes all attributes as a matter of course. Multi-class issues are solved using pairwise classification.

Analysis Metrics:

$$\text{Accuracy by class} = \frac{(\text{TP} + \text{TN})}{N} * 100, \tag{2}$$

$$\text{FPR} = \text{FAR} = \frac{\text{FP}}{\text{FP} + \text{TN}} \tag{3}$$

$$\text{TPR} = \text{Detection Rate} = \frac{\text{TP}}{\text{TP} + \text{FN}} \tag{4}$$

where N = total number of instances. For training set, N = 175,341.

TN = True Positives, TN = True Negatives, FP = False Positives, FN = False Negatives.

Let $A1$, $A2$, up to N times denote the actual target values and $P1$, $P2$, …, up to N times denote the predicted target values.

Mean absolute error is:

$$\text{MAE} = \frac{1}{N}\sum_{i=1}^{N} |Pi - Ai| \tag{5}$$

Root-mean-square error is:

$$\text{RMSE} = \sqrt{\frac{1}{N}\sum_{i=1}^{N} (Pi - Ai)^2} \tag{6}$$

Relative absolute error:

$$\text{RAE} = \frac{\sum_{i=1}^{N} |Pi - Ai|}{\sum_{i=1}^{N} |A - Ai|} \tag{7}$$

where A is a mean value of actual target values.

Root relative squared error:

$$\text{RRSE} = \sqrt{\frac{\sum_{i=1}^{N} (Pi - Ai)^2}{\sum_{i=1}^{N} (A - Ai)^2}} \tag{8}$$

Receiving Operating Characteristic (ROC) Area:

The ROC curve area is the region under the curve plotted with TPR against the FPR where TPR is on y-axis and FPR is on the x-axis.

Kappa Statistic:

It is a metric that compares an observed accuracy with an expected accuracy (random chance). The kappa statistic evaluates single classifier as well as classifiers among themselves. Furthermore, it takes into account random classifier that is why it is less misleading than simply using accuracy as a metric. The value of the kappa statistic lies between 0 and 1 ranges. 0 means totally disagree, and 1 means full agreement.

5 Experiment Setup

Whole work run by simulation on the computer was carried out using windows operating system, an Intel (R) Core (TM), i5-4200U CPU @ 1.60–2.30 GHz, 6 GB RAM. WEKA tool (WEKA 3.8) [25] is used to employ ML algorithms toward UNSW-NB15 dataset.

Table 3 UNSW_NB 15 dataset with all features (excluding id)

Classifier	No. of features	TPR	FPR	Accuracy (%)	Training time	Kappa statistic	MAE	RMSE
Naïve Bayes	44	0.723	0.023	72.30	0.22	0.6624	0.0567	0.2231
REPTree	44	0.874	0.026	87.36	18.04	0.8377	0.031	0.1282

6 Results and Discussion

In Table 3, an overview of the performance of predictor classifiers has shown. Tables 4 and 5 indicate the performance comparison of the several distinct attribute selection and attribute subset selection algorithms on 2 different predictor classifiers. Comparison shows that only the info-gain attribute evaluator with top 20 attributes could enhance the overall performance of REPTree in lesser time, whereas Chi-squared feature selection (with top 11 attributes) gives the best result for Naive Bayes classifier. Now, another comparison is needed to select one of the feature selection algorithms.

So a graph is shown in Fig. 2a which is for comparing the results of chi-square and info-gain to the results with full feature set with respect to both the predictor classifiers. This graph shows among the two only Info-gain classifier can improve performance for both the classifiers.

A quick overview of the performance of several well-known classifiers on the UNSW-NB15 dataset before feature selection is presented in Table 6. For the algorithms from Bayesian family (Naïve Bayes and online AODE), the dataset has been discretized before applying classifier.

Worthwhile to mention that IBK ($k = 5$), Bagging (on REPTree) and MultiBoost (Decision Stump as the weak learner) have been used here.

Then, the performance comparison of the same algorithms has been shown in Table 7 as well as in Fig. 2b.

The comparison shows that IBK has less training time but takes more time in predicting. Here, the other incremental classifier online AODE is the best according to time for multi-classification yet REPTree gives better detection accuracy and comparable result based on processing time. In Tables 8 and 9, detailed performance of REPTree is shown.

7 Conclusion and Future Work

Machine learning methodologies are by and large progressively embraced for digital security; subsequently, it is essential to assess when and which kind of algorithms can achieve adequate outcomes. Our analyzation of the research consists of 2

Table 4 Comparison of feature selection approaches with Naïve Bayes

FS algorithm	Search method	No. of features	TPR	FPR	Accuracy (%)	Training time	Kappa statistic	MAE	RMSE
Chi-square	Ranker	26	0.757	0.023	75.69	0.13	0.701	0.050	0.205
		24	0.978	0.004	76.10	0.05	0.706	0.049	0.2034
		20	0.778	0.022	77.75	0.05	0.725	0.046	0.191
		16	0.815	0.018	81.48	0.05	0.7695	0.040	0.1753
		11	0.841	0.017	84.07	0.03	0.8001	0.034	0.1605
		06	0.760	0.026	75.96	0.02	0.704	0.0503	0.1947
Info-gain	Ranker	26	0.756	0.023	75.58	0.06	0.7004	0.0504	0.2062
		24	0.761	0.023	76.10	0.06	0.7065	0.0494	0.2034
		20	0.784	0.021	78.38	0.06	0.7332	0.0447	0.189
		16	0.816	0.019	81.58	0.05	0.7705	0.0399	0.1745
		11	0.828	0.017	82.84	0.05	0.7853	0.0371	0.168
		06	0.831	0.018	83.10	0.03	0.7878	0.0368	0.1566
Gain-ratio	Ranker	26	0.733	0.024	73.26	0.14	0.673	0.0547	0.2143
		24	0.746	0.024	74.59	0.06	0.6883	0.0521	0.2057
		20	0.786	0.020	78.56	0.05	0.734	0.0471	0.1863
		16	0.807	0.020	80.67	0.05	0.758	0.043	0.1762
		11	0.778	0.024	77.82	0.03	0.7214	0.0528	0.1869
		06	0.743	0.036	74.26	0.03	0.6747	0.0595	0.193
OneR	Ranker	26	0.757	0.023	75.69	0.06	0.7017	0.050	0.2056
		24	0.764	0.023	76.37	0.05	0.7097	0.048	0.2014
		20	0.775	0.022	77.45	0.05	0.7223	0.0466	0.1952

(continued)

Table 4 (continued)

FS algorithm	Search method	No. of features	TPR	FPR	Accuracy (%)	Training time	Kappa statistic	MAE	RMSE
		16	0.779	0.021	77.89	0.05	0.7273	0.0457	0.1894
		11	0.769	0.025	76.93	0.03	0.715	0.048	0.1992
		06	0.760	0.026	75.96	0.03	0.704	0.0503	0.1947
Symmetrical uncertainty-based evaluation	Ranker	26	0.751	0.024	75.05	0.06	0.694	0.051	0.2083
		24	0.748	0.024	74.75	0.06	0.690	0.051	0.2094
		20	0.756	0.024	75.59	0.05	0.7003	0.0504	0.2045
		16	0.797	0.020	79.67	0.05	0.747	0.0420	0.1833
		11	0.829	0.018	82.89	0.03	0.785	0.0383	0.1681
		06	0.818	0.022	81.80	0.03	0.770	0.0410	0.1558
Significance-based evaluation	Ranker	26	0.758	0.022	75.78	0.06	0.7032	0.0497	0.2045
		24	0.769	0.021	76.93	0.06	0.7166	0.0474	0.198
		20	0.804	0.018	80.42	0.06	0.7571	0.0413	0.179
		16	0.821	0.016	82.07	0.06	0.7769	0.0384	0.1718
		11	0.812	0.016	81.24	0.03	0.7667	0.0402	0.1752
		06	0.831	0.018	83.10	0.03	0.7878	0.0368	0.1566
Consistency subset evaluator	Greedy step wise	28	0.754	0.024	75.37	0.08	0.6977	0.0507	0.2078
	Best first	27	0.756	0.024	75.61	0.07	0.700	0.050	0.2061
	Genetic	30	0.743	0.023	74.34	0.06	0.6859	0.052	0.2114
	PSO	35	0.738	0.023	73.80	0.23	0.6790	0.053	0.215
CFS subset evaluator	Greedy step wise	4	0.823	0.028	82.32	0.03	0.7753	0.442	0.1511
	Best first	4	0.823	0.028	82.32	0.03	0.775	0.442	0.1511
	Genetic	12	0.829	0.018	82.88	0.08	0.7857	0.039	0.1649

(continued)

Table 4 (continued)

FS algorithm	Search method	No. of features	TPR	FPR	Accuracy (%)	Training time	Kappa statistic	MAE	RMSE
	PSO	9	0.840	0.019	84.02	0.09	0.7987	0.035	0.1580

Table 5 Comparison of feature selection approaches with REPTree

FS algorithm	Search method	No. of features	TPR	FPR	Accuracy (%)	Training time	Kappa statistic	MAE	RMSE
Chi-square	Ranker	26	0.873	0.026	87.34	12.05	0.8376	0.0318	0.1283
		24	0.873	0.026	87.31	9.77	0.8371	0.0319	0.1285
		20	0.873	0.027	87.32	6.92	0.8372	0.0320	0.1286
		16	0.872	0.027	87.31	4.39	0.8351	0.0324	0.1295
		11	0.870	0.028	87.16	3.17	0.8325	0.0329	0.1305
		06	0.812	0.044	81.17	1.69	0.7582	0.0470	0.1569
Info-gain	Ranker	26	0.873	0.026	87.32	8.05	0.8373	0.0318	0.1283
		24	0.873	0.026	87.31	7.33	0.8371	0.0319	0.1284
		20	0.874	0.026	87.37	6.61	0.8379	0.0318	0.1282
		16	0.872	0.027	87.16	4.47	0.8351	0.0324	0.1295
		11	0.868	0.028	86.81	2.75	0.8306	0.0334	0.1314
		06	0.864	0.028	86.40	1.36	0.8252	0.0346	0.134
Gain-ratio	Ranker	26	0.870	0.027	87.02	8.78	0.8332	0.0325	0.1298
		24	0.870	0.027	86.96	6.96	0.8325	0.0326	0.1301
		20	0.867	0.028	86.72	5.46	0.8293	0.0338	0.1321
		16	0.865	0.028	86.54	4.16	0.8270	0.0345	0.1332
		11	0.806	0.037	80.61	3.02	0.7499	0.0501	0.1594
		06	0.763	0.049	76.33	1.06	0.6930	0.0655	0.181
OneR	Ranker	26	0.873	0.026	87.34	8.68	0.8376	0.0318	0.1283
		24	0.874	0.026	87.36	7.27	0.8378	0.0317	0.1282
		20	0.873	0.027	87.32	5.91	0.8371	0.0321	0.1286

(continued)

Table 5 (continued)

FS algorithm	Search method	No. of features	TPR	FPR	Accuracy (%)	Training time	Kappa statistic	MAE	RMSE
		16	0.871	0.027	87.09	4.56	0.8341	0.0327	0.1300
		11	0.823	0.041	82.25	3.48	0.7720	0.0445	0.1533
		06	0.812	0.044	81.17	1.73	0.7582	0.047	0.1569
Symmetrical uncertainty-based evaluation	Ranker	26	0.873	0.026	87.29	9.27	0.8368	0.0319	0.1286
		24	0.872	0.027	87.24	8.34	0.8362	0.0322	0.129
		20	0.870	0.027	87.01	6.36	0.8332	0.0325	0.1299
		16	0.872	0.027	87.16	5.03	0.8351	0.0322	0.1292
		11	0.867	0.028	86.73	3.03	0.8294	0.0338	0.132
		06	0.852	0.031	85.17	1.63	0.8092	0.0386	0.1402
Significance-based evaluation	Ranker	26	0.874	0.026	87.35	10.83	0.8377	0.0317	0.1282
		24	0.873	0.026	87.34	8.88	0.8375	0.0317	0.1282
		20	0.874	0.026	87.36	7.00	0.8378	0.0317	0.1281
		16	0.874	0.026	87.42	6.08	0.8386	0.0316	0.1279
		11	0.871	0.027	87.07	2.97	0.8338	0.0328	0.1301
		06	0.864	0.028	86.40	1.47	0.8252	0.0346	0.1340
Consistency subset evaluator	Greedy step wise	28	0.873	0.026	87.34	9.13	0.8375	0.0318	0.1284
	Best first	27	0.873	0.027	87.25	8.31	0.8363	0.0322	0.1291
	Genetic	30	0.873	0.026	87.33	10.63	0.8374	0.0318	0.1283
	PSO	35	0.874	0.026	87.37	13.47	0.8379	0.0317	0.1281
CFS subset evaluator	Greedy step wise	4	0.840	0.034	83.99	1.48	0.7935	0.0425	0.1469
	Best first	4	0.840	0.034	83.99	0.97	0.7935	0.0425	0.1469
	Genetic	30	0.863	0.028	86.306	4.31	0.8241	0.0346	0.1339

(continued)

Table 5 (continued)

FS algorithm	Search method	No. of features	TPR	FPR	Accuracy (%)	Training time	Kappa statistic	MAE	RMSE
	PSO	35	0.868	0.028	86.784	3.44	0.8302	0.0338	0.1318

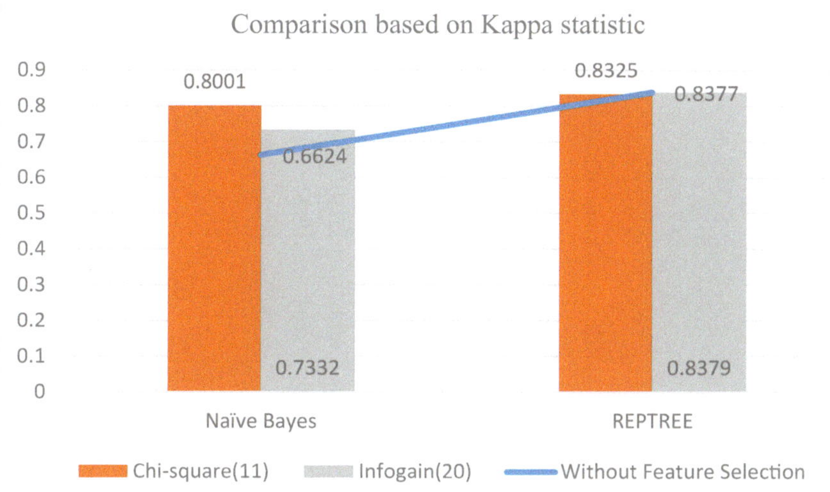

(a)

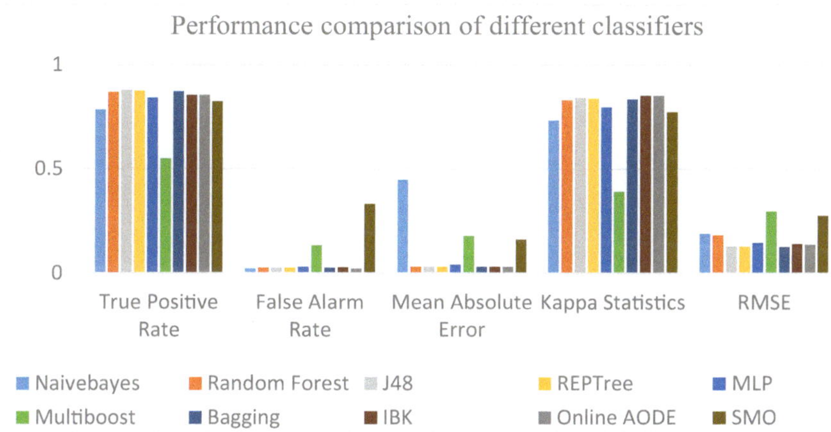

(b)

Fig. 2 **a** Comparison of 2 best feature selection classifiers based on Kappa statistic; **b** performance comparison of different classifiers

phases: before feature selection and after feature selection with the mainstream ML algorithms for intrusion detection issues, especially network intrusion detection.

At the end, it is concluded that with respect to the given performance metrics overall performance of REPTree is quite adaptable than others. The advantage of using REPTree is compared to other classifiers it gives low false alarm rates for

Table 6 Comparison of different classifiers before applying feature selection

Classifier	Naive Bayes	Random Forest	J48	REPtree	MLP	MultiBoost	Bagging	IBK	Online AODE	SMO
Acc. (%)	72.30	86.53	87.44	87.36	83.7	54.75	87.01	82.8	84.12	82.4616
Time (s)	0.22	277.88	45.46	18.04	167.4	14.94	173.75	0.06	2.19	11,700.06
MAE	0.0567	0.0333	0.030	0.031	0.041	0.178	0.03	0.04	0.0339	0.1616
RMSE	0.2231	0.1293	0.130	0.1282	0.1469	0.2969	0.1271	0.1499	0.1528	0.2747
RAE (%)	35.854	21.0836	19.47	20.1105	26.1355	112.6832	20.1797	25.6472	21.4759	102.2217
RRSE (%)	79.383	45.9997	46.41	45.6062	52.2646	105.6034	45.2169	53.3185	54.3629	97.7069
Kappa statistic	0.6624	0.8278	0.8387	0.837	0.79	0.390	0.83	0.78	0.799	0.7745
TPR	0.723	0.865	0.874	0.874	0.838	0.548	0.870	0.828	0.841	0.825
FPR	0.023	0.025	0.026	0.026	0.030	0.134	0.025	0.030	0.022	0.033

Table 7 Comparison of different classifiers after applying feature selection

Performance metrics	Naive Bayes	Random Forest	J48	REPTree	MLP	MultiBoost	Bagging	Online AODE	IBK	SMO
Acc. (%)	78.38	86.7	87.65	87.37	84.1	54.75	87.13	85.442	85.405	82.44
Time (s)	0.09	157.39	25.17	6.61	55.4	5.93	66.36	0.13	0.560	2729.73
TPR	0.784	0.868	0.877	0.874	0.841	0.548	0.871	0.854	0.854	0.824
FPR	0.021	0.025	0.026	0.026	0.030	0.134	0.025	0.027	0.021	0.033
MAE	0.0447	0.0317	0.030	0.0318	0.0414	0.1781	0.0317	0.0317	0.033	0.1616
RMSE	0.189	0.1281	0.128	0.1282	0.1461	0.2969	0.1267	0.1401	0.136	0.2748
RAE (%)	28.257	20.030	19.54	20.113	26.166	112.683	20.075	20.062	21.31	102.23
RRSE (%)	67.244	45.5787	45.54	45.6092	51.962	105.603	45.068	49.84	48.39	97.761
Kappa statistics	0.733	0.8306	0.841	0.8379	0.796	0.39	0.835	0.813	0.815	0.774

Table 8 Confusion matrix of REPTree

Classes	Normal	Backdoor	Analysis	Fuzzers	Shellcode	Reconnaissance	Exploits	DoS	Worms	Generic
Normal	56,000	0	0	0	0	0	0	0	0	0
Backdoor	0	185	15	13	18	11	1413	88	2	1
Analysis	0	17	462	3	0	0	1438	78	0	2
Fuzzers	0	5	27	16,167	113	19	1724	113	1	15
Shellcode	0	0	0	120	853	17	113	27	0	3
Reconnaissance	0	5	8	19	9	7918	2381	149	0	2
Exploits	0	20	103	443	157	548	31,001	1012	35	74
DoS	0	15	18	121	96	75	10,653	1255	1	30
Worms	0	0	0	4	1	1	36	2	80	6
Generic	0	8	5	42	13	6	561	79	2	39,284

Table 9 Accuracy by class with false alarm rate

Class	Normal	Backdoor	Analysis	Fuzzers	Shellcode	Reconnaissance	Exploits	DoS	Worms	Generic
Acc.	100	99.069	99.02	98.4	99.6	98.14	88.18	92.12	99.94	99.51
FAR	0.000	0.000	0.001	0.005	0.002	0.004	0.129	0.009	0.000	0.001

each class (except for Exploits) and it results in a progressively precise and simple classification tree, although it can have large amount of data.

This project till now was carried out to classify attacks into multiple classes for network anomaly detection framework utilizing machine learning and assess the performance of machine learning algorithms. This was done to analyze the behavior of the dataset to these classifiers.

Future research will be improvement of REPTree algorithm in lesser time complexity and higher accuracy.

References

1. Aydın M, Ali M, Zaim AH, Ceylan KG (2009) A hybrid intrusion detection system design for computer network security. Comput Electr Eng 517–526
2. McHugh J (2000) Testing intrusion detection systems: a critique of the 1998 and 1999 DARPA intrusion detection system evaluations as performed by Lincoln Laboratory. ACM Trans Inf Syst Secur 3(4):262–294. https://doi.org/10.1145/382912.382923
3. Stolfo SJ, Fan W, Lee W, Prodromidis A, Chan PK (2000) Cost-based modeling for fraud and intrusion detection: results from the JAM project. In: Proceedings of the IEEE DARPA information survivability conference and exposition, vol 2. USA, pp 130–144
4. Tavallaee M, Bagheri E, Lu W, Ghorbani AA (2009) A detailed analysis of the KDD CUP 99 data set. In: Proceedings of the 2009 IEEE symposium on computational intelligence in security and defense applications (CISDA 2009)
5. UNSW-NB15. Available: https://www.unsw.adfa.edu.au/unsw-canberracyber/cybersecurity/ADFA-NB15-Datasets/
6. Janarthanan T, Zargari S (2017) Feature selection in UNSW-NB15 and KDDCUP'99 datasets. In: IEEE 26th international symposium on industrial electronics (ISIE). IEEE
7. Kumar K, Kumar G, Kumar Y (2013) Feature selection approach for intrusion detection system. Int J Adv Trends Comput Sci Eng (IJATCSE) 2(5):47–53
8. Mogal DG, Ghungrad SR, Bhusare BB (2017) NIDS using machine learning classifiers on UNSW-NB15 and KDDCUP99 datasets. Int J Adv Res Comput Commun Eng 6(4)
9. Ben Amor N, Benferhat S, Elouedi Z (2004) Naive Bayes vs. decision trees in intrusion detection systems. In: Proceedings of the 2004 ACM symposium on applied computing, pp 420–424
10. Breiman L (2001) Random forests. Statistics Department, University of California, Berkeley
11. Zhang J, Zulkernine M (2006) A hybrid network intrusion detection technique using random forests. In: Proceedings of the first international conference on availability, reliability and security, pp 262–269
12. Hu W, Hu W, Maybank S (2008) AdaBoost-based algorithm for network intrusion detection. IEEE Trans Syst Man Cybern Part B Cybern 38(2):577–583
13. Mukkamala S, Janoski G, Sung A (2002) Intrusion detection: support vector machines and neural networks. In: Proceedings of the IEEE international joint conference on neural networks, pp 1702–1707
14. Sammany M, Sharawi M, El-Beltagy M, Saroit I (2007) Artificial neural networks architecture for intrusion detection systems and classification of attacks
15. Duque S, Bin Omar MN (2015) Using data mining algorithms for developing a model for intrusion detection system (IDS). Procedia Comput Sci 61:46–51
16. Nawir M, Amir A, Yaakob N, Lynn OB (2018) Multi-classification of UNSW-NB15 dataset for network anomaly detection system. J Theoret Appl Inf Technol 96(15)
17. Belouch M et al (2017) A two-stage classifier approach using RepTree algorithm for network intrusion detection. Int J Adv Comput Sci Appl 8(6):389–394

18. Quinlan JR (1993) C4.5: programs for machine learning, vol 1. Morgan Kaufmann
19. Breiman L (1996) Bagging predictors. Mach Learn 24(2):123–140
20. Efron B, Tibshirani RJ (1993) An introduction to the bootstrap. J Am Stat Assoc 89(428):436. https://doi.org/10.2307/2291020
21. Webb GI (2000) MultiBoosting: a technique for combining boosting and wagging. Mach Learn 40(2)
22. Aha DW, Kibler D (1991) Instance-based learning algorithms. Mach Learn 6:37–66
23. Webb G, Boughton J, Wang Z (2005) Not so Naive Bayes: aggregating one-dependence estimators. Mach Learn 58(1):5–24
24. Elomaa T, Kääriäinen M (2001) An analysis of reduced error pruning. J Artif Intell Res 15:163–187
25. Weka 3 homepage. https://www.cs.waikato.ac.nz/ml/weka/downloading.html. Accessed 2019/1/2

Chapter 41
Heterogeneous Resource Provisioning for Workflow-Based Applications Using AHP in Cloud Computing

S. M. Jaybhaye and V. Z. Attar

1 Introduction

Analytic hierarchy process (AHP) is an approach for multi-objective decision analysis that makes use of relative things. The strength of this method lies in utilizing insight-based soft information from the user in the form of relative values. As a result of its potential and capacity to support in decision making, it has pulled the attention of researchers from different fields. It is applied to many multifaceted problems with different criteria and domains. This approach is used to give best possible judgment and priorities to the decision factors based on user's experience.

A hierarchy is developed in which highest priority is given to the main objective. The next lower priority is given to the next important criteria and so on. The bottom of hierarchy is occupied by the available alternatives. Once the hierarchy is established, comparison matrix is formulated and comparisons of lower level criteria with respect to the property of upper levels are made. Different importance criteria are used for resource allocation to the tasks. Pairwise comparison is done between the criteria, and tasks are mapped to the heterogeneous virtual machine instances (VMIs) which are available in the cloud. Here, different VMIs are considered which are taken from the AWS cloud. If user's higher priority is time, then according to decision-making system, all resources should be selected.

S. M. Jaybhaye (✉) · V. Z. Attar
Department of Computer and Information Technology, College of Engineering Pune, Pune 411005, India
e-mail: smjaybhaye@gmail.com

V. Z. Attar
e-mail: vahida.comp@coep.ac.in

© Springer Nature Singapore Pte Ltd. 2021
S. D. Purohit et al. (eds.), *Proceedings of International Conference on Communication and Computational Technologies*, Algorithms for Intelligent Systems,
https://doi.org/10.1007/978-981-15-5077-5_41

1.1 Cloud Computing and Its Services

Cloud computing provides wide range of services varying from Software as Service, Platform as a Service, Infrastructure as a Service, Storage as a Service, and many more. Due to the characteristics of cloud computing environment, user can access services flexibly and efficiently. Resource mapping is one of the major and important issues; as if it is not done properly, it results into more execution time and more cost. Despite the seeming advances made in cloud computing provisioning, many major obstacles remain as yet basically unsolved. Some of the issues are

1. Scalability and heterogeneity of modern IT environments
2. Resource management and user behavior
3. Dynamic environment of cloud.

According to the applications requirement and user defined constraints, tasks should be mapped and scheduled on the suitable resource. Static scheduling approaches are not sufficient so dynamic scheduling, heuristic searches, evolutionary approaches are used to handle mapping and scheduling.

1.2 Heterogeneous Resources in Cloud

Cloud computing is remotely accessible service which provides cloud-based resources to its consumer. Cloud-based resources can be physical or virtual (software based). Physical server, storage device, and network device are physical resources, and virtual server, virtual storage, and virtual networks are software-based virtual resources. Infrastructure as a Services (IaaS) uses virtualization technology for resource provisioning.

Virtual machine instances are the images of physical machines which are described with its specification and configurations such as number of microprocessor (CPU) cycles, memory access, and network bandwidth. VMIs are the containers of assigned resources.

Heterogeneous VMIs allow user to receive different quality of service. The role of quality and service performance is crucial aspects in application execution. Cloud provider must contribute special interest while designing and implementing the IaaS.

1.3 Resource Allocation

Cloud resources are categorized into software resources and hardware resources. Examples for software resources are services and platform. Examples of hardware resources are processor, network, and storage (RAM and database). Infrastructure as a Service (IaaS) provides a logical resources from the available resource pool.

Table 1 Amazon EC2 instances

Instance name	Memory (GiB)	Clock speed (GHz)	Compute units (ECU)	vCPUs	Cost ($/h)
t3.small	2.0	2.3	0	1	0.020800
a1.medium	2.0	2.3	0	1	0.025500
r4.16xlarge	488.0	2.3	195	164	4.256000

It also provides a virtual machine instance, virtual network, virtual IP address, and virtual memory. VMIs are available in different sizes and specifications, and so its pricing is also changing. When user submits large application to the cloud, it must be assigned to the suitable VMI based on user's criteria. To meet customer demands and resource allocation is one of the challenging tasks in cloud computing. Dynamic nature of user's behavior and changes in application requirement makes this problem of allocation more complex. Limited resource pool, heterogeneity, and environmental problems are some of the issues that need to be considered while designing a new model which will improve utilization and will meet users' requirements. Table 1 shows some AWS EC2 instances.

1.4 Scientific Workflow Applications

A workflow is represented in Fig. 1 using directed acyclic graph. Figure 1 shows four layers, and task is represented as a node. Task number and its size are represented in circle. Tasks are dependent on each other, and they are shown using directed arrows. This type of inputs is given to schedule task using heterogeneous virtual machine instances (VMI) in the cloud. Here, the assumed environment for cloud is Amazon Web services. It provides many different types of VMIs. As per the applications' requirement, resource selection must be done. After workflow execution the output obtained as task execution time and cost for VMI is given in Table 2.

2 Literature Survey

A lot of literature is available on AHP that deals with the mathematics of the method. Gawali and Shinde [1] developed ranking of task by the modified analytic hierarchy process (MAHP). Other scheduling techniques and methods are used in combination of MAHP for resource allocation. Data size and time required to execute for Cyber-shake are more, and it is used to analyze results. Priorities are assigned on length and execution time of incoming tasks. After checking the availability of Virtual Machine's task is allocated. If that machine becomes overloaded and the other machine is free then the allocated task can be divided and executed on other machines. MAHP checks

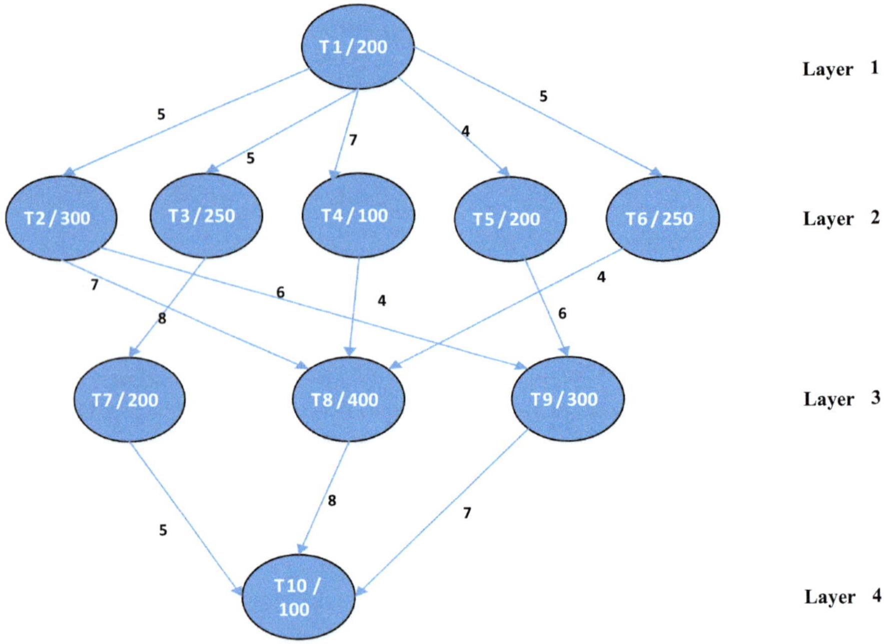

Fig. 1 Basic structure of workflow application using DAG

Table 2 Task execution time and cost required for VMI

Task ID	Execution time			Cost		
	VMI-1 Small	VMI-2 Medium	VMI-3 High	VMI-1 Small	VMI-2 Medium	VMI-3 High
T1	2.00	1.50	1.00	0.0416	0.03825	4.256
T2	3.00	2.25	1.50	0.0624	0.057375	6.384
T3	2.50	1.88	1.25	0.052	0.047813	5.32
T4	1.00	0.75	0.50	0.0208	0.019125	2.128
T5	2.00	1.50	1.00	0.0416	0.03825	4.256
T6	2.50	1.88	1.25	0.052	0.047813	5.32
T7	2.00	1.50	1.00	0.0416	0.03825	4.256
T8	4.00	3.00	2.00	0.0832	0.0765	8.512
T9	3.00	2.25	1.50	0.0624	0.057375	6.384
T10	1.00	0.75	0.50	0.0208	0.019125	2.128

availability of virtual machine and overloading and performs divide and conquer of task and then aggregation of results. It helps to allocate resources efficiently and to improve utilization of resources. Whaiduzzaman et al. [2] used attributes and aspects of different multi-criteria decision analysis for service selection. Nayak and Tripathy [3] used VIKOR method to optimize and ranking, based on deadline and duration of execution time. This method finds the idle task which can be backfilled. Zeng [4] considered rewards approach for fast and effective scheduling. Deadline, decay, penalty, time, and bottom line are some of the parameters used by researchers. Paired comparison matrix is formed for deadline, reparation duty, and profit objectives.

AHP Scheduler suggested by Revathy and Sekar [5] solves the important issue of resource optimization by reduction in time and effective allocation. Cloud Simulator tool simulates and executes task for different number of cloudlets. Priority values are assigned to the task based on ranking of users, computation time, storage, and number of cores which are the different criteria considered based on user's requirements of quality of service. Modified AHP is compared with FCFS and Round Robin. Scheduling algorithms can have its own requirement; Nayak and Tripathy [6] designed scheduler to take best possible lease. Given approach by the authors also minimizes lease rejection. Deadline, duration, and start time are the different criteria of lease which are considered for its selection. Ergu et al. [7] proposed reliability, cost, time, and network bandwidth criteria for resource allocation. A pairwise comparison of AHP shows improved consistency ratio and good results for dynamic resource allocation. Ebadifard and Babamir [8] used black hole algorithm for optimum workflow scheduling. Single-objective black hole algorithm is converted into multi-objective by using the AHP relationship for scheduling. It uses Pareto domination relationship for user references. It gives optimal solution by increasing the search space of the problem.

Saaty [9] discussed significance of eigenvector for multi-objective decision approach. A review of such developments is taken by Emrouznejad and Marra [10] for papers of past 38 years, i.e., (1979–2017) duration. Researchers [10] had identified combined approach with AHP and TOPSIS, Data Envelopment-AHP (DEAHP), SWOT analysis, Quality Function Deployment (QFD) which improves performance of decision making.

Furthermore, Annette et al. [11] had shown that AHP gives good solutions for many levels of hierarchy with subattributes for animation. This application requires high computing resources for rendering which can be easily available from the cloud services. Simple additive weighting (SAW) and AHP are applied and compared for rendering process. AHP approach is used for selection of data platform by Lnenicka [12]. Many big data platforms are available in the market such as Apache Hadoop, Storm and Spark, HPCC, Cloudera, Amazon Kinesis. Most suitable platform for the required application is selected using AHP. Technical, social, and cost policies are three categories for criteria selection. Singh et al. [13] suggested multi-criteria preference synthesis (MPS) method in combination of AHP for resource allocation which gives priority to the tasks. Java-based AHP tool is implemented by using JTree class [14] for alternative ranking. Liu and Zhang [15] used combination of AHP and evolutionary approaches. Kumar and Kaur [16] used this approach which meets the user's requirements while giving the ranking because it uses the crowing distance (CD) factor. It is calculated for every criterion and applied for ranking.

AHP method application summary for best selection of services, resources, or methods in different environment is given in Table 3. Overall analytical hierarchy process used for task mapping based on the user QoS requirements is shown in Fig. 2.

3 Quality of Service Function

End user of cloud computing services requires cost, performance time, reliability, scalability, resource utilization, availability, agility, and security.

These QoS attributes are required to the customer for selecting a service provider and estimation of budget [17]. Response time and performance of any application on cloud are known for its variability because it escalates and reduces every time due to dynamic environment of cloud.

1. Cost: Many services are available in cloud computing. Service provider and users both are concerned in the cost which varies as per to service charges. They cost differently depending on the usage. Pricing model also majorly contributes for cost calculation. On demand, reserved instances, and spot instances are some of the examples o pricing models.

$$\text{Total Cost} = \Sigma(\text{Execution Time} \times \text{Price}) + \text{Transfer Cost}$$

2. Time: Communication time, execution time, throughput, latency, and response time are some of the parameters which affect the calculation of total execution time. Total execution time required to execute the workflow is also called as makespan.

$$\text{Total Execution Time} = \Sigma\text{Execution Time} + \text{Transfer Time}$$

3. Reliability: The service should be provided as per the requirement of users. Fault management, fault tolerance, efficiency, user experience levels, and safety are some of the parameters which contribute to reliability.

$$\begin{aligned}&\text{Mean Time between Failure (MTBF)}\\&\quad = \text{Total working time of VMI/Number of failures within given time}\end{aligned}$$

4. Resource Utilization: Resource consumption and techniques like monitoring utilization are used to identify over-provisioning or underperformance. It is a ratio of actual time required for the application to execute it on the resource to the total uptime of the resource.
5. Availability: It is the time extent to till that service should be provided. Downtime of resources or service is important factor which affects the service. If services are

Table 3 Summary of the AHP methods used for best selection

Method name	Contribution	Application
Modified heuristic algorithm using AHP and BATS bandwidth aware divisible scheduling + BAR	Efficient resource allocation, maximum utilization of resource	Scientific workflow scheduling
Review of multi-criteria decision analysis (MCDA) multi-criteria methods	Summary of MCDA techniques with their unique features and comparison, review of and logical grouping of method	Service selection for cloud
Multi-criteria decision-making (MCDM) method	Identification and ranking of best ideal task Improved utilization and reduces task rejection	Task scheduling
Scheduling based on AHP	Dynamic task scheduling, increased service providers' revenue	Task scheduling
Resource allocation in cloud	Effective resource allocation based on users criteria	Task scheduling
Backfilling algorithm using AHP	AHP is used to select best possible lease for scheduling Performance of back filling algorithm is improved and rejection of lease is reduced	Deadline sensitive leases scheduling
Task-oriented resource allocation in cloud computing model name	Identification of the inconsistent elements, improve the consistency ratio	Task scheduling and resource allocation in cloud computing
A multi-objective workflow scheduling method by using the black hole algorithm	Reduced makespan, improved resource utilization as well as improved benefits for service provider	Workflow scheduling
AHP model for APS	Big data analytics platform selection	Businesses and public sector
Multi-criteria preference synthesis (MPS) method	Consistent priority providing to the users' tasks for violations	Resource allocation in cloud
jAHP (Java based application)	Multiple criteria decision making	Problem solving distributed places
AHP decision-making for cloud service composition	Combined approach for multi-objective evolutionary approaches	Best service selection

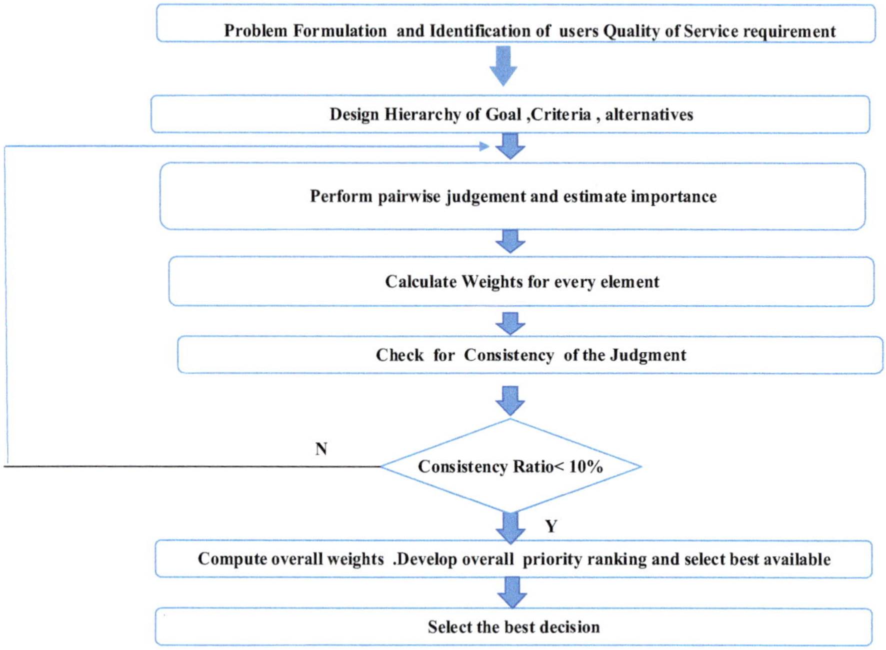

Fig. 2 Flowchart for analytical hierarchy process

provided twenty-four by seven, then those services are called as highly available (HA) services.

Following are the advantages of AHP application for resource allocation based on the different preferences.

1. Improved judgment and understanding through repetition
2. Easy, flexible model for resource selection
3. Solves complex decision-making problems
4. Deals with interdependent criteria of quality of service
5. Determines priorities
6. Overall estimation of the requirement for available alternatives.

4 Methodology

In the present work, AHP method is used for obtaining the weights for the attributes and ranking the alternatives for the execution of the task. The task is analyzed by constructing hierarchy of attributes which consist of time (Ti), cost (C), utilization (U), reliability (R), and availability (A) as main attributes for execution of task as shown in Fig. 3. The alternative choices are on the basis of different configurations of virtual machine instances as multi-criteria for decision.

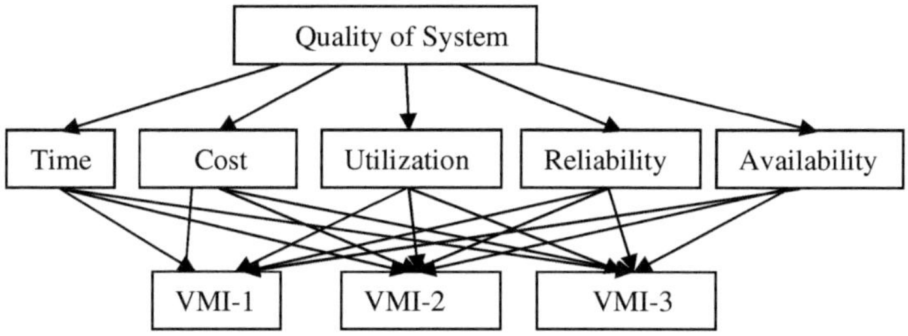

Fig. 3 Hierarchy for decision making through AHP

Table 4 Pairwise comparison of the quality of service for context 1

	Time	Cost	Utilization	Reliability	Availability
Time (T)	1	2	4	7	9
Cost (C)	0.5	1	3	4	6
Utilization (U)	0.25	0.33	1	2	3
Reliability (R)	0.14	0.25	0.5	1	2
Availability (A)	0.11	0.16	0.33	0.5	1

Two different contexts are used to allocate resources in cloud. In context 1, time (T), cost (C), utilization (U), reliability (R), and availability (A) are the criteria used to give the preferences. Most important parameters are arranged according to their percentage contribution in decision making. Various criteria such as memory are the user's requirement. Intermediate values of Satty's scale from 1 to 9 are used for decision making. Results must be based on criteria given by the users. Ranking is given to the criteria based on judgment of decision maker.

Table 4 shows pairwise comparison of the quality of services done and weightages assigned. When criteria are compared with itself, then weightage is 1 so all diagonal values are 1. It gives choices of the criteria in row to criteria of the column. If choice is twice, it indicates that column be 0.5 preferable for option row. Consistency ratio is degree of coordination between the criteria. CR < 10% must be satisfied; if not, the weight matrix must be modified till it satisfies.

The following steps are used to obtain consistency ratio

Step 1: Matrix A1 = Initial Comparison Matrix

	T	C	U	R	A	GM = (Pai)^0.2	Wt.
T	1	2	4	7	9	3.471252	0.480936
C	0.5	1	3	4	6	2.047673	0.283701
U	0.25	0.333333	1	2	3	0.870551	0.120613
R	0.142857	0.25	0.5	1	2	0.513533	0.071149
A	0.111111	0.166667	0.333333	0.5	1	0.314696	0.043601
						$\sum$GM = 7.217704	

Step 2: Matrix A2 = Wt. Matrix
Step 3: Matrix A3 = [A1] × [A2]

2.42124
1.432209
0.608514
0.358287
0.2201

Step 4: Matrix A4 = [A3]/[A2]

5.034435
5.048296
5.045169
5.035725
5.04811
λAvg = 5.042347

Step 5: CI = [λAvg − *N*]/*N* − 1 = 0.010587
Step 6: CR = CI/RI = 0.009.
Hence for context 1, consistency ratio is 0.0095 which is acceptable.

Table 5 shows multi-criteria decision making (MCDM) criteria and alternatives which indicate that how well each attribute serves with VMIs alternatives. The judgments are entered using the fundamental scale of the AHP method suggested by Saaty [18].

Table 6 gives ranking of alternatives. It is calculated by multiplying relative normalized weight (Wt) of each attribute with its corresponding normalized weight for each alternative and summing over the attributes for each alternative.

Table 5 Multi-criteria decision making

	Time	Cost	Utilization	Reliability	Availability
VMI1	0.601649	0.5452	0.157057	0.387371	0.58154987
VMI2	0.144623	0.165118	0.249312	0.443428	0.30899643
VMI3	0.253727	0.289683	0.593631	0.169201	0.1094537

Table 6 Ranking of alternatives

Ranking of alternatives		
VMI1	0.515889	1
VMI2	0.191491	3
VMI3	0.280582	2

Table 7 Priority for the QoS context 1

Quality of service	% in ranking	Ranking in sequence
Time	48.09	1
Cost	28.37	2
Utilization	12.06	3
Reliability	7.11	4
Availability	4.36	5

The priorities calculated are shown in Table 7. Comparison matrices are used to calculate the final priorities for available alternatives.

In context 2, time, cost, memory, utilization, availability, and reliability are the criteria used to give the preferences. As compared to context 1, a memory criterion is added to show that how priorities change if multiple factors are added to take decisions. The pairwise comparison of the quality of service for context 2 is shown in Table 8, and priority for the QoS context 2 is shown in Table 9. The consistency ratio obtained is 2.2%, number of comparisons are 15, and principal eigenvalue is 6. 263. It shows that the pairwise comparison matrix is valid and consistent.

Table 8 Pairwise comparison of the quality of service for context 2

	1 (Time)	2 (Cost)	3 (Memory)	4 (Utilization)	5 (Reliability)	6 (Availability)
1	1	3.00	3.00	5.00	9.00	9.00
2	0.33	1	2.00	3.00	5.00	7.00
3	0.33	0.50	1	2.00	3.00	4.00
4	0.20	0.33	0.50	1	3.00	3.00
5	0.11	0.20	0.33	0.33	1	2.00
6	0.11	0.14	0.25	0.33	0.50	1

Table 9 Priority for the QoS context 2

Quality of service	% in ranking	Ranking in sequence
Time	45.1	1
Cost	23.1	2
Memory	14.2	3
Utilization	9.4	4
Reliability	4.7	5
Availability	3.4	6

5 Result Analysis

The modular design approach on the basis of attributes and alternatives is presented.

Weightages and priority are assigned for the QoS, and ranking is done as shown in Table 7. Time is the most contributing factor with 48.09% in first context and 45.1% in second context compared to cost, utilization, reliability, and availability. Weightage of each criterion is changing based on user's priority as shown for two different contexts. Hence, change in alternative criteria will change design of scheduling approach. Overall ranking of available VMIs alternatives is done based on the weightages of individual QoS. Ranking is given as VMI1 = 0.515889, VMI3 = 0.280582, VMI2 = 0.191491, respectively. Wide ranges of flexibility in alternatives are considered which help in upscaling or downscaling of the application. Future research can be carried out regarding this considering subcriteria of resource scheduling based on (i) virtual machine: load, time, cost, speed; (ii) policy-based: security, processor; (iii) utility-based: response time, profit, application, user satisfaction; (iv) application-based: large, real, data-intensive, shared database; (v) SLA-based: response and throughput.

6 Conclusion

In the present paper, heterogeneous resource ranking for workflow-based applications using AHP in cloud computing method is proposed to perform multi-criteria decision system for resource allocation based on quality of service in cloud computing. This study also reveals the most contributing factor to design scheduling algorithm. Contributing factors are given by the weightage of the criteria. The simulation results are performed by varying the number of criteria of quality of service.

References

1. Gawali MB, Shinde SK (2018) Task scheduling and resource allocation in cloud computing using a heuristic approach. J Cloud Comput Adv Syst Appl 7(1):1–16
2. Whaiduzzaman M, Gani A, Anuar NB (2014) Cloud service selection using multicriteria decision analysis. Sci World J 2014, Article ID459375. https://doi.org/10.1155/2014/459375
3. Nayak SC, Tripathy C (2018) Deadline based task scheduling using multi-criteria decision-making in cloud environment. Ain Shams Eng J 9(4):3315–3324
4. Zeng W (2013) Task scheduling algorithm based on the analytic hierarchy process under cloud computing environment. J Converg Inf Technol (JCIT) 8(7):114
5. Revathy C, Sekar G (2018) Analytic hierarchy process for resource allocation in cloud environment. J Cyber Secur Mobil 7(1):25–38
6. Nayak SC, Tripathy C (2018) Deadline sensitive lease scheduling in cloud computing environment using AHP. J King Saud Univ Comput Inf Sci 30:152–163
7. Ergu D, Kou G, Peng Y, Shi Y (2011) The analytic hierarchy process: task scheduling and resource allocation in cloud computing environment. J Supercomput 64(3):835–848. https://doi.org/10.1007/s11227-011-0625-1

8. Ebadifard F, Babamir SM (2018) Optimal workflow scheduling in cloud computing using AHP based multi objective black hole algorithm, pp 36–42. CEUR-WS.org/vol-2145/p06.pdf
9. Saaty TL (2003) Decision-making with the AHP: why is the principal eigenvector necessary. Eur J Oper Res 145(1):85–91
10. Emrouznejad A, Marra M (2017) The state of the art development of AHP (1979–2017): a literature review with a social network analysis. Int J Prod Res 55(22):6653–6675
11. Annette R, Banu A, Chandran PS (2016) Comparison of multi criteria decision making algorithms for ranking cloud render farm services. Indian J Sci Technol 9(31). https://doi.org/10.17485/ijst/2016/v9i31/93467
12. Lnenicka M (2015) AHP model for the big data analytics platform selection. Acta Inform Prag 4(2):108–121. https://doi.org/10.18267/j.aip.64
13. Singh A, Dutta K, Singh A (2014) Allocation in cloud computing environment using AHP technique. Int J Cloud Appl Comput 4(1):33–44
14. Chou C-H, Charleston (2013) Design and implementation of jAHP: a Java-based analytic hierarchy process application. Int J Comput Appl 62(15):35–41
15. Liu L, Zhang M (2015) Multi-objective optimization model with AHP decision-making for cloud service composition. KSII Trans Internet Inf Syst 9(9):3293–3311
16. Kumar R, Kaur K (2018) Optimized algorithm: maxchild on workflow scheduling. Int J Appl Eng Res 13(12):10458–10461
17. Gupta S et al (2019) Tier application in multi-cloud databases to improve security and service availability. In: Handbook of research on cloud computing and big data applications in IoT. IGI Global, pp 82–93
18. Saaty TL (2008) Decision making with the analytic hierarchy process. Int J Serv Sci 1(1):83–98

Chapter 42
Contemplation and Analysis of MIMO-OFDM Technology with the Augmentation of Training-Based Channel Estimation Techniques

Meenal Parihar and Asif Sayeed Khan

1 Introduction

MIMO channel model introduces spatial multiplexing where several parallel streams of data are transmitted through multiple antennas at the transmitter and received through multiple antennas at the receiver to increase diversity at the receiver and also to mitigate fading in the wireless channel. In a single carrier system, several symbols are transmitted [1] over a single carrier for high data rate in a MIMO channel model. Let us suppose the broadband wireless channel has a bandwidth $B = 10$ MHz. The symbol time 'Ts' for a single carrier system is given as: $Ts = 1/B = 1/10$ MHz $= 0.1$ μs. The delay spread 'Td' in a wireless channel is between 2 and 3 μs. It is observed that $Ts << Td$, and with the increase in channel bandwidth, data rate increases and symbol time decreases further which is the main cause of inter-symbol interference in a frequency selective channel which leads to signal distortion and incorrect signal reception, and hence, there is a need to overcome the problem of ISI. In a multicarrier system, the channel bandwidth 'B' is divided into several sub-bands. Let us suppose the bandwidth 'B' is divided into 'M' number of sub-bands. Hence, the bandwidth of each sub-band is equal to 'B/M'. Each symbol occupies a unique sub-band with a distinct subcarrier frequency. Since $B = 10$ MHz and let $M = 1000$. Therefore, the bandwidth of each sub-band is given as: 10 MHz/1000 = 10 kHz. Now, the symbol time $Ts = 1/10$ kHz $= 0.1$ ms. Hence, in a multicarrier system $Ts >> Td$. The symbol time has been increased by dividing the entire bandwidth into 1000 number of sub-bands which is the also basic concept of OFDM. The subcarriers are orthogonal to each other in order to avoid interference from the adjacent symbol

M. Parihar (✉) · A. S. Khan
Department of Electronics and Communication Engineering, Government Women Engineering College Ajmer, Ajmer, India
e-mail: meenalparihar8@gmail.com

A. S. Khan
e-mail: asif@gweca.ac.in

© Springer Nature Singapore Pte Ltd. 2021

S. D. Purohit et al. (eds.), *Proceedings of International Conference on Communication and Computational Technologies*, Algorithms for Intelligent Systems,
https://doi.org/10.1007/978-981-15-5077-5_42

[1, 2]. The main advantage of using OFDM in a MIMO system than a multicarrier technique is the reduced hardware complexity. In a multicarrier system, one needs oscillators and band-limited filters. In case if there are one thousand subcarriers, then in a multicarrier system one needs one thousand oscillators to generate the subcarrier frequencies along with an equal number of band-limited filters at the transmitter. For detection of the subcarriers, matched filters are also required at the receiver section. OFDM rather works on computational algorithms which has a simplified system design. It works on DFT/IDFT or FFT/IFFT algorithms. Fast Fourier transform and inverse fast Fourier transform make the calculation much easier and hence can be preferred over discrete Fourier transform and inverse discrete Fourier transform (DFT/IDFT) [1, 2].

1.1 Mathematical Model of a Multicarrier Modulation (MCM) System

The fundamental frequency in a multicarrier system is given as: $F_0 = B/M$. Let $\{X_0, X_1, X_2, \ldots, X_{M-1}\}$ be the symbols transmitted on 'M' subcarriers, then the frequency of the kth subcarrier is given by $kF_0 = k \cdot B/M$ and the symbol transmitted on the kth subcarrier is given by: X_k. The overall transmitted signal over the wireless channel is given as: $x_k(t) = \sum_{k=0}^{M-1} X_k e^{j2\pi k F_0 t}$ where $e^{j2\pi k F_0 t}$ is the kth subcarrier and kF_0 is the frequency of the kth subcarrier. The signal $x_k(t)$ is generated [1, 2] using 'M' number of oscillators at the transmitter. The net received composite time domain signal at the receiver without noise is given by:

$$y(t) = \sum_{k=0}^{M-1} X_k e^{j2\pi k F_0 t} \tag{1.1}$$

Equation (1.1) is similar to a Fourier series expansion, and X_l is the lth Fourier coefficient. At [3] the receiver to extract the symbol 'X_l' on the lth subcarrier, the mathematical equations are given as follows:

$$X_l = F_0 \int_0^{1/F_0} e^{-j2\pi l F_0 t} \cdot y(t) dt \tag{1.2}$$

$$\Rightarrow X_l = F_0 \int_0^{1/F_0} e^{-j2\pi l F_0 t} \cdot \sum_{k=0}^{M-1} X_k e^{j2\pi k F_0 t} \tag{1.3}$$

$$\Rightarrow X_l = \sum_{k=0}^{M-1} X_k F_0 \int_0^{1/F_0} e^{j2\pi(k-l) F_0 t} dt \tag{1.4}$$

$1/F_0$ is the fundamental time period in the above integral.

$$\because \mathrm{e}^{\mathrm{j}2\pi(k-l)F_0 t} = \begin{cases} 1, & k = l \\ 0, & k \neq l \end{cases} \tag{1.5}$$

$$\therefore \text{RHS} = \sum_{k=0}^{M-1} X_k \delta(k-l) = X_l \Leftrightarrow k = l \tag{1.6}$$

Hence, from Eqs. (1.5) and (1.6), it is observed that the orthogonality principle has been maintained and hence the symbol 'X_l' is extracted precisely. The same set of equations is followed to extract all the symbols transmitted on 'M' subcarriers and termed as coherent demodulation at the receiver. By correlating with $\mathrm{e}^{\mathrm{j}2\pi l F_0 t}$ or matched filtering with $\mathrm{e}^{\mathrm{j}2\pi(k-l)F_0 t}$, one can extract the symbol on the lth subcarrier and this is the principle of multicarrier modulation (MCM) method [3, 4].

1.2 Mathematical Model of Orthogonal Frequency Division Multiplexing (OFDM) System

However, in an MCM system to generate the composite signal $x_k(t)$, one needs to tune the oscillators at the subcarrier frequencies with a spacing of 'B/M' in order to maintain the orthogonality among the subcarriers which is a very challenging task. OFDM system overcomes the problem of using large number of oscillators and filters. In the initial system, the signal is bandlimited to 'B' with [4] the maximum frequency ranging between $-B/2 \leq F_{\max} \leq B/2$. The samples are taken at nyquist sampling rate defined as: $R = 2F_{\max} = 2 \cdot B/2 = B$ and the sampling interval is given by: $T = 1/B$.

$$\because x(t) = \sum_{k=0}^{M-1} X_k \mathrm{e}^{\mathrm{j}2\pi k F_0 t} \tag{1.7}$$

The lth sampling instant $= lT = l/B$

$$x(l) = x(lt) = \sum_{k=0}^{M-1} X_k \mathrm{e}^{\mathrm{j}2\pi k F_0 l t} \tag{1.8}$$

$$x(l) = \sum_{k=0}^{M-1} X_k \mathrm{e}^{\mathrm{j}2\pi k B/M \cdot l \cdot 1/B} = \sum_{k=0}^{M-1} X_k \mathrm{e}^{\mathrm{j}2\pi k L/M} = X_l = \text{IDFT/IFFT} \tag{1.9}$$

$x(l)$ is the lth IFFT/IDFT point of the symbols $\{X_0, X_1, X_2, \ldots, X_{M-1}\}$, and the samples $\{x(0), x(1), x(2), \ldots, x(m-1)\}$ are taken at nyquist sampling rate and

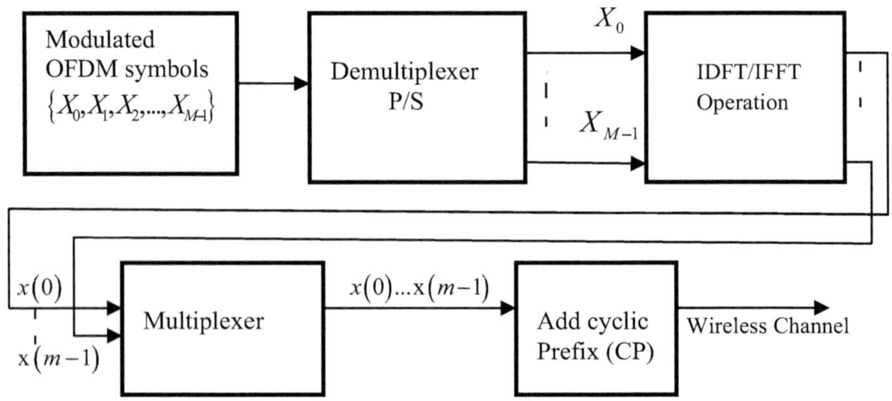

Fig. 1 MIMO-OFDM block diagram of transmitter section

generated with IFFT/FFT or IDFT/DFT [4, 5] which is the key principle of MIMO-OFDM. At the receiver, the symbols are extracted with DFT/FFT algorithms. Since OFDM does not use bandlimited filters and uses mathematical algorithms, the spectrum of subcarriers is overlapping for spectral efficiency. The addition of cyclic prefix (CP) is the most important criteria in the OFDM system as it takes 'L' number of samples from the tail of the OFDM signal and prefixes it at the front end. Hence, the total length of the OFDM block becomes '$M + L$'. It removes the interblock [5] interference (IBI) between two different OFDM blocks. With the addition of CP, the output at the receiver in time domain is given by circular convolution and multiplication in the frequency domain. The frequency domain response is given as $Y(k) = H(k) \cdot X(k), k = 0, 1 \ldots M - 1$. Each subcarrier experiences a flat fading channel response. The OFDM system removes the ISI by converting the frequency selective channel into 'M' number of flat fading channels [5, 6] (Figs. 1 and 2).

2 Channel Estimation Techniques in MIMO-OFDM System

In an OFDM system at the transmitter end, the message bits are modulated through various modulation techniques like PSK/QAM. The message symbols are then converted into time domain by IFFT operation [6] and then after adding cyclic prefix finally sent through a wireless channel. The received time domain signal is usually distorted by the non-ideal channel characteristics, and here we need the channel estimation in order to correctly detect the transmitted signal. Channel estimation measures the channel attenuations 'h' from the received observations 'y' and known transmitted signal 'x' at the receiver. In pilot-aided or training-based channel estimation techniques where pilot symbols known at the receiver are transmitted at the beginning of the preamble or scattered throughout the preamble in order to estimate

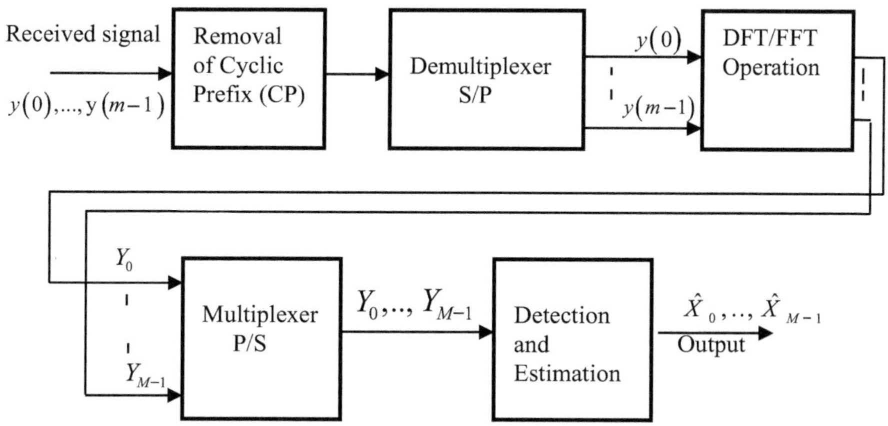

Fig. 2 MIMO-OFDM block diagram of receiver section

the channel characteristics via these known symbols. Block type and comb type are the most common pilot arrangement methods. In block type, for each OFDM symbol block, the pilot tones are transmitted for all the subcarriers. They are periodically transmitted among several OFDM symbol blocks within a time period to track the time-varying channel characteristics. Block type is basically considered for slow fading frequency selective channels where it is assumed that the channel remains constant up to few OFDM symbol blocks. In comb-type pilot arrangement, for each OFDM symbol block, the pilot tones are inserted periodically among the subcarriers and not for all the subcarriers within the same OFDM symbol block. The pilot tones are scattered periodically in the OFDM symbol, and they are transmitted for all the OFDM blocks but not for all the subcarriers in a single OFDM symbol and only for specific subcarriers. Comb type is beneficial when the channel variations are fast and they can change even within the adjacent OFDM symbol blocks. It is a frequency domain interpolation method where channel is estimated along the frequency axis [6].

In **block-type** pilot arrangement, S_t is the period of pilot symbol in time domain along the time axis as shown in Fig. 3a. The pilot symbol must be as kept as frequently as the coherence time in order to track the time varying channel characteristics. In **comb-type** pilot arrangement, S_f is the period of pilot symbol along the frequency axis as shown in Fig. 3b. The pilot symbol must be kept as frequently as the coherence bandwidth [6].

2.1 Training-Based Channel Estimation Techniques

The training or pilot symbols for 'M' subcarriers can be represented in the form of diagonal matrix as follows:

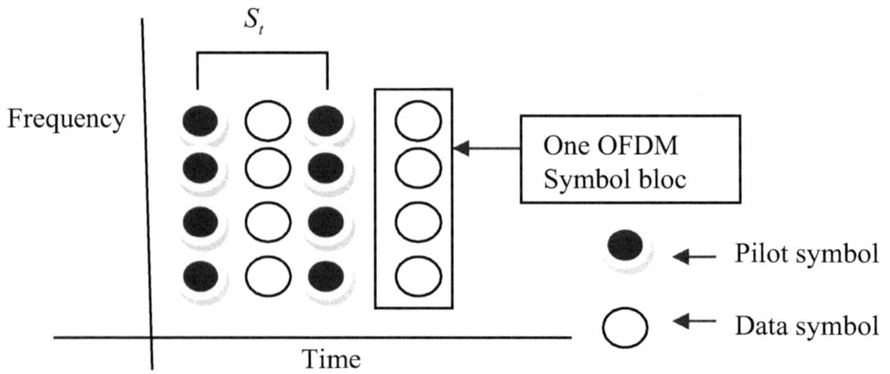

a. Block type

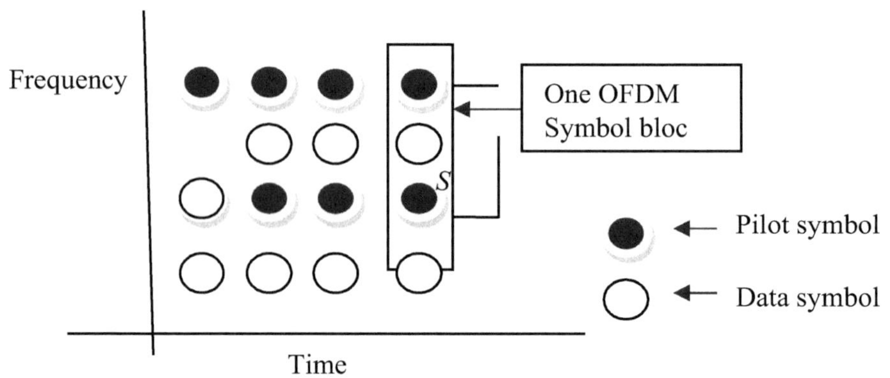

b. Comb type

Fig. 3 Block-type and comb-type pilot symbol arrangement in OFDM transmitted symbols blocks

$$X = \begin{bmatrix} X(0) & 0 & 0 & \dots & 0 & \dots \\ 0 & X(1) & 0 & \dots & 0 & \dots \\ 0 & 0 & X(2) & \dots & 0 & \\ \dots & \dots & \dots & \dots\dots & & \dots \\ \dots & \dots & \dots & \dots\dots & & \dots \\ 0 & 0 & 0 & \dots & & X(M-1) \end{bmatrix} \tag{2.1}$$

$X(k)$ denotes the symbol on the kth subcarrier with mean $E\{X(k)\} = 0$ and variance $\text{Var}\{X(k)\} = \sigma_X^2$, $k = 0, 1, \dots, M-1$. We represent X as diagonal matrix since the '*M*' subcarriers are orthogonal with respect to each other. The received symbol can be represented in matrix form as follows:

$$Y \triangleq \begin{bmatrix} Y(0) \\ Y(1) \\ \cdot \\ \cdot \\ \cdot \\ Y(M-1) \end{bmatrix} = \begin{bmatrix} X(0) & 0 & 0 & \dots & 0 & \dots \\ 0 & X(1) & 0 & \dots & 0 & \dots \\ 0 & 0 & X(2) & \dots & 0 & \\ \dots & \dots & \dots & \dots\dots & & \dots \\ \dots & \dots & \dots & \dots\dots & & \dots \\ 0 & 0 & 0 & \dots & & X(M-1) \end{bmatrix} \begin{bmatrix} H(0) \\ H(1) \\ \cdot \\ \cdot \\ \cdot \\ H(M-1) \end{bmatrix} \begin{bmatrix} W(0) \\ W(1) \\ \cdot \\ \cdot \\ \cdot \\ W(M-1) \end{bmatrix} \tag{2.2}$$

$$\therefore Y = XH + W \tag{2.3}$$

where H is the channel vector, W is the noise vector with mean $E\{W(K)\} = 0$ and variance $\text{Var}\{W(k)\} = \sigma_W^2, k = 0, 1, \dots, M-1$. The three basic channel estimation techniques are given as follows:

Least square (LS) channel estimation The least square channel estimation technique finds the minimum channel estimation $\hat{H}_{\text{LS}}$ of the channel matrix H by minimizing the cost function as expressed [7] by the following mathematical expressions given below:

$$J\left(\hat{H}\right) = \left\| Y - X\hat{H}_{\text{LS}} \right\|^2 \tag{2.4}$$

$$\because \hat{H}_{\text{LS}} = X^{-1}Y \tag{2.5}$$

The mean square error (MSE) for LS estimate is given by:

$$\text{MSE}_{\text{LS}} = \left\{ E\left(H - \hat{H}_{\text{LS}}\right)^H \left(H - \hat{H}_{\text{LS}}\right) \right\} \tag{2.6}$$

$$\therefore \text{MSE}_{\text{LS}} = \frac{\sigma_W^2}{\sigma_X^2} \tag{2.7}$$

Without going into the details of the derivation of MSE, the final solution is given by Eq. (2.7) which is inversely proportional to the signal-to-noise ratio. LS estimation leads to the major problem of noise enhancement when the channel is in deep null as concluded from Eq. (2.7). It is preferred due to its low complexity since it does not consider the knowledge of channel statistics but does not give optimum results in case of mean square error; hence, better techniques are implemented where effect of noise on the signal is considered thoroughly.

Minimum mean square error (MMSE) channel estimation MMSE technique gives optimum results in case of mean square error [8] since it considers channel correlation function. The LS channel estimate is given as: $\hat{H}_{\text{LS}} = X^{-1}Y = \tilde{H}$, the MMSE channel estimate is defined as:

$$\hat{H}_{\text{MMSE}} = V\tilde{H} \tag{2.8}$$

Hence, the mean square error of MMSE channel estimate is given as:

$$J\left(\hat{H}_{\text{MMSE}}\right) = E\{\|e\|^2\} = E\left\{\left\|H - \hat{H}_{\text{MMSE}}\right\|^2\right\} \tag{2.9}$$

From Eq. (2.9), it is observed that the MMSE channel estimation technique minimizes the mean square error by finding a better linear estimate in terms of the weight function 'V' where $V = R_{H\tilde{H}} \cdot R_{\tilde{H}\tilde{H}}$. '$R_{H\tilde{H}}$' is cross-correlation matrix between the true channel matrix H and the estimated channel matrix $\tilde{H}$ in frequency domain and $R_{\tilde{H}\tilde{H}}$ is the autocorrelation matrix of estimated channel matrix. On solving Eqs. (2.8) and (2.9), the solution for MMSE channel estimator is as follows:

$$\hat{H}_{\text{MMSE}} = R_{H\tilde{H}}\left(R_{HH} + \frac{\sigma_W^2}{\sigma_X^2} \cdot I\right)^{-1} \cdot \tilde{H} \tag{2.10}$$

The MMSE channel estimator gives much better results as compared to LS channel estimator in terms of MSE especially when signal-to-noise ratio is low as it considers the effect of noise as shown from Eq. (2.10) at the cost of computational complexity since the calculations [8] include matrix inversion of channel matrix X which is quite tedious.

Least square discrete Fourier transform (LS-DFT) channel estimation The least square channel estimation technique is easy to implement without any computational complexity with the major drawback of noise enhancement. This problem can be overcome with the augmentation of discrete Fourier transform and minimize the mean square error equivalent to MMSE channel estimation technique. DFT-based channel estimation allows the reduction of noise component. This is done by converting the frequency domain LS channel estimates at the receiver into time domain by IDFT operation and applying a smoothing filter in the time domain. DFT-based channel estimation is done in time domain as most of the signal power is concentrated to relatively few samples [8, 9]. At the transmitter after adding cyclic prefix, the time domain output at the receiver is given by circular convolution between channel impulse response and the transmitted signal, and hence, orthogonality is preserved among the subcarriers. However, this is the reason that the channel impulse response is concentrated on the first samples of cyclic prefix (CP) and filter is applied in time domain of length CP [10]. Let us assume that the length of CP be from 0 to $L - 1$. These samples have both signal power and noise. The samples from $L - 1$ to $M\text{-}1$ have only noise component. The filter accepts the samples in the CP with higher signal power than noise power, and only the 'L' samples are converted into frequency domain by DFT operation. Hence, the noise is reduced in the time domain.

The mathematical [9] expression for LS-DFT channel estimation is given as follows:

$$\text{IDFT}\{H(k)\} = h[m] + w[m] \triangleq \hat{h}[m] \quad m = 0, \ldots, M-1 \tag{2.11}$$

$$\hat{h}_{\text{DFT}}[m] = \begin{cases} h[m] + w[m] & m = 0, \ldots, L-1 \\ 0 & \text{otherwise} \end{cases} \tag{2.12}$$

The DFT-based channel estimation by converting the 'L' samples into frequency domain is given as follows:

$$\hat{H}_{\text{DFT}}(k) = \text{DFT}\left\{\hat{h}_{\text{DFT}}[m]\right\} \tag{2.13}$$

hence the noise reduction in LS estimation method can be achieved with the addition of DFT-based channel estimation with simple mathematical calculations as discussed briefly [9].

3 MATLAB Simulated Results for Training-Based Channel Estimation Techniques in MIMO-OFDM

From Fig. 4, we observe that DFT-based channel estimation technique reduces the effect of noise on the received samples. The mean square error (MSE) is reduced in least square channel estimation with DFT-based channel estimation, and from

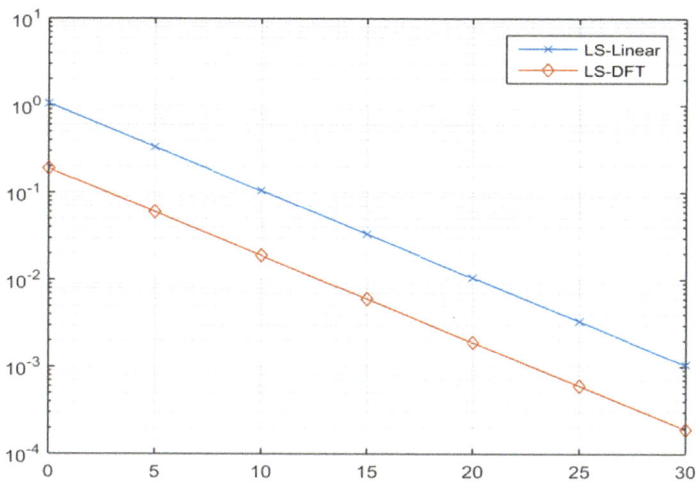

Fig. 4 Comparison of LS and LS-DFT channel estimation techniques with linear interpolation

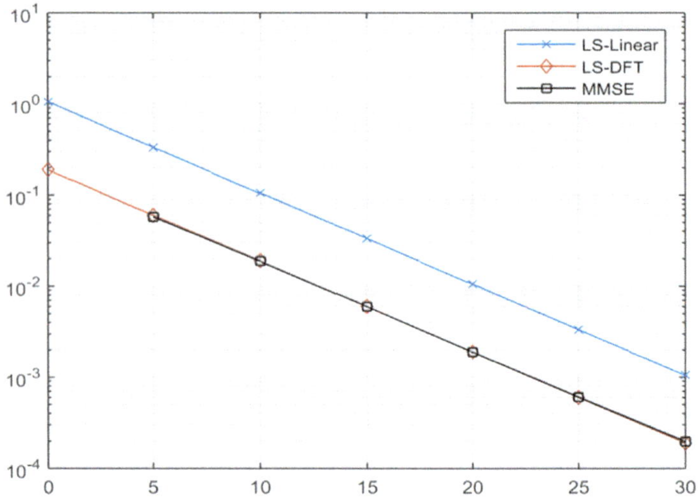

Fig. 5 Comparison of training-based channel estimation techniques in MIMO-OFDM

Fig. 5, we observe that the MSE performance of LS-DFT and MMSE are equal and the curves of both the estimation techniques are completely overlapping.

4 Conclusion

MIMO-OFDM reduces the inter-symbol interference with the implementation of simple IDFT/IFFT- and DFT/FFT-based algorithm at the transmitter and receiver section and avoids the use of filters and oscillators. Channel estimation by LS estimation technique is the simplest one to implement but has the disadvantage of noise enhancement. MMSE considers the effect of noise by estimating the channel statistics via covariance and cross-covariance channel matrices and reduces the mean square error at the cost of computational complexity. Hence, the performance of LS channel estimation technique with DFT-based channel estimator has reduced computational complexity and the MSE performance of LS-DFT is equivalent to that of MMSE channel estimation technique. Thus, it is preferable to implement DFT technique rather than MMSE technique to improve the performance of LS technique since DFT is easier to implement than the MMSE channel estimation technique.

References

1. Cho YS (2010) MIMO-OFDM wireless communication with MATLAB. Wiley
2. Li P (2018) A research on transmission performance of MIMO-OFDM wireless channel. In: International conference on communication, signal processing and systems. Springer, pp 308–316
3. Guo S (2016) Robust capacity maximization transceiver design for MIMO-OFDM systems. Sci China Inf Sci
4. Dheergha Rao K (2015) Channel coding techniques for wireless communications. Springer
5. Umesha GB (2018) Comb type pilot arrangement based channel estimation for spatial multiplexing MIMO systems. In: 2018 IRJET
6. Watanabe K (2019) Modified pilot selection for channel estimation of systematic polar coded MIMO-OFDM. ICT Express
7. Abdelhamid RIADI (2019) Performance of massive MIMO-OFDM system with M-QAM modulation based on LS channel estimation. researchgate.net
8. Koni S (2018) Performance evaluation of channel estimation methods in OFDM systems. Majlesi J Telecommun
9. Al-Salihi H (2018) DFT based channel estimation techniques for massive-MIMO systems. In: IEEE conference, June 2018
10. Singh V et al (2019) Source redundancy management and host intrusion detection in wireless sensor networks. Recent Patents Comput Sci 12(1)

Chapter 43
Gaussian Distribution Operator Based Spider Monkey Optimization Algorithm to Improve Global Search Ability

Himani and S. K. Agrawal

1 Introduction

Large number of optimization techniques have been introduced widely over the previous decennaries, but the performance of some of the optimization mechanisms are not effective in finding the global optima of the complex optimization functions. One of the most efficient optimization algorithms is Nature-Inspired Algorithm (NIA), this approach is based on the natural phenomenon. Several types of NIAs are swarm intelligence, evolutionary algorithms, bio-inspired algorithms, genetic algorithms, etc. Swarm optimization techniques are also known as Swarm Intelligence (SI), which is applied to a broad range of complex problems in order to produce the best optimal solution by imitating the cumulative behavior of natural swarm group. Some of the SI techniques are Spider Money Optimization, Artificial Bee Colony Optimization (ABC), Firefly Optimization, Bacterial Foraging Optimization and Particle Swarm Optimization algorithm (PSO), these techniques have been designed to solve composite functions in engineering field, transportation area, finance, warehousing business, real-life problems, etc. [1, 2]. One of the most influential SI optimization techniques is Spider Monkey Optimization (SMO) approach, which makes the use of foraging characteristics of spider monkeys to obtain the finest optimum solution of the given complex constrained or differential function. Bansal et al. derived this algorithm in 2014 [2]. The dominance of SMO algorithm is due to its capability to manage stagnation or early convergence over the canonical approaches like ABC, PSO, and Genetic Algorithms (GA), but SMO demands adjustment of meta-heuristic operations to a greater extent. Thus, in order to settle

Himani (✉) · S. K. Agrawal
Department of Electronics and Communication Engineering, Government Women Engineering College Ajmer, Ajmer, India
e-mail: himmiiim27@gmail.com

S. K. Agrawal
e-mail: skagarwal5@rediffmail.com

© Springer Nature Singapore Pte Ltd. 2021
S. D. Purohit et al. (eds.), *Proceedings of International Conference on Communication and Computational Technologies*, Algorithms for Intelligent Systems,
https://doi.org/10.1007/978-981-15-5077-5_43

these disputes and to improve the efficiency and impulsive convergence a variant of SMO is presented in this work by introducing an additional Gaussian operator in the solution search equation of basic SMO [3–5].

The rest of the paper is composed as follows: The basic synopsis of SMO algorithm is given in Sects. 2 and 3 give the improvements in the different stages of fundamental SMO. Experimental results for several benchmark functions are shown with discussion in Sect. 4. Lastly, the conclusion and future scope of the proposed paper has been given in Sect. 5.

2 Spider Monkey Optimization Algorithm

Spider Monkey Optimization algorithm was proposed by Bansal et al. in 2014, motivation of this approach is from the intelligent foraging characteristics of spider monkeys which belong to the FFSS system. In this system, a female monkey constantly instructs a crowd of generally 40–50 spider monkeys, which split into tiny troops in order to seek good food in sufficient quantity. In which each of the troop is further commanded by female, again and again, this process repeats itself until the fulfilment of objective (food) [6–8]. There are seven significant stages in this mechanism described as follows:

Initialization of Spider Monkeys Initially, SMO produces a population of M spider monkeys that are uniformly distributed. S_i is generated as follows:

$$S_{ij} = S_{\text{min}_j} + \alpha(0,\ 1) \times \left(S_{\text{max}_j} - S_{\text{min}_j}\right) \tag{1}$$

Here S_i illustrates the ith spider monkey in the crowd, S_{min_j} and S_{max_j} are lower and upper limits of the search area in jth dimension and $\alpha(0,\ 1)$ is a random number uniformly assigned in the range (0, 1).

Local Leader Phase (LLP) In this stage, each spider monkey renews its position, which depends on its fitness value. If the fitness of the current position is inferior to new position, then it is updated by new position and rejects the past one. The equation which is used to update position is as follows:

$$S\text{new}_{ij} = S_{ij} + \alpha(0,\ 1) \times \left(L_{\text{local},\ kj} - S_{ij}\right) + \alpha(-1,\ 1) \times \left(S_{rj} - S_{ij}\right) \tag{2}$$

Here first term in the equation is for persistence, second term is for local leader influence and third term represents the social influence. S_{ij} shows the jth dimension of ith spider monkey, $L_{\text{local},\ kj}$ denotes the kth local leader of that troop and S_{rj} is rth spider monkey elected randomly within kth group in jth dimension such that $r! = i$ and $\alpha(-1,\ 1)$ is an arbitrary number uniformly assigned in the range $(-1, 1)$.

Global Leader Phase (GLP) In GLP the spider monkey's position came to be renovated on the basis of the knowledge of previous states, using the position update equation as follows:

$$Snew_{ij} = S_{ij} + \alpha(0,\ 1) \times \left(L_{\text{global},\ j} - \text{SM}_{ij}\right) + \alpha(-1,\ 1) \times \left(S_{rj} - S_{ij}\right) \tag{3}$$

Here, location of group leader is $L_{\text{global},\ j}$ in jth dimension and $j \in \{1,\ 2,\ 3, \ldots,\ D\}$ is the index chosen arbitrarily. The quality of the given function is depicted by fitness value of that swarm. The equation of probability according to which solution gets updated is given as follows:

$$P_i = 0.9 * \frac{\text{fitness}_i}{\text{maxfit}} + 0.1 \tag{4}$$

Global Leader Learning (GLL) Phase Greedy selection approach is applied in this stage, to update the position of global leader. The spider money having the superior fitness value is elected as global ruler. The global limit count is increased by one if no change is there.

Local Leader Learning (LLL) Phase This stage of SMO is the same as above where greedy selection approach is applied to update the position of global leader in each local group. The spider monkey having the best fitness value in each group is elected as local ruler. The local limit count is increased by one, if no change is there.

Local Leader Decision (LLD) Phase If the position of the local leader is not rejuvenated up-to a predetermined threshold better known as Local Leader Limit, then in this stage a verdict is appropriated relating to the local leader position. Local leader position will be initialized randomly if no change occurs, or it can be decided according to the following equation:

$$Snew_{ij} = S_{ij} + \alpha(0,\ 1) \times \left(L_{\text{global},\ j} - S_{ij}\right) + \alpha(0,\ 1) \times \left(S_{rj} - L_{\text{local},\ kj}\right) \tag{5}$$

Global Leader Decision (GLD) Phase The position of the global leader is controlled in this stage, if it is not updated to a predetermined threshold (Global Leader Limit) then global ruler will split the whole crowd in tiny troops until the maximum no. of groups are generated. In GLD phase, LLL stage is initiated to choose the local leader in each of new group. In case if maximum no. of groups are formed but still there is no update in the global leader position then global ruler will merge all the troops to produce one group. Thus this mechanism is based on FFSS system [9–11].

3 Gaussian Spider Monkey Optimization Algorithm

In swarm intelligence field, SMO is quite a new algorithm with good convergence speed and balanced exploration and exploitation processes. Exploration and exploitation are two key features in SMO technique, exploration is related to autonomous or independent search while exploitation uses the prevailing knowledge to search food [12, 13]. In presented work, a new variant of SMO is proposed in order to improve the global search ability and convergence speed by making the use of regular Gaussian distribution operators [4, 8, 14].

Improved Local Leader Phase (LLP) This stage is the same as given in basic SMO except for that one distribution operator, which is introduced in the solution search equation. Thus the modified position update equation is defined as follows:

$$\begin{aligned} Snew_{ij} = \; & S_{ij} + \alpha(0,\ 1) \times \left(L_{\text{local},\,kj} - S_{ij}\right) \times 2y \\ & + \alpha(-1,\ 1) \times \left(S_{rj} - S_{ij}\right) \times 2y \end{aligned} \tag{6}$$

All the parameter in the above equation is same as given in Eq. (2) and y is defined as follows:

$$y = 0.5 - 0.25 \times \left(\frac{\text{Current\_Iteration}}{\text{Maximum\_iterations}}\right) \tag{7}$$

Improved Global Leader Phase (GLP) In GLP the spider monkey's position came to be updated according to the knowledge of previous states, using the position update equation as follows:

$$\begin{aligned} Snew_{ij} = \; & S_{ij} + \alpha(0,\ 1) \times \left(L_{\text{global},\,j} - S_{ij}\right) \times 2y \\ & + \alpha(-1,\ 1) \times \left(S_{rj} - S_{ij}\right) \times 2y \end{aligned} \tag{8}$$

Here, location of group leader is $L_{\text{global},\,j}$ in jth dimension and $j \in \{1,\ 2,\ 3, \ldots, D\}$ is the index chosen arbitrarily and y is the same as depicted in Eq. (7). The equation of probability according to which solution gets updated is the same as Eq. (4).

Improved Local Leader Decision (LLD) Phase Improved LLD phase is also the same as represented in SMO except one Gaussian operator is added in the position update equation. Improved Local leader position will be initialized randomly if no change takes place, or it can be decided according to following equation:

$$Snew_{ij} = S_{ij} + \alpha(0,\ 1) \times \left(L_{\text{global},\ j} - S_{ij}\right) \times 2y + \alpha(0,\ 1) \times \left(S_{rj} - L_{\text{local},\ kj}\right) \times 2y \quad (9)$$

Here y can be obtained by regular Gaussian distribution operator, the same as given in Eq. (7). Rest of the phases of GSMO are exactly the same as that of for fundamental SMO [15, 16].

4 Simulation Results and Discussion

In order to validate the effectiveness and optimization performance of the two algorithms, SMO algorithm, and GSMO algorithm, several standard optimization test functions have selected. These test functions have been used to analyze the performance of GSMO algorithm. Comparison of optimal values for these algorithms are shown in Figs. 1, 2, and 3 and in Table 1.

Test Functions Studied

The first standard objective function is Beale function, which is given by:

$$f(x,\ y) = (1.5 - x + xy)^2 + \left(2.25 - x + xy^2\right)^2 + \left(2.625 - x + xy^3\right)^2$$

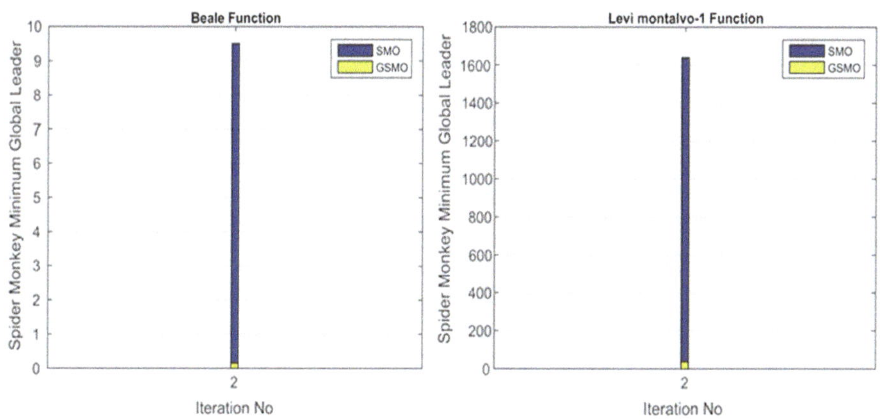

Fig. 1 Comparison of optimal values of Beale and Levi Montalvo-1 function for SMO and GSMO algorithms

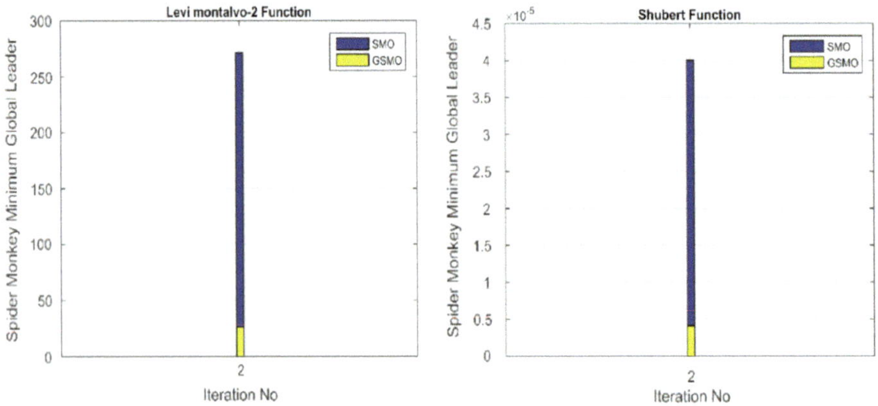

Fig. 2 Comparison of optimal values of Levi Montalvo-2 and Shubert function for SMO and GSMO algorithms

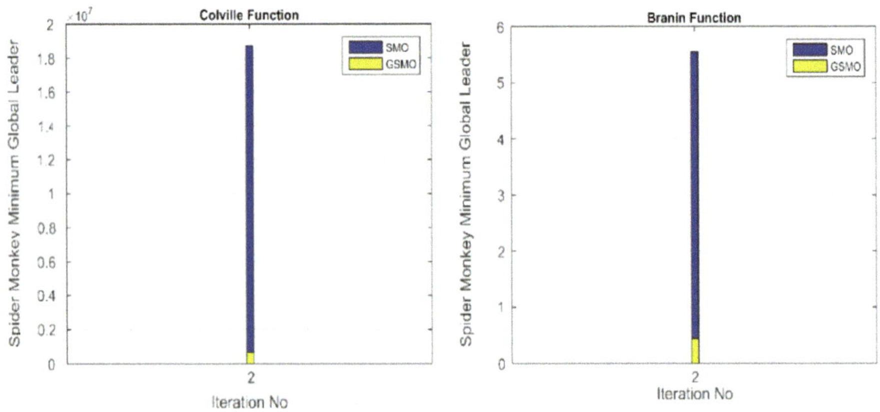

Fig. 3 Comparison of optimal values of Colville and Branin function for SMO and GSMO algorithms

Table 1 Optimization results for various benchmark function

Algorithms	SMO	GSMO
Optimal value for Beale function	9.513347	0.16062183
Optimal value for Levi Montalvo-1 function	1639.793963	38.94541030
Optimal value for Levi Montalvo-2 function	271.409329	26.44240288
Optimal value for Shubert function	0.00004004	0.00000415
Optimal value for Colville function	187,210,59.970139	671,282.85557
Optimal value for Branin function	5.54764495	0.44807023

The function is continuous and is defined in two-dimensional space. The function is multimodal and can be defined on any input domain but it is usually evaluated on $x \in [-4.5,\ 4.5]$ for all $i = 1,\ 2$. The function has one global minimum at: $f(x^*) = 0$ at $x^* = (3,\ 0.5)$.

Second test function is Levi Montalvo-1 presented by the following equation:

$$f(x) = \sin^2(\pi\omega_1) + \sum_{i=1}^{d-1}(\omega_i - 1)^2\left[1 + 10\sin^2(\pi\omega_i + 1)\right] + (\omega_d - 1)^2\left[1 + \sin^2(2\pi\omega_d)\right]$$

where $\omega_i = 1 + \frac{x_i - 1}{4}$ for all $i = 1,\ 2, \ldots,\ d$. The function is usually evaluated on the hypercube $x_i \in [-10,\ 10]$ for all $i = 1,\ 2\ , \ldots,\ d$ and global optima is $f(x^*) = 0$ at $x^* = (1,\ 1,\ \ldots,\ 1)$.

Third objective function is Levi Montalvo-2 given as follows:

$$(x,\ y) = \sin^2(3\pi x) + (x - 1)^2\left(1 + \sin^2(3\pi y)\right) + (y - 1)^2\left(1 + \sin^2(2\pi y)\right)$$

The function is continuous but not convex and is defined in two-dimensional space. It is multimodal, differentiable and is non-separable. The function can be defined on any input domain but it is usually evaluated on $x \in [-10,\ 10]$ and $y \in [-10,\ 10]$. The function has one global minimum at $f(x^*) = 0$ at $x^* = (1,\ 1)$.

Shubert function is the fourth optimization test function and is defined on 2-dimensional space presented as shown below:

$$f(x) = \left(\sum_{i=1}^{5} i\cos((i + 1)x_1 + 1\right)\left(\sum_{i=1}^{5} i\cos((i + 1)x_2 + 1\right)$$

The Shubert function has several local minima and many global minima. It is usually evaluated on the square $x_i \in [-10,\ 10]$ for all $x = 1,\ 2$. Although this may be restricted to the square $x_i \in [-5.12,\ 5.12]$ for all $i = 1,\ 2$. And global optima is $f(x^*) = -186.739$.

Next benchmark test function is Colville function is usually evaluated on the hypercube $x_i \in [-10,\ 10]$ for all $i = 1,\ 2,\ 3,\ 4$. And global optima is $f(x^*) = 0$ at $x^* = (1,\ 1,\ 1,\ 1)$. Function is shown below:

$$f(x) = 100\left(x_1^2 - x_2\right)^2 + (x_1 - 1)^2 + (x_3 - 1)^2 + 90\left(x_3^2 - x_4\right)^2 + 10.1\left((x_2 - 1)^2 + (x_4 - 1)^2\right) + 19.8(x_2 - 1)(x_4 - 1)$$

Branin function is the last objective function that has taken to analyze the performance of GSMO. This is a multimodal minimization problem defined as follows:

$$f(x) = \left(-1.275\frac{x_1^2}{\pi^2} + 5\frac{x_1}{\pi} + x_2 - 6\right)^2 + \left(10 - \frac{5}{4\pi}\right)\cos(x_1) + 10$$

Here, $x_1 \in [-5, 10]$ and $x_2 \in [0, 15]$, global optima is $f(x_i) = 0.397887357729738$ for $x = [-\pi, 12.275]$ or $x = [9.42478, 2.475]$ or $x = [\pi, 2.275]$.

Simulation setup The simulation results of SMO and GSMO are compared to justify that GSMO is a competed part of algorithms based on swarm intelligence. Arrangement of parameters used is given as follows:

Total number of Swarm or Spider Monkeys = 200;
Global Leader Limit = 50; Local Leader Limit = 30;
Maximum no. of groups = 20; and we have considered all the problems only for two iterations as shown in Figs. 1, 2, and 3.

Rest of the settings are taken from primitive works [4, 11, 15, 17].

5 Conclusion

The work has attempted to present a new technique called GSMO and the mainspring of the proposed paper is an astute foraging behavior of Spider monkeys. In the presented work it has been shown that the perseverance, persistence, adaptability, and efficiency of the GSMO algorithm is superior to that of the SMO algorithm, with the help of several benchmark optimization functions. The convergence speed of the GSMO algorithm is also increased with the introduction of the regular Gaussian distribution operator in the solution search equation of the SMO algorithm. In the forthcoming attempt, identification of disputes for which GSMO algorithm is functioning better or average than other nature-inspired algorithms and GSMO will also be applicable for real-time optimization problems.

References

1. Bonabeau E, Dorigo M, Theraulaz G (1999) Swarm intelligence: from natural to artificial systems, vol 1. Oxford University Press
2. Bansal JC, Sharma H, Jadon SS, Clerc M (2014) Spider monkey optimization algorithm for numerical optimization. Memet. comput 6(1):31–47
3. Devi RV, Sathya SS (2017) Monkey behavior based algorithms—a survey. Int J Intell Syst Appl 12:67–86
4. dos Santos Coelho L, Alotto P (2011) Gaussian Artificial bee colony algorithm approach applied to loney's solenoid benchmark problem. IEEE Transections Magn 47:1326–1329
5. Sharma A, Sharma H, Bhargava A, Sharma N (2016) Power law-based local search in spider monkey optimization for lower order system modelling. Int J Syst Sci, 1–11
6. Sharma A, Sharma A, Panigrahi BK, Kiran D, Kumar R (2016) Ageist spider monkey optimization algorithm. Swarm Evol Comput 28:58–77

7. Gupta K, Deep K (2016) Investigation of suitable perturbation rate scheme for spider monkey optimization algorithm. In: Proceedings of fifth international conference on soft computing for problem solving, vol 437, Springer, Singapore, pp 839–850
8. Gupta K, Deep K, Bansal JC (2016) Improving the local search ability of spider monkey optimization algorithm using quadratic approximation for unconstrained optimization. Comput Intell 0(0)
9. Swami V, Kumar S, Jain S (2017) An improved spider monkey optimization algorithm. Soft computing: theories and applications, advances in intelligent systems and computing, vol 583, pp 73–81
10. Kumar S, Sharma VK, Kumari R (2014) Self-adaptive spider monkey optimization algorithm for engineering optimization problems. Int J Inf Commun Comput Technol 2:96–107
11. Kumar S, Kumari R (2014) Modified position update in spider monkey optimization algorithm. International Journal of Emerging Technologies in Computational and Applied Sciences (IJETCAS). Citeseer
12. Pal SS, Kumar S, Kashyap M, Choudhary Y, Bhattacharya M (2016) Multi-level thresholding segmentation approach based on spider monkey optimization algorithm. In: Proceedings of the second international conference on computer and communication technologies, Springer, India, pp 273–287
13. Al-Azza AA, Al-Jodah AA, Harackiewicz FJ (2016) Spider monkey optimization (SMO): a novel optimization technique in electromagnetics. In: 2016 IEEE radio and wireless symposium (RWS), pp 238–240
14. Kaur A, Sharma H, Sharma N (2017) Disruption operator-based spider monkey optimization algorithm. In: International conference on computing, communication and automation (ICCCA-2017), pp 216–221
15. Hazratia G, Shannab H, Shannac N, Bansal JC (2016) Modified spider monkey optimization. In: 6 international workshop on computational intelligence (IWCI), pp 209–214
16. Kumar S, Kumari R, Sharma VK (2015) Fitness based position update in spider monkey optimization algorithm. Procedia Comput Sci 62:442–449
17. Panwar D, Tomar P, Singh V (2018) Hybridization of Cuckoo-ACO algorithm for test case prioritization. J Stat Manag Syst 21(4):539–546

Chapter 44
Dynamic Segment Array and Search

Ajay Kumar Mehta

1 Introduction

Dynamic Segment Array and Search (DSAS) is a complete form of Dynamic Segment array (DSA) and Dynamic Segment search (DSS). Dynamic Segment array (DSA) is a Segment base Array in which numerical data of similar type and length can be stored at run time.

Array is divided in multiple segment of equal size with an extra common reserve memory space. When data is initialized to array the data item stored in a specific segment of an array by applying a predefined condition to frequently changing digit (FCD) of number. The main advantage of this type of segmentation of array is minimization of data searching time. This Array can be initialized at run time (Table 1).

We scan from left side which digit is frequently changeable.

We find that,

In 5th position each number has digit 6 (i.e. fixed).

In 4th position numbers have digit 9, 4, 3 and 5 respectively (i.e. 1st frequently changeable digit (FCD) of numbers).

Hence, we get FCD place position = 4.

We scan from left side which digit is frequently changeable.

Now we find that FCD place position = 2.

Here, 5th position is fixed.

4th position is fixed.

3rd position is fixed.

We will use this FCD to storing data item into a specific array segment.

A. K. Mehta (✉)
Sri Balaji College of Engineering and Technology, Benad Road, Jaipur, India
e-mail: lasvi@rediffmail.com

© Springer Nature Singapore Pte Ltd. 2021

S. D. Purohit et al. (eds.), *Proceedings of International Conference on Communication and Computational Technologies*, Algorithms for Intelligent Systems,
https://doi.org/10.1007/978-981-15-5077-5_44

Example 1: Suppose we have group of 10 numerical data items to store in an Array as given below:

Table 1 Group of 10 numerical data items

54321	← Place Position of digits from left(i.e. FCD)
69432	
64782	
63900	
65640	
62514	
63584	
67895	
68521	
69852	
69231	

1.1 Dynamic Segment Array

An Array can be divided into multiple parts by using an appropriate algorithm. The main advantages of divide an array into multiple parts is

1. Data can be stored independently in a specific segment so that as and when a data is needed it can be retrieved by the user without searching whole array.
2. It decreases searching time up to 80% in worst cases (w.r.t. linear search).

We will divide an array into two parts:

i. Main Part
ii. Reserve Part.
i. Main Part: It is the major part of whole array it occupies 90% part of Max size of an array. This part is sub-divided into multiple segments (mostly 11 segment)
ii. Reserve Part: It is the Common part of all segments, when a segment completely filled by data items it transfers the control to reserve part of the array to store next data item for this segment. The logic behind prefixing of Dynamic word With Segment Array is initialization of this array Dynamically at run time, because this Array is most useful when we use a large volume of data such as roll number of students in any competitive examination.

Role of Frequently Changeable Digit Frequently Changeable Digit (FCD) play a key role to store data at a specific segment of Array. Each segment is labeled by integer value. Whole Array is divided into 11 segments in which first segment is labeled with integer 0, second segment is labeled with integer 1 and so on for 10 segments. 11th segment or last segment is reserve segment used as a common place for storing data after overflow of any particular segment. As per Table 2 if we want to insert 20 data in a Dynamic Segment Array the FCD will decide the proper label of segment.

Example 2: Suppose we have group of 20 numerical data items to store in an Array as given below:

Table 2 Group of 20 numerical data items

54321	**54321**	← Place Position of digits from left (i.e. FCD)
10042	10012	
10032	10092	
10002	10062	
10093	10083	
10023	10079	
10042	10062	
10039	10059	
10002	10002	
10073	10023	
10093	10003	

For Example: we have a Dynamic Segment Array(DSA) of MAXSIZE 22. Each segment will have MAXSIZE/2 memory location out of total 22 memory location of the array, this segment size is also denoted by segment_size. Size of Reserve part of the Array will also have 2 location e.g. reserve_part = maxsize-10*segment_size.

We have got the FCD place position = 2 for all the data of Table 2, Now we scan the data of Table 2 from left to right to store data in DSA.

First data is 10042, FCD indicating the place position of digits.

here Digit 4 is the FCD of the number, it will search label 4 to store in Dynamic Segment Array. The first location of segment 4 or label 4 is vacant now data will be stored at this location. Otherwise it searches vacant location at this label.

Next data is 10032 hence fcd is 3 data will be stored at 1st memory location of label 3.

Next data is 10002 hence fcd is 0 data will be stored at 1st memory location of label 0.

Next data is 10093 hence fcd is 9 data will be stored at 1st memory location of label 9.

Next data is 10023 hence fcd is 2 data will be stored at 1st memory location of label 2.

Next data is 10042 hence fcd is 4 data will be stored at 1st memory location of label 4.

Next data is 10039 hence fcd is 6 data will be stored at 2nd memory location of label 3.
Next data is 10002 hence fcd is 0 data will be stored at 2nd memory location of label 0.

And so on

3rd last data is 10002 hence fcd is 0 it should be stored at label 0 but segment 0 is completely full now data will be stored at 1st location of reserve part.

2nd last data is 10023 hence fcd is 2, data item will be stored at 2nd location of label 2.

Last data is 10003 hence fcd is 0, data will be stored at 2nd location of reserve part.

2 Dynamic Segment Search

This a searching technique used to retrieve a data item from Dynamic Segment Array searching is based on fcd of searching data. This technique is most useful when we are dealing with large volume of data items and where time complexity is important factor. For example, if we want to retrieve data item 10003 from Table 3, at first we find the fcd of searching data items which is 0, hence searching starts from 1st location of label 0 but data is not there now looks for 2nd location of same segment or label now data is also not found in segment 0. After search process complete in segment 0, The control is transferred to the reserve part of the Array. Search begins with the 1st location of the reserve part of Array data item is not found in this location now searches 2nd location of the reserve part of the Array. Here we get the required data item 10003; hence search is successful. This is the worst case because data item is found in last location of reserve part of the Array.

2.1 Time Complexity

Time Complexity [4] of any Searching algorithm [3] is very important it shows the overall efficiency of an algorithm. Now we consider Table 3 Dynamic Segment Search begins searches at 1st location of segment/label 0 it makes two comparison on segment 0 the time taken in searching of data item in segment 0 is $2/n$ and after segment 0 the searches begins in reserve part, there are also two comparison on reserve part, the searching is complete at 2nd comparison the time taken in reserve part is $2/n$. the data item is found at last location of reserve part of the array, thus this is the worst case for searching of data item. The total time taken in worst case is: Time taken in a particular segment (here segment/label 0) + time taken in reserve part.

i.e. Time taken in worst case: $T(n) = n/11 + n/11 + c$.

Table 3 Dynamic segment array

10002	Label 0
10002	
10013	Label 1
10012	
10023	Label 2
10023	
10032	Label 3
10039	
10042	Label 4
10043	
10059	Label 5
–	
10062	Label 6
10062	
10073	Label 7
10079	
10083	Label 8
–	
10093	Label 9
10092	
10002	Label reserve
10003	

or Time taken in worst case: $22/11 + 22/11 + c$.
or Time taken in worst case: $2n/11 + c$ or $4 + c$ (in this case).

Where c is constant value, stand for amount of time taken to find fcd or equal to number of digits in a data item.

In best case required data item will be found at 1st location of any segment the total time taken is:

$$T(1) + c \quad \text{or} \quad O(1) + c$$

In average case the data may be found anywhere between lower bound to upper bound of the array, hence total time taken is:

$$T(n/11)/ + c$$

Summary of time complexity The time complexity [5, 6] of dynamic search array is very less w.r.t. Linear search [4] for any unsorted numerical array the following table give the summary (Table 4).

Table 4 Time complexity of dynamic segment search

Time complexity	AT best case	AT average case	AT worst case
	$T(1) + c$	$T(n/11) + c$	$T(2n/11) + c$
	O(1)	O(n)	O(n)

Implementation The both technique Dynamic Segment Array and Dynamic Segment search can be implement by any programming language like C, C++, [7, 8] java, Python. I have used C++ to check the algorithm and code [9] is successfully executed in turbo C++ compiler on windows 7 professional operating system [10].

Important uses Both techniques (DSA and DSS) is interrelated, if DSA technique is used in any application to store the key element, then only DSS technique can access the key element. Other existed searching techniques [4] does not work on this Array.

This Dynamic Segment array can be used dynamically for store large volumes of values such as roll numbers of any competitive Examination. At the time of finding result of examination the Dynamic Segment Search technique can be used to retrieve the record very quickly.

3 Algorithm to Insert a Data in DSA

```
Step 1:   Begin
Step 2:   Declare:
          (A)  An ARRAY of MAXSIZE
                 (B)  Variable FCD
             SET SEGMENT_SIZE=(MAX SIZE)/11;
                SET RESERVE_SIZE=MAXSIZE-I_index*10;
                Initialize FCD Dynamically or Statically

Step 3:                    START LOOP-1

          (A)  INPUT DATA to store in ARRAY
               Check the Total   digits in number is >=FCD
                    START  LOOP-2
          (B)  Extract each digit of data and keep it in a small
               Array (i.e. ext_arr [])
          (C)  Repeat the LOOP-2 until each digit is stored in
               array
                       END of  LOOP-2
Step 4:
          (A)  Set the Condition statement by using FCD
                  Switch (ext_arr[FCD])
            (B)   Set the Case Label :( e.g. 0,1,2……9,10)
          (C)   Set the index/Lower Bound of an Array  for each
                case: (e.g. I=0,1*i_index,2*I_index,9*I _index, -
                     10*I_index)
          (D)   In matching Case label:
                        Check ,
                     IF I<=Segment _upper _limit
                        STORE the DATA in Segment by using
                        ARR[I]=DATA
               RETURN
                             ELSE
                          GOTO RESERVE PART  of ARRAY
                 CHECK ,IF I <MAX SIZE OF ARRAY
          Store DATA in ARRAY by using
                   ARR[I]=DATA
          UPDATE  I
          CHECK ,
                 IF I ==MAX
          PRINT OVERFLOW
           RETURN

Step 5:   Repeat step 3& 4 until all DATA are stored
                                   END of LOOP-1
Step 6:   END
```

4 Algorithm of Dynamic Segment Search

```
Step 1:  Begin
Step 2:
         (A) Read the KEY element
                           START  LOOP-1
         (B) By using LOOP extract each digit of
             KEY element
                            END LOOP-1
         (C) Check the Total digits of KEY element  must.
             not be smaller than FCD
Step 3:  (A) Select a particular array segment conjunction with
                unique case label utililizing FCD value as
                argument of conditional statement(i.e. switch
                mod[FCD])
         (B) SET I=lower Bound of Particular Segment
                              START LOOP –2
         (C) Traverse the array from Lower bound of segment to Upper bound of
             the  segment compare each data stored in this segment with KEY
             element
             by incrementing I each iteration
                IF,A MATCH IS FOUND
                PRINT “DATA”
                ELSE
             GOTO RESERVE_PART OF ARRAY
                               END LOOP-2
Step 4:  (A) SET I=lower Bound of Particular Segment
                              START LOOP -3
         (B) Traverse the array from Lower bound of RESERVE_PART to Up-
                             Per bound of RESERVE_PART.compare each
             data stored in this  Segment with KEY element, by
              incrementing I each iteration.
               IF,A MATCH IS FOUND
                PRINT “DATA”
               ELSE
             PRINT “SEARCH IS COMPLETED ITEM
                     NOT FOUND”
            Return -1
                               END LOOP -3
Step 6:     END
```

5 Conclusion

Before writing this thesis I studied Properties of array, sorting technique and searching algorithm, I observed that An array is always used sequentially to store data and search data which takes more time when we work with large volume of data.

To reduce searching time of data retrieval, many sorting algorithm is introduced, which takes extra time to sort an array. There are no any special algorithms which can retrieve the data item from unsorted Array in less time. My study concludes that each algorithm has its own advantage and disadvantage. After studying and considering above algorithms we underline a problem, that we cannot store data in a segment of array, if we success to divide a array in subgroups and put data on subgroup, then time complexity can be minimize to retrieve the data item from a particular segment of Array.

And finally I designed the algorithm by which we can store the data at a particular segment and also we can retrieve the data from a particular segment by keeping other segments out from the comparison.

Acknowledgements It is with great pleasure that I take this opportunity to bow my head in respect and gratitude for all those who helped us in making this research successfully.

I would like to express my sincere gratitude to supervisor Mr. Rahul Sharma for his Valuable support, comments and help throughout this work.

We wish to pay our sincerest gratitude to our esteemed teacher of CSE department SBCET, Jaipur and Mr. Amit Mishra Assistant professor of CS department of SBCET, Jaipur. For their invaluable support and impulsive suggestion for successful completion of our Report. No one can replace what he has done for mean the way he has helped me in bringing out this research. I will always be indebted to him all my long life.

We would like to give special thanks and heartily cheers to all my group member and fellows who are directly and indirectly involved in the project and were a source of inspiration and encouragement throughout the course of our Thesis preparation.

References

1. Samanta D (2009) Classic data structures, 2nd edn. Prentice Hall India Learning Private limited, Delhi
2. Pancholi H, Singhal R, Tulli S (2018) Data structure algorithm using C, 2nd edn., Genius Publication, Jaipur
3. Paira S, Chandra S, Aalam S, Patra S (2017) A brief study and analysis of different searching algorithms. In: 2nd ICECCT 2017, vol 4, Coimbatore, pp 944–948
4. LSA homepage. https://www.studytonight.com/data-structures/linear-search-algorithm
5. Best, worst and average case, homepage. https://en.wikipedia.org/wiki/Best,_worst_and_average_case
6. Pandey K, Pradhan N (2014) A Comparison and selection on basic type of searching algorithm in data structure. IJCSMC, vol 3, pp 751–758, Secunderabad
7. Shrivastava S, Shrivastava D (2011) Data structures through C in depth, 2nd edn., BPB Publications, Delhi
8. Langsam Y, Augenstein M, Tenenbaum A (2015) Data structures using C & C++, 2nd edn., Pearson Education India, Chennai
9. Knuth D (1998) The art of computer programming, 2nd edn. vol 3. Stanford University, Addison-Wesley

10. Saini GL, Deepak P, Vijander S (2014) Modeling, implementation and performance investigation of efficient-AODV routing protocol in MANET using data rate. IJERSTE 3(1): 76–82
11. Jain S, Singh S (2016) Data structures made simple. BPB Publication, Delhi

Chapter 45
A Comprehensive Survey and New Investigation on Sleep Disorder Detection Using EEG Signal

Santosh Kumar Satapathy and D. Loganathan

1 Introduction

Sleep is the primary function of the brain and plays a vital role in quality of life, good health, protection of mental and physical health, of human being throughout their life. Sleep is a reversible state in which the eyes are closed and several nervous system centers are inactive. Hence, sleep renders the individual either partially or completely unconscious and makes the brain a less complicated network. Nowadays, there are many people who suffer from sleep disorders like physiological and neurological disorders such as apnea, insomnia, and narcolepsy, which has great socioeconomic and public life concern. Humans spend around one-third of their lives sleeping, and conditions such as obstructive sleep apnea and insomnia are frequent and can severely affect physical health. Sleep analysis is used for diagnosis and treatment of such type of disorders. The analysis of sleep can be done by studying the electroencephalogram (EEG) signals, which is usually based on periods of 30 s length segments called epoch, and this method of sleep staging is also called polysomnography (PSG), which is a common procedure for the detection and diagnosis of sleep disorders. Sleep study is normally done based on one of two available standards: the Rechtschaffen and Kales (R&K) [1] and the new standard developed by the American Academy of Sleep Medicine (AASM) [2]. Human sleep is characterized by five stages such as awake (W), non-REM1, Non-REM2, Non-REM3, and rapid eye movement (REM) sleep. Characteristics of sleep are described as below.

S. K. Satapathy (✉) · D. Loganathan
Pondicherry Engineering College, Puducherry, India
e-mail: santosh.satapathy@pec.edu

© Springer Nature Singapore Pte Ltd. 2021

S. D. Purohit et al. (eds.), *Proceedings of International Conference on Communication and Computational Technologies*, Algorithms for Intelligent Systems,
https://doi.org/10.1007/978-981-15-5077-5_45

1.1 Wake Stage

The wake stage is characterized by low amplitude and mixed frequency of EEG signal. Generally, wake stage is seen in the beginning of the sleep and it can be defined as a transition stage from the full alertness state to the half-sleepy state. This stage consists of alpha and beta activities which are not similar to the slow activities seen in SWS, along with the eye movements and high muscle tone.

1.2 NREM Stage 1

In Stage 1 of NREM, the EEG signal has the highest amplitude nearly 50–70 μV with frequency ranges of 2–7 Hz. This stage acts as a transition stage between wakefulness and sleep, which usually lasts between 1 and 5 min. This stage consists of a low-voltage EEG tracing with well-defined alpha and theta activity. Stage N1 constitutes 2–5% of total sleep time and exhibits the absence of sleep spindles complexes, and REM sleep waves [3].

1.3 NREM Stage 2

N1 stage is followed by the Stage 2 which is also called as deeper sleep stage, which lasts for approximately 10–20 min. Stage N2 is act as a baseline of sleep, and it is characterized by the occurrence of sleep spindles and k-complexes. This relatively large waveform has duration of between 0.5 and 1.5 s duration and generally larger than 75 μV amplitude and a relatively low-voltage, mixed-frequency EEG background [4].

1.4 NREM Stage 3

Stage 3 is referred to a period during which at least 20% and not more than 50% of the sleep consists of EEG waves with frequencies of 2 Hz or smaller and amplitudes of more than 75 mV (delta waves). This stage normally appears only in the first one-third of the sleep episode, and it usually compromises 4–6% of whole sleep time. Stage 4 is similar to Stage 3, except that delta waves cover 50% or more of the record. Sleep Stage 4 usually represents 12–15% of total sleep time. Stages 3 and 4 together are also known as—deep sleep or, slow wave sleep (SWS), and this is the most restorative part of sleep. These two stages merely represent the same behavior; hence, according to the new standard developed by the American Academy of Sleep Medicine (AASM), both are combined to represent single stage as NREM 3.

1.5 REM Stage

REM is the sleep stage in which dreaming occurs and it constitutes up to 20–25% of total sleep. It is well characterized by the rapid eye movements under the closed eyelids and has low-voltage EEG patterns. In this sleep stage, the brain activity of human beings is reversed from Stage 4 to a Stage 1. The REM stage contains low-voltage, mixed-frequency EEG signal with saw-tooth wave-like pattern, EMG with low-like pattern, EMG with low amplitude, and high-amplitude EOG signal from both eyes [5].

Monitoring of these different stages through use of recorded signals from neural, respiratory, and cardiac activity during sleep can provide an evaluation of sleep in patients suffering from sleep disorders. An accurate and efficient automatic sleep stage classification system supports in order to improve the diagnosis of any type sleep related disorders. Nowadays, many biomedical signals such as EEG, ECG, EMG, and EOG offer useful details for clinical setups that are used in identifying sleep disorders with the EEG signal serving as the most effective signal in sleep stage classification [6].

Recently, many studies show that electroencephalogram (EEG) signals become widely used in sleep recording and sleep stage classification. It can be calculated by placing the electrodes either at C3-A2 or C4-A1 sites on the head of a subject and records the EEG signals from the brain, whereas classification of sleep stages of NREM and REM is mainly done by Rechtschaffen and Kales (R&K) rules to determine sleep characteristics [7, 8].

2 Electroencephalogram

The human brain is the command center for the human nervous system and that consists of millions of neurons interconnected by axons and dendrites. It receives signals from the body's sensory organs and outputs information to the muscles. The human brain has the same basic structure as other mammal brains but is larger in relation to body size than any other brains.

According to [7–9], the human brain is one of the largest and most complex organs in the human body. The brain is made up of many specialized areas that work together. The brain can be divided into three primary structures: the cerebrum, cerebellum, and brain stem. The brain is also divided into several lobes as:

- The frontal lobes are responsible for problem solving and judgment and motor function.
- The parietal lobes manage sensation, handwriting, and body position.
- The temporal lobes are involved with memory and hearing.
- The occipital lobes contain the brains visual processing system.

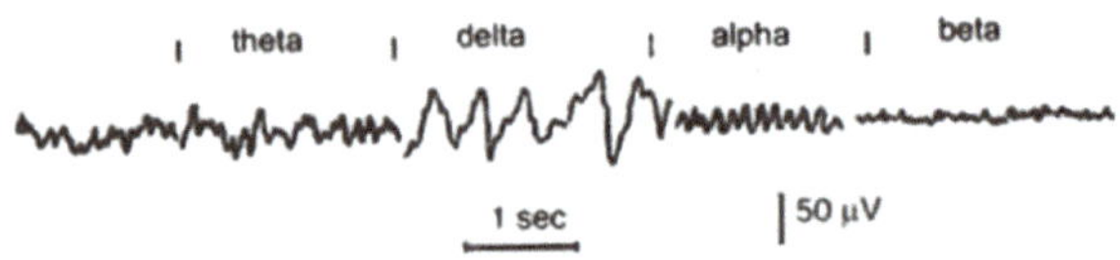

Fig. 1 Illustration of the sleep states of theta, delta, alpha, beta rhythms, respectively

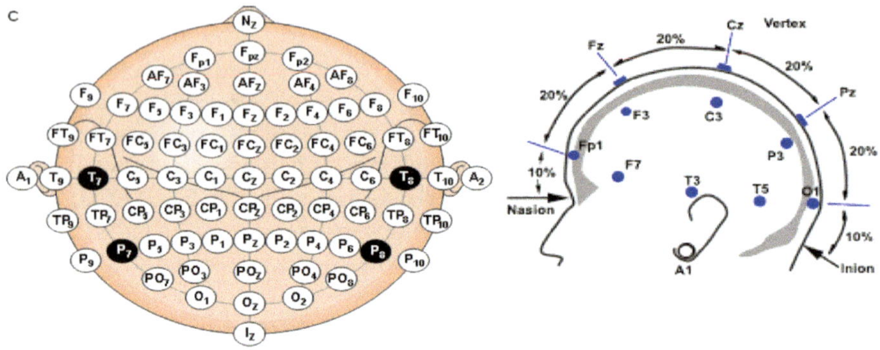

Fig. 2 EEG electrode placement system

Various methods have been developed to measure the signal activity generated in the human brain, including EEG, MEG, functional magnetic resonance imaging (f-MRI), and function near-infrared spectroscopy (fNIRS), and position emission tomography (PET) [10–13]. Out of all these signals, EEG is one of the strongest biomedical signals that has significance and high practical value for applications in clinical neurology [14]. EEG is a non-invasive method for measuring the electrical activity of the cerebral cortex [15–18].

The recorded EEG signals are characterized by frequency, patterns, and amplitude. The information related to sleep states in EEG mainly focused on rhythm (1–4 Hz), rhythm (4–8 Hz), rhythm (8–13 Hz), and rhythm (14–30 Hz) [1] (Fig. 1).

EEG is measured through multiple electrodes that are placed in different scalp at multiple locations according the international 10/20 placement system [19–23]. This system was extended to the 10/10 system with more electrodes since the multi-channel EEG hardware system systems have been developed [24, 25, 26] (Fig. 2).

3 Review of Existing Research

Radha et al. [27] considered six different EEG signals and various features, such as time domain, frequency domain, and nonlinear features. In this work, additionally SVM and random forest (RF) were considered for classification of five sleep stages. The important work of this research work was to develop an online algorithm for an automatic sleep stage classification using a single EEG channel. The results showed that best performance occurred by considered frontal EEG signals. The authors found the accuracy results of 6-stages, 5-stages, 4-stages, 3-stages, and 2-stages as 85.57%, 86.53%, 87.49%, 89.77%, and 95.05%, respectively.

Hassan and Bhuiyan [28] proposed a single-channel-EEG-based method for sleep stage scoring with using complete ensemble empirical mode decomposition (EMD). In this study, the author has used bagging technique which was applied to classify the sleep stages. This work achieved an accuracy of 86.89%, 90.69%, 92.14%, 94.10%, and 99.48% for 6-Class, 5-Class, 4-Class, 3-Class, and 2-Class, respectively.

Lan et al. [29] proposed multi-channel PSG data to develop an automatic classification of sleep states. In this research, the authors focused to extract the spectral features based on fast Fourier transform (FFT). In this work, for classification of the sleep stages, a rule-based DT classifier achieved an accuracy of 84%.

In [30], the authors proposed a work to distinguish between the drowsy and wake states through EEG signals. In this proposed work, authors suggested three channels of EEG waveforms. The results of the drowsiness detection approach indicated a high accuracy and precision of 98.01% and 97.91%, respectively.

Zhu et al. [31] developed sleep stage scoring system based on single channel of EEG signal, and also used a multi-class SVM as the classifier. In this approach, the authors have extracted nine graph domain features from visibility graph and a horizontal visibility graph. The accuracy of the classification of six sleep stages achieved 87.5% using the SVM classifier.

Hsu et al. [32], in this work, have classified five sleep stages by considering on six energy features from a single EEG channel. In this study, the authors used Elman recurrent neural classifier and achieved classification accuracy as 87.2%.

In [33], the authors proposed PSG signals, from which he was extracted time and frequency domain that included two EEG channel, two EOG channels, and one EMG channel for automatic sleep stage scoring. In this proposed work, the authors has used a Dendrogram-SVM (DSVM) and achieved accuracy as 92%.

Krakovská and Mezeiová [34] addressed an ASSC technique, where he has extracted the 14 features of PSG signals and used 6 EEG channels, 2 EOG channels, and 1 EMG channel, and in this study, the authors has used the quadratic discriminant analysis for classification of sleep stages and the results confirmed an accuracy of 81%.

In [35], the authors have considered the single-channel EEG signal; in this approach, the researchers obtained from a total of 39 features and it is related to time domain and frequency domain. In this study for classification purpose, binary SVM is used for classification of sleep stages and the algorithm was capable of achieving an accuracy of 95.88%.

In the study of Phan et al. [36], a single-channel of EEG data was introduced and additionally a machine learning approach was adopted. In this study, the researchers used a two-step classification based on the KNN method to first classify wake/sleep and then four sleep stages consisting of wake, Stage 1 + REM, Stage 2 and SWS. The achieved accuracy was 98 and 95%.

In [37], the authors have used a single-channel EEG signal and extracted autoregressive coefficient from the EEG signal to classify both sleep and wake stages with an accuracy of 91%. In this study for classification, a partial least squares regression (PLSR) classifier is introduced.

In [38], the researcher has proposed a FCM algorithm which was used to detect sleep disorder, which was able to distinguish the wake stage as well as Stages 1–4 and REM sleep stages, using single-channel EEG signals. The results showed that the algorithm could provide an accuracy of 92%.

Cic et al. [39] used EMD method from that generated time-frequency features and applying the generalized zero crossing (GZC) method on the obtained intrinsic mode functions (IMF) for sleep stage classification. In this study, the authors have used a single EEG channel, and for classification purpose, SVM classifier is used. The approach is performed with an overall 90% accuracy.

Chen et al. [40] proposed another sleep stage classification study to estimate different sleep stages including wake stage, Stage 1 and Stage 2. In this study, four recorded EEG signals were used. The proposed method achieved 80% accuracy, and for classification of sleep stages, hopfield neural network (HNN) classifier was used.

Another sleep stage classification study was conducted by Aboalayon et al. [41] proposed to differentiate between awake and sleep Stage 1. In this work, SVM classifier is used for classification between human awake from Stage 1. The experimental results indicate that the proposed methodology achieves the average classification accuracy of 92.5%.

Mora et al. [42] reported that an efficient differentiation of human sleep stages is very important for the treatment of sleep disorders such as apnea, insomnia, and narcolepsy.

Doroshenkov et al. [43] have developed a classification algorithm based on hidden Markov models using only EEG signals.

Tzimourta et al. [44] have developed an EEG-based method in which 5-class automatic sleep staging is represented classification of sleep stages. FIR filters are applied to the EEG signals, extracting the EEG rhythms. The results showed that random forest classifiers achieved best accuracy 75% in case of classification of different sleep stages.

4 Signal Analysis Methods

Various methods are used for acquiring EEG signals. They are differing in the way the electrodes placed. The methods can be categorized as follows:

1. Invasive: Invasive EEG recordings are those recordings that are captured with electrodes that are placed on the surface or within the depth of the brain [45].
2. Non-invasive: Non-invasive EEG recordings are those that are captured with electrodes that are placed on the scalp rather than placing it on the surface or within the depth of the brain.

4.1 Signal Preprocessing

In biomedical signal processing, it is quite difficult to determine the noise, artifacts, and any other trends present in the raw signals, and before its further processing toward analysis of feature extraction, we need to reduce all these noise part from the raw signal. Normally, EEG recordings have a wide variety of artifacts, from that some due to some technical origin and others have a physiological origin which mixed with the brain signals [46]. For that reason, we need to use preprocessing methods before feature extraction from EEG signal analysis without losing any important relevant information [47]. As per the different review on literature on sleep stage detection reported that employ of preprocessing techniques before extracting the features from the signal. In some cases selective digital filters like low-pass, high-pass, band-pass, and band-stop filters have been widely used in artifact detection and removal.

4.2 Feature Extraction

Feature extraction is the most important procedure for any type EEG-related analysis. As per our literature survey, normally sleep stage detection schemes are employed feature extraction algorithms. As per the literature survey, the sleep stage-related features can be meticulously categorized into four main groups: time, frequency, time-frequency domain, and non-linear features. As per our literature survey of different contributed research work, most of the authors have used the methods for feature extraction as (i) standard statistics, (ii) non-parametric statistics of the spectral domain, and (iii) wavelet transforms [48, 49].

4.3 Feature Selection

In the process of feature selection, we are trying to avoid feature redundancy and noisy data that affect the classification process and also try to reduce the high dimensional feature vector representation to low vector representation. To achieve this, we have applied dimensionality reduction and feature selection algorithms and generate the most discriminate features and try to keep the dimensionality of features as low as possible [50]. To give the best performances, a feature selection algorithm is applied to reduce the complexity of the classifier and select the best of significant features and those features are input to the classifier [51]. As our survey, most of the researchers used the principal component analysis and sequential selection methods for selecting the features for classifying [52]. In our experimental work, we have used the sequential selection method for selecting the features for classifying.

4.4 Classification

Based on our review on different previous researches on ASSC, most of the authors use the different classification algorithms. Approximately 34% of the ASSC methods use classification schemes that are based on SVM classifiers, 19% based on NN classifiers, 12% based on LDA, 9% based on KNN, 5% based on DT, 5% based on NB classifiers, and less than 5% based on other types, such as HMM, Adaboost, Bagging, and K-means.

5 Discussion and Results

In this study, we have considered two channel of EEG signal as F3-A2 and O1-A2. The data used for this evaluation of the proposed scheme are taken from the Sleep Medicine Centre of the Hospital of Coimbra University (CHUC), which is publicly available; basically, this dataset is named as ISRUC-SLEEP and that was created for sleep researchers. This dataset recorded information from human adults including healthy subjects and subjects with sleep disorder with under medication [53]. This EEG signal were filtered and segmented 10 s epochs. In this study, we have tried to differentiate between wake stage and Stage 1 of sleep; for each subject, we have recorded 8 h recoding throughout the night; in this time duration both the channels recorded brain signal activities from subject; for each subject, we have extracted 850 epochs and 6000 samples. In this work, we have computed to estimate the power spectral density (PSD) from the recorded 10 s epochs and work we have estimate the power spectrum of the 10 s epoch with help of Welch's method and the period gram method or study is based on the available guidelines for sleep stage classification, the AASM standard, which includes five-state sleep stages: wake, Stage 1, Stage 2, Stage 3 and REM. In this study, we have considered 10 patients and their one recording to be considered. From each channel, we have considered 23 features and one response value. The chosen classifier applied to this proposed method in this study includes as SVM, KNN, DT, and ENSEMBLE which are among the most important classifiers which was also earlier used in automatic sleep stage; in this research work, we have input the two acquired channel like F3-A2 and O1-A2, and for each classifier, the overall accuracy is mentioned in Table 1 (Figs. 3, 4, 5 and 6).

Table 1 Results of sleep staging for the 4 classifiers in terms of accuracy

	SVM (%)	DT (%)	KNN (%)	ENSEMBLE (%)
F3-A2 channel				
Accuracy	71.8	64.3	59.8	71.8
O1-A2 channel				
Accuracy	57.8	57.7	57.7	53.8

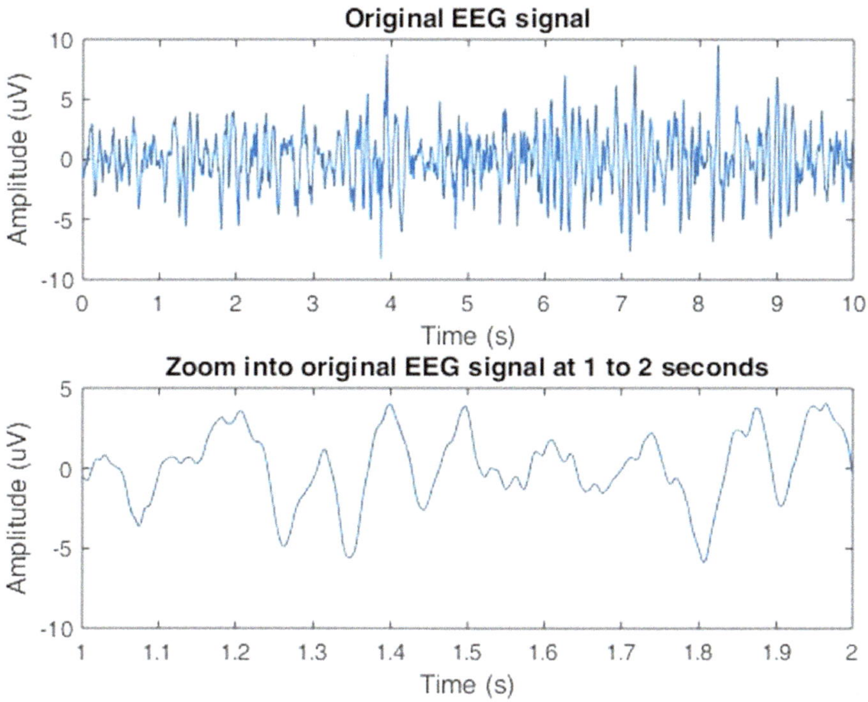

Fig. 3 Original input EEG signal

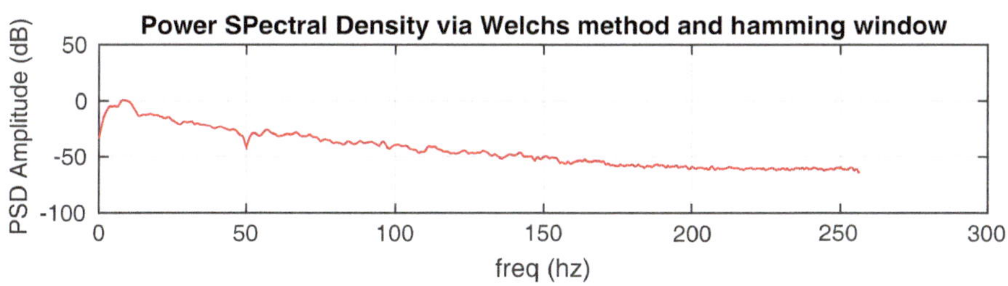

Fig. 4 PSD of a EEG signal using Welch method

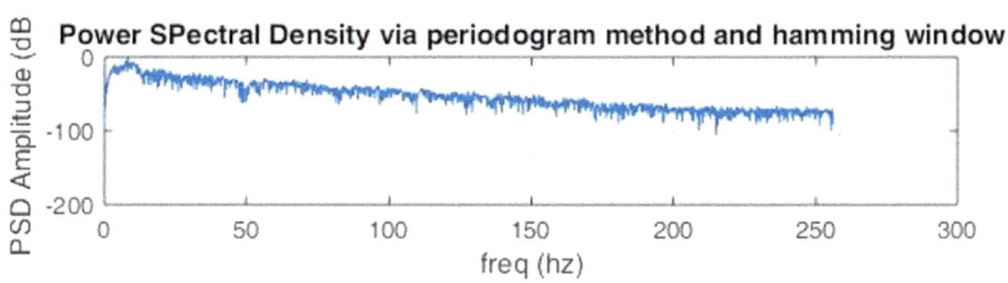

Fig. 5 PSD of a EEG signal using period gram method

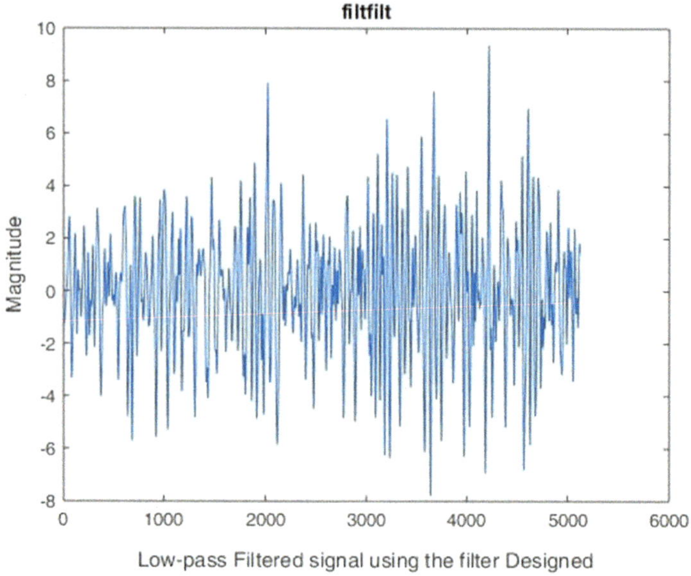

Fig. 6 Filtered EEG signal

The best classification accuracy obtained from F3-A2 channel through SVM and Ensemble classifiers achieved with 71.8% to differentiate between wake stage and Stage 1 of sleep. Similarly for O1-A2 channel, we have achieved best accuracy from SVM classifier with 57.8% in related to differentiate between wake stage and Stage 1 of sleep (Figs. 7 and 8).

6 Conclusion

This work provided a brief survey on sleep studies and a comprehensive survey in related to identify the different sleep stages from EEG channel and subchannel from different frequency sub-bands though automatic manner. The ASSC analysis has been divided into three essential parts including: feature extraction, feature selection, and classification. This research gives a brief idea to the researchers in subject to signal processing and analysis and discusses their performances and efficiency. On these selected features, base classifier such as SVM, DT and KNN algorithms are applied. According to the result given in classification Table 1, it could be inferred that SVM and Ensemble of classification models generates highest classification result as compared to other base classifiers.

From Table 1, we can say that our SVM and Ensemble method gives 71.8% average accuracy for F3-A2 channel which is better than all other two base classifier for the same data set and for O1-A2 channel; we can say that SVM method gives

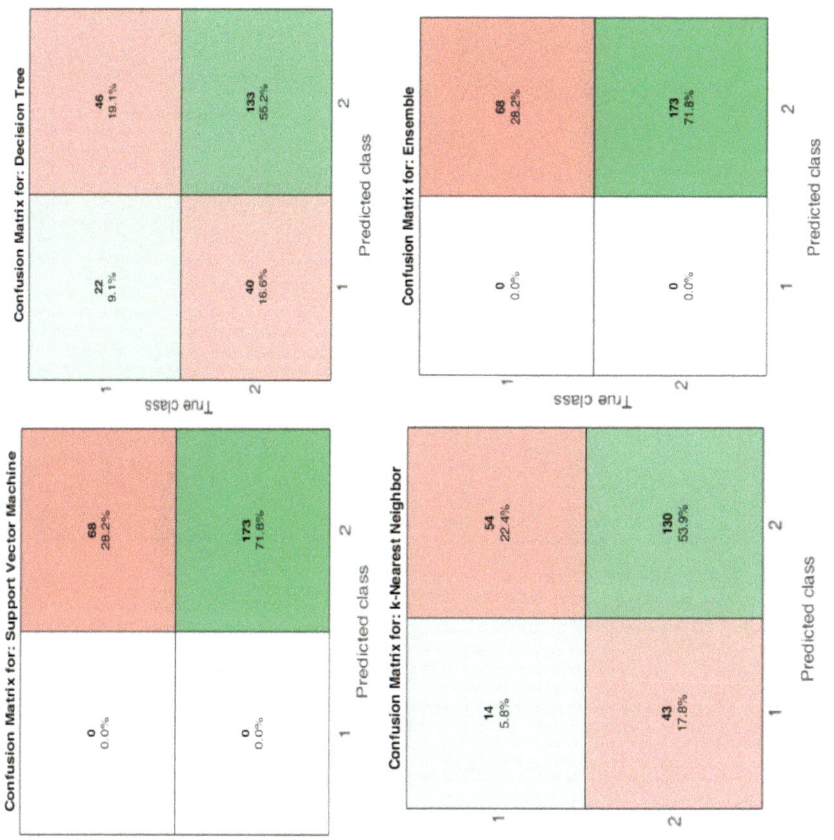

Fig. 7 Confusion matrix for F3-A2 channel

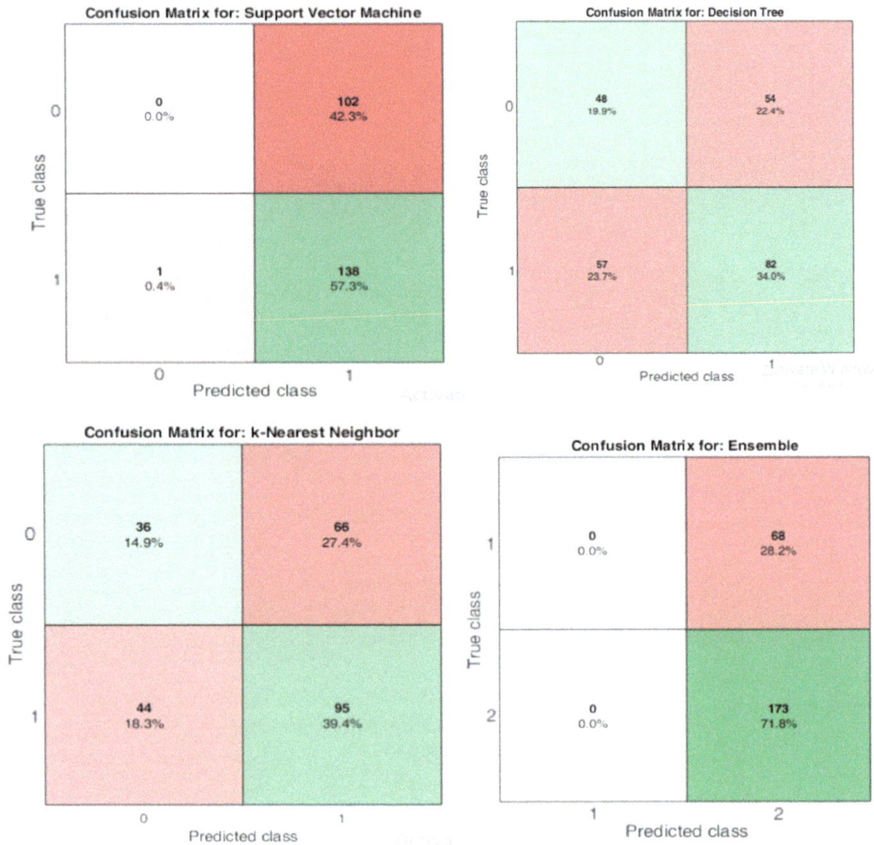

Fig. 8 Confusion matrix for O1-A2 channel

57.8% average classification accuracy in order to differentiate between wake stage and Stage 1 of sleep.

Therefore, our simpler and more feasible scheme makes our approach more attractive for easy implementation in any embedded microcontroller device to identify certain patterns of various sleep disorders in near real time.

In our study, we have only used two EEG channel F3-A2 and O1-A2 for experimentation, in which we have observed that somehow F3-A2 channel accuracy more in related to classification between wake and Stage 1 of sleep; in future, we will apply more channels of EEG signal and again we will improve the classification accuracy in regarding automatic sleep stage classification.

References

1. EEG signals using fractal dimension: a nonlinear approach (2014) In: Proceedings of the IEEE international conference on smart instrumentation, measurement and applications (ICSIMA), Kuala Lumpur, 25–25 Nov 2014, pp 1–5
2. Rechtschaffen A, Kales A (1968) A manual of standardized terminology, techniques and scoring system for sleep stages of human subjects. No. 204. National Institutes of Health Publications; Brain Information Service/Brain Research Institute, Los Angeles; and U.S. Government Printing Office, Washington
3. Iber A-IS, Chesson A, Quan SF (2007) The AASM manual for the scoring of sleep and associated events: rules, terminology, and technical specifications, 1st edn. American Academy of Sleep Medicine, Westchester
4. Sanei S, Chambers J (2007) EEG signal processing. Wiley, England. ISBN-13 978-0-470-02581-9
5. Erdamar A, Duman F, Yetkin S (2012) A wavelet and teager energy operator based method for automatic detection of K-complex in sleep EEG. Expert Syst Appl 39:1284–1290
6. Cvetkovic D, Übeyli ED, Cosic I (2008) Wavelet transform feature extraction from human PPG, ECG, and EEG signal responses to ELF PEMF exposures: a pilot study. Digit Signal Process 18:861–874
7. Ebrahimi F, Mikaeili M, Estrada E, Nazeran H (2008) Automatic sleep stage classification based on EEG signals by using neural networks and wavelet packet coefficients. In: 30th annual international conference of the IEEE engineering in medicine and biology society, 2008. EMBS 2008, 20–25 Aug 2008, pp 1151, 1154
8. Khalifa W, Salem A, Roushdy M, Revett K (2012) A survey of EEG based user authentication schemes. In: Proceedings of the 8th international conference on informatics and systems (INFOS), Cairo, 14–16 May 2012
9. Ahmad RF, Malik AS, Kamel N, Amin H, Zafar R, Qayyum A, Reza F (2016) Discriminating the different human brain states
10. Aqil M, Hong K-S, Jeong M-Y, Ge SS (2012) Detection of event-related hemodynamic response to neuroactivation by dynamic modelling of brain activity. Neuroimaging 63:553–568
11. Hu X-S, Hong K-S, Ge SS (2012) FNIRS-based online deception decoding. J Neural Eng 9:26012–26019
12. Naseer N, Hong K-S (2015) FNIRS-based brain-computer interfaces: a review. Front Hum Neurosci 9:3
13. Khan MJ, Hong K-S (2015) Passive BCI based on drowsiness detection: an fNIRS study. Biomed Opt Express 6:4063–4078
14. Liu Y, Yan L, Zeng B, Wang W (2010) Automatic sleep stage scoring using Hilbert–Huang transform with BP neural network. In: Proceedings of the 4th international conference on bioinformatics and biomedical engineering (iCBBE), Chengdu, 18–20 June 2010, pp 1–4
15. Rao R, Derakhshani R (2005) A comparison of EEG pre-processing methods using time delay neural networks. In: Proceedings of the 2nd IEEE EMBS international conference on neural engineering, Arlington, 16–19 Mar 2005, pp 262–264
16. Roy V, Shukla S (2015) A survey on artifacts detection techniques for electro-encephalography (EEG) signals. Int J Multimed Ubiquitous Eng 10:425–442
17. Mantri S, Patil V, Mitkar R (2012) EEG based emotional distress analysis—a survey. Int J Eng Dev 4:24–28
18. Padmanabh ML, Shastri RK, Biradar D (2015) EEG signal processing techniques for mental task classification. Int J Adv Comput Electron Technol 2:66–73
19. Cić M, Šoda J, Bonković M (2013) Automatic classification of infant sleep based on instantaneous frequencies in a single-channel EEG signal. Comput Biol Med 43:2110–2117
20. See AR, Liang C-K (2011) A study on sleep EEG using sample entropy and power spectrum analysis. In: Proceedings of the defence science research conference and expo (DSR), Singapore, 3–5 Aug 2011, pp 1–4

21. Gorur D, Halici U, Aydin H, Ongun G, Ozgen F, Leblebicioglu K (2002) Sleep spindles detection using short time fourier transform and neural networks. In: Proceedings of the international joint conference on neural networks, Honolulu, 12–17 May 2002, pp 1631–1636
22. Abeyratne UR, Swarnkar V, Rathnayake SI, Hukins C (2007) Sleep-stage and event dependency of brain asynchrony as manifested through surface EEG. In: Proceedings of the 29th IEEE annual international conference of the engineering in medicine and biology society, Lyon, 22–26 Aug 2007, pp 709–712
23. Huang L, Sun Q, Cheng J (2003) Novel method of fast automated discrimination of sleep stages. In: Proceedings of the 25th IEEE annual international conference of the engineering in medicine and biology society, Cancun, 17–21 Sept 2003, pp 2273–2276
24. Jurcak V, Tsuzuki D, Dan I (2007) 10/20, 10/10, and 10/5 systems revisited: their validity as relative head-surface-based positioning systems. Neuroimage 34:1600–1611
25. Agrawal SS, Mane VM (2009) Bio-medical electronics with oral question and answers. Nirali Prakashan
26. Berry RB, Wagner MH (2014) Sleep medicine pearls. Elsevier Health Sciences
27. Radha M, Garcia-Molina G, Poel M, Tononi G (2014) Comparison of feature and classifier algorithms for online automatic sleep staging based on a single EEG signal. In: Proceedings of the 36th IEEE annual international conference of engineering in medicine and biology society, Chicago, 26–30 Aug 2014, pp 1876–1880
28. Hassan AR, Bhuiyan MIH (2016) Computer-aided sleep staging using complete ensemble empirical mode decomposition with adaptive noise and bootstrap aggregating. Biomed Signal Process Control 24:1–10
29. Lan K-C, Chang D-W, Kuo C-E, Wei M-Z, Li Y-H, Shaw F-Z, Liang S-F (2015) Using off-the-shelf lossy compression for wireless home sleep staging. J Neurosci Methods 246:142–152
30. Yu S, Li P, Lin H, Rohani E, Choi G, Shao B, Wang Q (2013) Support vector machine based detection of drowsiness using minimum EEG features. In: Proceedings of the IEEE international conference on social computing (SocialCom), Alexandria, 8–14 Sept 2013, pp 827–835
31. Zhu G, Li Y, Wen PP (2014) Analysis and classification of sleep stages based on difference visibility graphs from a single-channel EEG signal. IEEE J Biomed Health Inform 18:1813–1821
32. Hsu Y-L, Yang Y-T, Wang J-S, Hsu C-Y (2013) Automatic sleep stage recurrent neural classifier using energy features of EEG signals. Neurocomputing 104:105–114
33. Lajnef T, Chaibi S, Ruby P, Aguera P-E, Eichenlaub J-B, Samet M, Kachouri A, Jerbi (2015) Learning machines and sleeping brains: automatic sleep stage classification using decision-tree multi-class support vector machines. J Neurosci Methods 250:94–105
34. Krakovská A, Mezeiová K (2011) Automatic sleep scoring: a search for an optimal combination of measures. Artif Intell Med 53:25–33
35. Koley B, Dey D (2012) An ensemble system for automatic sleep stage classification using single channel EEG signal. Comput Biol Med 42:1186–1195
36. Phan H, Do Q, Do T-L, Vu D-L (2013) Metric learning for automatic sleep stage classification. In: Proceedings of the 35th IEEE annual international conference of the engineering in medicine and biology society (EMBC), Osaka, 3–7 July 2013, pp 5025–5028
37. Kayikcioglu T, Maleki M, Eroglu K (2015) Fast and accurate PLS-based classification of EEG sleep using single channel data. Expert Syst Appl 42:7825–7830
38. Obayya M, Abou-Chadi F (2014) Automatic classification of sleep stages using EEG records based on fuzzy c-means (FCM) algorithm. In: Proceedings of the 31st national radio science conference (NRSC), Cairo, 28–30 Apr 2014, pp 265–272
39. Cić M, Šoda J, Bonković M (2013) Automatic classification of infant sleep based on instantaneous frequencies in a single-channel EEG signal. Comput Biol Med 43:2110–2117
40. Chen X, Wang B, Wang X (2013) Automatic sleep stage classification for daytime nap based on hopfield neural network. In: Proceedings of the Chinese control and decision conference (CCDC), Guiyang, 25–27 May 2013, pp 2671–2674
41. Aboalayon KAI, Ocbagabir HT, Faezipour M (2014) Efficient sleep stage classification based on EEG signals. IEEE. ISBN 978-1-4577-1343-9/12

42. Mora AM, Fernandes CM, Herrera LJ, Castillo PA, Merelo J-J, Rojas F, Rosa AC (2010) Sleeping with ants, SVMs, multilayer perceptrons and SOMs. In: 2010 10th international conference on intelligent systems design and applications (ISDA), 29 Nov 2010–1 Dec 2010, pp 126, 131
43. Doroshenkov L, Konyshev V, Selishchev S (2007) Classification of human sleep stages based on EEG processing using hidden Markov models. Biomed Eng 41(1):25–28
44. Tzimourta KD, Tsilimbaris A, Tzioukalia K, Tzallas AT, Tsipouras MG, Astrakas LG, Giannakeas N (2018) EEG-based automatic sleep stage classification. Biomed J Sci Tech Res (BJSTR) 7(4)
45. https://my.clevelandclinic.org/health/diagnostics/17144-invasive-eeg-monitoring
46. Motamedi-Fakhr S, Moshrefi-Torbati M, Hill M, Hill CM, White PR (2014) Signal processing techniques applied to human sleep EEG signals—a review. Biomed Signal Process Control 10:21–33
47. Rao R, Derakhshani R (2005) A comparison of EEG preprocessing methods using time delay neural networks. In: Proceedings of the 2nd IEEE EMBS international conference on neural engineering, Arlington, 16–19 Mar 2005, pp 262–264
48. Motamedi-Fakhr S, Moshrefi-Torbati M, Hill M, Hill CM, White PR (2014) Signal processing techniques applied to human sleep EEG signals—a review. Biomed Signal Process Control 10:21–33
49. Lakshmi MR, Prasad DT, Prakash DVC (2014) Survey on EEG signal processing methods. Int J Adv Res Comput Sci Softw Eng 4:84–91
50. Lan T (2011) Feature extraction feature selection and dimensionality reduction techniques for brain computer interface. Ph.D. thesis, Oregon Health & Science University, Portland
51. Koley B, Dey D (2012) An ensemble system for automatic sleep stage classification using single channel EEG signal. Comput Biol Med 42:1186–1195
52. Khatwani P, Tiwari A (2013) A survey on different noise removal techniques of EEG signals. Int J Adv Res Comput Commun Eng 2:1091–1095
53. Sirvan K, Sousa T, Santos JM, Nunes U (2016) ISRUC-sleep: a comprehensive public dataset for sleep researchers. Comput Methods Programs Biomed 124:180–192

Chapter 46
Security Issues in Cloud Computing: A Review

Alok Bansal, Amit Kumar Bairwa, and Saroj Hiranwal

1 Introduction

Cloud computing is a model that empowers everywhere, provides favorable, on-demand system access to a common pool of configurable assets (e.g. systems, applications, administrations, servers, and capacity), quickly delivery [1]. Security issues and difficulties then emerged since there is an absence of mindfulness while the clients are utilizing the cloud administration that is given by the cloud specialist co-op like AWS, Google Platform, IBM Azure, MongoDB [2]. To accomplish five fundamental attributes of distributed computing, it gives three sorts of administration models: Software as a Service (SaaS), Platform as a Service (PaaS), and Infrastructure as a Service (IaaS) [3].

Cloud computing satisfies five characteristics; they are: 1. on-demand service, 2. resource pooling, 3. access network, 4. elasticity, and 5. measured service. The rapidly expanding world of cloud computing is rife with 'SaaS', 'PaaS', and 'IaaS' which are three of the most commonly used acronyms in the cloud sector. These three terms distinguish between the primary cloud computing service models: 'Software as a Service', 'Platform as a Service', and 'Infrastructure as a Service' (Fig. 1).

A. Bansal (✉) · A. K. Bairwa · S. Hiranwal
Rajasthan Institute of Engineering and Technology, Jaipur, India
e-mail: alok241986@gmail.com

S. Hiranwal
e-mail: ersaroj_hiranwal@rediffmail.com

A. K. Bairwa
Manipal University Jaipur, Rajasthan, India
e-mail: amitbairwa@gmail.com

© Springer Nature Singapore Pte Ltd. 2021

S. D. Purohit et al. (eds.), *Proceedings of International Conference on Communication and Computational Technologies*, Algorithms for Intelligent Systems,
https://doi.org/10.1007/978-981-15-5077-5_46

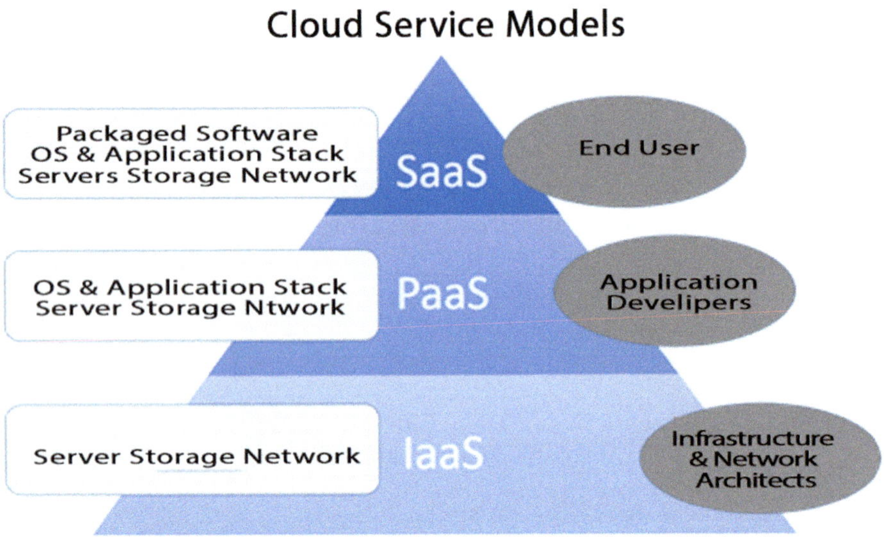

Fig. 1 Cloud computing service models

1.1 IaaS (Infrastructure as a Service)

Lowest-level cloud service paradigm is IaaS; pre-configured hardware resources are provided to users through a virtual interface by IaaS. IaaS can provide extra storage for corporate data backups, network bandwidth for a company Web site server, or it can even enable access to high power computing. Popular IaaS offerings like Amazon EC2, IBM SoftLayer, and Google's Compute Engine (GCE) are silently powering a huge portion of the backbone of the Internet (Fig. 2).

1.2 PaaS (Platform as a Service)

PaaS is a cloud service model where the cloud is used to deliver a platform to users from which they can develop, initialize, and manage applications. PaaS is sometimes called 'middleware', referring to how it conceptually sits between SaaS and IaaS. Platforms like Google's App Engine, Apache's Stratos, and IBM BlueMix are popular PaaS.

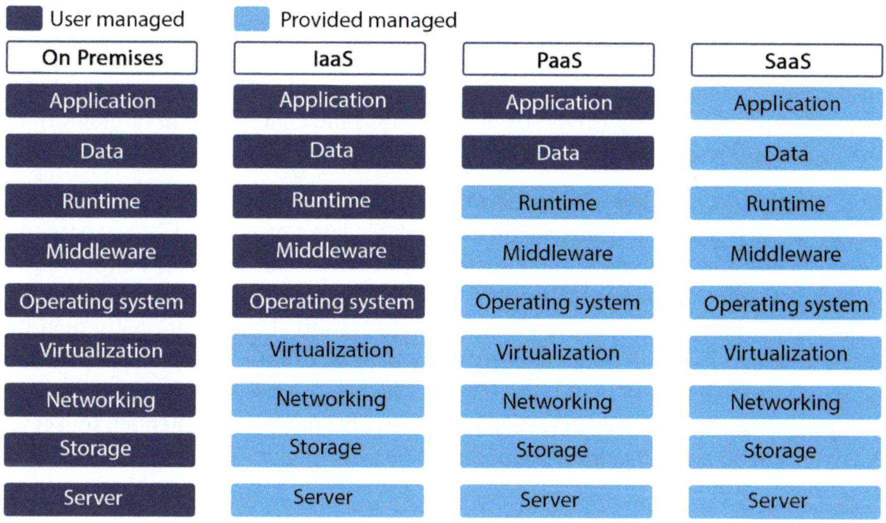

Fig. 2 Difference between IaaS, PaaS, and SaaS platforms

1.3 SaaS (Software as a Service)

'On-demand software', SaaS, is a software licensing and delivery model, fully functional and complete software product which is delivered to users over the Web on a subscription basis. SaaS offerings are typically accessed by end users through a Web browser. Popular products like Office 365 and Salesforce have thrust SaaS offerings to the many organizations.

2 Issues and Challenges

There are various issues in cloud computing because it incorporates extremely advanced components e.g. systems, databases, working systems, transactions to the executives, virtualization, asset booking, simultaneous control, and memory. It is important to take all the issues seriously because the cloud solution provider must guarantee that clients are not facing any major issues like information theft and loss. Which may cause an unusual loss relying upon the affectability of the information placed elsewhere in the cloud. There are few issues to be examined.

Network and Data Security in distributed computing have a few features, for example, information secrecy, accessibility, trustworthiness, reinforcement, and debacle recuperation.

Data Confidentiality: It is a key issue to be viewed as when redistributing profoundly delicate information to the cloud. Approaches ought to be set up to keep unapproved clients from inducing anything from the data being put away in the cloud database. Data privacy is accomplished through encryption of data. Different encryption calculations have been proposed, for example, Rivest Shamir Adelman (RSA), Triple Data Encryption Standard (3DES), and homomorphic encryption. In homomorphic encryption real issues are it's computational complexity and higher cost.

Data Availability: A few circumstances exist in which information accessibility cannot be ensured. For instance, in unavoidable circumstances like catastrophic events, it is critical to know whether the information can be used, confirmed, or recouped by the information proprietors. The cloud clients must know about the safety efforts being taken by the cloud specialist co-op. They additionally read the print of the service-level agreement (SLA) which went into with the specialist co-ops. High accessibility in cloud foundations can be accomplished by planning deficiency tolerant cloud frameworks. Cloud frameworks ought to be intended for server disappointment, zone disappointment, and cloud disappointment.

Data Integrity: Information Integrity can be defined as the trustworthiness and dependability of the information. It's one of the key issues in cloud-based systems. In distributed computing, the respectability of information stockpiling is an essential and significant prerequisite. The respectability of information demonstrates its consistency, normality, and legitimacy. At the point when cloud administration necessities increment, the CSP may need to scale up their capacity frameworks and this may prompt high shot of information misfortune, information defilement, circle disappointment, hub disappointment, or equipment disappointment. A cloud inspector freely assesses cloud administrations and cloud framework. Occasional outsider evaluating systems could be started to confirm information uprightness.

Backup and Disaster Recovery: It is fundamental for CSPs to give a reinforcement and debacle recuperation plan for information security, recuperation, flexibility of server farm, information accessibility and to permit business congruity after system or framework disappointments. The cloud clients must be educated about the backup type and prerequisites and what catastrophe recuperation plan is accessible.

2.1 Answers for Cloud Security Issues

Cloud computing enables the client to effortlessly recoil the activity and furthermore help to spare expense. Be that as it may, with the expanded reception rate of the cloud administration, the security issues and hazard have been expanded also [4]. So as to make distributed computing a superior alternative to build the client stockpiling limit and spare their private data safely, there are not many arrangements.

1. **Trusted cloud service provider**

Most of the cloud solution providers have highly effective methodologies to protect the information in the cloud. Entrenched and experienced cloud specialist co-op is increasingly trusty and better decision. Plus, the gauges and guidelines of the cloud specialist organization are likewise significant. Instances of believed mists specialist organizations are Amazon Web Services (AWS), IBM, Google, and Microsoft [5]. They asserted that the application analyzes the expense and execution of cloud specialist organizations and guarantees decency, consistence and delegates while constraining estimation cost structure.

2. **Facilities for recovery**

Cloud service provider should assume the liability to recuperate the information of clients if there is any information misfortune because of specific issues. Cloud specialist organization should ensure that they have legitimate reinforcement and can recover and recoup the private information of the clients that may be expensive. The cloud specialist organizations can likewise execute the accompanying answer to guarantee information recuperation.

(a) Using quickest plate innovation in occasion of fiasco for replication of information in peril.
(b) Changing grimy page edge.
(c) Expectation and substitution of unsafe gadgets.

3. **Use cloud service wisely**

The information put away in the cloud ought to be classified, and even the cloud service provider ought not to approach those data. The information put away in the cloud ought to be very much encoded to guarantee the security of the client's data. Any individual who needs access to the information in the cloud ought to request the authorization of the clients before doing.

4. **Vulnerability shielding**

The cloud specialist co-op should check the defenselessness of their cloud administration much of the time and consistently update and keep up the cloud to restrain the conceivable passage and decrease the danger of assault of the cloud by the programmers. The cloud specialist organization may likewise utilize the intrusion detection system (IDS) to ensure that the cloud administration is secure and safe.

5. **Data storage guidelines**

The cloud condition guarantees the security of the information put away in the cloud. The clients must comprehend the idea of the information stockpiling the guidelines which is trailed by the cloud specialist organization. Cloud service provider gives security arrangement according to the guideline, for example, HIPAA, DSS, PCI, and EU information insurance laws.

6. **Security check events**

The clients have contract with the cloud specialist co-op, so the clients can guarantee if any mishaps or ruptures of the delicate information put away in the cloud. The clients must have concurrence with the cloud specialist organization before utilizing the cloud administrations given by that specific cloud specialist organization. The clients ought to guarantee that the cloud specialist co-op gives enough insights concerning satisfaction of guarantees, break remediation and revealing possibility.

7. **Identification administration and confirmation**

If client needs to get to the information put away in the cloud, they should be verified by utilizing the username and secret word as well as the advanced information. Staggered validation method presented can likewise be executed in distributed computing. The system creates secret key in a few levels before the client can get to the cloud administrations. Wrong validation of the data at the time of development may create a serious security breach. Moreover, the proposed method can likewise be connected in distributed computing because the newly developed method also helps to detect the man in the middle attack. Spillage versatile validation can likewise be used so as to improve the security of the cloud administrations.

8. **Access control**

The cloud specialist organization should set up the information access control with rights, and the clients who access the information ought to be checked by the cloud specialist co-op. The CSP must guarantee that solitary approved clients may approach the information put away in cloud [6]. The technique can lessen the danger of the information access by the unapproved clients and give a much secure condition to store delicate information. The outsider inspecting can likewise be one of the choices to guarantee information respectability of the capacity in the cloud.

The examining system ought to have the accompanying properties:

(a) Confidentiality: Auditing conventions keep the client's information secret from inspector.
(b) Dynamic reviewing: Auditing convention bolster updates the information in cloud.
(c) Batch inspecting: Auditing convention bolster bunch reviews the mists and different clients.

9. **Enterprise infrastructure**

Client must verify the information when they need to keep in the cloud framework. The CSP gives a foundation which encourages clients to introduce and arrange equipment segments like firewalls, server, intermediary server, and switches.

3 Conclusions

Cloud computing is a model that paces up and increments the adaptability of information of the executives with decreased expense. The security turns into a worry to everybody who uses cloud administrations. There are is numerous opportunities in the field of security while there is continuous development or improvement in the security models of the cloud development environment. It is normal that this paper gives some premise or establishment with respect to issues and difficulties in distributed computing. Distributed computing administration models like Software as a Service (SaaS), Platform as a Service (PaaS), and Infrastructure as a Service (IaaS) are presented in this paper and security issues of cloud. Cloud development is continuously growing day by day by lots of cloud professionals. In terms of security, lots of issues are still unresolved.

References

1. Armbrust M, Fox A, Griffith R, Joseph AD, Katz R, Konwinski A, Lee G, Patterson D, Rabkin A, Stoica I, Zaharia M (2010) A view of cloud computing. Commun ACM Mag 53(4):50–58
2. Ashraf I (2014) An overview of service model of cloud computing. Int J Multidiscip Curr Res 2:779–783
3. Tari Z, Yi X, Premarathne US, Bertok P, Khalil I (2015) Security and privacy in cloud computing: vision, trends, and challenges. IEEE Cloud Comput 2(2):30–38
4. Mell P, Grance T (2011) The NIST definition of cloud computing. Retrieved from https://doi.org/10.6028/NIST.SP.800-145
5. Jain S, Kumar R, Kumawat S, Jangir SK (2014) An analysis of security and privacy issues, challenges with possible solution in cloud computing. In: Proceedings of the national conference on computational and mathematical sciences (COMPUTATIA-IV), pp 1–7
6. Gupta S et al (2019) Tier application in multi-cloud databases to improve security and service availability. In: Handbook of research on cloud computing and big data applications in IoT. IGI Global, pp 82–93

Chapter 47
A Conceptual Framework for Sensitive Big Data Publishing

Nancy Victor and Daphne Lopez

1 Introduction

Data that focuses mainly on 3 Vs—Volume, Velocity and Variety—is termed as big data [1]. The major sources of big data include healthcare data sets, genetic data sets, demographic data sets, search logs, etc. Big data analytics refers to the technique of analyzing big data sets to uncover the information which is hidden in these sets. The major benefits of big data analytics include performing risk analysis, creating new revenue streams, offering tailored health care, etc. But, one of the major concerns with big data is preserving the privacy of data that is published.

The traditional mechanism for privacy preservation includes removing Personally Identifying Information (PII) from the data. PII mainly includes the data which uniquely identifies an individual, such as personal ID. However, this may not be sufficient for effective preservation of privacy [2]. The identity of the individual can still be found out by linking the published data set with some background knowledge about the individual. Various anonymization, anatomization and perturbation approaches have been developed with respect to the re-identification challenge in published data sets. A survey of various big data privacy models and the challenges concerned has been studied by researchers worldwide [3, 4].

One of the major sources of big data include social networking sites such as Twitter. Privacy preservation in social network data is not an easy task because the data will be usually in unstructured formats. A common question can arise that if the users are not concerned about their privacy, who else cares. But, the important thing to be noted is that users are not aware of the privacy risks hidden behind their

N. Victor (✉) · D. Lopez
School of Information Technology and Engineering, Vellore Institute of Technology, Vellore, Tamil Nadu, India
e-mail: nancyvictor@vit.ac.in

D. Lopez
e-mail: daphnelopez@vit.ac.in

© Springer Nature Singapore Pte Ltd. 2021

S. D. Purohit et al. (eds.), *Proceedings of International Conference on Communication and Computational Technologies*, Algorithms for Intelligent Systems,
https://doi.org/10.1007/978-981-15-5077-5_47

Facebook posts or tweets. There can be a situation where you may not be aware that you are continuously being followed by someone who can possible harm you by knowing about your day-to-day activities. A system which publishes social network data for public use should make sure that no information can be obtained about a particular individual from the published information or even by linking the data with some external information. This is possible only by applying some anonymization techniques before sharing the data for public use. In order to identify the sensitive information from unstructured data like tweets, natural language processing techniques can be used.

One of the main issues with publishing tweets includes leaked tweets. These are the tweets that were actually published from a private account but would have been re-tweeted or copy and pasted, finally resulting in making the private information public. The other issue with tweets is when the user unknowingly discloses some information. This mainly includes enabling the location services instead of turning it off. The main reason is because the users are not aware of the problems hidden behind their public posts or tweets.

As we are dealing with big data sets, there should be some efficient mechanism to handle both batch and streaming data that are produced from various data sources. Lambda architecture is one of the best architectures which defines how historical data and near real-time data can be combined together to solve a variety of use cases. This paper proposes a privacy framework that can be embedded in the Lambda architecture exclusively for dealing with unstructured social network data.

2 Related Work

Various architectures have been developed for working with big data. Apache Kafka is a distributed messaging architecture which includes the producer, consumer, message and message storage servers called brokers [5]. The Samza architecture includes three layers such as streaming, execution and processing layers [6]. Lambda architecture has three major components—batch layer, serving layer and speed layer— for processing batch and streaming data [7]. As Lambda architecture helps for processing different types of data, this will be the best suited for big data applications. Section 2.1 explains about Lambda architecture and Sect. 2.2 focuses on NLP techniques.

2.1 The Lambda Architecture

The Lambda architecture was proposed by Nathan Marz, based on the three design principles for big data processing: human fault tolerance, data immutability and re-computation. Figure 1 shows the basic structure of the Lambda architecture [7].

The architecture has three vital components:

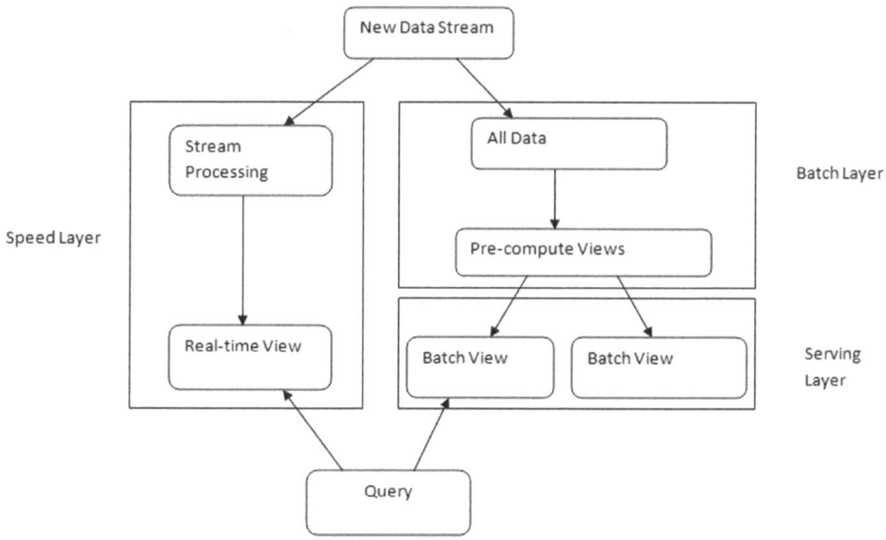

Fig. 1 Lambda architecture

1. Batch Layer
2. Serving Layer
3. Speed Layer.

Batch Layer: This layer focuses on managing the ineradicable, append-only raw data set (the master data set) and computing arbitrary views from the entire data set. This layer makes use of Apache Hadoop framework where HDFS is used for master data set management and MapReduce computation model is used for computing the views.

Serving Layer: All the batch views will be indexed in serving layer so that queries can be fired at any point of time. This layer gets updated whenever the batch layer finishes the computation of batch views. Cloudera Impala can be used in this layer.

Speed Layer: The need of speed layer is to work on real-time data so that they can be computed using functions on arbitrary data. This updates the real-time view (incremental updating) instead of re-computing the views. Storm is used for computing the real-time views. As the speed layer requires both random reads and writes, HBase is used to increment the real-time views in a persistent manner.

Every time a new data stream arrives at the system, it will be given into the batch layer and speed layer. Batch layer takes care of the master data set, and the current data also will be appended along with that, so that a new set of batch views will be produced. These views will be indexed in the serving layer. The current data stream will be taken by the speed layer and creates an incremental view of the most recent data. Both the batch views and real-time views are merged to give response to the query fired by the analyst/researcher.

2.2 NLP Techniques

Natural language processing techniques enable computers to analyze human/natural language and derive meaningful information. This way of human–computer interaction proves to be efficient in the areas of automatic text summarization, sentiment analysis, named-entity recognition, optical character recognition, etc. Named-entity recognition (NER) refers to the technique of categorizing the entities into some classes such as names and locations.

The major approaches used for NER include rule-based and statistics-based approaches. Every NER technique focuses on (1) building the knowledge base and (2) identifying and classifying the entities based on the knowledge base created.

Supervised learning works by labeling all the data available, and the approaches include conditional random field classifiers, Naive Bayes classifiers, support vector machines, etc. Unsupervised learning works by making use of the input data for identifying the hidden structures behind the data based on their similarity. The major approaches for unsupervised learning include various techniques such as k-means clustering. Semi-supervised learning makes use of a combination of supervised and unsupervised techniques for learning.

3 Privacy-Preserving Data Publishing Using NLP Techniques

Social networking sites such as Twitter can be considered as one of the major big data sources. As the tweets produced are unstructured, the traditional anonymization techniques [8–11] proposed for relational data sets cannot work efficiently with unstructured data. Named-entity recognition plays a major role here by extracting the entities of interest according to the application under consideration. These entities can be further used for anonymization, as these will fall into one of the attribute categories such as explicit identifiers, quasi-identifiers, sensitive attributes or non-sensitive attributes. Based on the category, appropriate anonymization techniques can be applied.

3.1 The Proposed Algorithm

De-identification of healthcare data such as pathology reports has been studied and surveyed by various medical researchers. As the structure of tweets is almost similar to the pathology reports from a hospital, the work can be extended for anonymizing medical data. The input to the system is the unstructured data, and the result is the anonymized version of the input. The system works incrementally by adding up the information in recent tweets with that of the historical tweets before publishing them to the analyst/researcher.

The steps for privacy-preserving unstructured data publishing are given below:
Input: Unstructured data (Tweets)
Output: Anonymized data.

Steps:

1. Input the document which is to be anonymized.
2. Use named-entity recognition techniques to classify the attributes.
    (a) All explicit identifiers are replaced with level 1 technique in the anonymization ladder. This includes either the removal or replacement of data with more general values such as "*".
    (b) All sensitive attributes are retained as such because this is the data that the user wants.
    (c) All non-sensitive attributes are removed completely.
    (d) All quasi-identifiers (QI) are retained for applying the appropriate anonymization technique.
    (e) Generate the data as QI $<QI_1, QI_2, \ldots QI_n>$ SA $<SA_1, SA_2, \ldots SA_n>$, where QI is the quasi-identifier and SA is the sensitive attribute. The data in the QI is added with the existing QI data set to generate the anonymized values.
3. Check whether the data set conforms to the basic anonymization principles. If yes, go to Step 4. Else, anonymize the data set.
4. Generate MQ $<MQ_1, MQ_2, \ldots MQ_n>$ SA $<SA_1, SA_2, \ldots SA_n>$, where MQ is the modified quasi-identifier.
5. Add noise parameter for achieving better privacy, if needed and output the result as:
   $<MQ + x>$ $<SA>$, where "x" is the noise which is added to the result.

3.2 The Proposed Conceptual Framework

Figure 2 depicts the proposed conceptual framework for big data publishing without compromising privacy. As the basic architecture is modeled for processing batch and real-time data, the input data that is fed into the system can take any formats.

Whenever private data is shared by the organization, the primary level of anonymization is from the organization's end by removing the explicit identifiers such as name and ID number. This will be in turn fed into the various layers.

The named entities present in the data, which is supplied to the speed layer, will be extracted using NER techniques. The final list of named entities and the data in the batch layer will be compared and anonymized in the serving layer depending upon the level of anonymization required.

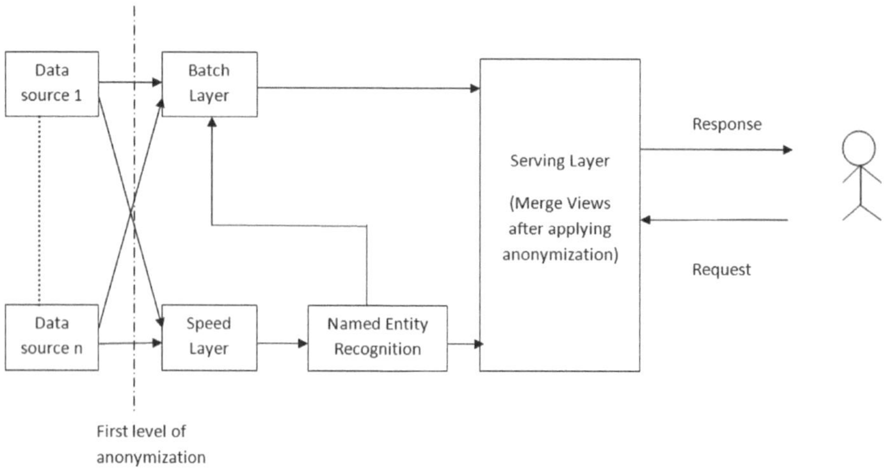

Fig. 2 Proposed framework for big data publishing

4 Experimental Results

4.1 Extracting and Analyzing Tweets

The tweets are extracted using the Twitter API. Creating an application in Twitter developers results in generating the consumer key and secret used for extracting tweets related to a particular keyword. The keyword used here is "AIDS", as this is one of the sensitive information available. Using "searchTwitter" command, the related tweets are retrieved. This is converted to a data frame and is saved as a text file. Figure 3 shows the resultant tweets provided.

```
text
Overwatch quick play is so aids, everyone wants to do their own thing
Walking the Walk for AIDS https://t.co/O3nWsraC33 #STD #HIV
Never before has US been involved in so many places, making such impact: Ebola, AIDS, Afghanistan unity gov't, S. China sea #IdeasForum
RT @ASAPYams first symptom of aids is wearing bootcut jeans
Uma em cada dez crianças possui defesas naturais contra Aids https://t.co/JgP93uqyvn https://t.co/REDnNXMySp
#HowToStaySkinny #AMemoir1) Don't eat.2) Ask @god for cancer or AIDS.3) Repeat Numbers 1 &amp; 2.
Didn't think Fifa soundtracks could get anymore Aids but this year's is a special effort
Like my vape smells  like strawberry smoothies and you dodge that cloud like it's AIDS. Cigs are cancer and you just like... aight.
People who scored D in HIV/AIDS and also Communication Skills in first year. Hi. ??
Cannabis Stock Leader:  Modern Mobility Aids Inc (OTO:MDRM) Gains 72.73% at Midday on September 29 https://t.co/UQqURpRqDh
One in 10 children has 'Aids defence' - BBC News https://t.co/5e7dHIsMDa
#WeirdOneLetterOffBrandsBand-AIDS
@SundayShoutsFC AIDS
#Stigmabase | IT -  Aids, 2 sieropositivi su 3 con più di 50 anni entro 2030 - Le persone con Hiv cambiano volto:... https://t.co/6V5Y3asnxT
Health &amp; Wellness Update: Local HIV/AIDS Service Organizations https://t.co/dtawgA01JG
RT @easterncomd: Army unit provides training aids to primary school in Chandel District of Manipur https://t.co/MVuYAnaSMD @adgpi https://t...
RT @USAIDKenya: "@PEPFAR has worked successfully to improve the quality of human resources for health in delivering #HIV/AIDS services," Am...
Health &amp; Wellness Update: Local HIV/AIDS Service Organizations https://t.co/Pp1u7WIube
BBC News - One in 10 children has 'Aids defence' https://t.co/9YeuIs4dQV
Help Tim raise $1,000 for the amazing work of AIDS Walk Los Angeles! https://t.co/BOBp6D9BV0HELP ME TODAY AND DONATE PLEASE,
```

Fig. 3 Tweets related to the keyword "AIDS"

Fig. 4 Word cloud

A wordcloud is generated from these tweets to visualize what people are tweeting about "AIDS". Figure 4 shows the wordcloud created.

This visualization helps us to identify the most important terms related to the keyword "AIDS".

4.2 Applying NER Techniques

The input for the anonymization algorithm is the unstructured data. In order to process the tweets, it has been converted to a text file. Figure 5 shows a part of the corresponding text file for the tweets generated. Named-entity recognition works by extracting "named entities" related to some predefined group of classifiers.

OpenNLP is a package used for processing natural language text for applying machine learning techniques. Various NER models have been created already for working with OpenNLP package. But, this works only for some predefined set of categories such as "person names", "organization names" and "location". Using these predefined set of classes, the person names, organization names and locations

```
1) Don't eat.
2) Ask @god for cancer or AIDS.
3) Repeat Numbers 1 &amp; 2.",FALSE,0,NA,2
"7","Didn't think Fifa soundtracks could g
"8","Like my vape smells  like strawberry
"9","People who scored D in HIV/AIDS and a
"10","Cannabis Stock Leader:  Modern Mobil
"11","One in 10 children has 'Aids defence
"12","#WeirdOneLetterOffBrands
Band-AIDS",FALSE,0,NA,2016-09-29 09:27:27,
"13","@SundayShoutsFC AIDS",FALSE,0,"Sunda
"14","#Stigmabase | IT -  Aids, 2 sieropos
"15","Health &amp; Wellness Update: Local
"16","RT @easterncomd: Army unit provides
"17","RT @USAIDKenya: "@PEPFAR has worked
"18","Health &amp; Wellness Update: Local
"19","BBC News - One in 10 children has 'A
"20","Help Tim raise $1,000 for the amazin
HELP ME TODAY AND DONATE PLEASE,",FALSE,0,
```

Fig. 5 Tweets in text file

```
> l
[1] "Afghanistan"          "China"                 "Manchester"          "Chandel District of"
> o
[1] "Mobility Aids Inc ("                     "BBC News"
[3] "White Privilege Blvd"                    "Health"
[5] "Local HIV/AIDS Service Organizations" "Army"
[7] "Health"                                  "Local HIV/AIDS Service Organizations"
[9] "AIDS"
```

Fig. 6 Entities extracted using NER techniques in OpenNLP

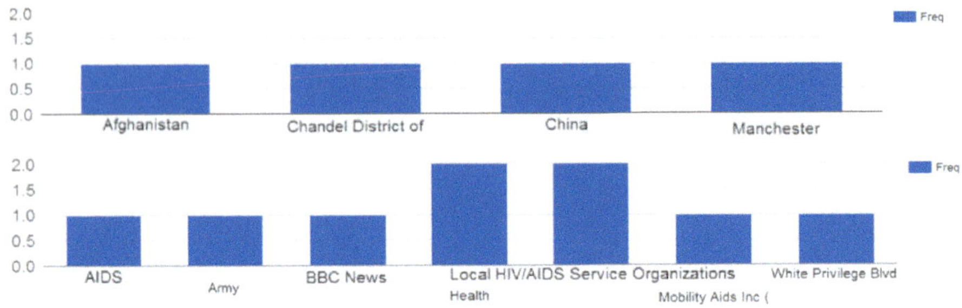

Fig. 7 Frequency of extracted entities

have been extracted using named-entity recognition. The results are shown in Fig. 6. Custom name finder models also can be created using OpenNLP models.

The frequency of each entity is shown in Fig. 7. Only twenty tweets related to "AIDS" are retrieved for this implementation. This is to show that even with very less amount of publicly shared information, the frequency of certain information that is shared will be more. This can be combined with other publicly available information to link and analyze sensitive information.

4.3 Anonymizing the Data

The data to be anonymized is found out by combining the information that is extracted using NER techniques with the data that is already available, the historical data. It can be seen that even the information about some persons is published along with the tweets. Any system which publishes the data for public/research use should make sure that no explicit identifiers are shared for public use [12]. For anonymizing the data, the explicit identifiers, quasi-identifiers and sensitive information should be identified first. Each identifier should be considered as a different set of classifier. As it is very clear that all the person names are explicit identifiers, this should be completely removed and should be replaced with level 1 technique in the anonymization ladder. This involves replacing the values with the more general "*". Quasi-identifiers are transformed using generalization mechanisms [13]. Figure 8 depicts the generalization hierarchy developed for the attribute "work class".

Fig. 8 A sample generalization hierarchy

Level-0	Level-1	Level-2
Private	Non-Government	*
Self-emp-not-inc	Non-Government	*
Self-emp-inc	Non-Government	*
Federal-gov	Government	*
Local-gov	Government	*
State-gov	Government	*
Without-pay	Unemployed	*
Never-worked	Unemployed	*

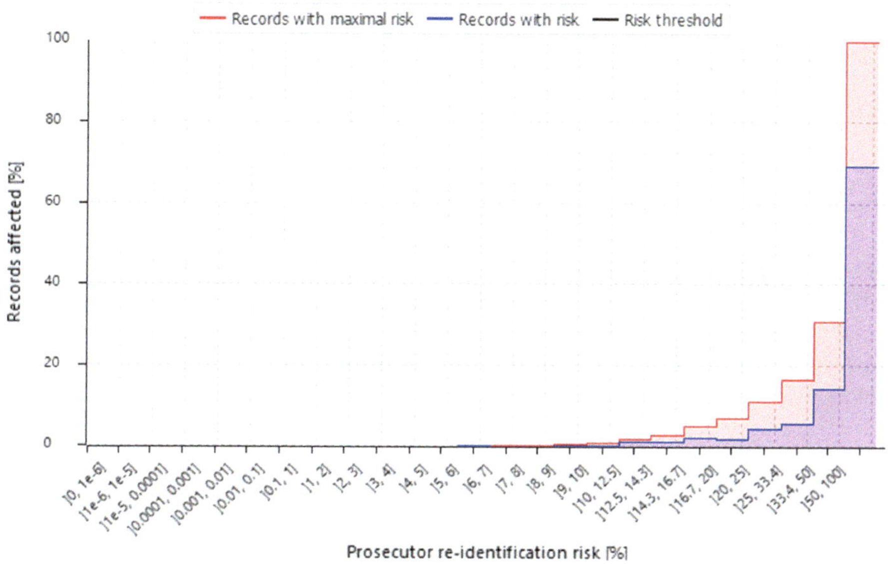

Fig. 9 Distribution of risk (without anonymization)

Once the generalization hierarchy is set for all QIs, any anonymization technique can be applied for anonymizing the data. k-anonymization technique is applied here for anonymizing the data. The distribution of risk without and with anonymization is given in Figs. 9 and 10 respectively.

5 Conclusion

This paper presents a conceptual framework for big data publishing by preserving the privacy of the published information. This is accomplished by integrating natural

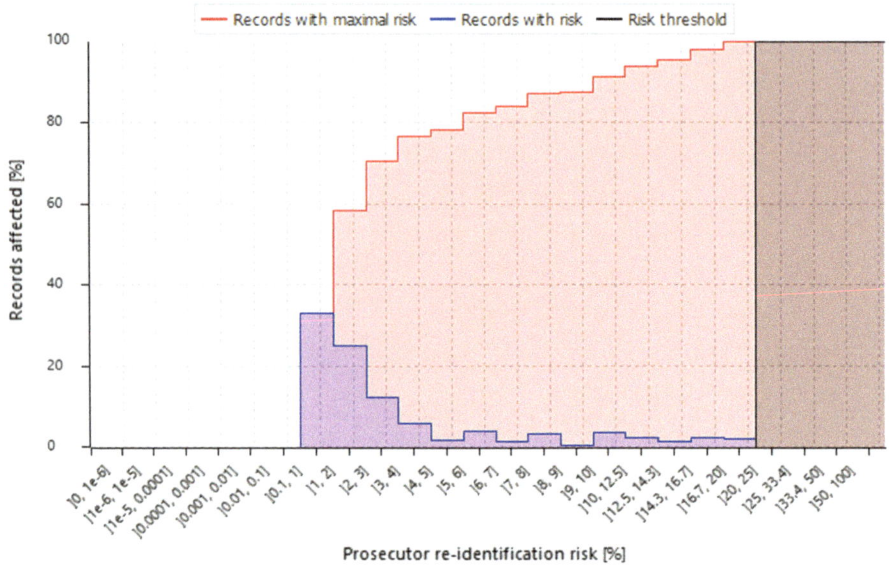

Fig. 10 Distribution of risk (with anonymization)

language processing techniques with the traditional anonymization algorithms. This framework is ideal for big data sets as it can process structured, unstructured and semi-structured data. Major applications focus on publishing medical data, social network data and other sensitive contents. The work can be further extended by making use of neural network architectures.

References

1. Katal A, Wazid M, Goudar RH (2013) Big data: issues, challenges, tools and good practices. In: 2013 sixth international conference on contemporary computing (IC3). IEEE
2. Fung B et al (2010) Privacy-preserving data publishing: a survey of recent developments. ACM Comput Surv (CSUR) 42(4):14
3. Victor N, Lopez D, Abawajy JH (2016) Privacy models for big data: a survey. Int J Big Data Intell 3(1):61–75
4. Victor N, Lopez D (2018) Privacy preserving big data publishing: challenges, techniques, and architectures. In: HCI challenges and privacy preservation in big data security. IGI Global, pp 47–70
5. http://kafka.apache.org/
6. http://samza.apache.org/
7. Marz N, Warren J (2013) Big data: principles and best practices of scalable realtime data systems. Manning Publications
8. Sweeney L (2002) k-anonymity: a model for protecting privacy. Int J Uncertain Fuzziness Knowl-Based Syst 10(5):557–570
9. Machanavajjhala A et al (2007) l-diversity: privacy beyond k-anonymity. ACM Trans Knowl Discov Data (TKDD) 1(1):3

10. Li N, Li T, Venkatasubramanian S (2007) t-closeness: privacy beyond k-anonymity and l-diversity. In: IEEE 23rd international conference on data engineering. ICDE 2007. IEEE
11. Dwork C (2011) Differential privacy. In: Encyclopedia of cryptography and security. Springer US, pp 338–340
12. Gupta S et al (2019) Tier application in multi-cloud databases to improve security and service availability. In: Handbook of research on cloud computing and big data applications in IoT. IGI Global, pp 82–93
13. Prasser F, Kohlmayer F, Kuhn KA (2014) A benchmark of globally-optimal anonymization methods for biomedical data. In: Proceedings of the 27th IEEE international symposium on computer-based medical systems, New York City

Chapter 48
Verifiable Color Progressive Visual Cryptography with Cheating Detection

Sachin Kumar Singh, Manish Yadav, Abhishek Chaudhary, and Umang Singhal

1 Introduction

Expand headway in advances has prompted the far-reaching utilization of computerized media in our everyday lives. Transmission and recovery of advanced media are conceivable now nearly in commotion-free structure over the remote information correspondence arrange. Computerized media can be effectively caught, dispersed, in some cases even adjusted and copied without any such unlawful activities. Consequently, a genuine worry as different security-related issues, for example, information misappropriation, unlawful information use, get to control and so on are imagined which thus request some type of assurance and security in information transmission [1, 2]. To offer security in several techniques such as cryptography (encryption), data hiding, secret sharing schemes, etc., are developed [3, 4]. In visual cryptography, secret sharing splits the secret data into several noise like parts called shares. A gathering of approved members holding these shares can just peruse/recover the secret message, in this manner fills the need of access control in one manner, and in the other manner, it keeps away from purposeful or coincidental information misfortune [5].

Secret image sharing (SIS) scheme is a procedure of sharing a secret picture into few commotions like shadow/share images in the manner that whenever shared images are joined in a certain way, the secret picture can be decoded [6]. These shares

S. K. Singh (✉) · M. Yadav · A. Chaudhary · U. Singhal
AKTU, Lucknow, India
e-mail: sachin.ls@outlook.com

M. Yadav
e-mail: manish.it41@gmail.com

A. Chaudhary
e-mail: recs.cse1604@gmail.com

U. Singhal
e-mail: umangsinghal96@gmail.com

© Springer Nature Singapore Pte Ltd. 2021

S. D. Purohit et al. (eds.), *Proceedings of International Conference on Communication and Computational Technologies*, Algorithms for Intelligent Systems,
https://doi.org/10.1007/978-981-15-5077-5_48

are disseminated among the gathering of legitimate members. Just an approved subset of participants can decipher the secret picture through participation. The prohibited subset of members neglects to recover any data about the secret picture. All the approved and the illegal subset of clients are the individuals from the entrance structure of the SIS scheme [7]. The idea of secret sharing was first independently explored and detailed by Blakley [8] and Shamir [9].

In (t, n) threshold VCS, we will construct a threshold progressive VCS that encodes a secret image to n shares and posses the decoding property that in the event that the number of shares pooled is not as much as threshold t, then the stacking of the shares pooled uncovers nothing about the secret picture, but on the off chance that the number of shares pooled surpasses threshold t and increases, the stacking uncovers the secret picture progressively. It means that the number of shares stacked together after t increases, the quality of the reconstructing also increases.

2 XOR-Based Progressive VCS Without Pixel Expansion

This section discusses a progressive VCS without pixel expansion, which is based on the classical XOR-based (2, 2)-threshold VCS whose basis matrices are

$$C_0 = \begin{Bmatrix} 1\,0 \\ 1\,0 \end{Bmatrix}, \quad C_1 = \begin{Bmatrix} 1\,0 \\ 0\,1 \end{Bmatrix}$$

The construction of the extended basis matrices of the XOR-based (2, n)-threshold progressive VCS is almost the same as the OR-based (2, n)-threshold progressive VCS, where $n \geq 2$, reconstruct the secret image progressively. The encoding algorithm is also analogous that of the OR-based but while decoding of an XOR-based scheme uses the XORing instead of the stacking (ORing) (Fig. 1).

(a) (b) (c) (d)

Fig. 1 Figure shows (2, 4)-threshold progressive VCS **a** shows a share; **b**, **c** and **d** show images obtained from stacking t shares, $t = 2$, 3 and 4 (all shares)

3 Our Approach

Our work can be divided into the following. We firstly extended classical visual cryptography for color images. We have built up an encoding plan to change a picture from color space into monochrome spaces/grayscale channels, progressive VCS is applied on these monochrome spaces, and progressive shares of these grayscale channels are generated, called as primitive shares. This primitive share from each monochromatic space/grayscale channel is stacked to create final shares, which are also progressive in nature enables perfect restoration of the original image. Second, we implemented an cheating detection scheme which is based on Diagonal Parity and Gama Weight. This scheme generates an authentication matrix which consists of information generated by Diagonal Parity and Gama Weight. This matrix is used in transaction with its particular share for examining the integrity and to authenticate the share.

4 Extension of Progressive VCS to Color Images

Digital halftoning has been extensively used. For visual cryptography, digital halftoning converts the grayscale image into a monochrome image. Then, we will get the binary image on which original visual cryptography technique can be applied. But, there is a loss on quality.

Encoding is the stage that is used to create shares. This is divided into the following steps:

1. *Color halftoning*: One could do color halftoning first followed by the splitting then do the grayscale channels.

$$I \rightarrow \text{Color Halftoning } I_{\text{htone}} \rightarrow \text{split RGB } [I_R, I_G, I_B]$$

2. *Scheme application*: A VCS scheme will be deployed to each of the grayscale channel individually, generating shares corresponding to each grayscale channel. These shares are knows as primitive shares. The thing to be emphasized is that these shares are getting generated through a progressive VCS(t, n) scheme, so theses shares posses a progressive nature.

$$\begin{aligned} I_R &\rightarrow [S_{R1}, S_{R2}, \ldots, S_{Rn}] \\ I_G &\rightarrow [S_{G1}, S_{G2}, \ldots, S_{Gn}] \\ I_B &\rightarrow [S_{B1}, S_{B2}, \ldots, S_{Bn}] \end{aligned}$$

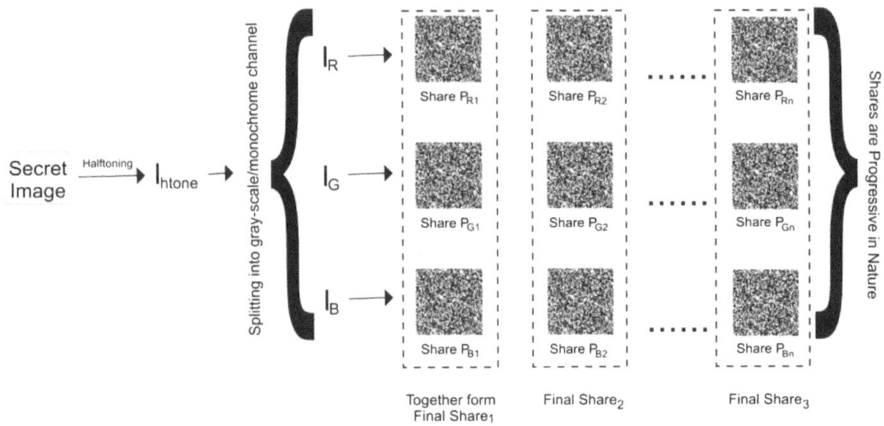

Fig. 2 Proposed Progressive VCS scheme decoding is the stage for the reconstruction of the original secret image

3. *Final share generation*: Final shares are generated with the primitive share generated for each grayscale channel, and this will be done through stacking. Three primitive shares will be stacked together belonging to each grayscale channel that will lead to generation of final share. As the primitive shares are generated through progressive scheme and exhibit progressive nature, the same is inherited by the final shares also. Hence, the final share generated by stacking also exhibits progressive nature,

The number of share generated for each grayscale channel depends on the number final share, i.e., n in progressive VCS(t, n) (Fig. 2).

Decoding is the stage for the reconstruction of the original secret image. All the shares consist of stacked primitive shares, and each share belongs to a certain grayscale channel. During reconstruction, the grayscale channel primitive share should be XORed with the same grayscale channel [10, 11]. As the number of share increases and after the threshold is attained the quality of the secret start enhancing.

5 Diagonal Parity and Gama Weight

This is the cheating detection method we are going to for detection of cheating in VCS. In a two-dimensional diagonal parity-check, the image bits are organized in a matrix consisting of rows and columns. For all the possible diagonals, one parity-check bit is calculated. The total number of diagonal in any $n \times m$ image is $n + m - 1$. The diagonal of the matrix will be traversed in two ways. One from left top corner to right bottom corner and the second will be from right top corner to left bottom corner because of this two different way traversal total number of parity bits calculated will be $2(n + m - 1)$ in a $n \times m$ matrix. During the parity checking,

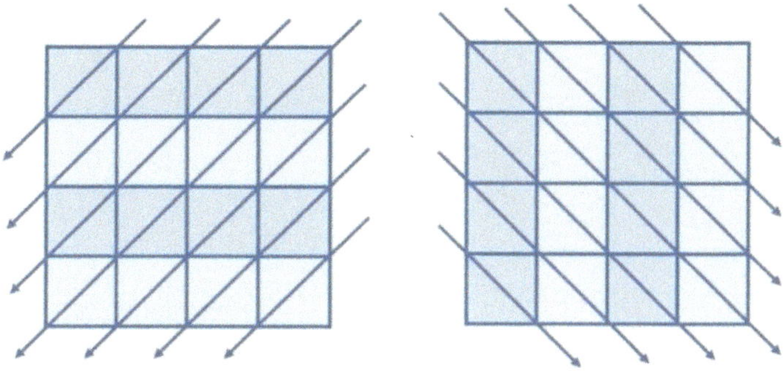

Fig. 3 Diagonal Parity

count has been maintained of the occurrence of 1, if this count is even the parity if concluded 1 else 0 (In halftone, we will use this by counting the number of 255). This diagonal parity can be applied to generated shares and can be used for the error checking. The integrity of the share after any transaction can be tested, the originality can be verified, and hence, the influence of intruder and cheating can be detected (Fig. 3).

As seen from the figure, diagonal is drawn in two possible directions for covering all the possible diagonals. The first can be termed as diagonal form left top corner to right bottom corner (D_{LR}), and the second one will be from right top corner to left bottom corner (D_{RL}).

But if we talk about the accuracy and error detection capacity of two-dimensional diagonal parity, it can detect all single cheating errors and detect two and three cheating errors that occur anywhere in the matrix. However, only some patterns with four and its multiple cheating errors cannot be detected (Fig. 4).

Due to the existence of these undetected errors, only diagonal parity will not be able to detect cheating error in the share successfully. So for that, we are going to use Gama weight.

For each diagonal, a Sigma value is maintained. Sigma value actually adds up the difference in location of the first occurrence of 1 and the other occurrences of 1. (Location is specified as if element is at index (1, 5), its location value will be 15, and indexing method should be same). When first 1 occurs in the diagonal, then its location has been recorded and further used to compute the value of sigma. In halftone, we will use this by recording the location of 255. This weight will be calculated for both the directions of the diagonal as shown in figure. So corresponding to one share, there will be two Gama weight calculated, the first which are diagonals from left top corner to right bottom corner (Gama_{LR}) and the second one which are from right top corner to left bottom corner (Gama_{RL}).

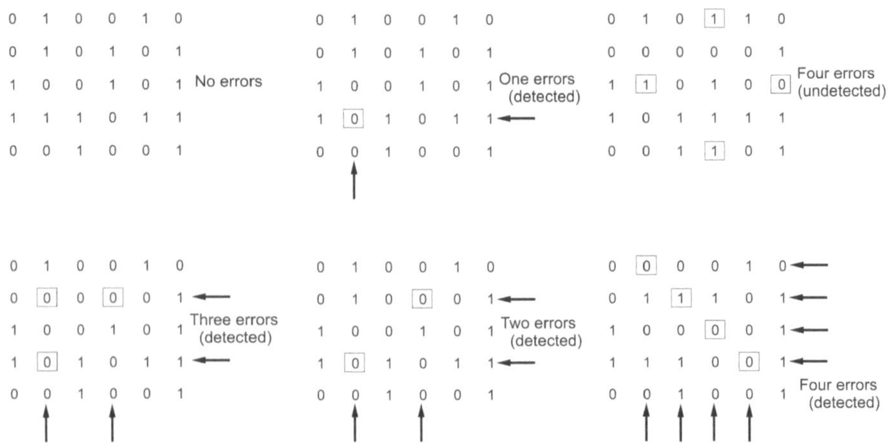

Fig. 4 Errors case in Diagonal Parity

With the help of sigma, Gama weight is calculated corresponding to each diagonal, which is calculated as if the value of a particular diagonal parity is corresponding to a diagonal is 1, then the corresponding sigma for the diagonal will be added to Gama if the Diagonal Parity is zero sigma value of that diagonal will be subtracted from Gama. The algorithm is as follows.

Algorithm: Compute_Gama()

Input—Matrix of image binary or halftoned and Parity of each diagonal as Parity_i.

Output—Gama Weight

```
for each Diagonal D_i:
  for each element in D_i:
    base_loc = location of first occurrence of 1 or 255
    if(Di[element] = = 1 or 255) then
      sigma=sigma+abs(base - element_location)
    else
      continue
    // forends
  // forends
if Parity_i == 1 or 255 then
  Gama = Gama + sigma of Diagonal D_i
else
  Gama = Gama - sigma of Diagonal D_i
```

In the figures below, graph has been plotted corresponding to the Gama values calculated on all possible combination of 3 × 3 and 4 × 4 binary matrix.

These figures show the variation in the value of Gama from one another and give a clear idea about the distribution of the data. The measure shows the heterogeneity of the distribution of the Gama Weight, so using Gama Weight and Diagonal Parity cheating activity can be detected because different images will not have both same.

6 Combining Progressive VCS and Cheating Detection Method

In this paper, we have discussed Color Progressive VCS. We did it through making the generated primitive shares' progressive while generating shares of grayscale channels. Due to the presence of progressive nature in primitive shares, the shares which are further generated by stacking them like a color image is also progressive. The decoding of original secret from these share have been discussed earlier in the paper. This is a way how Color Progressive VCS can be achieved. Now, we come to the cheating process, as shown in the Sect. 5, it has been shown that how the Diagonal Parity and Gama Weight can be calculated for the share generated. Figs. 5 and 6 in Sect. 5 also show the efficiency and ability to generate discreet values for different combination is reasonably good.

In practical implementation, we can integrate both these process. In Progressive VCS(t, n), a total of n number of final shares will be generated, and t is required for the reconstruction. We will generate n share by the proposed Progressive VCS, and in this, every share will consist of three grayscaled primitive shares which are stacked together. For each of the share, we will calculate the set of Diagonal Parity which will be two D_{LR} and D_{RL} and Gama weight which will be also two Gama_{LR} and Gama_{RL}. For final share, there will be six set Diagonal Parity of size $n + m - 1$ each, and we will have six Gama Weight corresponding to the three primitive shares in each final share. We will use an additional array of size $12 \times (n + m - 1)$ for storing the six set of Diagonal Parity and Gama Weights. This matrix will be generated for all the n

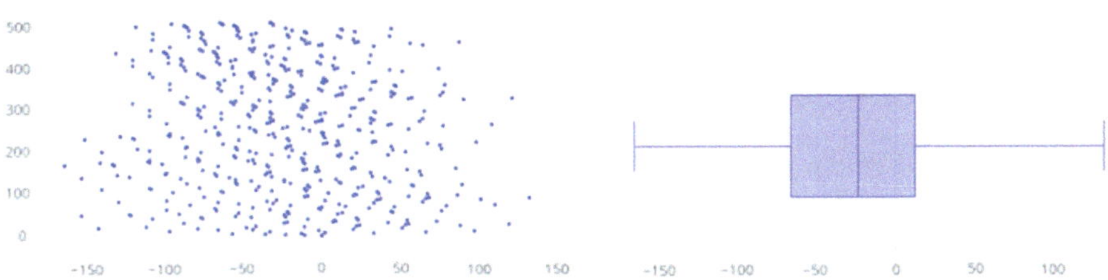

Fig. 5 Scatter plot and box plot for values of Gama for all possible 3×3 binary matrix

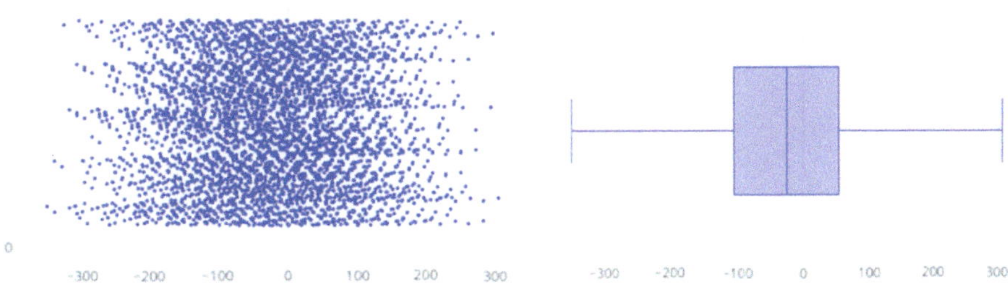

Fig. 6 Scatter plot and box plot for values of Gama for all possible 4×4 binary matrix

share in our Progressive VCS(t, n). This matrix will be binded or transmitted with the Progressive share itself. Whoever receives a share will be able to verify the integrity and will be able to authenticate the received share by calculating and comparing the set of Diagonal Parity and Gama Weight.

7 Experimental Result

In this section, we experimentally examine the cheating detection method and proposed Progressive VCS (Fig. 7).

As it can be seen in the Figure, there is an original secret(a), and after going through some cheating process in transit, the reconstructed secret(b) is fake. With our proposed cheating detection method, the Gama Weight and Diagonal Parity should be calculated for both the secret and compared for the sake of authenticity and verification.

This method is implemented and executed in MATLAB. The Diagonal Parity and Gama Weight for the Original Image are as follows:

$$D_{LR} = [0100001010101010101\ 101\ 10 \\ 1010101000110001010101010]$$

$$D_{RL} = [010001\ 101010101010100\ 01\ 11\ 1\ 0 \\ 1010100111000101010101010]$$

$$\text{Gama}_{LR} = 490\ \text{Gama}_{RL} = 3521$$

The Diagonal Parity and Gama Weight for the original image are as follows:

$$D_{LR} = [0001010101011101010010110 \\ 011010101010101010101010]$$

$$D_{RL} = [010001011101010000010000 \\ 1100111101010101010101010]$$

Fig. 7 **a** is the original secret, and **b** is the fake secret after cheating

$$\text{Gama}_{LR} = 805\ \text{Gama}_{RL} = -8027$$

As seen from the experiment result, the Diagonal Parity as well as the Gama Weight are different for both the images, and hence, the secret reconstructed is not original.

We are applying our proposed VCS to the secret image (Figs. 8 and 12a).

First of all, its halftoning is performed on the secret (shown in Fig. 12b).Then, the halftoned secret is splitted into monochrome/grayscale channel. If visualization of these grayscale channels is performed, grayscale images will be produced (Fig. 9).

The loss in quality is due to halftoning. Figure 10 is used for better understanding and visualization of the splitting in RGB (the Red, Green and Blue images are produced by the help of a pure black share and concatenating it with the RGB channels) (Fig. 10).

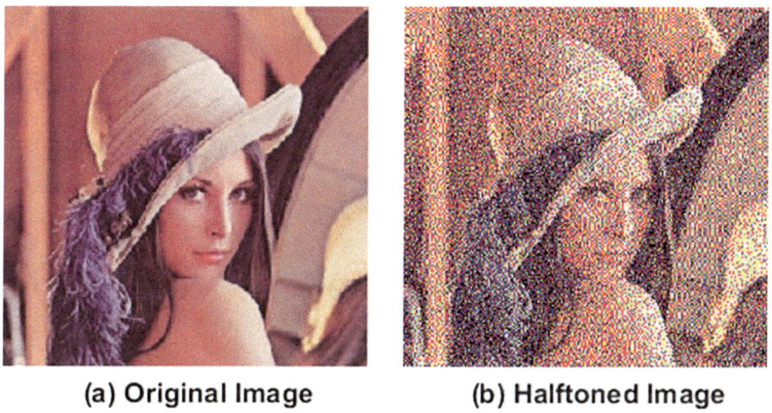

Fig. 8 Original and halftone image

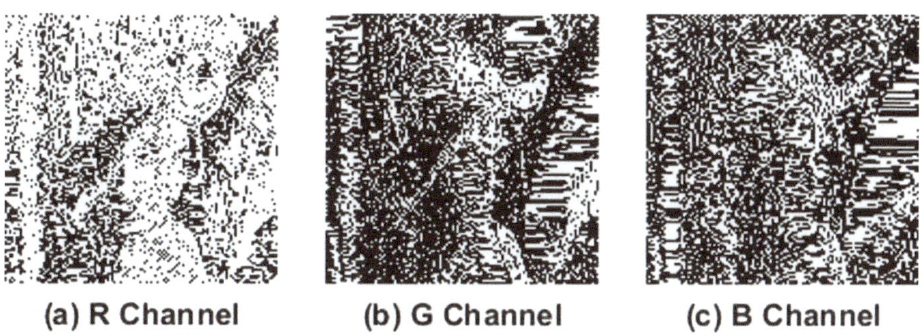

Fig. 9 All the three channel (RGB) in grayscale due to split

Fig. 10 The red, green and blue images are produced by the help of a pure black share and concatenating it with the monochromatic channels

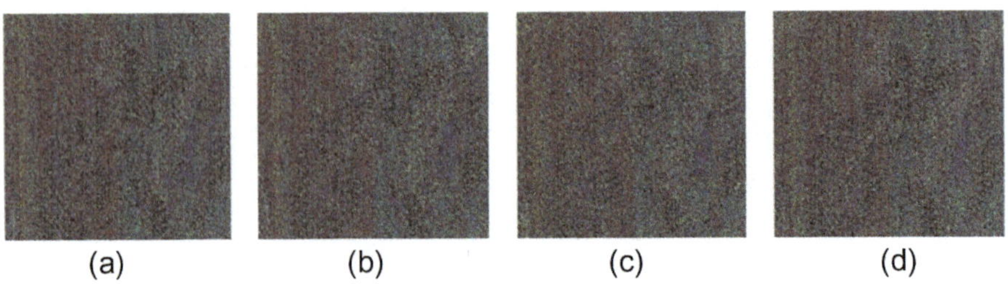

Fig. 11 (3, 4)-threshold Progressive VCS shows the final shares; **a**, **b**, **c** and **d**

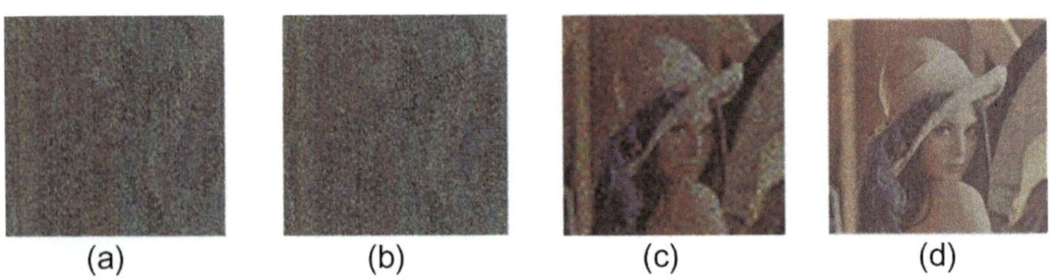

Fig. 12 (2, 4)-threshold Progressive VCS shows the resulting images obtained from stacking t shares, $t = 1, 2, 3$ and 4 (all shares)

Progressive share of each of the channel is produced (VCS(3, 4) is used in the experiment) which are primitive shares. Primitive share one from each channel is stacked together to form final share (Fig. 11).

In this paper, we have presented XOR-based progressive VCS scheme with no pixel expansion, which is driven by primitive shares generated for each monochromatic channel. For detecting the cheating, Gama Weight and Diagonal Parity are used. The Diagonal Parity and discreet value of Gama Weight ensure unique combination of both for different secret and prevent cheating.

References

1. Hou YC, Quan ZY, Tsai CF (2015) A privilege-based visual secret sharing model. J Visual Commun Image Represent 33:358–367
2. Bhattacharjee T, Maity SP (2016) An image-in-image communication scheme using secret sharing and m-ary spread spectrum watermarking. Microsyst Technol 1–14. https://doi.org/10.1007/s00542-016-3104-z
3. Wei SC, Hou YC, Lu YC (2015) A technique for sharing a digital image. Comput Stand Interfaces 40:53–61
4. Liu L, Wang A, Chang CC, Li Z (2014) A novel real-time and progressive secret image sharing with flexible shadows based on compressive sensing. Signal Process 29(1):128–134
5. Ou D, Sun W (2014) Reversible AMBTC-based secret sharing scheme with abilities of two decryptions. J Visual Commun Image Represent 25(5):1222–1239
6. Yang CN, Chen CH, Cai SR (2016) Enhanced boolean-based multi secret image sharing scheme. J Syst Softw 116:22–34
7. Jin J, Hong Wu Z (2012) A secret image sharing based on neighbourhood configurations of 2-d cellular automata. Opt Laser Technol 44(3):538–548
8. Blakley G (1979) Safeguarding cryptographic keys. In: Proceedings of the 1979 AFIPS National Computer Conference, pp 313–317
9. Shamir A (1979) How to share a secret. Commun ACM 1979 22(11):612–613
10. Kumar A, et al. (2019) An enhanced quantum key distribution protocol for security authentication. J Discrete Math Sci Crypt 22(4):499–507
11. Kumar A, et al. (2019) An improved quantum key distribution protocol for verification. J Discrete Math Sci Crypt 22(4):491–498

Chapter 49
2-Terminal Reliability Analysis of Single Fault-Tolerant Multistage Interconnection Networks

Amit Prakash and Dilip Kumar Yadav

1 Introduction

The advancements of high-performance, cost-effective, and fault-tolerant multistage interconnection networks (MINs) are essential for efficient multiprocessor system in the field of parallel processing. Besides, it is equally challenging to assure their high reliability.

The design of MIN comprises highly fault-tolerant capability, number of redundant and disjoint paths, hardware cost, routing, and rerouting behavior which provides strong re-readability under any switch failures [1]. MIN should tolerate the multiple switch failures i.e., if a processing element (PE) fails, the residual PEs can communicate without any interruption and reliability should be higher to determine the performance of the network [2].

The fault-tolerant MIN provides multiple disjoint paths between any source–destination node pairs [3]. Fault-tolerance and reliability of MINs with disjoints and redundant paths improve the performance and reliability [4].

The design like Shuffle exchange network was introduced by increasing a greater number of stages in shuffle exchange, but they have lower 2-terminal reliability (2-TR) due to the increased number of components and complexity [5, 6].

In this paper, reliability block diagram (RBD) of MINs based on improved sum of disjoint products—multiple variable inversion technique (SDP-MVI) is used for the evaluation of 2-TR, with an assumption that all fault occurs independently, links are reliable, and switches are either working or failed.

A. Prakash
Department of Electronics and Communication Engineering, National Institute of Technology Jamshedpur, Jamshedpur, Jharkhand, India

D. K. Yadav (✉)
Department of Computer Applications, National Institute of Technology Jamshedpur, Jamshedpur, Jharkhand, India
e-mail: dkyadav.ca@nitjsr.ac.in

© Springer Nature Singapore Pte Ltd. 2021

S. D. Purohit et al. (eds.), *Proceedings of International Conference on Communication and Computational Technologies*, Algorithms for Intelligent Systems,
https://doi.org/10.1007/978-981-15-5077-5_49

The paper is ordered as follows: Sect. 1 states the introduction and need of MINs; in Sect. 2.1, the 2-terminal reliability has been estimated for various MINs; Sect. 2 presents performance analysis; and Sect. 3 concludes the paper, followed by the references.

1.1 Shuffle Exchange Network (SEN)

Single switch/link failure network, where each stage has $N/2$ switching elements (SEs) of size 2×2, and network complexity of $((N/2) * \log_2 N)$ [6, 7], is shown in Fig. 1. RBD of SEN is presented in Fig. 2.

Reliability expression for SEN: $R = p_2 p_7 p_{10}$

Here, 'p_i' is the reliability of ith SE and by assuming identical SEs, with reliability $p_i = r$,

2-TR (SEN)$= r^3$

Paths (2, 3; $\delta = 1$) or (010, 011): {2–7–10}

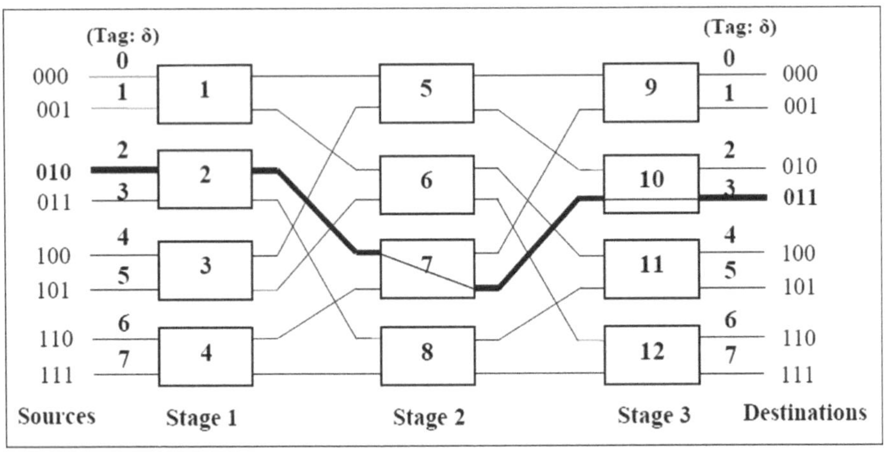

Fig. 1 SEN: 2-TR paths from $S_{2\,(\delta = 2,\,3)}$ to $D_{10\,(\delta = 2,\,3)}$

Fig. 2 RBD of SEN (S_2–D_{10})

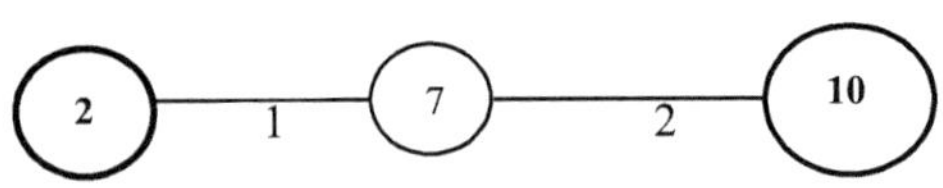

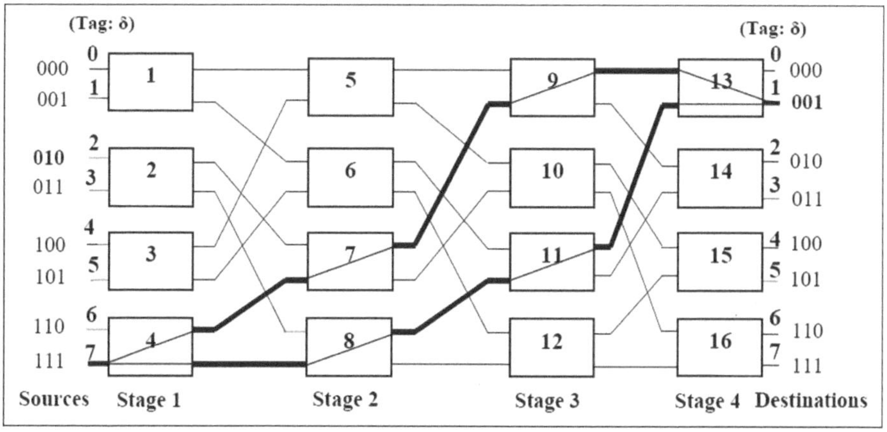

Fig. 3 SEN+: 2-TR paths from $S_{4\ (\delta = 6,\ 7)}$ to $D_{13\ (\delta = 0,\ 1)}$

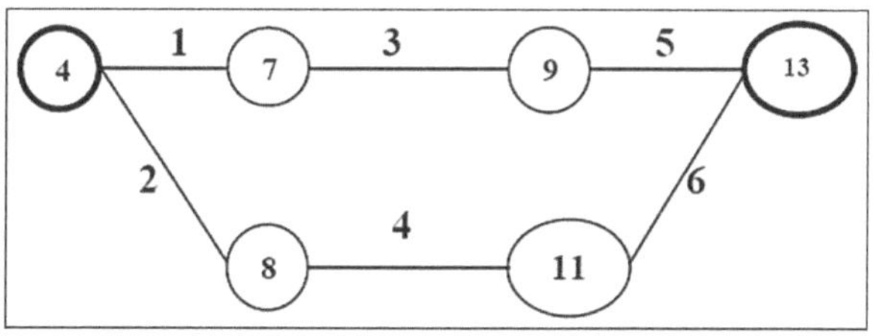

Fig. 4 RBD of SEN+ for S_4–D_{13}

1.2 Shuffle Exchange Network with One Additional Stage (SEN+)

Stage of size $N \times N$ is added having $N/2$ SEs of size 2×2 [8]. It has ($\log_2 N + 1$) stage, network complexity is $((N/2) * ((\log_2 N) + 1))$ [5, 7], and the model is shown in Fig. 3. RBD of SEN+ is presented in Fig. 4.

Reliability expression for SEN+: $R = p_4 p_7 p_9 p_{13} + p_4 p_8 p_{11} p_{13} (1 - p_7 p_9)$

2-TR (SEN+)$= r^4\ (2 - r^2)$

Paths (6, 7; $\delta = 1$) or (111, 001): {4–7–9–13}, {4–8–11–13}

1.3 Shuffle Exchange Network with Two Additional Stages (SEN+2)

It consists of $N/2$ SEs per stage, ($\log_2 N + 2$) stages having network complexity of $((N/2) * ((\log_2 N) + 2))$ [5, 7] and the model is shown in Fig. 5. RBD of SEN+2 is presented in Fig. 6.

Reliability expression for SEN+2:

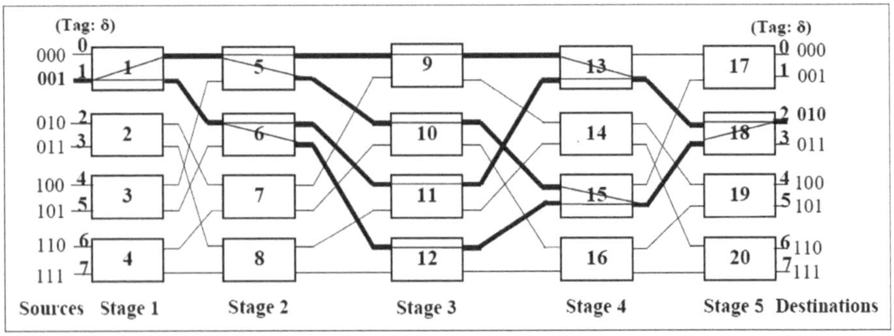

Fig. 5 SEN+2: 2-TR paths from $S_{1(\delta = 0,\ 1)}$ to $D_{18\ (\delta = 2,\ 3)}$

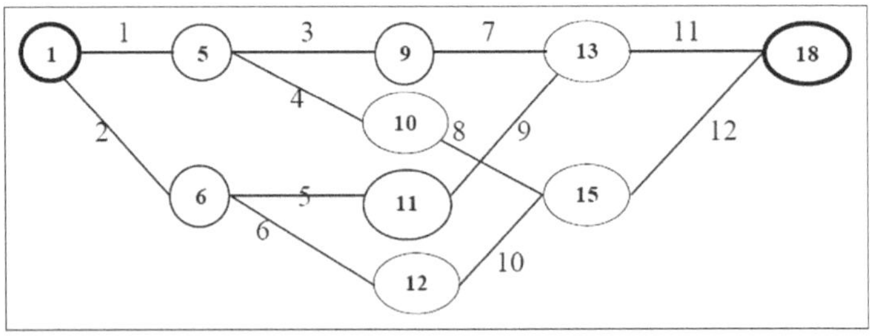

Fig. 6 RBD of SEN+2 for S_1–D_{18}

$$
\begin{aligned}
R = {} & p_1 p_5 p_9 p_{13} p_{18} + p_1 p_5 p_{10} p_{15} p_{18}(1 - p_9 p_{13}) + p_1 p_6 p_{11} p_{13} p_{18}(1 - p_5) \\
& + p_1 p_5 p_6 p_{11} p_{13} p_{18}(1 - p_{10} p_{15})(1 - p_9) + p_1 p_6 p_{12} p_{15} p_{18}(1 - p_5)\,(1 - p_{11} p_{13}) \\
& + p_1 p_5 p_6 p_{12} p_{15} p_{18}(1 - p_{10})(1 - p_{13}) \\
& + p_1 p_5 p_6 p_{12} p_{13} p_{15} p_{18}(1 - p_9)\,(1 - p_{10})(1 - p_{11})
\end{aligned}
$$

2-TR (SEN+2)= $r^5\ (4 - r^5 + 4r^4 - 2r^3 - 4r^2)$

Paths (1, 2; $\delta = 1$) or (001, 010): {1–5–9–13–18}, {1–5–10–15–18}, {1–6–11–13–18}, {1–6–12–15–18}

2 Performance Analysis

Performance analysis of MIN is a very critical issue in parallel processing systems. Performance of MIN can be analyzed from different indices like 2-TR, network complexity, fault-tolerance, and hardware cost.

2.1 2-Terminal Reliability Comparison

For source to destination node pairs, the evaluation of MINs for 2-terminal reliability are presented in Table 1 and Fig. 7.

2.2 Network Complexity

It is the function of *N* (switch input/output size) which provides information regarding number of SEs in the network and its connectivity pattern [6]. SEN, SEN+, and SEN+2 have lower 2-TR at relatively lower network complexity. The network complexity comparison is shown in Table 2.

2.3 Hardware Cost

Various MIN comparisons based on hardware cost are illustrated in Table 3.

2.4 Comparison Summary of MINs

The summary of various MIN which is self-explanatory is provided in Table 4 for ease of understanding the characteristics.

3 Conclusion

2-TR reliability of MIN is evaluated for different tag values and validated. In any switch/link failure state, the MIN design provides full access and dynamic rerouting property. The performance analysis based on various parameters signifies fault-tolerant, cost-effective, reroutable, affordable, and ideal for multiprocessor interconnection network.

2-TR increases exponentially for switching element reliability values 0.5–0.9 and linearly for 0.9–0.99. Hardware cost of SEN+2 model is slightly higher than that of other existing models, where the additional SEs in SEN+2 do not increase the terminal reliability, rather these added links make the possibility of high network system failure.

This concludes that the better design of MINs should have the provision of more disjoint paths, higher terminal reliability, low hardware cost, and high rerouting capability.

Table 1 2-TR evaluation of MINs at SE reliability value (0.50, 0.60, 0.70, 0.80, 0.90, 0.91, 0.92, 0.93, 0.94, 0.95, 0.96, 0.97, 0.98, and 0.99) for $N = 8$

Model	Switching element reliability													
	0.50	0.60	0.70	0.80	0.90	0.91	0.92	0.93	0.94	0.95	0.96	0.97	0.98	0.99
	Terminal reliability (2-TR)													
SEN	0.12500	0.21600	0.34300	0.51200	0.72900	0.75357	0.77869	0.80436	0.83058	0.85738	0.88474	0.91267	0.94119	0.970299
SEN+	0.10938	0.21254	0.36255	0.55706	0.78076	0.80363	0.82643	0.84911	0.87163	0.89392	0.91594	0.93761	0.95889	0.979712
SEN+2	0.08496	0.18481	0.33956	0.54484	0.78884	0.81088	0.83276	0.85445	0.87593	0.89719	0.91823	0.93902	0.95957	0.979896

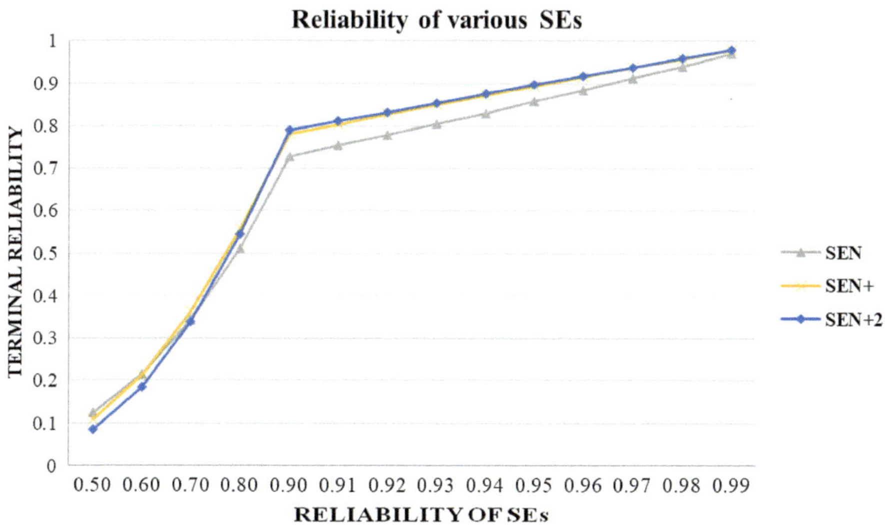

Fig. 7 2-TR evaluation of MINs

Table 2 Network complexity comparison

Model	Tag number (δ)	2-TR expressions	Network complexity	2-TR ($r = 0.9$)
SEN [6, 7]	0, 1, 2, 3, 4, 5, 6, 7	r^3	12	0.72900
SEN+ [5, 7]	0, 1, 2, 3, 4, 5, 6, 7	$r^4 (2 - r^2)$	16	0.78076
SEN+2 [5, 7]	0, 1, 2, 3, 4, 5, 6, 7	$r^5 (4 - r^5 + 4r^4 - 2r^3 - 4r^2)$	20	0.78884

Table 3 Cost of various MINs for $N = 8$

Stages	SEN	SEN+	SEN+2
1	$2 \times 2 \times 4$	$2 \times 2 \times 4$	$2 \times 2 \times 4$
2	$2 \times 2 \times 4$	$2 \times 2 \times 4$	$2 \times 2 \times 4$
3	$2 \times 2 \times 4$	$2 \times 2 \times 4$	$2 \times 2 \times 4$
4	–	$2 \times 2 \times 4$	$2 \times 2 \times 4$
5	–	–	$2 \times 2 \times 4$
Cost	48	64	80
No. of SEs	12	16	20

Researchers should attempt for obtaining reliable and fault-tolerant MIN and methods to enumerate the precise reliability of MINs for complex parallel systems.

Table 4 Comparison summary of MINs

Sl. No.	Type of comparison	SEN	SEN+	SEN+2
1.	No. of stages ($N = 2^n$)	$\log_2 N$	$\log_2 N + 1$	$\log_2 N + 2$
2.	Switches	$N/2 * (\log_2 N)$	$(N/2) * (\log_2 N + 2)$	$(N/2) * (\log_2 N + 1)$
3.	Size of switches	2×2	2×2	2×2
4.	For 8×8 size	3 stages 12 SEs 32 links	4 stages 16 SEs 40 links	5 stages 20 SEs 48 links
5.	Fault-tolerance	No	Yes	Yes
6.	Fault-tolerance paths	–	Disjoint path	Disjoint path
7.	Disjoint	0	2	2
8.	Multiplexer/de-multiplexer	–	–	–
9.	Hardware cost	48	64	80
10.	2-TR	r^3	$r^4\,(2 - r^2)$	$r^2[1 - (1 - (r^2(1 - (1 - r)^2)))^2]$

Acknowledgements The authors acknowledge the National Institute of Technology, Jamshedpur, India, for providing the research opportunity and facilities.

References

1. Abd-El-Barr M (2007) Design and analysis of reliable and fault-tolerant computer systems. Imperial College Press, London
2. Prakash A, Yadav DK, Choubey A (2020) A survey of multistage interconnection networks. Recent Adv Electr Electron Eng Bentham Science 13(2):165–183
3. Rajkumar S, Goyal NK (2016) Review of multistage interconnection networks reliability and fault tolerance. IETE Tech Rev 33(3):223–230
4. Goyal NK, Rajkumar S (2017) Multi-source multi-terminal reliability evaluation of interconnection networks. Microsys Technol 23(1):255–274
5. Gunawan I (2008) Reliability analysis of shuffle-exchange network systems. Reliab Eng Syst Saf 93(2):271–276
6. Blake JT, Trivedi KS (1988) Reliability of shuffle-exchange networks and its variants. IEEE, TH-0209-7/88
7. Bistouni F, Jahanshahi M (2014) Analyzing the reliability of shuffle-exchange networks using reliability block diagrams. Reliab Eng Syst Saf 132:97–106
8. Blake JT, Trivedi KS (1989) Multistage interconnection network reliability. IEEE, Trans. Comput 38(11)

Chapter 50
Predicting English Premier League Matches Using Classification and Regression

Tuhin Kundu, Akash Roy Choudhury, and Sruti Rai

1 Introduction

The English Premier League (EPL) is the highest level of the England football system which is contested by 20 clubs and operated by the system of promotion and relegation with the English Football League. Each team in the EPL plays every other team in the league twice (at home and away stadiums), thus essentially the total number of football matches conducted in a single season stands at $380(=2 *^{20} C_2)$. A single season usually runs from August to May in the following year, except for exceptional circumstances. The EPL is the most watched sports league in the world, broadcasted in 212 territories to 643 million homes and a potential television audience of 4.7 billion people approximately. The league generates revenue in the region of €2.2 billion in international and domestic television rights alone with overall revenue from multiple sources such as broadcast, merchandising, and tickets peaking at €5.8 billion in the 2016/17 season.

Sports gambling in the United Kingdom has been legalized and has a long history of being controlled, but has been relaxed more recently. The online sports betting market in the United Kingdom is approximately worth £650 million with a compounding annual growth rate of 7% from 2009 to 2012. The total online gambling population in the United Kingdom is estimated at 2.1 million customers. These statistics showcase the huge popularity of online sports gambling which makes us consider the betting

T. Kundu · A. R. Choudhury · S. Rai (✉)
Jalpaiguri Government Engineering College, Jalpaiguri, West Bengal, India
e-mail: sr1914@cse.jgec.ac.in

T. Kundu
e-mail: tuhinkundu@outlook.com

A. R. Choudhury
e-mail: arc1909@cse.jgec.ac.in

© Springer Nature Singapore Pte Ltd. 2021

S. D. Purohit et al. (eds.), *Proceedings of International Conference on Communication and Computational Technologies*, Algorithms for Intelligent Systems,
https://doi.org/10.1007/978-981-15-5077-5_50

statistics available across the entirety of 12 seasons to train models capable of classifying the outcome of a given football competitive match in the EPL between any two given teams.

The rest of the paper is organized as follows: Sect. 2 provides a brief review of the literature to understand the problem and to build the model. Section 3 describes the proposed methodology on the dataset that is used, the preprocessing steps done on it, and how the models are built. In Sect. 4, classifiers performance is investigated in terms of training accuracy and testing accuracy. Finally, the article is concluded with future work in Sect. 5.

2 Literature Review

Research for predicting the result of a football match started as early as 1977 by Stefani [25] which used a model called the least squared model that used a goal scoring distribution matrix to rate the strengths of both home and away teams. Min et al. [15] used an intricate framework to predict the results of football matches. This intricate framework, also known as the FRES system, used time-series method to generate more accurate and realistic predictions. The system consisted of two elements: Bayesian network element and rules-based theorem, which needed adequate expert knowledge for it to run well. Ulmer et al. [26] used historical match data to create a feature set to include gameday data and the current team form and used five different supervised learning classifiers to predict match outcomes.

Baio et al. [2] suggested a Bayesian hierarchical model by considering the number of goals scored by both teams in each match which in turn is used to develop the prediction model to predict match results. Constantinou et al. [5] developed a prediction model called pi-rating using various kinds of statistics for the EPL 2010/11 season which included match outcomes such as home win, away win, or draw, with the inclusion of objective and subjective information such as team form and strength, fatigue, and psychological impact. Joseph et al. [12] used Bayesian networks to forecast results. Bayesian networks provided means of displaying, representing, and predicting the expert knowledge in the game of football while also showing that it outperformed other supervised classification algorithms such as decision tree learner, Naive Bayes, and k-nearest neighbor. They used attributes such as home advantage, presence and absence of important players, and opposition team quality based on the league table position. Thus, we can see that Bayesian network has been extensively used for the problem of football match prediction.

Predictive models for football matches in the English Premier League have been developed previously. Schumaker et al. [23] gathered tweet sentiments related to football and used it to predict outcomes and use as a wagering decision system. Rosenberg et al. [21] used a swarm intelligence model that amplified predictive accuracy when games were considered together as real-time swarms. Ryan et al. [22] analyzed the influence of maturity status and investigated the relationships between physical tests. Angelini et al. [1] proposed a PARX model to model the goal distribution of a football team and improve forecast performance. Koopman et al. [14] proposed a statistical model for forecasting of football match results and use the statistical framework to improve bookmaker's odds.

3 Proposed Methodology

3.1 *Supervised Machine Learning Algorithms*

Linear Regression Let us take into assumption that there are a total of $n \geq t+1$ observational points on a t-vector of input variables, $x' = (x_1 \ldots x_i)$ and a scalar response, y, such that $j = 1 \ldots n$ where jth response is determined by Eq. (1).

$$y_i = \beta_0 + \sum_{i=1}^{t} \beta_i x_{ij} + e_j \tag{1}$$

where e_j is the residual which is independently and identically distributed, with unknown variance, σ^2 and mean zero. The inputs x_{ij} are considered to be specified design variables, but in specific cases, it is more appropriate to use them as random variables and assume joint distribution on y and x, for suppose in multivariate norm. Implicit note is considered that the assumption $x_1 \ldots x_t$ include all relevant variables although extraneous variables is included. Equation (2) is expressed in a matrix as

$$Y = X\beta + e \tag{2}$$

where Y is the vector of n size containing observed responses, X is the design matrix with dimensional size $n \times (t+1)$ as defined by Eq. (1), where it is assumed that $t+1$ is the rank and β is the $(t+1)$ vector containing unknown regression coefficients [11, 24].

Gradient Boosting Regressor Small trees, where each tree has a high bias, sequentially add up for gradient boosting regressor trees (GBRT). A new tree is to be added with each new iteration which is focused on the documents which are responsible for the remaining (currently) regression error.

Let $T(x_i)$ of sample x_i be the sample prediction. Assume a convex, differentiable, and continuous loss function $L(T(x_i), \ldots, T(x_n))$ which attains minimum $T(x_i) = y_i$ for all values of x_i. With every new iteration, the existing classifier $T(.)$ is added with a new tree $h(.)$. The first-order Taylor expansion $L(T + h_i)$ is used to calculate the best $h(.)$ with respect to $h_t(.)$, which is minimized [16].

Instance space $x_1, \ldots, x_n$ is used to calculate the gradient descent by GBRT. Equation (3) showcases the gradient step used to predict $T(x_i)$ in every iteration.

$$T(x_i) \leftarrow T(x_i) - \alpha \frac{L}{T(x_i)} \tag{3}$$

where learning rate is denoted by $\alpha > 0$. Prediction of regression tree $h_t(x_i)$ is used to be approximated along with the negative gradient as shown in Eq. (4).

$$h_t \approx -\arg\min_{h \in T_d} \sum_{i=1}^{n} (h_t(x_i) - r_i)^2, \quad \text{where } r_i : \frac{\partial L}{\partial T(x_i)} \tag{4}$$

Sequential Minimal Optimization Support vector machine (SVM) is an algorithm specifically utilized for binary classification tasks and uses an ensemble of SVMs in a stacked structure which is used for multiple class classification tasks [6]. Sequential minimal optimization (SMO) is a form of SVM which has been optimized by John Platt's SMO algorithm [13], which uses either the Gaussian or the polynomial kernel function. SMO breaks down the optimization problem in SVM into the smallest possible subproblems which are solved analytically on an independent basis [3], whereas it uses the concept of pairwise coupling to tackle multi-class problems, thus widening its usefulness as a multi-class classifier [10, 27].

AdaBoost Freund and Schapire [7] proposed the AdaBoost algorithm which converts a weak classifier into a highly accurate prediction rule by learning the weak classifier sequentially and combining it with the weak hypothesis linearly [4]. The input dataset is denoted by Eq. (5).

$$S = \{(x_1, y_1), \ldots, (x_m, y_m)\}, y \in \{-1, +1\} \tag{5}$$

where x_i represents ith vector and y_i would represent the target class labels. AdaBoost repeatedly invokes the weak learning classifier in a series representing the time intervals $t = 1, 2, \ldots, T$ with each sample from the training set having equal initialized weight as shown in Eq. (6).

$$\omega_{t,i} = \left\{ \begin{array}{l} \frac{1}{2M}, \; y_i = -1 \\ \\ \frac{1}{2L}, \; y_i = 1 \end{array} \right\} \tag{6}$$

where M denotes all the 'false' samples and L denotes all the 'true' samples. T iterations were used to train using the training set where the target is to find the optimal classifier h_i which is optimized to be the strongest classifier. The realization of the strongest classifier is identified by updating and manipulating the weights of the classifier by increasing or decreasing it after each iteration of the training round of the training set. The classifier is updated according to the weights given by Eq. (7).

$$D_{t+1}(i) = \frac{D_t(i)e^{-\alpha_i y_i h_i(x_i)}}{Z_t} = \frac{e^{-\sum_{j=1}^{t} \alpha_i y_i h_i(x_i)}}{L\prod_{j=1}^{t} Z_j} = \frac{e^{-mrg(x_i, y_i, f_i)}}{L\prod_{j=1}^{t} Z_j} \tag{7}$$

where h_i is the base classifier, Z stands for the normalization factor, α_i is the parameter that plays a role in the minimization of h_t, and $mrg(x_i, y_i, f_i)$ represents the functional margin of a point (x_i, y_i) in respect to the function [18].

Bayesian Networks The Bayesian network is basically a probabilistic network that consists primarily of two segments, that is, the dependency structure and local probability models [17, 19]. The dependency structure signifies the relation between the variables which is known by constructing directed edges between the various variables while keeping the network free from directed cycles. Every variable is dependent on their parents which makes inclusion of other variables possible in an empty set as shown in Eq. (8).

$$p(x_1, \ldots, x_n) = \prod_{i=1}^{n} p(x_i | Pa(x_i)) \quad (8)$$

where $Pa(x_i)$ are supposed to be the parents of x_i. The local probability models specify the dependency of the variable with respect to their parents. Discrete-valued Bayesian networks are used; thus, the local probability models are showcased by conditional probability tables (CPTs) [8].

3.2 *Correlation-Based Feature Selection (CFS)*

CFS considers a function as an input and uses it with a search algorithm to justify the given feature set. CFS heuristics include the quality of feature subsets to predict class labels by considering individual features with the level of intercorrelation between them. Equation (9) conveys the heuristics of the algorithm.

$$G_s = \frac{k\bar{r_d}}{\sqrt{k + k(k-1)\bar{r_d}}} \quad (9)$$

The numerator of the heuristic is an indication of the predictive prowess of the feature set within the class, whereas the denominator signifies the redundancy among the feature set. The heuristic quality or goodness should remove the irrelevant features which are poor predictors of the class in question.

Continuous features present in the feature set are binned by converting them to nominal. If X and Y are considered to be discrete random variables with respective ranges R_x and R_y, entropy of Y before and after observing variable X is provided by Eqs. (10) and (11) [9].

$$H(Y) = -\sum_{y \in R_y} p(y) \log(p(y)) \quad (10)$$

$$H(Y|X) = -\sum_{x \in R_x} p(x) \sum_{y \in R_y} p(y|x) \log(p(y|x)) \quad (11)$$

3.3 Dataset Used

The historical match and bookie odds datasets were collected for the seasons between 2005–2006 and 2016–2017, both inclusive. A single season in the EPL hosts to 380 ($=2 *^{20} C_2$) matches; thus, 12 seasons hosts a total of 4560 ($=380 \times 12$) matches. The historical real-time data stored was obtained along with the betting datasets to obtain the bookie odds of some of biggest gambling companies such as Bet365, Ladbrokes, Interwetten, Willian Hill, Betbrain, among others.

One of the main challenges of the predictive model is to account for the nature of randomness of the dataset. EPL is one of the most unpredictable football leagues in the world. It is primarily because, several times, a team from the bottom half of the league table defeats its opponent from the top half of the league table. This complicates the task of predicting the match outcomes as there is always a certain probabilistic chance for a team to win a match. In the 12 seasons of EPL statistics that was considered for our model, it is observed that among the total 4560 games played, the home team had won 2127 of them, 1291 games had been won by the away teams, and rest of the 1142 games resulted into a draw. In terms of percentage, it makes up 46.7% won by the home team, 28.3% by the away team, and 25% of the matches resulted into a draw. With the statistical data, the entropy is calculated as 0.96 in Eq. (12). The value of 0.96 is very close to an entropy value of 1, which signifies pure randomness.

$$\text{Entropy} = -(0.47 * \log_3(0.47) + 0.28 * \log_3(0.28) + 0.25 * \log_3(0.25)) = 0.96 \tag{12}$$

Four models are considered to search for the optimal predictive model. The entire experiment is divided into four tasks. Football match results are classified into three classes which are home team win, away team win, and draw. To achieve the target of optimal classification into the three classes, classification as well as regressional algorithms are used.

3.4 Data Preprocessing

Data obtained of all 12 seasons of the EPL were merged using the feature labels into a single dataset. Any feature with a large number of missing data was removed in the preprocessing step, because of the randomness of the historical data in the datasets.

This single dataset created is taken as input for both training and testing purposes. The dataset consisted of the following historical match statistics:

- Home and away teams
- Full-time home and away team goals
- Half-time home and away team goals
- Full- and half-time results

- Home and away team shots
- Home and away team shots on target
- Home and away team corners
- Home and away team fouls committed
- Home and away team yellow cards
- Home and away team red cards.

The betting odds statistics obtained and used in our model are as follows.

- Away win, home win, and draw odds for Bet365, Bet&Win, Interwetten, Ladbrokes, VC Bet, William Hill.
- Betbrain odds for maximum and average away win, home win, and draw.
- Betbrain maximum and average over and under 2.5 goals.

3.5 Classification Model

The purpose of our classification model is to classify the outcomes of football matches on the basis of the analytical match and betting odds statistical features. Before building the model, data is collected and preprocessed. The preprocessed data is divided into training and testing sets and used as input by the predictive model. The training set is used to train the model, which is tested later on against the testing set.

The first task of the model is to predict match outcomes directly from the historical match statistical features. The second task included using the betting odds statistical features to predict match results and the examination of accuracy records for the betting companies. The primary reason of examining the accuracy records is to find which company's betting statistical features are the most accurate. For this purpose, we used correlation-based feature selection (CFS) on the betting statistical features. The company whose average correlation coefficient associated with the target class is the highest among all is the company to provide the best odds. The third task is to combine the historical match statistical features and betting odds statistical features for the purpose of classification of the match outcomes. At this stage, we checked if betting statistical features could also be included in our classification models to attain our target classes. It was then compared to see if betting odds are to be considered as features to better the purpose of classification. The fourth task included choosing the appropriate features from the historical match statistical features and using their correlation coefficient to remove redundant features. The model is showcased via flowchart as shown in Fig. 1.

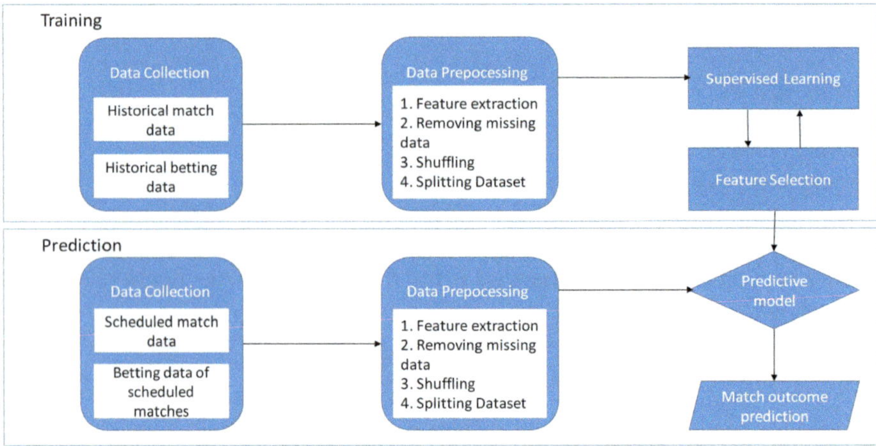

Fig. 1 Flowchart of the classification model developed

3.6 Classification via Regression Model

This model included the task to predict the full-time home team and away team goals using the supervised regression algorithms to determine the match outcomes on the basis of the number of goals scored.

This model used regression to predict the goals scored by the home and away teams in a particular match, using which match outcome is decided. The match outcomes predicted by the regression algorithms are used to calculate the predictive accuracy of the regressional model. Further, the model included correct outcomes even if the teams scored a different number of goals than the number of goals scored in the actual match. For example, if in a game between Manchester City and Manchester United at Manchester City's home ground, Manchester United won 2-1, and if the regression model predicted the goals wrongly to be 3-2, it still is able to predict that Manchester United scored more goals than Manchester City and emerge victorious. Thus, it is able to classify that match outcome accurately even if the number of goals predicted turned out to be incorrect.

Another task included is defining features present in the match statistical data as proactive features, for both the proposed models. Proactive features included non-redundant features important for the classification task. Among the statistical data, the dataset also consisted of non-proactive features such as full-time home goals, full-time away goals, half-time home goals, half-time away goals, half-time result, and full-time result. For the classification models, full-time home and away team goals cannot be considered as proactive features as predictive model would come down to a simple task of calculating the goal difference and getting the result without making any predictions. Also, if the half-time home team and away team goals are considered as proactive features, the predictive model would only function specifically for only the second half (45 min) of each match, instead of the full 90 min of the match. So, the four non-proactive features and half-time results are removed, but it kept the full-time result only as class labels. The class labels are actual match outcomes using which the predictive accuracy of the models are calculated by comparing the predictive and actual match outcomes.

The full-time away goals and full-time home goals are used as class labels in the regressional model. This is because we are predicting the match outcomes according to the predicted scores calculated by the regressional model. The goal difference calculated on the predicted scores is used to predict the match outcome, which then is compared with the actual match outcome. In our model, we are not predicting the result of the game directly but the goals scored by the home and away teams at the end of the match. Then, we differentiated the predicted goals scored by the home and away team using which we predicted the match result by calculating the goal difference.

4 Results

For the purpose of training and testing, the previously mentioned supervised models are used. *K* folds cross-validation is considered as the model validation technique for the two models proposed. *K* is taken as 5 with a 75–25% training–testing dataset split. The 75–25% training–testing dataset split includes match and betting statistical data for nine seasons, 2005–2006 to 2013–2014 as the training set, while the statistical data from three seasons, 2014–2015 to 2016–2017 as the testing set. Also, the training dataset is used as our test set to calculate the training accuracy, used to validate our models against overfitting and underfitting.

At first, the historical match statistical data and predicted the match outcome are used. The accuracies for the models are given in Table 1, where it is observed that SMO provides the best comparative accuracy of 58.8% among the classification algorithms.

Later, the betting odds statistical data is considered to predict the match result. The accuracies observed are given in Table 2.

Standalone accuracy results using betting statistical features are mediocre as given in Table 2 without the match statistical features. Finally, the combination of both historical match statistical features and betting odds statistical features is taken to check whether an optimal model is reached. The accuracies obtained by us are given in Table 3.

Table 1 Results using match statistical data

Classifier name	Training accuracy (%)	Cross-validation accuracy (%)	75–25% split accuracy (%)
SMO	59.6	57.5	58.8
AdaBoost	51.5	49.2	52.2
Bayes net	55.8	53.8	54.2

Table 2 Results using betting odds statistical data

Classifier name	Training accuracy (%)	Cross-validation accuracy (%)	75–25% split accuracy (%)
SMO	54.2	53.0	53.3
AdaBoost	51.6	52.0	52.9
Bayes net	50.6	49.3	49.8

Table 3 Results using match and betting odds statistical data

Classifier name	Training accuracy (%)	Cross-validation accuracy (%)	75–25% split accuracy (%)
SMO	61.4	58.0	59.0
AdaBoost	51.7	51.7	53.0
Bayes net	51.4	50.1	51.0

Table 4 Results for gambling company's correlation coefficient

Gambling company	Correlation coefficient
Interwetten	0.2305
Bet&Win	0.2298
Bet365	0.2255
Ladbrokes	0.2254
VC bet	0.2245
William Hill	0.2223

The correlation-based feature selection algorithm is applied on the betting odds statistical features to infer which company's betting statistical features were more important than the others in our feature set. The results of which are given in Table 4 where Interwetten has the highest correlation coefficient of 0.23 among others.

The correlation-based feature selection algorithm is applied on the historical match statistical features as well to see which features are the most important in our classification model. The result is given in Table 5. The results clearly indicate that the shots on target for the home and away teams contribute significantly to the match outcome having a correlation coefficient of 0.22 and 0.20, respectively.

Using the correlation coefficients, it is chosen to remove six features with least coefficient values. In spite of having poor correlation coefficient for the home team and away team features, they are used in the process of calculating the final accuracy of our models. The top nine features with the most correlation coefficient values are taken for our classification model and performed the experiments again. The results of which are given in Table 6.

Table 5 Results for historical match features correlation coefficient

Gambling company	Correlation coefficient
Home team shots on target	0.2215
Away team shots on target	0.2009
Away team shots	0.1622
Home team shots	0.1546
Home team red cards	0.0931
Home team yellow cards	0.0883
Away team red cards	0.0745
Home team	0.0394
Away team	0.0375
Away team fouls committed	0.0369
Home team fouls committed	0.0355
Away team yellow cards	0.0321
Home team corners	0.0300
Away team corners	0.0220
Month of game played	0.0187

Table 6 Results after considering feature selection on match statistical data

Classifier name	Training accuracy (%)	Cross-validation accuracy (%)	75–25% split accuracy (%)
SMO	59.6	57.3	58.5
AdaBoost	48.1	48.2	48.3
Bayes net	56.6	55.0	55.6

Table 7 Full-time home and away goal prediction accuracy

Regressor name	Full-time home goal training accuracy (%)	Full-time away goal training accuracy (%)	Full-time home goal testing accuracy (%)	Full-time away goal testing accuracy (%)
Linear regressor	43.3	49.2	42.5	46.0
Gradient boosting regressor	45.1	52.7	41.2	48.4

The proposed regressional models used linear regression and GBRT to predict the home and away team goals using the historical match statistics. The results of which are given in Table 7.

As discussed in the previous section, the full-time home and away team goals are predicted to calculate the match outcomes on the basis of goal difference. The results of training and testing accuracies of the full-time match result of the proposed regressional models are given in Table 8.

Table 8 Accuracy for full-time result

Regressor name	Training accuracy (%)	Testing accuracy (%)
Linear regressor	95.1	95.1
Gradient boosting regressor	98.3	97.4

The high accuracy of 97.4% using GBRT, despite the high entropy value, is attributed to regression models considering winning and loosing streaks of the home and away teams. Also, conventionally strong teams, such as Manchester City, Manchester United, Chelsea, Liverpool, and Arsenal, are going to always suffer a low number of losses every season except any other calamatical ambiguity. Such conventionally strong teams are expected to score a set number of goals at least in their home matches which provides them a huge advantage over other teams which makes the task of predicting the number of goals scored by the home team easier. This is why the classification via linear regression model achieves a very high accuracy despite a large number of wrong predictions in the number of goals a team scored, as it manages to point out the sentiment present about the match outcome in terms of home win, away win, and draw with a relative high accuracy on the basis of historical statistical data.

5 Conclusion and Future Work

During our experiments, various strategies are undertaken, and several models are applied to see their predictive accuracy on EPL matches. It is observed that among the three classification algorithms, SMO provided the best accuracy while the models are also evaluated for any overfitting or underfitting. The best classification testing accuracy observed is of 58.8% using SMO with match statistical features. The accuracy of all the classification models are mediocre primarily due to the randomness or entropy of the match statistical data. As previously discussed, the entropy value of all the matches in the 12 seasons was 0.96, indicating a value near pure randomness.

Razali et al. [20] in their work stated an accuracy of 79.47% using the same dataset taking only three seasons of statistical data for testing and training sets. The model included a single season of match statistical features in each of their predictive models. Every year, the teams on the top of the league table play well consistently, and the teams on the bottom of the table played poorly on a consistent basis which creates a pattern for the purpose of classification. But, when match statistical data of 12 years is taken into account, different teams have different league positions every season, thus vastly increasing the entropy of the match statistical data. This randomness nullifies the pattern that emerges in the model of Razali et al., which in turn reduces the predictive accuracy of our model.

By applying correlation-based feature selection algorithm on the betting statistical data, it is observed that the value of all the betting companies are different by a small margin. Yet in the last 12 seasons, Interwetten provided the most reliable betting odds among others. Calculating in the same way for historical match data, it is also observed that for the classification model, shots on target was most important feature followed by the total number of shots taken by each team, which is very understandable as the two features have a direct correlation in terms of goals scored in a football match in most cases.

From the results of the regressional models, it is observed that the predictive accuracy of the goals scored by each and every team is low, but the results of classification of match outcome using the predicted goals are very high. The model does not predict how many goals were scored by each team very accurately but is able to predict the goal difference between the teams, thus providing the correct match outcome 97.4% times using GBRT.

The future course of work is very promising when the problem of predictive statistics is similarly applied in other sports given statistical data is available. Betting odds improve the application of artificially intelligent algorithms that have a significant impact in the football gambling business. Inherent player statistics, team winning streaks, home team arena attendance, player form and passing attributes, club transfer policy and business, youth-level infrastructure investment, and scouting network are just some of the features that should be used to make the football match prediction model more accurate. Thus, predictive football analytics find their usage with football pundits, betting markets, and the viewers around the world of this sport.

References

1. Angelini G, De Angelis L (2017) Parx model for football match predictions. J Forecast 36(7):795–807
2. Baio G, Blangiardo M (2010) Bayesian hierarchical model for the prediction of football results. J Appl Stat 37(2):253–264
3. Chen L, Chu C, Feng K (2016) Predicting the types of metabolic pathway of compounds using molecular fragments and sequential minimal optimization. Comb Chem High Throughput Screen 19(2):136–143
4. Choi S, Kim YJ, Briceno S, Mavris D (2016) Prediction of weather-induced airline delays based on machine learning algorithms. In: 2016 IEEE/AIAA 35th digital avionics systems conference (DASC). IEEE, pp 1–6
5. Constantinou AC, Fenton NE, Neil M (2012) Pi-football: a bayesian network model for forecasting association football match outcomes. Knowl Based Syst 36:322–339
6. Cortes C, Vapnik V (1995) Support-vector networks. Mach Learn 20(3):273–297
7. Freund Y, Schapire RE (1997) A decision-theoretic generalization of on-line learning and an application to boosting. J Comput Syst Sci 55(1):119–139
8. Gevaert O, Smet FD, Timmerman D, Moreau Y, Moor BD (2006) Predicting the prognosis of breast cancer by integrating clinical and microarray data with bayesian networks. Bioinformatics 22(14):e184–e190
9. Hall MA, Smith LA (1998) Practical feature subset selection for machine learning. In: McDonald C (Ed) Computer Science '98 Proceedings of the 21st Australasian Computer Science Conference ACSC'98, Perth, 4–6 February, 1998. Springer, Berlin, pp 181–191

10. Hastie T, Tibshirani R (1998) Classification by pairwise coupling. In: Advances in neural information processing systems, pp 507–513
11. Hocking R (1976) Theanalysis and selection of variables in linear regression. Biometrica 32:1–49
12. Joseph A, Fenton NE, Neil M (2006) Predicting football results using bayesian nets and other machine learning techniques. Knowl Based Syst 19(7):544–553
13. Keerthi SS, Shevade SK, Bhattacharyya C, Murthy KRK (2001) Improvements to platt's SMO algorithm for SVM classifier design. Neural Comput 13(3):637–649
14. Koopman SJ, Lit R (2015) A dynamic bivariate poisson model for analysing and forecasting match results in the english premier league. J R Stat Soc Ser A Stat Soc 178(1):167–186
15. Min B, Kim J, Choe C, Eom H, McKay RB (2008) A compound framework for sports results prediction: a football case study. Knowl Based Syst 21(7):551–562
16. Mohan A, Chen Z, Weinberger K (2011) Web-search ranking with initialized gradient boosted regression trees. In: Proceedings of the learning to rank challenge, pp 77–89
17. Neapolitan RE et al (2004) Learning bayesian networks, vol 38. Pearson Prentice Hall, Upper Saddle River
18. Niu B, Cai YD, Lu WC, Li GZ, Chou KC (2006) Predicting protein structural class with AdaBoost learner. Protein Pept Lett 13(5):489–492
19. Pearl J (2014) Probabilistic reasoning in intelligent systems: networks of plausible inference. Elsevier
20. Razali N, Mustapha A, Yatim FA, Ab Aziz R (2017) Predicting football matches results using bayesian networks for english premier league (EPL). In: IOP conference series: materials science and engineering, vol 226. IOP Publishing, p 012099
21. Rosenberg L, Pescetelli N (2017) Amplifying prediction accuracy using swarm AI. In: 2017 intelligent systems conference (IntelliSys). IEEE, pp 61–65
22. Ryan D, McCall A, Fitzpatrick G, Hennessy L, Meyer T, McCunn R (2018) The influence of maturity status on movement quality among english premier league academy soccer players. Sport Perf Sci Rep 32:1–3
23. Schumaker RP, Jarmoszko AT, Labedz CS Jr (2016) Predicting wins and spread in the premier league using a sentiment analysis of twitter. Decis Supp Syst 88:76–84
24. Seber GA, Lee AJ (2012) Linear regression analysis, vol 329. Wiley
25. Stefani RT (1977) Football and basketball predictions using least squares. IEEE Trans Syst Man Cybernet 7:117–121
26. Ulmer B, Fernandez M, Peterson M (2013) Predicting soccer match results in the english premier league. Ph.D. thesis, Doctoral dissertation, Ph.D. dissertation. Stanford
27. Xu Z, Dai M, Meng D (2009) Fast and efficient strategies for model selection of gaussian support vector machine. IEEE Trans Syst Man Cybernet Part B Cybernet 39(5):1292–1307

Chapter 51
Wavelet Transform-Based Artificial Intelligence Scheme for the Detection of Transforming Faults, Multi-location Faults and Cross-Country Faults in Wind Farm-Integrated Double-Circuit Transmission Line

Deepika Sharma, Gaurav Kapoor, and Shilpa Agarwal

1 Introduction

The perfect fault detection is the significant feature of any protection technique employed for the protection of transmission lines. Over the years, researchers have been in the search to identify quick and perfect detection of faults in the transmission lines using numerous protection techniques, so that the faulted section/zone can be secluded from credible critical effects caused by the fault. Accordingly, it is very decisive to have a competent protection technique for detecting the faults in transmission lines. Many protection techniques have been proposed in recent years, and the brief review of them is described hereafter. In [1], a combination of S-transform and FFNN has been used for the protection of distribution grids. Db4WT has been applied for the protection of fifteen-phase transmission line [2]. In [3], a fault detection technique using WHT has been proposed for SCCTL protection. HHT has been used for SPTL protection [4]. In [5], WT has been used for fault detection in DCTL. In [6], HHT has been used. In [7, 8], wavelet transform has been applied for detecting faults in series compensation-based three-phase and six-phase transmission lines, respectively. In [9, 10], Hilbert-Huang transform has been employed for fault recognition in wind-connected series compensation-based transmission line and six-phase transmission

D. Sharma · S. Agarwal
Department of Electrical Engineering, Sobhasaria Engineering College, Sikar, India
e-mail: dipikaas06@gmail.com

S. Agarwal
e-mail: hodee@secs.ac.in

G. Kapoor (✉)
Department of Electrical Engineering, Modi Institute of Technology, Kota, India
e-mail: gaurav.kapoor019@gmail.com

© Springer Nature Singapore Pte Ltd. 2021
S. D. Purohit et al. (eds.), *Proceedings of International Conference on Communication and Computational Technologies*, Algorithms for Intelligent Systems,
https://doi.org/10.1007/978-981-15-5077-5_51

line, respectively. Further, in [11, 12], wavelet-based schemes were developed for the protection of twelve-phase and three-phase transmission lines, respectively. In [13], a mathematical morphological gradient-based technique was designed for double-circuit transmission line protection. In [14, 15], the faults were detected in double circuit transmission line using wavelet transform. In [16, 17], wavelet transform and mathematical morphological gradient were used for detecting/classifying faults in six-phase transmission line, respectively.

In this work, the Db5WT is used and is executed for fault detection in the WFIDCTL. Such type of study has not been reported thus far to the top of the information of the authors. The results show that the Db5WT expertly detects the faults. The consistency of Db5WT is not susceptible to the fault factors variation.

This paper is prepared as: Sect. 2 presents the details of WFIDCTL. Section 3 describes the fault detection process. Section 4 presents the results. Section 5 ends the paper.

2 DCTL Integrated with Wind Farm

The design of WFIDCTL is exemplified in Fig. 1. The WFIDCTL is connected to a 400 kV, 50 Hz source at the sending end. The WFIDCTL is 200 km long in length. Further, the WFIDCTL is divided into two zones. Each zone has a length of 100 km. For the measurement of fault currents of both circuits, the current measurement blocks are connected at bus-1. The simulation model of a 400 kV WFIDCTL is designed using MATLAB.

The currents of circuit-1 and circuit-2 for no-fault are shown in Fig. 2. Figure 3 depicts the Db5WT coefficients of circuit-1 currents for no-fault. Figure 4 shows the Db5WT coefficients of circuit-2 currents for no-fault. The results of Db5WT for no-fault are reported in Table 1.

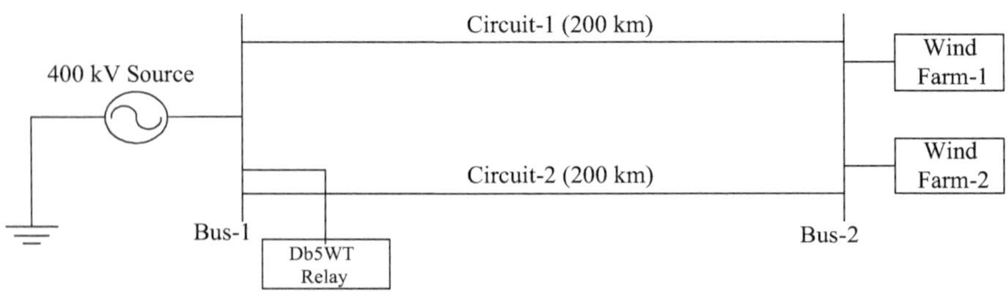

Fig. 1 Demonstration of WFIDCTL

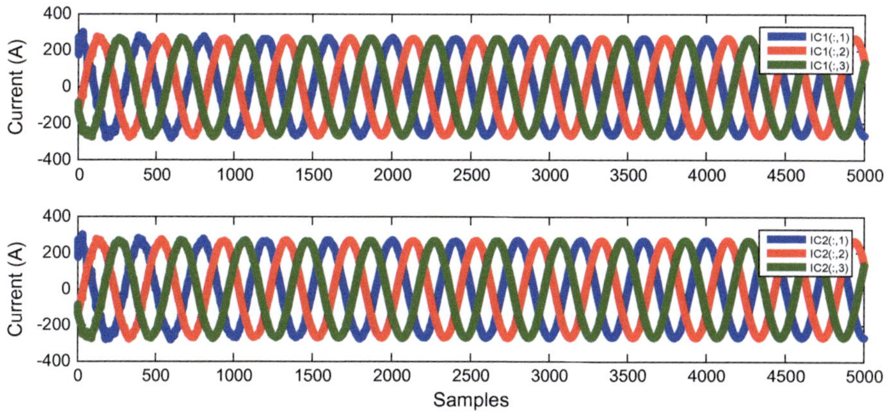

Fig. 2 Circuit-1 and circuit-2 currents for no-fault

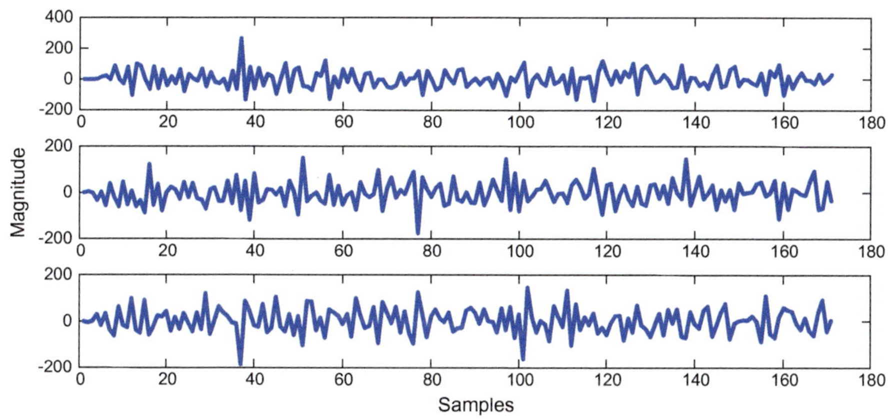

Fig. 3 Db5WT coefficients of circuit-1 currents for no-fault

3 The Process for Daubechies-5 Wavelet Transform (Db5WT)

The process for Db5WT is shown in Fig. 5, and the steps are exemplified beneath.

Step 1 Simulate the WFIDCTL for triggering faults and generate the post-fault currents for both circuits.

Step 2 Use Db5WT to inspect the post-fault currents for characteristics retrieval and evaluate the range of Db5WT coefficients.

Step 3 The phase will be asserted as the faulted phase if its Db5WT coefficient has the larger amplitude as compared to the healthy phase.

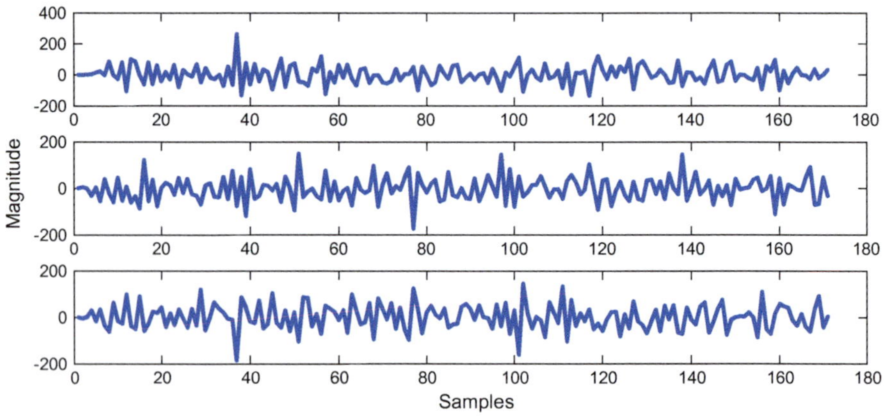

Fig. 4 Db5WT coefficients of circuit-2 currents for no-fault

Table 1 Results of Db5WT for no-fault

Db5WT coefficients					
Phase-A1	Phase-B1	Phase-C1	Phase-A2	Phase-B2	Phase-C2
262.8732	149.5768	146.3137	262.8732	149.5768	146.3137

4 Performance Assessment

The simulation studies have been carried out for the converting faults, multi-position faults and cross-country faults with the objective of verifying the feasibility of the Db5WT. In the separate subcategories, the outcomes of the work are investigated.

4.1 *The Response of Db5WT for Converting Faults*

The Db5WT has been tested for the converting faults. Figure 6 exemplifies the circuit-1 and circuit-2 currents of the WFIDCTL when initially an A2B2G fault at 60% from bus-1 at 0.05 s is firstly converted into the A2B2C2G fault at 0.2 s which is then again converted into the A2C2G fault at 0.35 s among $R_F = 20\ \Omega$ and $R_G = 20\ \Omega$. Figures 7 and 8, respectively, depict the Db5WT coefficients of the currents of circuit-1 and circuit-2. Table 2 reports the results for different converting faults. It can be viewed from Table 2 that the Db5WT perfectly detects all types of converting faults.

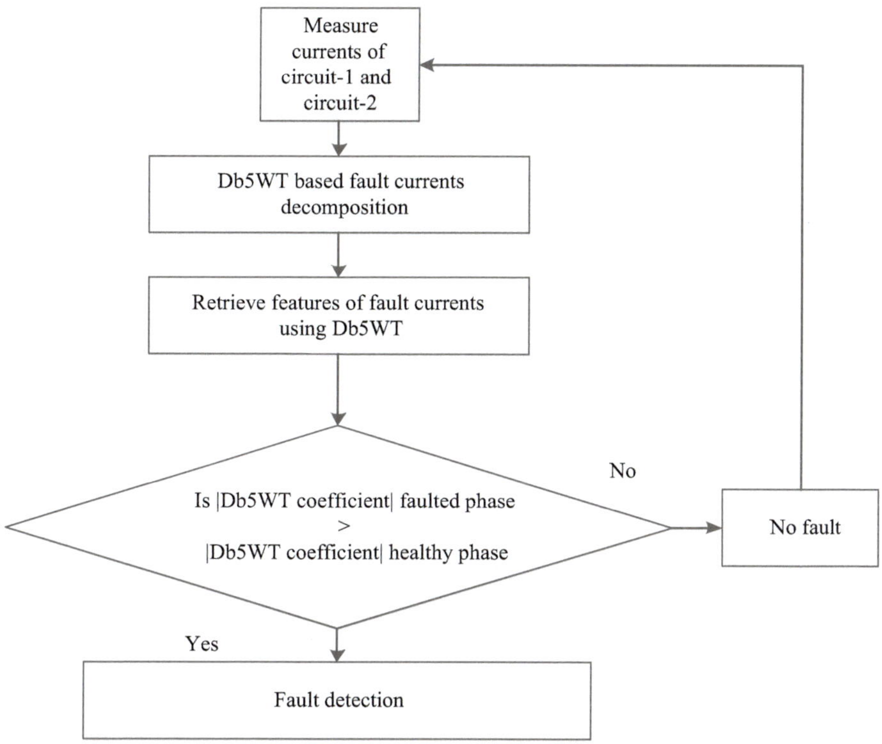

Fig. 5 Process for Db5WT

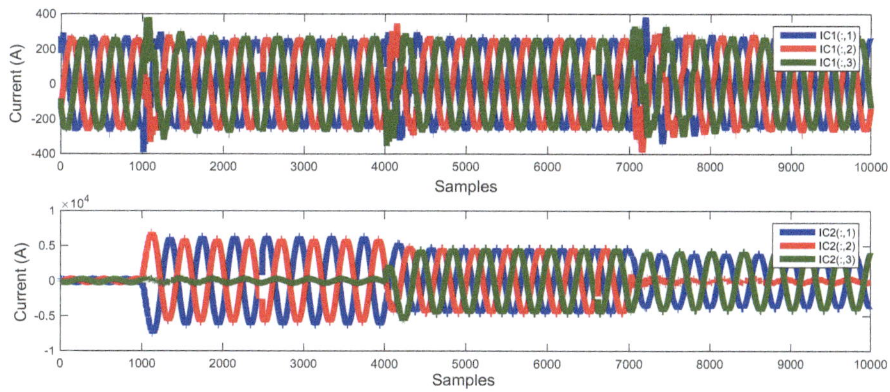

Fig. 6 Circuit-1 and circuit-2 currents when an A2B2G fault at 0.05 s is converted into A2B2C2G fault at 0.2 s which is again converted into A2C2G fault at 0.35 s at 60% from bus-1 among $R_F = 20\ \Omega$ and $R_G = 20\ \Omega$

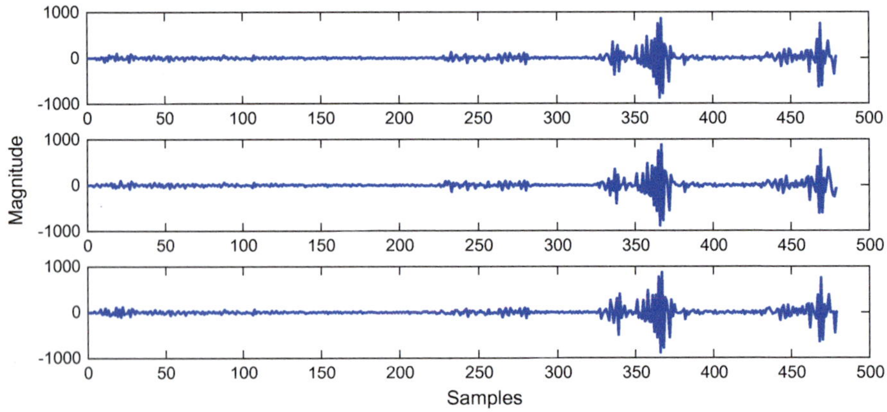

Fig. 7 Db5WT coefficients of circuit-1 currents for a converting fault

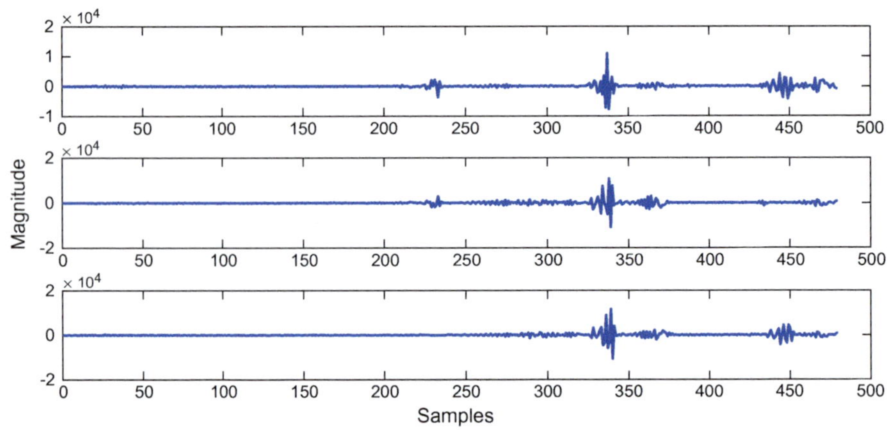

Fig. 8 Db5WT coefficients of circuit-2 currents for a converting fault

4.2 *The Response of Db5WT for Multi-position Faults*

The Db5WT is investigated for different multi-position faults. Figure 9 depicts the currents of circuit-1 and circuit-2 when the WFIDCTL is simulated for the A1B1G fault at 37.5% and A2B2C2G fault at 62.5% at 0.1075 s among $R_F = 8\ \Omega$ and $R_G = 10\ \Omega$. The Db5WT coefficients of circuit-1 and circuit-2 currents are shown in Figs. 10 and 11, respectively. Table 3 reports the results for the multi-position faults. It is seen from Table 3 that the Db5WT performs well for the detection of multi-position faults.

Table 2 Results of Db5WT for different converting faults

Fault with time (s)	Converted fault with time (s)	Db5WT coefficients					
		Phase-A1	Phase-B1	Phase-C1	Phase-A2	Phase-B2	Phase-C2
A1G (0.05)	A1B1G (0.2)	$1.1563 * 10^4$	$1.0222 * 10^4$	275.1417	167.0730	208.8998	258.9736
B1G (0.0925)	A1B1C1G (0.167)	$1.6971 * 10^4$	$1.4277 * 10^4$	$1.2749 * 10^4$	368.5240	550.8031	352.5181
C2G (0.074)	A2C2G (0.174)	193.6862	326.0349	173.0082	$4.7498 * 10^3$	344.0925	$4.8035 * 10^3$
A2B2C2G (0.05)	B2G (0.25)	556.0868	537.2722	517.9296	$3.8759 * 10^3$	$5.8357 * 10^3$	$3.3673 * 10^3$
A2B2G (0.05)	A2B2C2G (0.2) To A2C2G (0.35)	857.3156	882.0287	873.4756	$1.0910 * 10^4$	$1.0777 * 10^4$	$1.1617 * 10^4$

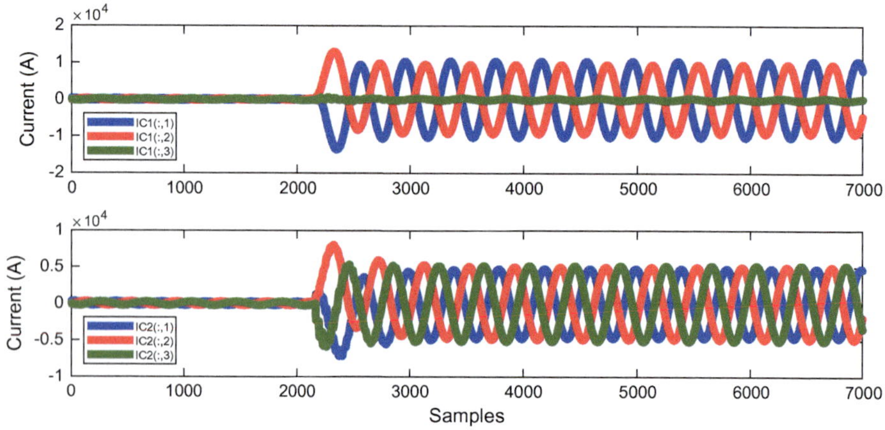

Fig. 9 Circuit-1 and circuit-2 currents for a multi-position fault A1B1G at 37.5% and A2B2C2G at 62.5% at 0.1075 s among $R_F = 8\ \Omega$ and $R_G = 10\ \Omega$

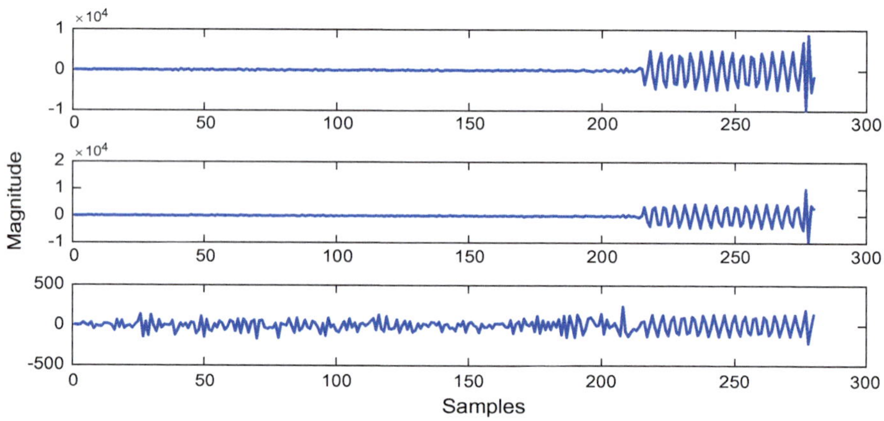

Fig. 10 Db5WT coefficients of circuit-1 currents for a multi-position fault A1B1G at 37.5% and A2B2C2G at 62.5% at 0.1075 s among $R_F = 8\ \Omega$ and $R_G = 10\ \Omega$

4.3 The Response of Db5WT for Cross-Country Faults

The Db5WT is checked for different cross-country faults. Figure 12 exemplifies the currents of circuit-1 and circuit-2 when the WFIDCTL is simulated for the cross-country A1G fault at 50% and B1C1G fault at 50% at 0.15 s among $R_F = 4\ \Omega$ and $R_G = 8\ \Omega$. The Db5WT coefficients of circuit-1 and circuit-2 currents are depicted in Figs. 13 and 14, correspondingly. Table 4 reports the results for different cross-country faults. It is examined from Table 4 that the Db5WT performs fine for the detection of cross-country faults.

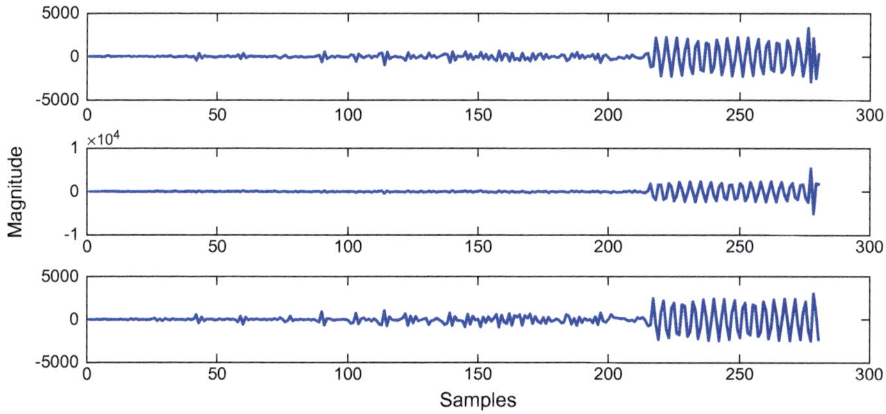

Fig. 11 Db5WT coefficients of circuit-2 currents for a multi-position fault A1B1G at 37.5% and A2B2C2G at 62.5% at 0.1075 s among $R_{\mathrm{F}} = 8\ \Omega$ and $R_{\mathrm{G}} = 10\ \Omega$

5 Conclusion

A Db5WT-based fault detection technique has been presented in this work for the protection of WFIDCTL. The Db5WT is exploited which decomposes the fault currents to estimate the Db5WT coefficients. The fault factors of WFIDCTL are varied. According to the results, it is discovered that the Db5WT efficiently detects the converting faults, multi-position faults and cross-country faults.

Table 3 Results of Db5WT for different multi-position faults

Fault-1 position (%)	Fault-2 position (%)	Db5WT coefficients					
		Phase-A1	Phase-B1	Phase-C1	Phase-A2	Phase-B2	Phase-C2
A1G (20)	C2G (80)	$3.5137 * 10^3$	143.3008	142.7626	169.4437	161.0264	733.1095
A1B1G (45)	A2C2G (55)	$1.0284 * 10^4$	$6.4017 * 10^3$	863.6864	$2.9318 * 10^3$	858.0003	$3.4845 * 10^3$
A1B1C1G (60)	A2B2C2G (40)	$2.3102 * 10^4$	$2.3151 * 10^4$	$2.3137 * 10^4$	$1.4747 * 10^4$	$1.4726 * 10^4$	$1.4740 * 10^4$
A1B1G (37.5)	A2B2C2G (62.5)	$8.7025 * 10^3$	$1.0074 * 10^4$	235.3219	$3.3063 * 10^3$	$5.3333 * 10^3$	$3.0050 * 10^3$
A2B2G (67.5)	C1G (32.5)	215.1429	379.8409	$2.3179 * 10^3$	$4.3500 * 10^3$	$2.4101 * 10^3$	575.6575

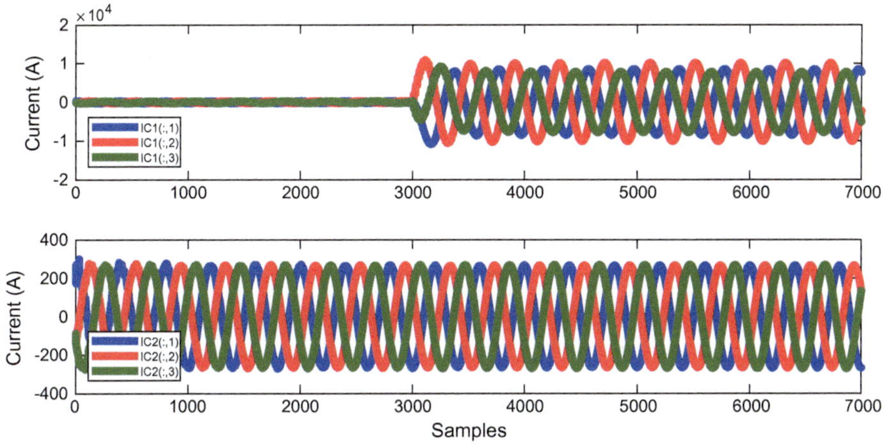

Fig. 12 Circuit-1 and circuit-2 currents for a cross-country fault A1G at 50% and B1C1G at 50% at 0.15 s among $R_F = 4\ \Omega$ and $R_G = 8\ \Omega$

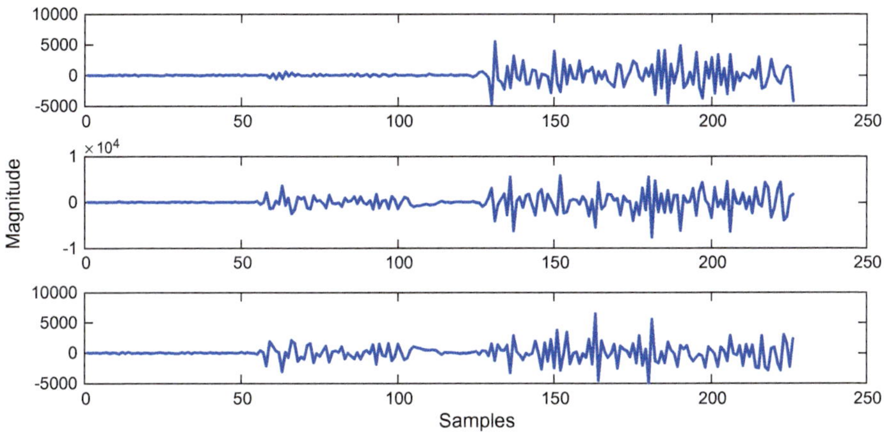

Fig. 13 Db5WT coefficients of circuit-1 currents for cross-country fault A1G at 50% and B1C1G at 50% at 0.15 s with $R_F = 4\ \Omega$ and $R_G = 8\ \Omega$

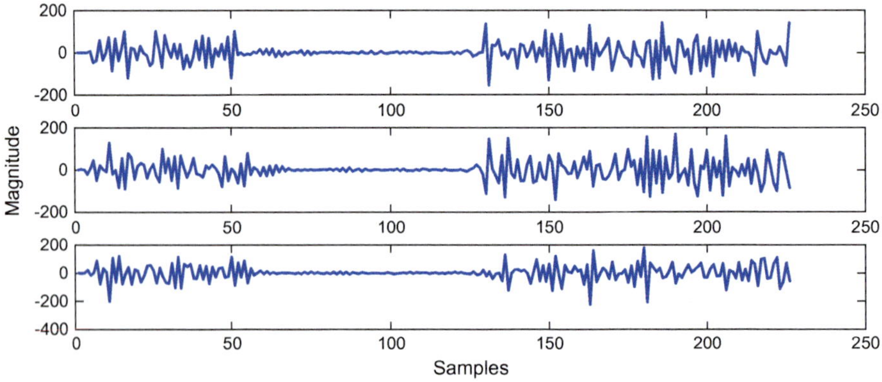

Fig. 14 Db5WT coefficients of circuit-2 currents for cross-country fault A1G at 50% and B1C1G at 50% at 0.15 s with $R_F = 4\ \Omega$ and $R_G = 8\ \Omega$

Table 4 Results of Db5WT for different cross-country faults

Fault-1 position (%)	Fault-2 position (%)	Db5WT coefficients					
		Phase-A1	Phase-B1	Phase-C1	Phase-A2	Phase-B2	Phase-C2
A1G (30)	B1G (70)	$1.0291 * 10^4$	$8.0801 * 10^3$	274.4176	143.3340	259.0455	157.2288
A1G (50)	B1C1G (50)	$5.5209 * 10^3$	$5.7609 * 10^3$	$6.4625 * 10^3$	142.3065	169.2909	177.3180
C2G (55)	A2G (45)	179.5677	308.7469	184.0779	$3.1036 * 10^3$	378.4034	$3.7690 * 10^3$
A2B2G (47.5)	C2G (52.5)	86.1705	162.8641	152.0568	$3.7427 * 10^3$	$3.8973 * 10^3$	$1.1973 * 10^3$
A2G (42.5)	B2G (57.5)	217.5079	218.6197	378.7452	$5.6364 * 10^3$	$2.8125 * 10^3$	395.3029

References

1. Shafiullah Md, Abido MA (2018) S-transform based FFNN approach for distribution grids fault detection and classification. IEEE Access 6:8080–8088
2. Kapoor G (2020) Fifteen phase transmission line protection using daubechies-4 wavelet transform. Int J Eng Sci Technol (IJEST-Africa) 12(1):1–14
3. Sharma P, Kapoor G, Ali S (2018) Fault detection on series capacitor compensated transmission line using Walsh Hadamard transform. In: Proceedings of IEEE international conference on computing, power and communication technologies (GUCON). IEEE, Greater Noida, pp 763–768
4. Kapoor G (2019) Detection and classification of single line to ground boundary faults in a 138 kV six phase transmission line using Hilbert Huang transform. i-manager's J Electr Eng 12(3):28–41
5. Gautam N, Kapoor G, Ali S (2018) Wavelet transform based technique for fault detection and classification in a 400 kV double circuit transmission line. Asian J Electr Sci 7(2):77–83
6. Kapoor G (2019) Detection and classification of three phase to ground faults in a 138 kV six-phase transmission line using Hilbert-Huang transform. JEA J Electr Eng 3(1):1–11
7. Sharma P, Ali S, Kapoor G (2018) Wavelet transform approach for fault detection of three phase transmission line compensated with series capacitor. Asian J Electr Sci 7(2):8–13
8. Kapoor G (2018) Fault detection of phase to phase fault in series capacitor compensated six-phase transmission line using wavelet transform. Jordan J Electr Eng 4(3):151–164
9. Sharma N, Ali S, Kapoor G (2018) Fault detection in wind farm integrated series capacitor compensated transmission line using Hilbert Huang transform. In: Proceedings of the IEEE international conference on computing, power and communication technologies (GUCON). IEEE, Greater Noida, pp 774–778
10. Kapoor G (2019) Detection and classification of four phase to ground faults in a 138 kV six phase transmission line using Hilbert Huang transform. Int J Eng Sci Technol (IJEST-Africa) 11(4):10–22
11. Kapoor G (2018) Wavelet transform based detection and classification of multi-location three phase to ground faults in twelve phase transmission line. Majlesi J Mechatron Syst 7(4):47–60
12. Rathore B, Shaik AG (2017) Wavelet-alienation based transmission line protection scheme. IET Gener Transm Distrib 11(4):995–1003
13. Kapoor G (2018) Mathematical morphology based fault detector for protection of double circuit transmission line. ICTACT J Microelectron 4(2):589–600
14. Gautam N, Ali S, Kapoor G (2018) Detection of fault in series capacitor compensated double circuit transmission line using wavelet transform. In: Proceedings of the IEEE international conference on computing, power and communication technologies (GUCON). IEEE, Greater Noida, pp 769–773
15. Kapoor G (2019) A protection technique for series capacitor compensated 400 kV double circuit transmission line based on wavelet transform including inter-circuit and cross-country faults. Int J Eng Sci Technol 11(2):1–20
16. Kapoor G (2018) Six-phase transmission line boundary protection using wavelet transform. In: Proceedings of the 8th IEEE India international conference on power electronics (IICPE). IEEE, Jaipur
17. Kapoor G (2018) Six phase transmission line boundary protection using mathematical morphology. In: Proceedings of the IEEE international conference on computing, power and communication technologies (GUCON). IEEE, Greater Noida, pp 857–861

Chapter 52
Performance Analysis of Dual-band, Triple-band and Multi-band Microstrip Antenna Using Metamaterial

Akshta Kaushal and Gaurav Bharadwaj

1 Introduction

Microstrip patch antenna is compatible that makes it unique by the attachment with planar monolithic microwave integrated circuit (MMIC) [1], easy fabrication, low cost, lightweight and low profile. Due to such features, these are the preferable and highly adaptable for wireless communication systems. Researchers suggest numerous topologies based on printed antenna to improve their performance. Changing the geometry of the rings is one of them which works at certain frequencies. In this work, several testings prove the credibility of microstrip patch antenna's dual-, triple- and multi-bands.

This work also includes the attractive properties of metamaterial which consists the concept of complementary split ring resonators (CSRRs). Metamaterial antenna improves the performance, such as by utilizing TL-MTM technique, it produces a multi-band response. It can also enhance the antenna parameters like gain and bandwidth. Here, this paper presents the outcomes attained by changing the orientation of rings at certain frequency or different frequencies with their enhanced bandwidth. Metamaterial also controls the polarization of antenna that is its main required feature [2–4].

This paper demonstrates the CSRRs based compact triple-band microstrip antenna at two unlike geometries. For WiMAX band having frequency range from 3.2 to 3.8 GHz, the conventional patch antenna is working at 3.5 GHz [5]. This patch antenna attached with two rectangular CSRRs consists of ground plane at suitable dimensions. Testing of all antennas has been performed on CST software that provides

A. Kaushal (✉) · G. Bharadwaj
Department of Electronics and Communication Engineering, Government Women Engineering College Ajmer, Ajmer, India
e-mail: akshtakaushal@gmail.com

G. Bharadwaj
e-mail: gauravb@gweca.ac.in

© Springer Nature Singapore Pte Ltd. 2021
S. D. Purohit et al. (eds.), *Proceedings of International Conference on Communication and Computational Technologies*, Algorithms for Intelligent Systems,
https://doi.org/10.1007/978-981-15-5077-5_52

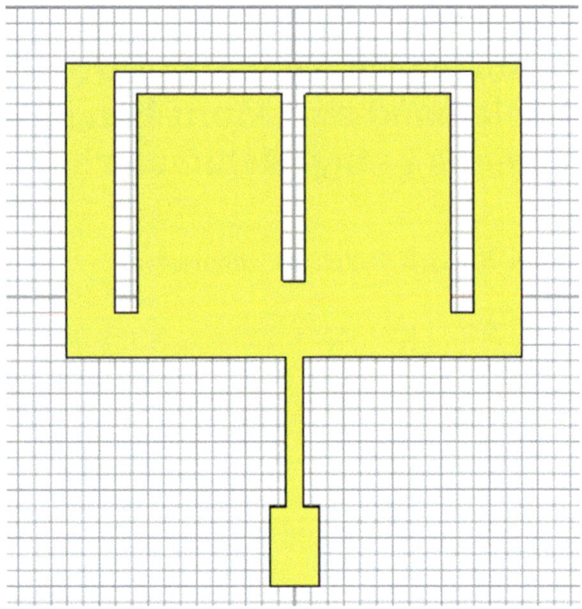

Fig. 1 Front view of E-slot microstrip patch antenna

the optimized outcomes as geometrical parameters. Different multi-band antenna designs support precise values at all resonant frequencies; i.e., specific resonator is denoted by specific resonant frequency. In addition, at different achieved frequencies, relatively high radiation characteristics make proposed designs a good option for WiMAX [6], UWB and radar sensor applications (Fig. 1).

Nature does not exhibit the several properties of engineered metamaterials. The different structures and properties of metamaterials highly affect the electromagnetic environment. In previous years, Silva et al. suggested a developed design of antenna that used resonating arrays where splitting resonator is attached in microstrip patch antenna. This design gave the theory of split ring resonator that changed the behavior of electromagnetic properties of antenna [7]. Furthermore, the work extended on behalf of the sides of the feed attached with the loading of a pair of split ring resonators [8]. After attaining certain results, extra pair of the antenna added in order to observe the behavior of multi-band antenna. Moreover, for WiMAX and WLAN applications, Raghavan et al. [9] developed an antenna with split ring resonator. Jackson performed the operation of the estimation of dielectric constant [10]. Kumar et al. gave circular split ring resonator with a dual-band antenna covering the bands of Bluetooth, WiMAX and Wi-Fi applications [11]. Siddiqui et al. [12] proposed a split ring resonator for UWB application located at the behind of the antenna. It provided multiple resonance due to the presence of several SRRs on the behind. Zheng suggested antenna applicable for multi-band areas. Observations have been made on covering the applications of Bluetooth and WiMAX through planar monopole antenna [13]. Yadav et al. [14] proposed triple-notch bands loaded antenna applicable for ultra-wideband range. Analysis had been based on CSRR on the patch,

S-shaped slot in the feed and mirror SRR. It resulted as circular patch antenna with a triple-notch band.

Yang [15] proposed a combination of CSRR- and SRR-based filtenna. His work was specifically for the lower cutoff frequency, i.e., 3.7 GHz, which clearly differentiates between CSRR and SRR. Madhav et al. gave the models of multiple antenna with metamaterial concept and defected ground [16–18]. The proposed antenna is operable for the bands WiMAX, LTE 33-37 and LTE 43-47.

2 Antenna Design

In this paper, a comparative analysis for dual-band antenna has been performed at the frequencies of 6.5 and 9.5 GHz and for triple-band antenna at the frequencies of 4.6, 6.6 and 9.5 GHz, whereas for multi-band antenna it is performed at the frequencies of 4.5, 6.6, 8.3 and 9.5 GHz. This paper presents the design of microstrip patch antenna having E-slot on patch and a pair of CSRRs on ground plane. A microstrip feedline is also connected with the rectangular patch. The geometric dimensions of the substrate are as the width and length of 36 mm each with its thickness of 1.6 mm (Fig. 2 and Table 1).

Fig. 2 Back view of multiband microstrip patch antenna

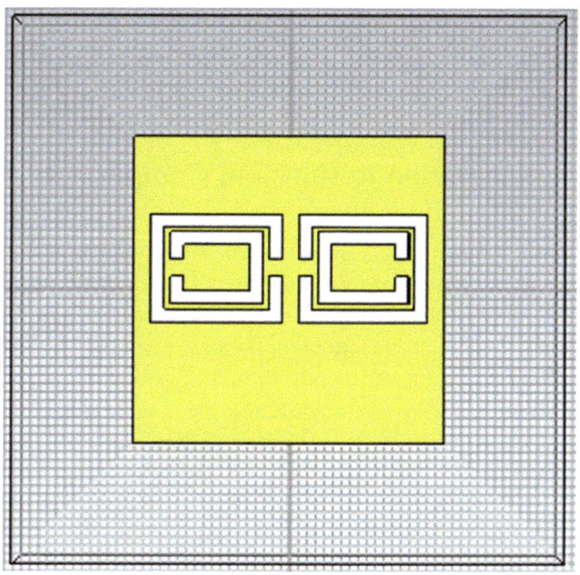

Table 1 Technical specifications of the proposed antennas

Parameter	Value (mm)
Substrate width sl	36
Substrate length sl	36
Substrate height	1.6
Patch width pw	28
Patch length pl	18.25
Patch depth *t*	1
E-slot width slw	22
E-slot length sll	15
Ground height	1
CSRR outer ring width srrW	15
CSRR outer ring length srrL	12.75
CSRR inner ring width	13.6
CSRR inner ring length	11.35
Split length in the ring	2
Gap between two CSRRs	2
Feedline width fw	3
Feedline height fl	5
Base feedline width fw1	1
Base feedline height fl1	9.25
Ground width	36
Ground length	36

3 Simulation Results and Comparative Analysis

Testing of novel designs of microstrip patch have been simulated and analyzed with the help of CST studio software. These results are investigated in between their obtained values of return loss versus frequency of antenna as shown in upcoming figures. The proposed antenna has resonant frequencies based on their bands. Tables 2, 3 and 4 illustrate the comparative analysis in between dual-, triple- and multi-bands of antenna with their gains and return losses. This comparison suggests the best application for each type of antenna.

Table 2 Dual-band antenna's return loss and gain

Frequency (GHz)	Return loss	Gain
6.517	−24.895	3.941
9.505	−32.806	−2.310

Table 3 Triple-band antenna's return loss and gain

Frequency (GHz)	Return loss	Gain
4.6	−33.546	−1.574
6.6	−22.105	4.417
9.5	−24.416	0.708

Table 4 Multi-band antenna's return loss and gain

Frequency (GHz)	Return loss	Gain
3.57	−6.343	−0.188
4.58	−18.406	−3.917
6.62	−19.647	3.705
8.38	−20.253	2.066
9.55	−17.918	−0.047

3.1 Comparison in Return Losses

See Figs. 3 and 4.

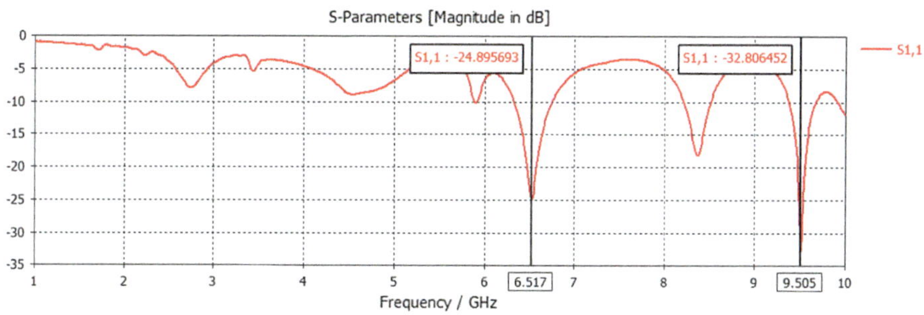

Fig. 3 Dual band antenna's return loss

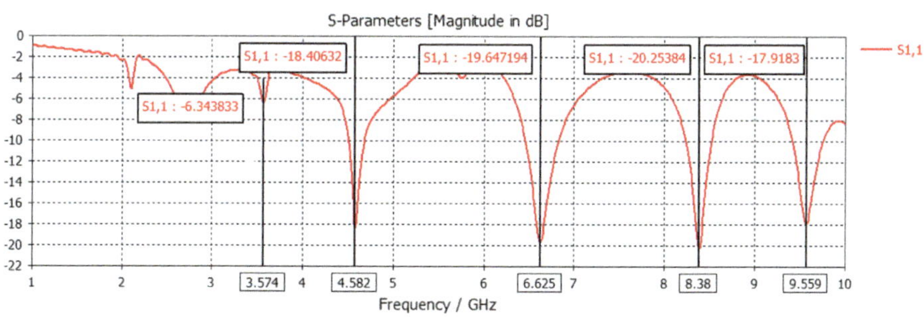

Fig. 4 Multi-band antenna's return loss

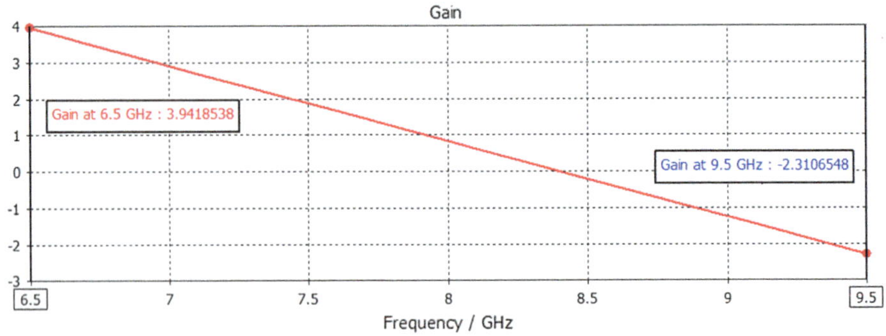

Fig. 5 Dual band antenna's gain

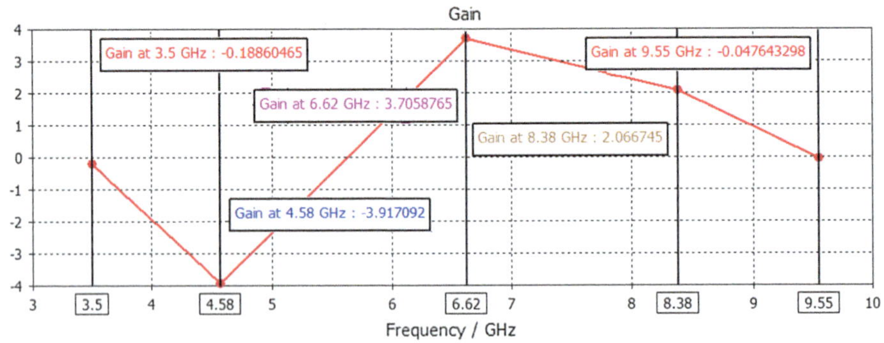

Fig. 6 Multi-band antenna's gain

3.2 Comparison in Antenna Gain

The performance analysis in terms of their gain has been presented in Figs. 5 and 6.

3.3 Comparison in Farfield Directivity Representations of Designed Antennas

Farfield representations of the band are illustrated in Figs. 7 and 8.

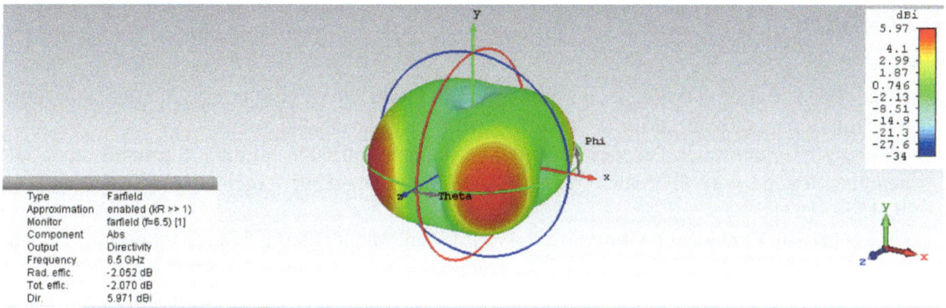

Fig. 7 Antenna's farfield directivity in 3-D form at 6.5 GHz

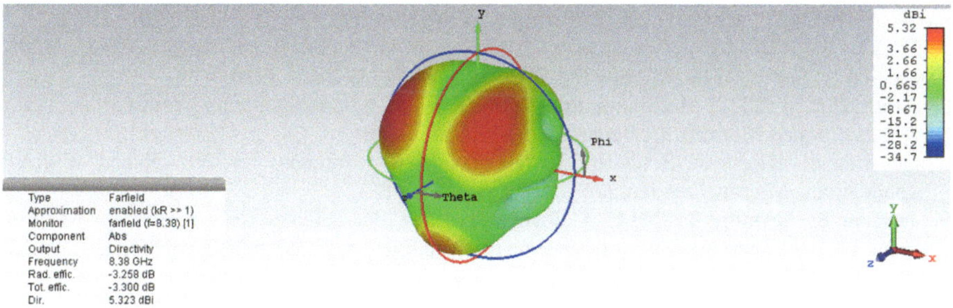

Fig. 8 Antenna's farfield directivity in 3-D form at 8.38 GHz

4 Conclusion

In this paper, the selection of appropriate antenna for certain applications is illustrated. It involves the change in the geometry or the direction of rings in order to attain the optimized performance of developed antennas. Comparative analysis between numerous antennas saves lot of efforts and time of the researchers for selection process. Certain frequencies are specified to perform definite tasks, such as 4.5 GHz is used for UWA applications while 9.5 GHz is excellent for radar-based applications. It makes simpler to select the number of bands at certain frequency to attain the maximum gain and minimum return loss. Mostly, metamaterial-based triple-band microstrip patch antenna is preferred for radar sensor, UWB and INSAT c-band receiver applications. The proposed antennas are economically cheap, compact, and can be easily fabricated.

References

1. Hasan MM, Faruque MRI, Islam SS, Islam MT (2016) A new compact double-negative miniaturized metamaterial for wideband operation. Materials 9(10):830
2. Martínez FJ, Zamora G, Paredes F, Martín F, Bonache J (2011) Multiband printed monopole antennas loaded with OCSRRs for PANs and WLANs. IEEE Antennas Wirel Propag Lett 10:1528–1531
3. Huang H, Liu Y, Zhang S, Gong S (2014) Multiband metamaterial-loaded monopole antenna for WLAN/WiMAX applications. IEEE Antennas Wirel Propag Lett 14:662–665
4. Wong KL, Chen LY (2014) Small-size LTE/WWAN tablet device antenna with two hybrid feeds. IEEE Trans Antennas Propag 62(6):2926–2934
5. Abed AT, Singh MSJ (2016) Slot antenna single layer fed by step impedance strip line for Wi-Fi and Wi-Max applications. Electron Lett 52(14):1196–1198
6. Nandi S, Mohan A (2017) CRLH unit cell loaded tri-band compact MIMO antenna for WLAN/WiMAX applications. IEEE Antennas Wirel Propag Lett 16:1816–1819
7. Da Silva JL et al (2015) Microstrip patch antenna project with split ring resonator periodically arrayed on the substrate. Microw Opt Technol Lett 57(12):2715–2720
8. Kim OD et al (2017) Multiband unidirectional planar monopole antenna with two split ring resonator pairs. Microw Opt Technol Lett 59(4):753–758
9. Rosaline SI, Raghavan S (2016) Design of split ring antennas for WLAN and WiMAX applications. Microw Opt Technol Lett 58(9):2117–2122
10. Jackson B, Jayanthy T (2014) A novel microstrip slot antenna for permittivity measurement. In: 2014 international conference on information communication and embedded systems (ICICES). IEEE
11. Kumar A, Naidu PV (2016) A compact O-shaped printed ACS fed monopole dual-band antenna for 2.4 GHz Bluetooth and 5 GHz WLAN/WiMAX applications. In: Progress in electromagnetic research symposium (PIERS). IEEE
12. Siddiqui JY, Saha C, Antar MMY (2014) Compact SRR loaded UWB circular monopole antenna with frequency notch characteristics. IEEE Trans Antennas Propag 62(8):4015–4020
13. Li WT et al (2012) Planar antenna for 3G/Bluetooth/WiMAX and UWB applications with dual band-notched characteristics. IEEE Antennas Wirel Propag Lett 11:61–64
14. Yadav S, Gautam A, Kanaujia B (2017) Design of dual band-notched lamp-shaped antenna with UWB characteristics. Int J Microw Wirel Technol 9(2):395–402
15. Yang Y, Zhu X, Karmakar NC (2012) Microstrip low pass filter based on split ring and complementary split ring resonators. Microw Opt Technol Lett 54(7):1723–1726
16. Babu MA (2015) Flared V-shape slotted monopole multiband antenna with metamaterial loading. Int J Commun Antenna Propag 5(2):93–97
17. Ramkiran DS (2017) Pentagonal shaped koch fractal monopole slot antenna for multiband applications. ARPN J Eng Appl Sci 12(15)
18. Khan H (2017) A CPW fed U-slot multiband fractal antenna. Int J Pure Appl Math 115(7):459–463

Chapter 53
Mustard and Cotton Waste-Based Biomass Gasifier Simulation by Using Aspen Plus

Rakesh Kumar Dabkeya and Mahendra Lalwani

1 Introduction

Aspen Plus software is used to develop a widespread process model for biomass gasification of atmospheric fixed bed reactor type. The investigational capability of the gasifier developed by the authors has a stipulation for appropriate cooling and filtration scheme to obtain adequate performance with low emissions [1, 2].

The model of gasifier is validated with the investigational data obtained from four different types of feedstock, i.e., neem wood, babool wood, mango wood and bagasse, etc. The model has fine predicted composition of H_2, CO and CO_2, whereas it has under predicted the CH_4. The conversion efficiency for babool wood is obtained higher than other wood due to its higher carbon and H_2 content and lesser amount of ash concentrations.

To find out the volumetric composition of different components available in biomass fuel, the ultimate and proximate analysis is performed. The result of this analysis is fed into the downdraft gasifier Aspen Plus model created with Aspen Plus software and performed the sensitivity analysis [3, 4].

2 Aspen Plus

Aspen Plus is used in the manufacturing chemical process modeling, simulation, sensitivity investigation, optimization and economic assessment. It delivers the

R. K. Dabkeya (✉) · M. Lalwani
Department of Renewable Energy, Department of Electrical Engineering, Rajasthan Technical University, Kota, India
e-mail: Rakeshkumar0707@gmail.com

M. Lalwani
e-mail: mlalwani@gmail.com

© Springer Nature Singapore Pte Ltd. 2021

S. D. Purohit et al. (eds.), *Proceedings of International Conference on Communication and Computational Technologies*, Algorithms for Intelligent Systems,
https://doi.org/10.1007/978-981-15-5077-5_53

comprehensive physical property representations and the library of component operation models, fast and reliable process simulation functions and forward-thinking calculation techniques. Engineers and researchers are capable to simulate genuine plant performance efficiently and precisely with the physical property database, and the operation models provided by Aspen Plus software thereby increase and improve the output production and reduce the expenditures [5, 6].

3 Model Validations

The different stages considered in Aspen Plus simulation include stream mixing, chemical equilibrium and heat exchange between streams, to improve the quality of producer gas [7, 8].

The following assumption was considered in modeling the gasification process:

1. The whole process is in steady state, and the reactions reach chemical equilibrium.
2. The heat exchange in a fixed bed is ideal, and it is isothermal in the same section.
3. The heat exchange occurs instantaneously at equilibrium with volatile products including mainly H_2, CO, CO_2, CH_4 and H_2O.
4. Tars are assumed to be negligible in the syngas.
5. Char only contains carbon and ash; ash is inert and does not participate in chemical reactions.

The model is validated by using the specifications of gasifier system, proximate and ultimate analysis of the biomass stocks, experimental setup parameters used in the simulation and reaction block description utilized in the simulation are given in Tables 1 and 2.

Table 1 Proximate and ultimate analysis of the biomass stocks [3]

Characteristic Proximate analysis (wt% dry basis)	Babul wood (Arcacia arabica)	Neem wood (Azadirachta indica)	Mango wood (Mangifera Indica)	Bagasse
Proximate analysis (wt% dry basis)				
Fixed carbon Content (FC)	16.38	16.35	16.34	15.8
Ash content (AC)	1.2266	1.35	1.36	2.8
Volatile Matter (VM)	82.40	82.35	82.30	80.7
Ultimate analysis (% dry basis)				
Carbon	50.24	50.20	50.18	46.2
Hydrogen	6.40	6.38	6.35	5.6
Oxygen	43.24	43.28	43.34	40.3
Nitrogen	0.12	0.14	0.13	0.2

Table 2 Experimental setup parameters used in the simulation [3]

Fixed bed reactor	
Temperature	750–875 °C
Pressure	1.05 bar
Air	
Temperature	65 °C
Flow rate	27.5–45 m^3/h

To validate the simulation results, experimental data from gasification of babul wood (Arcacia arabica), neem wood (Azadirachta indica) and mango wood (Mangifera indica) and bagasse waste in downdraft gasifier power plant were used [9–12].

The data for different fuel are collected from proximate and ultimate analysis, and after that, it is fed into the gasifier model which is developed on Aspen Plus software for the performance of experiment (sensitivity analysis) of the available literature [13]. The results after simulation of these data are compared with the results of above-mentioned paper and found that the results are better as compared to their practical work [3]. It can be seen in comparative study of results shown in Figs. 1, 2, 3 and 4 that the simulation results are better than the experimental results. The observation can be done that CO, H_2 and CH_4 have increased in volumetric percentage and the CO_2 content is reduced which is also a good significance for environmental issues solution which is shown in graphs below.

The producer gas composition acquired after simulation in Aspen Plus simulator software and experiment is shown in tables. The reasonable agreement with the investigational conclusions for all constituents with the exemption of CH_4 has shown by the model. The volumetric composition (%) of different components in babool, mango, neem wood and bagasse is shown in Figs. 1, 2, 3 and 4.

The model is under the anticipated composition of CH_4 that the oxygen in the presence of air reacts with the char. The under expectation resulted an error in the

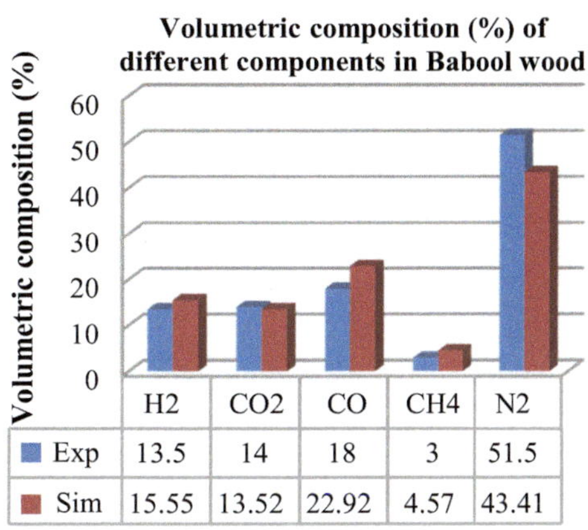

Fig. 1 Volumetric composition (%) of CO, H_2, CH_4, CO_2, N_2 in babool wood

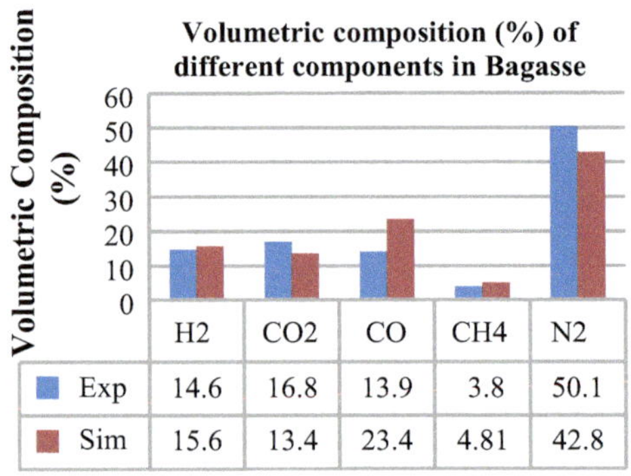

Fig. 2 Volumetric composition (%) of CO, H_2, CH_4, CO_2, N_2 in bagasse wood

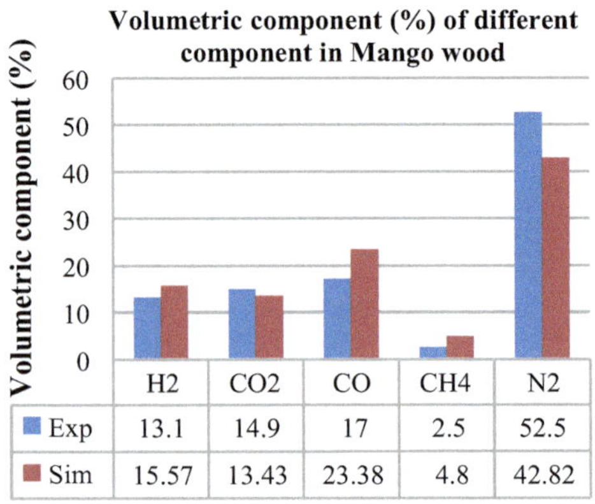

Fig. 3 Volumetric composition (%) of CO, H_2, CH_4, CO_2, N_2 in mango wood

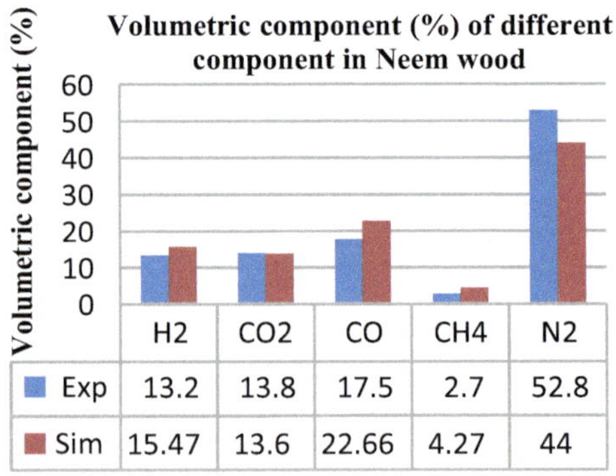

Fig. 4 Volumetric composition (%) of CO, H_2, CH_4, CO_2, N_2 in neem wood

calculation of the gas heating value and gas conversion efficiency. The pyrolysis foodstuffs are splintered in a province of high temperature and in the manifestation of O_2, so it is possible that selected amount of the CH_4 manufactured will go through combustion with O_2 [14–17].

The gas transformation efficiency of babul wood is greater than other types of biomass because of higher carbon content, more H_2 and less ash present in it. The proportion of overall combustible components like H_2, CO and CH_4 for babul wood is healthier than the neem wood, mango wood and bagasse, correspondingly. This is considerably healthier than that described by other plainclothes man in their research works [17–20].

4 Simulation Model

The Aspen Plus software is widely used for simulating the coal conversion, gasification, pyrolysis and other chemical-related procedures. Literatures like coal hydro gasification progressions, coal gasification simulation and integrated coal gasification combined cycle (IGCC) power plants have been previously published.

Fig. 5 shows the complete gasification simulation model developed for biomass gasification that has been used for gasification of mustard and cotton waste. The simulation model of gasification prepared in four zones, viz. drying, gasifier, heat exchanger and purification.

As mentioned above, the overall gasification process is simulated in four separate zones as shown in Figs. 6 and 7. In which, drying and gasifier zone are described as follows:

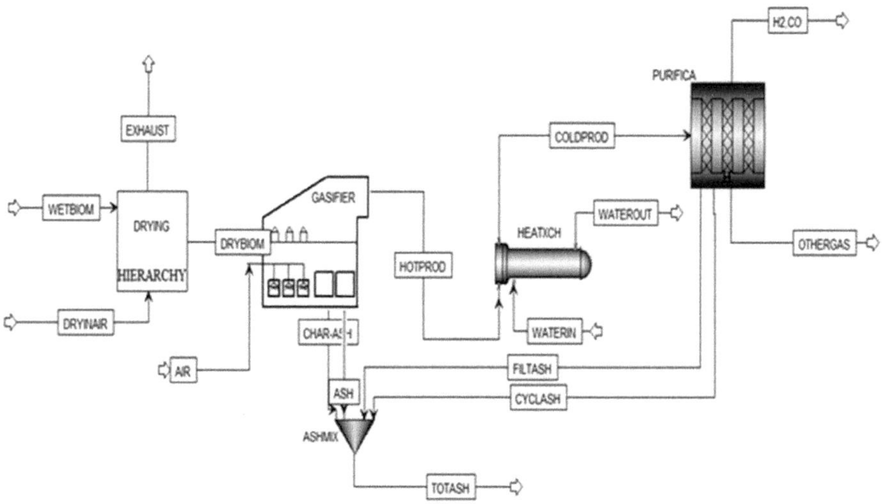

Fig. 5 Main flow sheet of simulation on Aspen Plus software

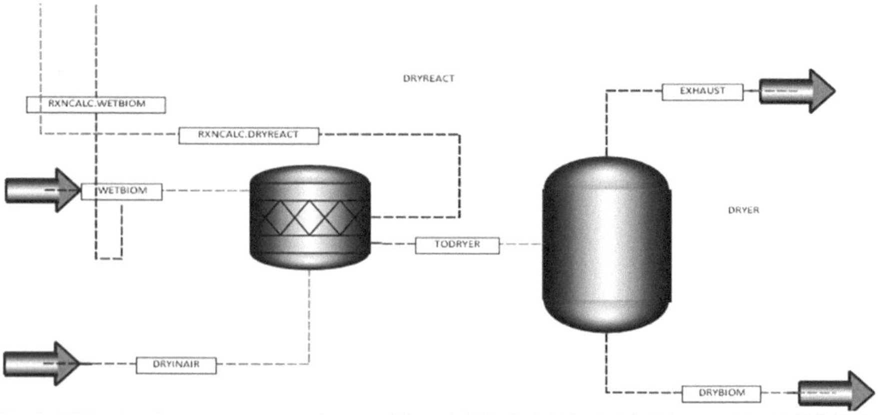

Fig. 6 Dryer hierarchy

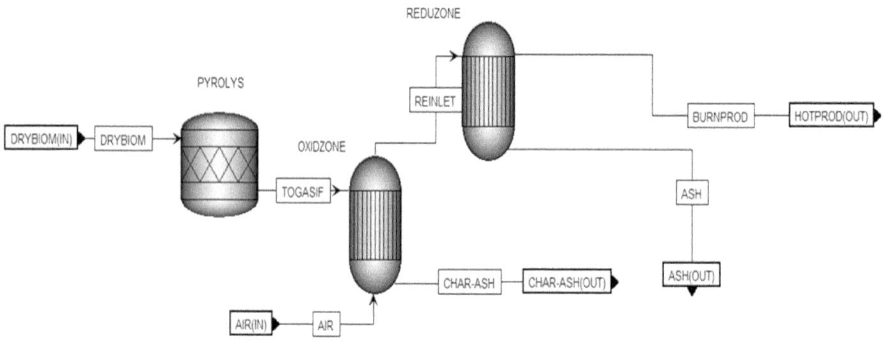

Fig. 7 Gasifier hierarchy

4.1 Drying Zone

Moisture content of the feedstock is an important factor to be able to stabilize a good combustion bed while having high moisture feed and to determine if the gasifier is capable to run in a steady-state condition for a long time. Also, the heating value of the gas produced depends on the moisture content of the feedstock. Moisture content can be determined on a dry basis as well as on a wet basis method. In this study, the dry basis method was used to calculate the moisture content as shown in equation below. The vaporization of water to steam requires a heat input of 1000 Btu/lb of water [6]. Energy which could be useful in steam production is diverted to drying the wood fuel. So, high moisture content reduces the thermal efficiency and results in low heating value of produced gas. Also, in downdraft gasifiers, high moisture contents give rise to low temperatures in the combustion zone which leads to high tar formation. Moisture content is calculated using the following equation:

$$\text{Moisture content} = \left[(\text{Wet weight} - \text{Dry weight})/\text{Dry weight}\right] \times 100$$

The modeling part of drying zone includes a yield reactor with a separator which removes water vapor. Free water is separated from the wet biomass. Water vapor along with dry biomass is sent to pyrolysis zone.

4.2 Pyrolysis Zone

Pyrolysis is where the volatile component vaporizes to a mixture of gases (devolatilization). The volatile vapor mainly consists of hydrogen, carbon monoxide, carbon dioxide, methane, hydrocarbon gases, tar and water vapor. As biomass has high volatile content, pyrolysis is an important step in biomass gasification. Remaining solid char and ash are also produced in this step. Primary products characterized by compounds evolved from cellulose, hemicellulose or lignin [15]. The pyrolysis zone in Aspen is modeled in three separate reactions:

- Devolatilization

Devolatilization is a one-step reaction modeled in a yield reactor which decomposes biomass fuels volatiles, char and ash.

- Primary pyrolysis

In this step, light gases are driven off along with tar from volatiles by the following reaction where primary tar is defined as C6.607H11.454O3.482 [19].

Volatile: 0.268 CO + 0.295 CO_2 + 0.094 CH_4 + 0.5 H_2 + 0.255 H_2O + 0.004 NH_3 + 0.0002 H_2S + 0.2 primary tar

Rate of this reaction is given by: Rp1 = 4.38 × 109 exp (−1.527 × 10 5/RTs) C_{volatile}.

- Secondary pyrolysis

Reaction below converts the primary tar to secondary tar and other products where secondary tar is defined to be pure benzene [15].

Primary tar: 0.261 secondary tar + 2.6 CO + 0.441 CO_2 + 0.983 CH_4 + 2.161 H_2 + 0.408 C_2H_4

Reaction rate for the secondary pyrolysis is given by: Rp2 = 4.28 × 10^6exp (−1.08 × 105/RTg) $C_{\text{primary tar}}$.

4.3 Combustion Zone

Char reactions are considered non-reacted shrinking core model which assumes char particles to be spherical; grains and solid-gas phase reaction take place on the external surface [15]. Combustion reactions are modeled with two types of reaction:

- Char oxidation reactions

$$C + O_2 = 2CO \qquad \Delta H = -123\,\text{kJ/mol}$$
$$C + O_2 = CO_2 \qquad \Delta H = -393.5\,\text{kJ/mol}$$

- Hydrogen combustion reaction

$$H_2 + 0.5\,O_2 = H_2O \qquad \Delta H = -242\,\text{kJ/mol}$$

In downdraft gasifiers, generally air fed into the combustion zone which has a large volume of nitrogen. This dilutes the syngas and reduces the concentration of hydrogen (H_2) and carbon monoxide (CO), which reduce syngas heat value [17]. For this reason, in simulation, the air has been replaced with oxygen which determines the product and temperature distribution of a gasification system.

4.4 Gasification Zone

Gasification zone is the most critical zone in a gasifier. The hot gases and carbon burnt go through a series of reduction reactions. Temperature in gasifier zone is less compared to combustion zone which is due to endothermic reactions. The hot syngas is cooled in heat exchanger and transferred to purification hierarchy and then separates the H_2 and CO. Important reduction reactions taking place in gasification zone are as follows:

Water Gas Reaction :

$$C + H_2O = CO + H_2 \qquad \Delta H = 118.5\,\text{kJ/mol}$$

Boudourd Reaction :

$$C + CO_2 = 2CO \qquad \Delta H = 159.9\,\text{kJ/mol}$$

Hydrogasification Reaction :

$$C + 2H_2 = CH_4 \qquad \Delta H = -87.5\,\text{kJ/mol}$$

Water Gas Shift Reaction :

$$CO + H_2O = CO_2 + H_2 \qquad \Delta H = -40.9\,\text{kJ/mol}$$

Biomass gasifier is called a device in which transformation of the biochemical fuels in gaseous forms takes place. After that, those products are used for different application. Gasifiers have three categories of applications, viz. biomass to electrical energy, heating mode of gasifier and hydrogen production.

5 Biomass Fuel Used

Agricultural residues can be described as by-product of agricultural production, and it contains husks, stalks, shells and straws. Agricultural waste or by-product comprises a most important fraction of the whole yearly production of agricultural production, and it is a significant source of energy for industrial and domestic purposes in various application [21–25].

5.1 Mustard Stalk

India grows the most mustard–rapeseed in world and stands at the second position in production after china. It was cultivated in 7.49 million hectare in India that is 33.8% of world and contains 16% of world total production. It is the second major significant oilseed later than groundnut. It supplies over 1.5 million tons of oil and 3–3.2 million tons of meal [18]. The mustard–rapeseed plant bears oily blackish to reddish brown seeds, spherical in shape. The oil extracted from the seeds has been utilized for various purposes for hundreds of years.

The straw and the stalk, other part than of the seed, constitute to about 70% of the total plant and are considered as agricultural waste after the removal of the seed for its utilization (Fig. 8). The agricultural waste of straw and stalk is usually being used to burn along with hay after drying for household cooking purpose, but this huge

Fig. 8 Images of the mustard waste for feedstock to biomass gasifier

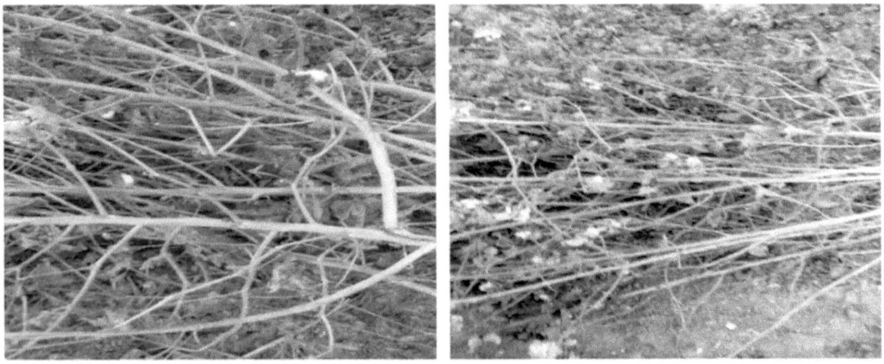

Fig. 9 Images of the cotton waste for feedstock to biomass gasifier

amount of generated waste can be utilized in a proper scientific manner for energy generation.

5.2 Cotton Waste

Cotton shell is one of the largest sources of biomass in India for agriculture residues. The scientific name of cotton is gossypium arboretum which is a species of native of cotton to Pakistan, India and other subtropical and tropical regions of the world. The fresh images of cotton waste are shown in Fig. 9 for feedstock utilization in biomass gasifier.

The cotton crop is being grown on the 9.2 million hectares, and it has a share of 27% in world. India has three zones for cotton cultivation mainly Northern zone, Southern zone and central zone, and its whole production consists of about 15.8 million bales. The seed is detached from cotton, and these are used to produce extremely thick oil of cotton seed. This oil can be put into direct burning purpose, and further, it can be transformed into bio-diesel. After the elimination of cotton and seeds, the shells are burnt in open atmosphere due to its low bulk density. Burning of shells has some adverse effect on the environment. So, appropriate recycling is important, and it should be acknowledged. The higher fraction of volatile matters gives positive response to energy conversion process.

6 Preliminary Analysis of the Biomass Samples

The following analyses have been carried out for the characterization of the different biomass samples.

Table 3 Proximate analysis

Name of content	VM	A	FC	Moisture
Cotton stalk	70.11	4.68	17.84	7.37
Mustard stalk	66.80	6.70	17.40	9.10
Mixture of both	68.455	5.42	17.62	8.505

- Proximate Analysis
- Ultimate Analysis.

The proximate analysis gives the values of volatile material, fixed carbon, ash content and moisture available in the waste of fuel utilized in the test. This test is performed by using the bomb calorimeter in the laboratory. The ultimate analysis gives the different components available in the fuel used for test in bomb calorimeter. This ultimate analysis results are carried out with the correlation given in paper [26, 27].

6.1 *Proximate Analysis*

Determination of moisture content, volatile matter, ash content and fixed carbon in the biomass sample is known as the proximate analysis. The proximate analysis results for different biomass samples give the results which are listed in Table 3.

6.2 *Ultimate Analysis*

Determination of total carbon, hydrogen, nitrogen, oxygen and sulfur percentages in the biomass sample is carried out by its ultimate analysis. With the ultimate analysis for all these biomass samples, the result is shown in Table 4 was obtained.

The samples of both mustard and cotton stalks were taken for the proximate analysis of both in bomb calorimeter, and this practical is performed at M.L.V. Textile and Engineering college, Bhilwara, Rajasthan (India), and the ultimate analysis is carried out by using correlation given in [26, 27]. The mixture sample of both the crop wastes in same proportion is taken for the proximate and ultimate analysis of

Table 4 Ultimate analysis

Name of content	HHV (MJ/kg)	C	H	N	O	S	Ash content
Cotton stalk	17.88	41.49	6.20	1.81	47.49	0.36	2.65
Mustard stalk	17.57	43.88	4.58	0.76	42.75	0.66	7.37
Mixture of both	17.725	42.685	5.39	1.285	45.12	0.51	5.01

mixture is also performed for checking the results of both and their mixture as shown in Tables 3 and 4.

This data of proximate and ultimate analysis are utilized for volumetric composition in Aspen plus model for getting the different components available in the cotton and mustard stalks and their mixture. The volumetric components contain the H_2, CO, CO_2, CH_4 and N_2, and these components contribute for the formation of methane and other useful biogases from feedstock's available as wastes in fields of Indian agriculture market. This ultimate analysis can also be found with the correlation [26, 27].

7 Result Discussion

The waste of the crops grown in fields of Rajasthan is mainly used for the testing of its volumetric content because they are making waste and environmental problems. These wastes of agricultural crops can be used as a fuel in biomass gasifier, and after that it will use like heating, methane production etc. The gas form of biomass can be used for cooking and other applications too.

The main waste of crops in India is discussed above in introduction, and various samples of crops have been investigated for solutions of those wastes. Here, in this thesis, the author has tried to get the mustard and cotton stems as fuel for the biomass, and these both wastes have higher calorific values and grows in Rajasthan in bulk amount, It is also available in bulk amount here in Rajasthan for utilization. These wastes of mustard have its stalks and husk as waste, and cotton uses its bollshell, stalks and husk for feedstock in biomass gasifier. Per year potential for mustard and cotton stalks is 8657.1 kilo tons per year and 52,936.5 kilo tons per year, respectively. By using this much amount of agricultural waste of farms and fields, a huge amount of energy can be generated or extracted through biomass gasifier, and the dependency on conventional fuels can be minimized. This change of dependency will reduce the greenhouse impact on environment and will also be a new source for energy and power generation for the rural and places where the grid availability is also a dream in villages of Rajasthan and rural India [8].

Sensitivity Analysis: Aspen Plus simulator is being used for sensitivity analysis performance which varies the parameters, i.e., temperature, air fuel ratio, fuel flow, pressure, to study their effect on composition of producer gases.

Fuel flow Rate: The biomass fuel flow in kg/hr, which is fed to the biomass gasifier is varied, and the change in component of producer gases is noted, and a graph is plotted between the component of producer and rate of fuel fed to gasifier for mustard, cotton and mixed fuel separately.

The change in the components (CO and H_2) of producer gases with change in rate of flow of biomass fed is plotted in a graph for mustard, cotton and mixed fuel separately which is shown in Figs. 10, 11 and 12.

The sensitivity analysis is performed for mass flow rate of the biomass between 50 and 500 kg/hr at 1000 points for getting the smooth curve between the given

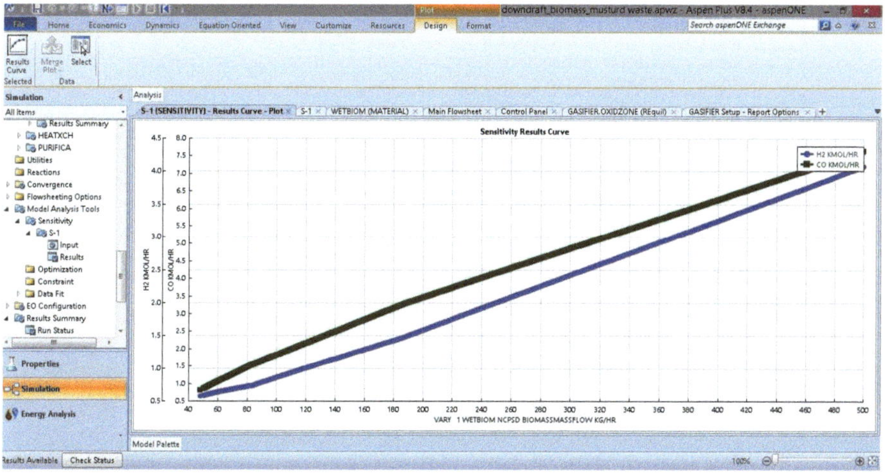

Fig. 10 Variation in component of CO and $H_{2\,mol}$ flow with wet biomass mass flow rate in kg/hr of mustard waste

Fig. 11 Variation in component of CO and $H_{2\,mol}$ flow with wet biomass mass flow rate in kg/hr of cotton waste

points. The plotted graphs are shown below for mustard waste, cotton waste and mixed waste of both the agricultural crops waste. The given graphs are plotted for H_2 and CO gases which are responsible for biogas production after the completion of gasification process.

The effect of variation in component of CO and $H_{2\,mol}$ flow with wet biomass mass flow rate in kg/hr of mustard waste, cotton waste and their mixed waste is plotted in figures. The mole flow of CO and H_2 is increasing with the increase in mass flow of

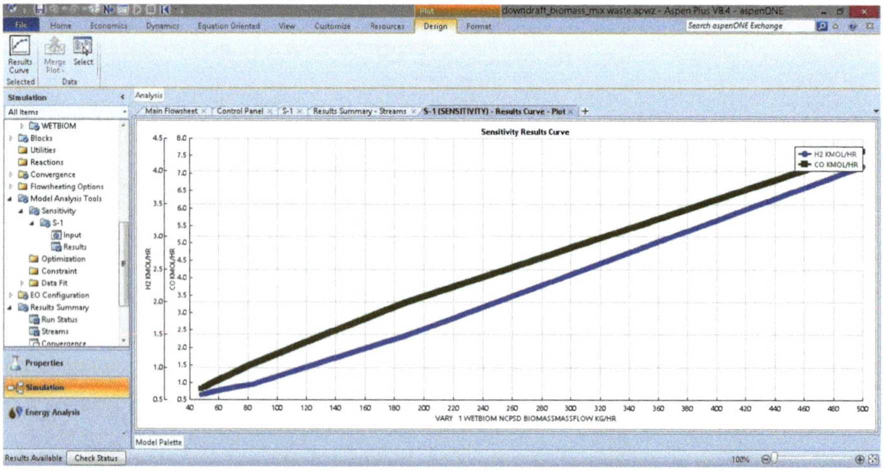

Fig. 12 Variation in component of CO and $H2_{mol}$ flow with wet biomass mass flow rate in kg/hr of cotton and mustard mixed biomass waste

biomass waste for gasification. The mustard and mixed biomass figures are showing that they are increasing parallel with each other of CO and H_2 components, but in cotton biomass, it can be noted that both components' graphs cross each other at around 400 kg/hr biomass flow rate. At 400 kg/h mass flow rate of cotton waste, the mole flow of CO increases than H_2 during the sensitivity analysis performance.

In this thesis study, a downdraft type of gasifier is modeled using the built in reactor models of Aspen Plus software process simulator taking the equilibrium reactions. This model is allowed to study the mass flow of wet biomass and sensitivity analysis of the given feedstock. In summary, it can be understood that the simulation and modeling of gasifier are the best way to analyze or predict the performance of unit and to study the effect of all working vital operation of gasifier and applied design parameters. The simulation result of modeled gasifier is shown with the change in component production mole flow with mass flow of feedstock. Both the components CO and H_2 are the main components for biofuel.

Acknowledgements The authors are highly grateful to faculty of MLV Textile and engineering college, Bhilwara, Rajasthan (India), for providing lab support to perform proximate and ultimate analysis.

References

1. Atnaw SM, Sulaiman SA, Yusup S (2011) A simulation study of downdraft gasification of oil-palm fronds using Aspen plus. J Appl Sci 11(11):1913–1920

2. Baredar P, Sethi VK, Pandey M (2010) Correlation analysis of small wind–solar–biomass hybrid energy system installed at RGTU Bhopal MP (India). Clean Technol Environ Policy 12(3):265–271
3. Keche AJ, Gaddale APR, Tated RG (2015) Simulation of biomass gasification in downdraft gasifier for different biomass fuels using Aspen plus. Clean Techn Environ Policy 17(2):465–473
4. Basavaraj RJ, Jayanti S (2014) Syngas-fueled, chemical-looping combustion-based power plant lay-out for clean energy generation. Clean Technol Environ Policy 17(1):237–247
5. Blasi C Di (2000) Dynamic behaviour of stratified downdraft gasifiers. Chem Engg Sci 55(1):2931–2944
6. Ewida KT, Salmawy H, El Atta N N, Mahmoud MM (2006) A sustainable approach to the recycling of rice straw through pelletization and controlled burning. Clean Technol Environ Policy 8(3):188–197
7. Zainal ZA, Rifau A, Quadir GA, Seetharamu KN (2002) Experimental investigation of a downdraft biomass gasifier. Biomass Bioenergy 23(4):283–289
8. Kumar A, Kumar N, Prashant B, Ashish S (2015) A review on biomass energy resources, potential, conversion and policy in India. Renew Sustain Energy Rev 45(1):530–553
9. Wang Z, Yang JN, Xiang Z, Li Y (2009) Syngas composition study. Front Energy Power Eng Chin 3(3):369–372
10. Liu Z, Han G (2015) Production of solid fuel biochar from waste biomass by low temperature pyrolysis. Fuel 158(1):159–165
11. Lago V, Greenhalf C, Briens C, Berruti F (2015) Mixing and operability characteristics of mechanically fluidized reactors for the pyrolysis of biomass. Powder Technol 274(1):205–212
12. Chawdhary R (2018) Biofuels from Indian lignocellulosic wastes through pyrolysis: a review with some case studies. Waste management and resource efficiency. Springer, Singapore, pp 131–146
13. Mamphweli NS, Meyer EL (2010) Evaluation of the conversion Efficiency of the 180N^{m3}/h Johansson Biomass Gasifier TM. Int J Energy Environ 1(1):113–120
14. Mansaray KG, Al-Taweel AM, Ghaly AE, Hamdullahpur F, Ugursal VI (2000) Mathematical modeling of a fluidized bed rice husk gasifier. Energy Sources 22(1):83–98
15. Manurung RK, Beenackers AACM (1993) Modelling and simulation of an open core downdraft moving bed rice husk gasifier. In: Bridgwater AV (ed) Advances in thermochemical biomass conversion. Blackie A&P, London, pp 288–309
16. Nikoo M, Mahinpey BN (2008) Simulation of biomass gasification in fluidized bed reactor using Aspen plus. Biomass Bioenergy 32(1):1245–1254
17. Ojha K (2010) Need of independent rural power producers in India—an overview. Clean Technol Environ Policy 12(1):495–501
18. Rajvanshi AK (1986) Alternative energy in agriculture. In: Yogi Goswami D (ed), edn 1, vol 2. CRC Press, pp 83–102
19. Ramzan N, Ashraf A, Naveed S, Malik A (2011) Simulation of hybrid biomass gasification using Aspen plus: a comparative performance analysis for food, municipal solid and poultry waste. Biomass Bioenergy 35(1):1–8
20. Rogel A, Aguillon J (2006) The 2D Eulerian approach of entrained flow and temperature in a biomass stratified downdraft gasifier. Am J Appl Sci 3(10):2068–2075
21. Sridhar G, Dasappa S, Sridhar HV, Paul PJ, Rajan NK (2005) Gaseous emissions using producer gas as fuel in reciprocating engines. SAE world congress detroit, michigan 2005:01–1732
22. Dhaked DK, Lalwani M (2017) A comprehensive review on a D-facts controller: enhanced power flow controller (EPFC). Int J Adv Eng Technol (IJAET) 10(1):84–92
23. Dhaked DK, Lalwani M (2017) Modeling and analysis of a D-facts device: enhanced power flow controller. Imanager's J Electr Eng (JEE) 11(1):41–49
24. Sikdar K (2013) Subhas, resurgence of fossil fuels? Clean Technol Environ Policy 15(1):203–204
25. Sansaniwal SK, Rosen MA, Tyagi SK (2017) Global challenges in the sustainable development of biomass gasification: an overview. Renew Sustain Energy Rev 80(1):23–43

26. Shen J, Zhu S, Liu X, Zhang H, Tan J (2010) The prediction of elemental composition of biomass based on proximate analysis. Energy Convers Manag 51(1):983–987
27. Jigisha Parikh, Channiwala SA, Ghosal GK (2007) A correlation for calculating elemental composition from proximate analysis of biomass materials. Fuel 86(1):1710–1719

Chapter 54
Study of Bond-Agent-Based Resource Discovery in Cloud Computing

Vijindra Rajendran and Prabakaran Mani

1 Introduction

Distributed system is a collection of independent computers (nodes) that appears to its users as a single coherent system. Our motive of research focus is on resource management in cloud. Resource management in cloud computing is not easy. Managing distributed heterogeneous resources causes some problems: problem of resource info management, no standard definition of resource requirements, and difficulty of guaranteeing compatibility of resource allocation [1]. Resource management further divides into two broad areas like resource allocation and scheduling. Resource allocation has five stages, i.e., resource discovery, filtering, verification, decision making, and job submission. For a successful resource allocation, an accurate resource discovery plays an important role. However, it is tough to produce excellent resources from varied organizations as a result of management policies and descriptions regarding varied resources square measure completely different in every organization.

These differences cause a problem about providing a uniformed view from various resources. User's requests are information of resource components which users want to use, such as CPUs, memory sizes, network bandwidth, or IP numbers, and a degree of similarity (DS) as a reference value to be used in comparing resources with user's requirements and to choose proper resources which are closely related user's requirements [1].

V. Rajendran (✉) · P. Mani
SRM University, Chennai, India
e-mail: vijindrarajendran@gmail.com

© Springer Nature Singapore Pte Ltd. 2021

S. D. Purohit et al. (eds.), *Proceedings of International Conference on Communication and Computational Technologies*, Algorithms for Intelligent Systems,
https://doi.org/10.1007/978-981-15-5077-5_54

2 Related Study

Kong et al. [2] proposed a general resource discovery system for open distributed processing. It uses DSTC architecture model to specify a resource discovery with the basic objects of objects, bindings, and interfaces. It is one of among the pioneer work for resource discovery in the field of distributed processing. Castano et al. [3] proposed a semantic approach based on: (i) *ontology* for a semantically rich description of the metadata characterizing the resources to be shared; (ii) *semantic matching techniques* for comparing ontological resource descriptions against a target resource request. The proposed H-MATCH algorithm is able to discover the location of compatible resources with respect to a target request, also in heterogeneous environments with lack of agreement on resource descriptions. Ranjan et al. [4] proposed a decentralized grid resource discovery system based on a spatial publish/subscribe index. It utilizes a distributed hash table (DHT) routing substrate for delegation of d-dimensional service messages with various parameters such as resource query rate, index load distribution, number of index messages generated, overlay routing hops, and system size. Yulan et al. [5] presented a hierarchical RD method to make use of attractive features of lumped and exhaustive methods. It proposes 3-layer resource organizational model with the parameters of extensibility and performance. Ranjan et al. [6] presented a tutorial about P2P-based RD in global grid. It investigates various decentralized RD techniques primarily driven by P2P network model. Also it provides complete state of art, resource taxonomy, existing works on RD in grids. Mohammed et al. [7] reviews the state of the art of RD techniques in grid environment. It separates the RD into 4 categories, i.e., centralized, hierarchical, P2P, and agent-based. And finally, HPPGrid model is proposed. Mahamat et al. [8] proposed a novel semantic-based scalable decentralized grid RD framework by integrating ontology, P2P networks, and intelligent agents. Dong et al. [9] proposed a novel hierarchical P2P-based grid model based on the existing grid resource discovery models. Leyli et al. [10] used a weighted tree structure for resource discovery in which the edges have weight. It uses only one bitmap for the identification of available resources in nodes and also resources of children and their descended nodes. Stelios et al. [11] explored decentralized distributed search protocols that transform nodes of large size distributed systems to act as both clients and servers and proposed an architecture that implies nodes orchestration based on previous resource requests, and we model a service meta-registry for inter-cloud systems to store relevant information to service submission. Sarhadi et al. [12] proposed dynamic structure of GIS based on peer-to-peer model using learning automata based on the parameters success rate, average number of hops/requirements, and average execution time. Chung et al. [13] proposed a direction-aware strategy which can alleviate the network traffic among unstructured information systems for distributed resource discovery service. Praveen et al. [14] proposed a distributed cloud architecture to make use of independent resources provided by the devices/users using multi-valued distributed hash tables for efficient resource discovery. Mollamotalebi et al. [15] focused on the resource discovery approaches used in grid computing environments along with their behaviours in different conditions. Barati et al. [16, 17] studied the problem of after the failure of a resource in a chain of resources made for a specific task in grid environment, discovering and finding a new resource that reconstructs the chain, and they define a new agent that is called task agent, and by proposing an algorithm for increasing the fault tolerance against probable failure of a resource in the resource chain. This work considers the resources of CPU, memory, disc capacity, network bandwidth, devices, software, etc. Zargar et al. [18] introduced for the resource discovery with implications for searching efficient resources using fuzzy logic and taboo table. Sim [19] proposed (1) an agent-based fog computing model, (2) a gossip-based resource discovery algorithm for propagating resource requests, (3) a reasoning technique for determining similarities between user requirements and resource profiles, and (4) a bargaining mechanism for dynamically pricing vertically arranged fog and cloud resources. Boukadi et al. [20] proposed a focused crawler for cloud service discovery (FC4DC). Its service-oriented architecture ensures an easier modification for rapid updating and better performance. Furthermore, the proposed crawler guarantees saving the search time and a better exploitation of the provider offerings thanks to a dedicated cloud service description ontology. Frederic et al. [21] surveyed the state of the art of cloud resource management for infrastructure as a service. They provided an overview of the recent research findings and technologies while focusing on the resource management techniques such as resource provisioning, resource discovery, resource monitoring, resource mapping, resource allocation, resource consolidation, resource modelling, and resource scheduling. Santos et al. [22] formulated the problem of resource allocation in cloud of sensors and proposed Zeus, a hybrid (partly decentralized) algorithm for solving it. Zeus has two key features. First, it is able to identify requests that are common for multiple applications and perform only once their required tasks, sharing the results among the applications for saving resources. Second, its hybrid approach leveraging edge computing makes Zeus scalable in terms of the number of VNs and applications supported, and suitable for meeting delay-sensitive applications. Sun et al. [23] focused on resource allocation in fog computing, proposed a fog computing structure and presented a crowd-funding algorithm to integrate spare resources in the network. Furthermore, to encourage more resource owners to share their resources with the resource pool and to supervise the resource supporters as they actively perform their tasks, they proposed an incentive mechanism in proposed algorithm. Faiez et al. [24] focused on heterogeneity, interoperability, integration, and portability issues in cloud computing and proposed Open Cloud Computing Interface (OCCI) as a community-based and open recommendation standard for managing any kind of cloud resources and also proposed the OCCIware approach, which represents the first approach to design, validate, generate, implement, deploy, execute, and supervise everything as a service with OCCI. Zarrin et al. [25] provided an investigation on the current state of resource discovery protocols, mechanisms, and platforms for large-scale distributed environments, focusing on the design aspects and also provided an overall guideline to design novel discovery approaches for distributed systems.

None of the previous studies and works investigate the resource discovery in terms of binding and independent message passing with agent technology. In this paper, we propose a bond-agent-based resource discovery mechanism for cloud computing.

3 Methodology

3.1 DSTC Architecture

Distributed applications within the DSTC design model are specified in terms of objects, their interface types, and bindings between the interfaces of objects.

In the DSTC architecture model, an object is an entity that encapsulates state, behaviour, and activity (i.e., the ability to take independent action). Objects move by exchanging powerfully written messages over a binding. Objects connect to a binding through their interfaces. A binding is an association between a set of objects that allows the objects to interact. Bindings are strongly typed—the binding type defines the roles of objects in a binding and the interaction that can occur between objects fulfilling those roles.

The fundamental unit of interaction during a binding could be a single, powerfully written message, though more complicated, high-level interactions are outlined, as an example, RPC or multicast.

An interface is instantiated to fulfil the role of an object in a particular binding. Interfaces are strongly typed—an interface type describes the possible structure and semantics for interactions of an object during a binding.

The server is responsible for accessing local resources, communicating with other remote servers to access their resources, and providing common services required by its clients. The client provides a mechanism for users to discover and access resources. The client can be further decomposed into two major components: a user part which provides an interface specifically designed for the end-user environment and an agent part, which provides user-specific services and communicates with the local server. Each server can support more than one agent, and each agent can support more than one user [2] (Fig. 1).

3.2 Bond-Agent-Based Resource Discovery

Bond is a Java-based distributed object system and agent framework, with an emphasis on flexibility and performance. It is composed of (a) a core containing the object model and message-oriented middleware, (b) a service layer containing distributed services like directory and persistent storage services, and (c) the agent framework, providing the essential tools for making autonomous network agents in

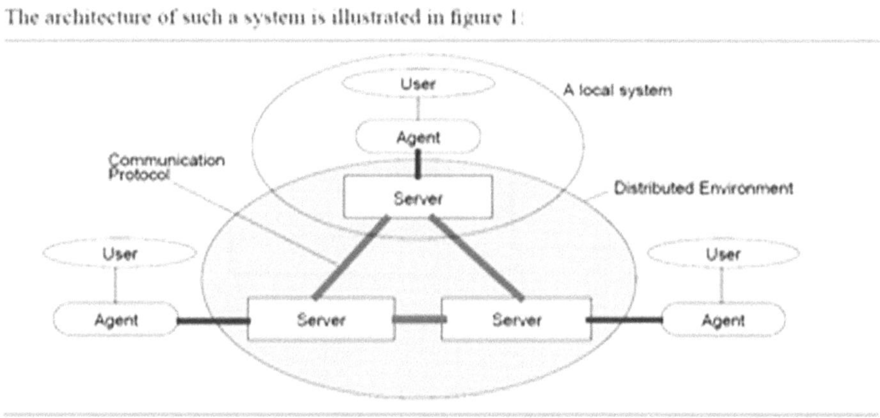

Fig. 1. General architecture of resource discovery system

conjunction with a information of unremarkably used methods which permit developers to assemble agents with no or less programming [26]. Bond Core, at the heart of the Bond system, there is a Java Bean-compatible component architecture.

Bond objects extend JavaBeans by permitting users to connect new properties to the item throughout runtime and supply a standardized API for accessing regular fields, dynamic properties, and JavaBeans-style setField/getField-defined virtual fields. This allows programmers the same flexibility as languages like Lisp or Scheme while maintaining the familiar Java programming syntax. Bond objects are network objects by default: They can be both senders and receivers of messages. [27]. No post-processing of the object code as in RMI or COREA-like stub generation is needed. Bond uses message passing, while RMI or COREA-based component architectures use remote method invocation. This approach has the advantage that there is no single point of failure, and the distributed awareness mechanism reconstitutes the network of directories even after catastrophic failures. [27]

In order to find suitable resources for every user, DSTC architecture is used as a reference model [28]. As like DSTC architecture, the proposed framework contains three basic objects, i.e., users, agent, and server (Fig. 2).

The interactions between the objects can be classified into user–agent binding, agent–server binding, and server–server binding which grouped into multi-party binding. The interface type used here is a user interface. Agent used here is a bond agent. Bond is a Java-based, open-source object-oriented middleware for network computing [27]. In this paper, we introduce an agent-based framework for resource discovery.

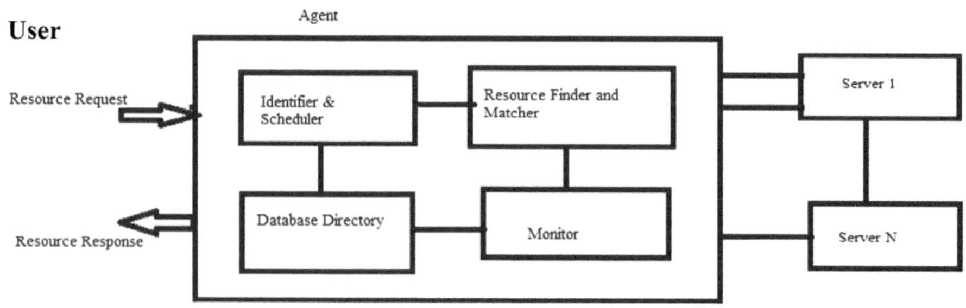

Fig. 2. Bond-agent-based framework for resource discovery

3.3 Resource Discovery Algorithm Formulation

Consider r is a request from the user i and R denotes all possible resources in the server S. Resource request from user i is denoted as

$r_i \in R$
Resource Request algorithm
Begin
For all $r \in R$ do
If $r_i \in AG$ &&match ($r_i = R$) then
Send r_i to the user from AG database
Else if
Advertise r_i to S then
If $r_i \in S$ &&match ($r_i = S$)
Send r_i to user through AG
Endif
Endif
For $i = 1$ to n do
Send r_i to i.S
Receive r_{resp}(accepted_rreq_list, rejected_rreq_list) from i.S
Endfor
Endfor
End

Resource Response Algorithm
Begin
For all $r_i \in S_j$ do
If r_i can be searched && r_i is not already accepted then
Add r_i to the accepted_rreq_list
Elseif r_i cannot be searchable then
Send $r_i$to $s_j + 1$
Increment $j + 1$ then goto step2
Repeat until time_frame >=10secs

If time_frame < 10secs
Add r_i to the rejected_rreq_list
Endif
Endfor
Send r_{resp}(accepted_rreq_list, rejected_rreq_list) to AG
……..
End

An agent component consists of four parts, i.e., identifier and scheduler (IS), resource finder and matcher (RFM), database, and monitor. All are interconnected together and interact using message passing. Once the user requests for a resource to the agent, the IS sends the request to RFM. If the requested resource is available in the agent's database, it immediately sends back to the user; if it is not available, RFM advertises the request to the connected servers using secure service discovery service protocol (SSDS). SSDS protocol is implemented on java and uses XML for service description and location. This protocol adds reliability, scalability and security features to the algorithm. If the server is busy due to network traffic or the resource is not available in the server, other connected servers will find the resource within a given time frame or else the advertisement will be rejected.

Performance measure can be measured by using
Time complexity—time taken to find a resource
Message complexity—number of messages sent to discover a resource.

4 Conclusion

Resource discovery plays a vital role in cloud resource allocation and scheduling. In this work, bond-agent-based resource discovery framework and algorithm is introduced in cloud computing. According to the literature study, it is the first step towards cloud resource discovery using DSTC architecture and bond agent. The future work focuses on the development of an algorithm which considers user-specific resource parameters and QoS factors using bond agent.

References

1. Yoo H, Hur C, Kim S, Kim Y (2009) An ontology-based resource selection service on science cloud. Int J Grid Distrib Comput 2(4)
2. Kong Q, Berry A (1995) A general resource discovery system for open distributed processing, open distributed processing, ISBN: 978-1-4757-6074-3
3. Castano S, Ferrara A, Montanelli S, Racca G (2004) Matching techniques for resource discovery in distributed systems using heterogeneous ontology descriptions. In: ITCC '04: Proceedings of the International Conference on Information Technology: Coding and Computing (ITCC'04) Volume 2 - Volume 2 April 2004

4. Ranjan R, Chan L, Harwood A, Buyya R, Karunasekera S (2007) A scalable, robust, and decentralised resource discovery service for large scale federated grids, 2007. https://www.gridbus.org
5. Yin Y, Cui H, Chen X (2007) The grid resource discovery method based on hierarchical model. Inf Technol J 6(7):1090–1094
6. Ranjan R, Harwood A, Buyya R (2007) Peer-to-peer-based resource discovery in global grids: a tutorial. IEEE Commun Surv Tutor 10(2):6–33
7. Bashir MB, Latiff MS, Ahmed AA, Coulibaly Y, Abdullah AH, Yousif A (2011) A hybrid resource discovery model for grid computing. Int J Grid Comput Appl 2(3):1
8. Hassan MI, Abdullah AB (2011) A new grid resource discovery framework. Int Arab J Inf Technol 8(1):99–107
9. Chen D, Chang G, Zheng X, Sun D, Li J, Wang X (2011) A novel p2p based grid resource discovery model. Journal of Networks 6(10):1390
10. Khanli LM, Kargar S, Niari AK A new approach to resource discovery in grid computing. www.intechopen.com
11. Sotiriadis S, Bessis N, Kuonen P (2012) Advancing inter-cloud resource discovery based on past service experiences of transient resource clustering. In: 2012 3rd international conference on emerging intelligent data and web technologies. IEEE, pp 38–45
12. Sarhadi A, Yousefi A, Broumandnia A (2012) New method for grid computing resource discovery with dynamic structure of peer-to-peer model based on learning automata. World Appl Sci J 19(1):153–158
13. Chung W-C, Hsu C-J, Lai K-C, Li K-C, Chung Y-C (2013) Direction-aware resource discovery in large-scale distributed computing environments. J Supercomput 66:229–248
14. Khethavath P, Thomas J, Chan-Tin E, Liu H (2013) Introducing a distributed cloud architecture with efficient resource discovery and optimal resource allocation. In: 2013 IEEE Ninth World Congress on Services, Santa Clara, CA, 2013, pp 386–392. https://doi.org/10.1109/SERVICES.2013.68
15. MollaMotalebi M, Maghami R, Ismail AS, Atahar Ahmed A (2013) Resource discovery approaches for grid environments. Int J Netw Commun 53–61
16. Barati Masoud, Lotfi Soheil, Rahmati Azizallah (2014) A fault tolerance algorithm for resource discovery in semantic grid computing using task agents. J Softw Eng Appl 7:256–263
17. Barati M, Alizadeh R (2014) A new method for resource discovery in a grid resource service chain. Int J Grid Distrib Comput 7(2):21–32
18. Nasrollahi AZ, Kazem AA (2016) Resource discovery in grid computing using fuzzy logic and tabu table. Int J Comput Sci Netw Secur (IJCSNS) 16(9):61
19. Sim KM (2018) Agent-based fog computing: gossiping, reasoning, and bargaining. IEEE Lett Comput Soc 1(2):21–24. https://doi.org/10.1109/locs.2018.2886828
20. Boukadi K, Rekik M, Rekik M et al (2018) FC4CD: a new SOA-based focused crawler for cloud service discovery. Computing 100:1081. https://doi.org/10.1007/s00607-018-0600-2
21. Nzanywayingoma F, Yang Y Efficient resource management techniques in cloud computing environment: a review and discussion. Int J Comput Appl 41(3):165–182. Doi: https://doi.org/10.1080/1206212x.2017.1416558
22. Santos IL, Pirmez L, Delicato FC, Oliveira GM, Farias CM, Khan SU, Zomaya AY (2019) Zeus: A resource allocation algorithm for the cloud of sensors. Future Gener Comput Syst 92:564–581
23. Sun Y, Zhang N (2017) A resource-sharing model based on a repeated game in fog computing. Saudi J Biol Sci 24(3):687–694
24. Zalila F, Challita S, Merle P (2019) Model-driven cloud resource management with OCCIware. Future Gener Comput Syst 99:260–277
25. Zarrin J, Aguiar RL, Barraca JP (2018) Resource discovery for distributed computing systems: a comprehensive survey. J Parallel Distrib Comput 113:127–166
26. Bölöni L, Jun K, Palacz K, Sion R, Marinescu DC (2000) The bond agent system and applications. Computer science technical reports

27. Jun K, Boloni L, Palacz K, Marinescu DC (1999) Agent-based resource discovery. Computer science technical reports. Paper 1464
28. Gupta S et al (2019) Tier application in multi-cloud databases to improve security and service availability. Handbook of research on cloud computing and big data applications in IoT. IGI Global, pp 82–93

Chapter 55
Bias Correction of GCM Data Using Quantile Mapping Technique

Shweta Panjwani, S. Naresh Kumar, and Laxmi Ahuja

1 Introduction

Global climate models (GCMs) generated climate data are used for different climate assessment studies in various areas like hydrology and socio-economic. These climate projections derived by these GCMs have lots of uncertainties [1–3]. Use of these bias climate change projections may result in uncertain predictions of climate change impacts studies [4, 5]. Since these models have lots of bias, it is advised that the bias-corrected data should be used as input for respective climate change studies [6].

There are various bias correction techniques which are used to bias correct the climate data with respect to observed data. Linear scaling, mean bias remove technique, regression and quantile mapping are some of the bias correction techniques. Several comparative studies are performed using different techniques for Rhine Basin [7], for hydrological impact studies over North America [8], for Indian summer monsoon [9] and for temperature and precipitation extremes [10]. From the above studies, quantile mapping technique found to be more suitable. But this technique has different methods for performing bias correction: linear and tricube. So, it is very difficult to perform the bias correction using both the methods at large-scale climate data for different climatic parameters. So, the main aim of this study is to compare both the methods of the quantile mapping bias correction technique, which is being performed on statistical tool R.

S. Panjwani (✉) · S. Naresh Kumar
Centre for Environment Science and Climate Resilient Agriculture, Indian Agricultural Research Institute, New Delhi, New Delhi 110 012, India
e-mail: shweta.panjwani537@gmail.com

L. Ahuja
Amity Institute of Information Technology, Amity University, Noida, UP, India

© Springer Nature Singapore Pte Ltd. 2021

S. D. Purohit et al. (eds.), *Proceedings of International Conference on Communication and Computational Technologies*, Algorithms for Intelligent Systems,
https://doi.org/10.1007/978-981-15-5077-5_55

2 Materials and Methods

The Indian Meteorological Department (IMD) daily data have been used to bias correct the CSIRO-MK3 climate model daily data (temperature and rainfall) for the time period 1976–2005 for Delhi location. The most popular method quantile mapping is used to bias correct the data [11, 12]. This method used the empirical probability distributions of observed and simulated values. This method assumes that both observed and simulated precipitation intensity distributions are nearly estimated by the gamma distribution. To correct the bias from the data, the inverse of cumulative distribution function (CDF) of observed values at the probability corresponding to the model output CDF at the particular value is calculated. Suppose, F_Y and $F_{\bar{F}}$ are CDF for observed data and ensemble mean of model, respectively, then the bias-corrected value Q will then be as follows:

$$Q = F_Y^{-1}(F_{\bar{F}}(\bar{F})) \tag{1}$$

Here, F^{-1} is an inverse of CDF.

Bias correction technique applied using Qmap package in R "fitQmapQUANT" method used to remove the bias, which estimates values of the empirical CDF in regularly spaced quantiles for observed and simulated long-term climate data. This approach uses two methods to estimates the value: linear and tricube. This approach used wet parameter especially for rainfall data. If value of wet parameter is TRUE, then the empirical probability of nonzero observations is calculated (obs $\geq$ 0) and the value of its corresponding model sets as a threshold and remaining other simulated models' values below this threshold are set to zero. [13, 14].

Daily model simulated temperature and rainfall data were bias corrected using linear and tricube approach. Then yearly average and sum were calculated for temperature and rainfall, respectively [15]. Finally, its parameter wise statistics (RMSE) were prepared to evaluate the overall performance of both the approaches.

3 Result and Discussion

Climate model generated temperature and rainfall data corrected with respect to the observed data using quantile mapping approach in R. The average temperature per year of the observed data varies between 29.5 and 32.5 °C but these ranges vary between 26.3 and 29.5 °C for simulated data(biased). After removing the bias using Qmap, these varied between 30.5 and 33 °C for both linear and tricube method (Fig. 1).

Similarly, the total rainfall per year measured in the observed data is around 200–1200 mm while total rainfall per year simulated by climate model is high (600–1200 mm). The bias-corrected rainfall data using both the method values ranges between 300–1150 mm (Fig. 2).

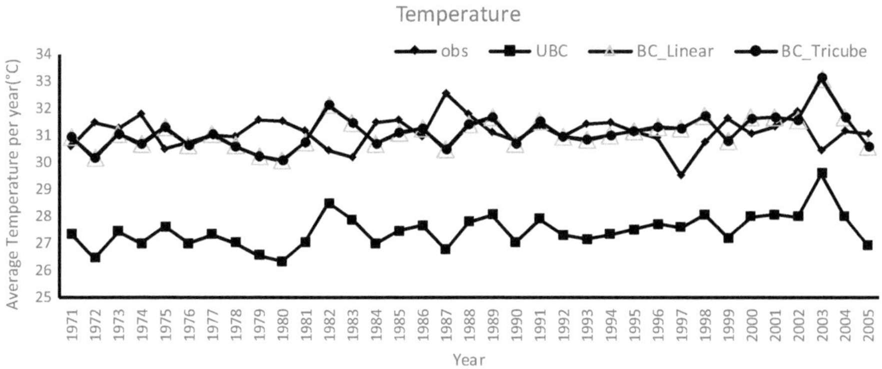

Fig. 1. Average temperature per year for Observed(obs), Simulated (UBC), Bias corrected (BC_Linear and BC_Tricube) data

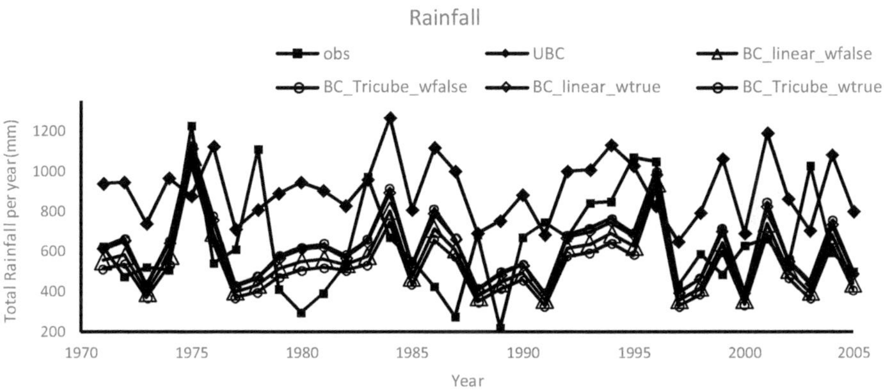

Fig. 2. Total rainfall per year for Observed (obs), Simulated (UBC), Bias corrected (BC_Linear and BC_Tricube with wet_true and wet_false combination) data

To evaluate the overall performance of all the methods (linear and tricube with all combinations) of quantile mapping, statistics for temperature and rainfall were examined. The mean temperature for observed and simulated data over 30 years' period is 31.12 °C and 27.12 °C, respectively. The value of average mean temperature data corrected using tricube method found the same as the observed data (31.12 °C). Similarly, total model simulated rainfall (31,725 mm) over 30 years' period is much higher than total observed rainfall (22,420 mm). But total rainfall over 30 years' period became much

nearer to the observed data after removing the bias using tricube method with wet_true combination (21,006 mm). As proceeding towards RMSE statistics, these values were found less with tricube with wet_false condition for temperature (3.99) and for rainfall (8.96) as compared to other methods (Table 1).

Table 1. Statistics of simulated and other bias correction methods with observed mean for temperature and rainfall

	Obs	Sim	Linear_wfalse	Tricube_wfalse	Linear_wtrue	Tricube_wtrue
Temperature						
Mean	31.12	27.46	31.09	31.12	–	–
RMSE	–	5.48	3.96	3.99	–	–
Rainfall						
Total	22,420	31,725	18,828	17,426	20,594	21,006
RMSE	–	10.09	9.12	8.96	9.47	9.61

4 Conclusion

Different bias correction techniques are used to remove bias from model-generated data by stakeholders. The performance of different methods of quantile mapping technique was evaluated to select the best technique, and their statistical parameter was compared. This study revealed that overall best method of quantile mapping for model-derived data is tricube with wet false condition. Use of accurate climate data is an important aspect of decision-making in various sectors which can be increased by applying this method.

References

1. Woldemeskel FM, Sharma A, Sivakumar B, Mehrotra R (2016) Quantification of precipitation and temperature uncertainties simulated by CMIP3 and CMIP5 models. J Geophys Res Atmos 121(1):3–17
2. Mandal S, Simonovic SP (2017) Quantification of uncertainty in the assessment of future streamflow under changing climate conditions. Hydrol Process 31(11):2076–2094
3. Shen M, Chen J, Zhuan M, Chen H, Xu CY, Xiong L (2018) Estimating uncertainty and its temporal variation related to global climate models in quantifying climate change impacts on hydrology. J Hydrol 556:10–24
4. Asseng S, Ewert F, Rosenzweig C, Jones JW, Hatfield JL, Ruane AC, … Brisson N (2013) Uncertainty in simulating wheat yields under climate change. Nat Clim Change 3(9):827
5. Raymundo R, Asseng S, Robertson R, Petsakos A, Hoogenboom G, Quiroz R, … Wolf J (2018) Climate change impact on global potato production. Euro J Agron 100:87–98
6. Panjwani, S., Kumar, S. N., Ahuja, L., & Islam, A. Prioritization of global climate models using fuzzy analytic hierarchy process and reliability index. Theoretical and Applied Climatology, 1–12 (2019)
7. Terink W, Hurkmans RTWL, Torfs PJJF, Uijlenhoet R (2010) Evaluation of a bias correction method applied to downscaled precipitation and temperature reanalysis data for the Rhine basin. Hydrol Earth Syst Sci 14(4):687–703
8. Chen J, Brissette FP, Chaumont D, Braun M (2013) Finding appropriate bias correction methods in downscaling precipitation for hydrologic impact studies over North America. Water Resour Res 49(7):4187–4205
9. Acharya N, Chattopadhyay S, Mohanty UC, Dash SK, Sahoo LN (2013) On the bias correction of general circulation model output for Indian summer monsoon. Meteorol Appl 20(3):349–356

10. Iizumi T, Takikawa H, Hirabayashi Y, Hanasaki N, Nishimori M (2017) Contributions of different bias-correction methods and reference meteorological forcing data sets to uncertainty in projected temperature and precipitation extremes. J Geophys Res Atmos 122(15):7800–7819
11. Wood AW, Maurer EP, Kumar A, Lettenmaier DP (2002) Long-range experimental hydrologic forecasting for the eastern United States. J Geophys Res (Atmos) 107:4429
12. Piani C, Haerter J, Coppola E (2010) Statistical bias correction for daily precipitation in regional climate models over Europe. Theor Appl Climatol 99(1–2):187–192
13. Gudmundsson L (2016) qmap: Statistical transformations for post-processing climate model output. R package version 1.0–4
14. Gudmundsson L, Bremnes JB, Haugen JE, Engen-Skaugen T (2012) Technical note: downscaling RCM precipitation to the station scale using statistical transformations - a comparison of methods. Hydrol Earth Syst Sci 16:3383–3390. https://doi.org/10.5194/hess-16-3383-2012
15. Pankaj A et al (2019) Sustainable smart-farming framework: smart farming. In: Smart farming technologies for sustainable agricultural development. IGI Global, pp 147–173

Chapter 56
A Collaborative Versioning Framework for Model-Based Version Control Systems

Jai Prakash Jyani and N. C. Barwar

1 Introduction

Software projects undergo constant evolution. Specifications undergo change, enhancements and customizations are made, bugs are fixed, and different software versions are created to suit different needs. Handling such changes for a large system is a significant and tedious task. Version control system (VCS) is an important tool in the area of software development which is mainly used for tracking and managing changes to the project file contents (mainly source code). They help developers to carry out parallel modifications in the software artifacts and facilitate them to share, collaborate and merge the changes [1]. If version control systems are not used, the developers face problems in communicating their changes and are forced to merge the changes manually which makes the changes untraceable. Version control systems store the changes in the server and allow the developers to coordinate their changes in a well-structured and traceable manner. Conflicts are detected and resolved properly, thus reducing the possibility of errors. The basic idea of version control systems is to keep the working copies (files on which developers work) separated from the master copies (files which are stored in the repository). The developers check out the working copies from the repository to their local system, make changes in the working copy and finally check in (commit) their changes back into the repository. Every time a developer commits a file, and the version control system creates a new version in the repository. In this way, the repository stores all the versions of a file. The basic diagram of version control system is shown in Fig. 1.

J. P. Jyani (✉) · N. C. Barwar
Department of Computer Science and Engineering, M.B.M. Engineering College, Jai Narain Vyas University, Jodhpur, India
e-mail: jai_jyani@yahoo.com

N. C. Barwar
e-mail: ncbarwar@jnvu.edu.in

© Springer Nature Singapore Pte Ltd. 2021

S. D. Purohit et al. (eds.), *Proceedings of International Conference on Communication and Computational Technologies*, Algorithms for Intelligent Systems,
https://doi.org/10.1007/978-981-15-5077-5_56

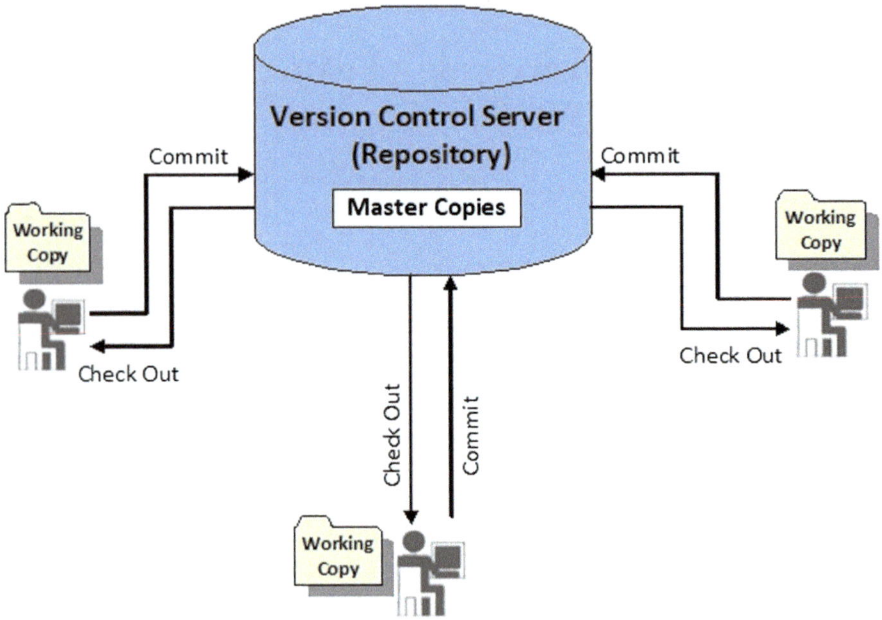

Fig. 1 Basic version control system

2 Types of Version Control Systems

Since the 1970s version control systems have been used in software development projects to archive and manage different versions of files, especially source code. A large number of VCSs have evolved over the period of time with different features and capabilities. The version control systems are based on either of the two fundamental architectures: Centralized (CVCS) or Decentralized/Distributed (DVCS).

2.1 Centralized Architecture

Earliest version control systems, i.e., SCCS and RCS were based on centralized client–server architecture. The versioning software runs in a central server where all the version control operations take place. As depicted in Fig. 2, the project artifacts reside in the repository on a central machine/server. The developer usually checks out the latest version of artifact from the repository onto his local machine (working copy), makes changes to the working copy and then commits those changes back to the central repository/server.

Users can check out either single file or the entire repository by specifying the version which is required (i.e., head or any other version). The repository can be

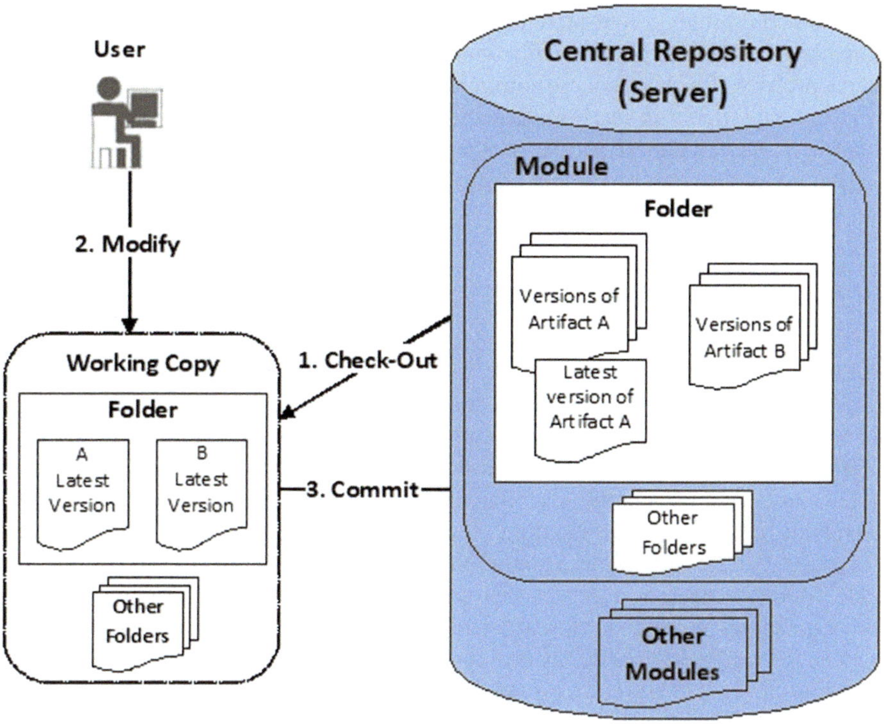

Fig. 2 Centralized version control system

accessed from anywhere through the network, i.e., LAN or WAN. The differences between any two versions of a document/file are called diff or delta.

2.2 Distributed Architecture

The distributed or decentralized version control systems work based on the peer-to-peer architecture and do not have any dependency on a central server. Before starting working on a project, every developer has to copy (clone) the entire master repository onto his local machine. Once the repository has been copied locally, the user can make changes in the local repository. In DVCS, the user is allowed to work in the local repository in an offline mode which improves the execution speed of the operations and reduces the unnecessary overhead. After the changes have been committed in the local repository, at a certain point of time, the local changes need to be shared with other developers and also pushed to the central (master) repository. This process is termed as publishing the changes from the local repository. At the time of publishing, conflicts may arise due to overlapped and contradicting changes performed by developers working simultaneously on the same project. Conflicts are resolved through proper resolution techniques, and efficient merging is applied thereafter. DVCS is highly suitable and widely used for the development of open source projects. The open source community involves big teams of developers which work collaboratively in large-scale projects and are geographically distributed all over the world. Git and Mercurial are the most popular and widely used DVCS. The distributed architecture of version control systems is shown in Fig. 3.

3 Model-Driven Software Development

Software modeling has today become an integral part of the software development process. It uses the principles of abstraction, modularity, problem partitioning, refactoring and reusability. The paradigm of model-based software development has revolutionized the traditional approaches used for developing high standard software. A system model represents the specification and abstract view of the actual system to be built. It helps in understanding the intricacies and functionality of the system under study, and a partial implementation can be carried out using the system model itself. Models represent different abstraction levels of the system such as conceptual, logical and physical. Different forms can be used to represent different kinds of models. Requirement models can be expressed using plain natural language. Unified modeling language (UML) is a powerful modeling tool to visualize the graphical components of software design models. Generative techniques can be used to directly convert the UML models into source code for high-level languages such as Java, PHP or .Net.

3.1 Version Control Mechanism in Model-Driven Software Development

Conventional version control systems like Subversion, CVS and VSS are generally used to record the changes to flat text-based files such as source code files and related technical documents, etc. In model-driven development approach, the software artifacts are not limited to only text files, but they usually constitute models like UML diagrams. The fundamental problem faced in using the conventional version control systems to maintain and manage the model versions is that the classical versioning methodology works only on plain, text-based files [2]. Models possess a graphical and hierarchical structure which cannot be handled by traditional VCSs. The models also evolve continuously throughout their development life cycle and require certain version control mechanism to monitor the collaborative evolution [3, 4].

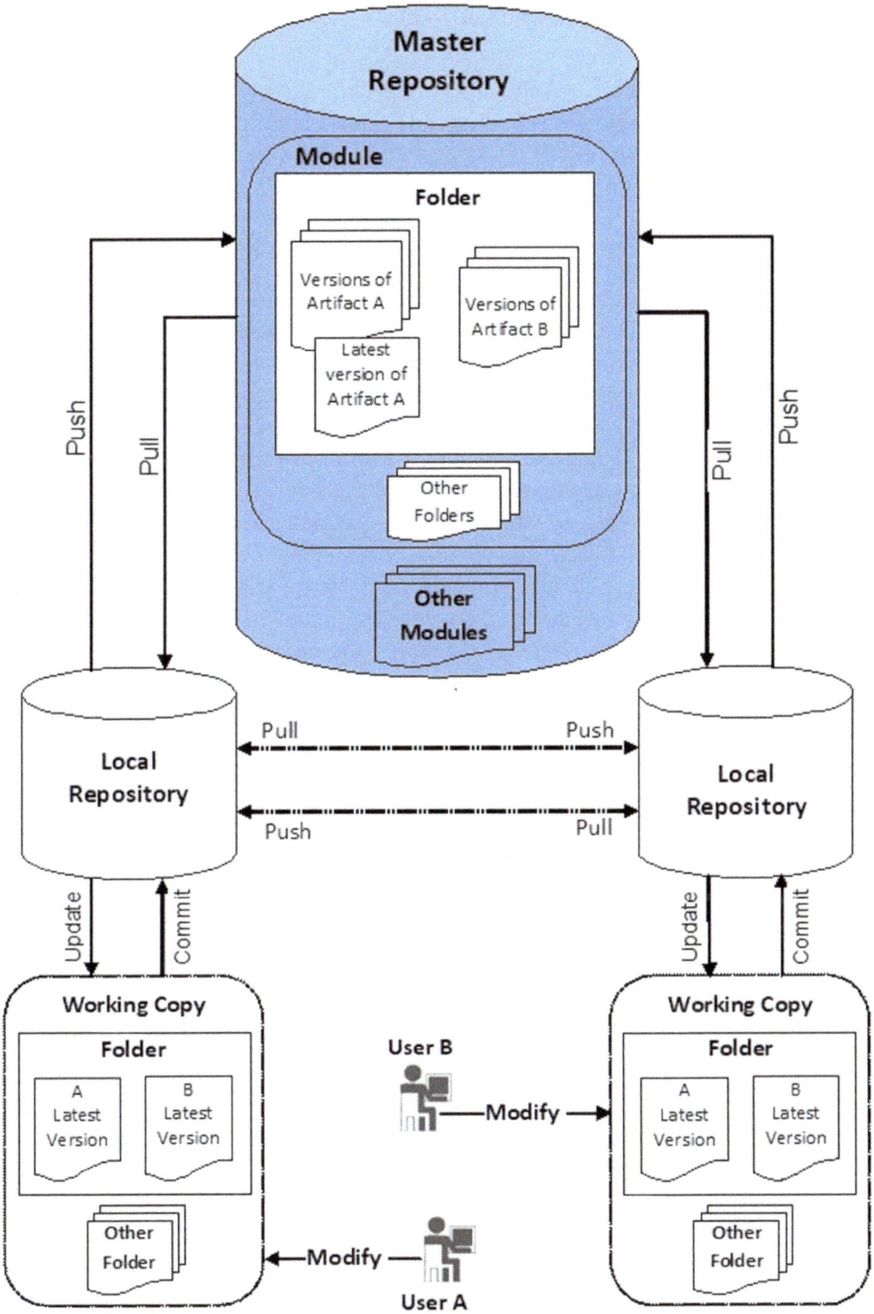

Fig. 3 Distributed version control system

3.2 Challenges in Model Versioning

In model-based software development, models form the essential component in the development lifecycle. However, models mature over a period of time where concurrent developers work on the same model and finally integrate their changes to a common repository. This triggers the need of using necessary and effective version control techniques for maintaining the collaborative model evolution [5]. The challenges faced in model-based version systems can be summarized as follows:

1. **Complexity in model representation**: Models are often rendered visually and offer a high abstraction level to simplify the development process. The graphical representation of models is generally described through UML diagrams, and maintaining the evolved versions of these complex diagrams is a tedious task.
2. **Tracking the model evolution**: During the development of models, different users may perform the operations, viz. addition, deletion and updation on the model. Since the model diagrams involve elements like classes, operations, attributes, relationships, etc., tracking the complete version history of a model is a challenge in model-based version control systems.
3. **Model comparison**: The task of model comparison consists of identifying the relationships and the differences between two models [6]. The graphical nature of models does not suit well for calculating the differences (deltas) among the model versions created due to parallel development. Some simplified storage mechanism is required for storing the models and subsequently perform the model comparison.
4. **Model merging**: Similar to source code files, models can also be developed collaboratively where more than one user can make changes to the same model. Parallel modifications in models can lead to conflicts. The conflicts should be properly detected and resolved to create consistent merged versions which can be finally committed in the repository.
5. **Conflict detection and resolution**: During model merging, situations may arise where a model change has been made in the same element by two developers working concurrently; say the same class name or attribute name has been changed. This may lead to a conflict during merging the two versions of the model. Once conflicts are detected, appropriate steps need to be taken to resolve the conflicts in order to receive a consistent merged version [7, 8].

4 Version Control Techniques for Models Using XML

Models are usually rendered in a graphical notation and may be persistently stored in an intermediate language like XML [9]. The XML representation of the file is then archived in the version repository. Through this technique, the intuitive hierarchy of a model is converted and replaced by XML nodes and tags, and it is possible to compare models via their XML representation. Thus, model versioning can be

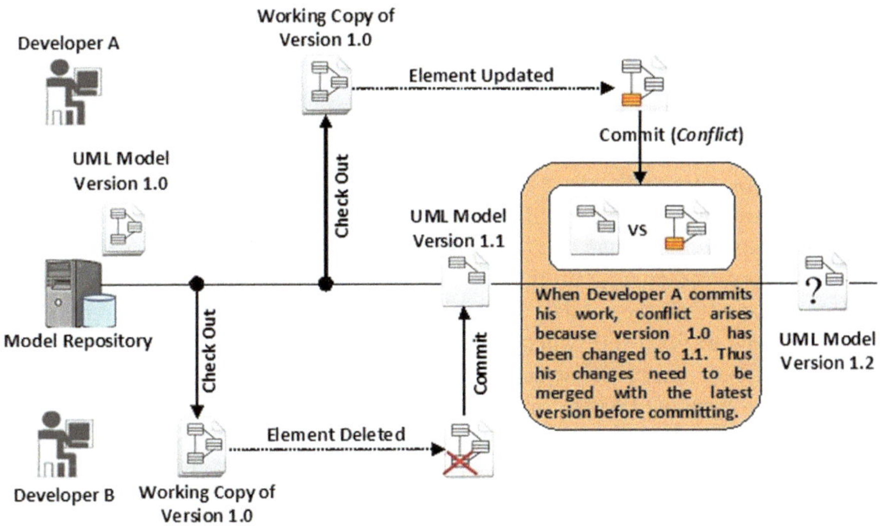

Fig. 4 Comparison of UML models

reduced to simply comparing the XML files for computation of model differences (delta) [10, 11]. As shown in Fig. 4, developer A and developer B work on the same UML model simultaneously. Both of them check out model version 1.0 from the shared repository and edit their local working copy. User B deletes certain element in the model and commits his changes into the repository, and thus, version 1.0 is updated to version 1.1. User A does not know about the changes committed by user B and performs updation on the old version, i.e., 1.0. Later on, user A tries to commit his modifications made in the local version 1.0, into the repository. This creates a conflict scenario because the same artifact has already been changed and committed in the repository by user B at some previous point of time.

Here, if proper conflict resolution mechanism is not applied, the commit by user A will discard the changes made earlier by user B and overwrite them [12, 13]. Thus, the developer who commits last has to perform merging to detect the conflicts and resolve them to create a merged artifact. This may be a time-consuming task so developers need certain version control tool to deal with it.

5 Proposed Model Versioning Tool

To achieve version control functionality for models such as UML class diagrams, a tool to simulate the model comparison and versioning features has been designed. The models will be represented in the form of XML which will conform to a standardized well-formed structure [14, 15]. The operations performed on the model, i.e., addition, deletion or updation will be stored in the XML so that the differences can be computed and proper version history can be maintained for the models.

The proposed model version control system (ModVCS) has been developed using the java swings platform. It allows the users to work collaboratively and concurrently on UML models like class diagrams and store these in a XML-based shared repository. The application provides version control features like model comparison for viewing the differences among different versions, merging the model changes done simultaneously by more than one user and effective conflict detection and resolution techniques to create consistent merged versions [16]. The user can view the entire version history of all the models stored in the model repository and compare any of the versions to find who, when and what changes were performed on a particular model. The hierarchical structure of the UML class diagram is serialized and converted to a structured XML format, and all the operations are performed on these XMLs to implement the model versioning functions.

5.1 Basic Design

The architecture diagram of XML-based model version control system is depicted in Fig. 5. The model engineer generates the UML class diagrams based on the technical system study. These UML diagrams are inputted into the model version control system. It is a login-based application designed for evolving models which simulates the functionality of basic version control system. The authorized users need to login

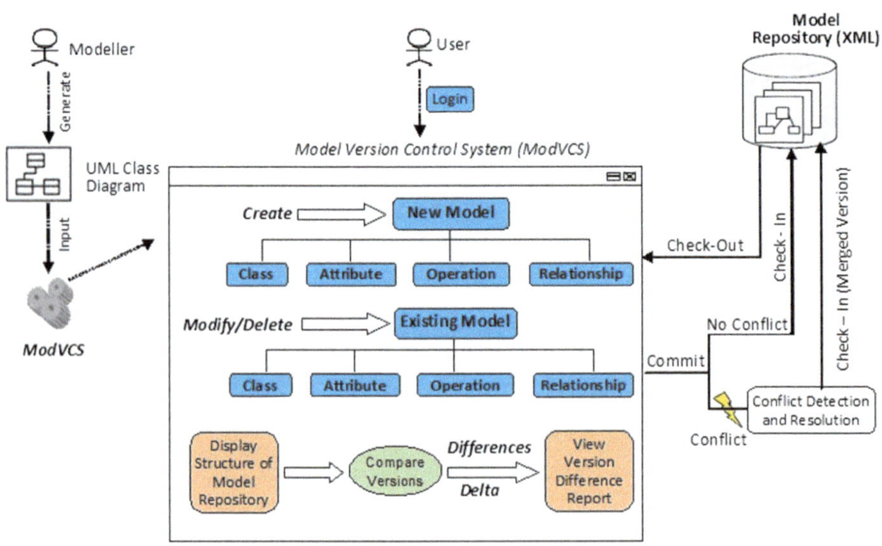

Fig. 5 Architecture of model version control system

into the application using their credentials. Once the user has successfully logged in, he gets access to the shared model repository.

The user can create new models or modify/delete the elements in the existing models. The basic versioning commands, i.e., checkout and checkin (commit) are used to edit the working copy of artifacts and save the changes back into the repository. The application provides the feature to display all the versions of a model in a tree structure. The user can compare the versions and view the difference report of model versions. The tool implements collaborative version control functionality for model development where multiple users can login into the system concurrently and perform modifications on the shared model repository. Effective and efficient conflict detection and resolution mechanism has been integrated in the tool to enable consistency and integrity for concurrent development. This takes care that proper merging techniques are applied in the case of conflicts and a consistent merged version is checked in without any data loss.

5.2 Assumptions and Limitations

The proposed tool has been designed to implement the version control functionality for UML class diagrams. It is assumed that the modeler has already designed the class diagram structure and is ready with it before using this tool. Simple and easy-to-use interface has been provided in the tool to add the different elements of the class diagram and their interrelationships. The tool does not provide the feature to draw and design the graphical components of the UML class diagram. Another limitation is that it has been carried out to manage the versioning of class diagrams only. Other UML diagrams such as use-case diagrams, activity diagrams, state machine diagrams, sequence diagrams, component diagrams, etc., are not handled by the designed tool currently and can be proposed as a future scope.

6 Implementation and Working

The primary goal of ModVCS application is to provide a comprehensive and easy versioning framework for model-based artifacts. The front end has been designed using java swings platform, and the back end consists of well-formed XML formats for storing the serialized class diagrams. Apart from this, the java-based World Wide Web Consortium (W3C) API for document object model (DOM) has been used for parsing the XML documents. The implementation and working of the model version control tool and its related features are described as follows:

6.1 *Serialization of Models (Class Diagram)*

UML class diagrams contain graphical notations for displaying the various elements (class, attributes, operations, etc.) and their interrelationships, and they are converted into a pre-defined XML format so that model comparison and delta (difference) calculation can be performed in easy manner. The XML representation of different elements of UML class diagram is presented below:

(1) **New Model**: The first step in the application is to add a new model. The user has to provide a valid model name and add the model to the repository. The XML representation of newly created model is

```
<?xml version="1.0" encoding="UTF-8"?>
<Model name="" createdBy="" createdOn="" versionNo="">
</Model>
```

(2) **New Class**: A new class can be added to the model by providing class name and class type. The XML representation of the new class is

```
<?xml version="1.0" encoding="UTF-8"?>
<Model name="" createdBy="" createdOn="" versionNo="">
    <Class name="" type="">
    </Class>
</Model>
```

(3) **New Attribute**: A new attribute can be added to the class by providing attribute name and attribute type. The XML representation of the newly created attribute is

```
<?xml version="1.0" encoding="UTF-8"?>
<Model name="" createdBy="" createdOn="" versionNo="">
   <Class name="" type="">
      <Attributes>
         <Attribute>
          <name>......</name>
          <type>......</type>
       </Attribute>
     </Attributes>
   </Class>
</Model>
```

<Attributes> tag is used to declare all the attributes in the class. <Attribute> represents an individual attribute with its name and type.

(4) **New Operation:** A new operation (method) can be added to the class by selecting the class in a model and providing the operation name along with appropriate method return type and mentioning comma-separated data types of input parameters to the method, if any exists. The XML representation of the newly created operation is

```
<?xml version="1.0" encoding="UTF-8"?>
<Model name="" createdBy="" createdOn="" versionNo="">
    <Class name="" type="">
        <Operations>
            <Operation>
                <name>……</name>
                <returnType>……</returnType>
                <parameters>……</parameters>
            </Operation>
        </Operations>
    </Class>
</Model>
```

<Operations> tag is used to declare all the operations in the class. <Operation> represents an individual operation with its name, return type and parameters.

(5) **New Relationship**: A model or class diagram may consist of multiple classes which might be interrelated to each other in some manner. The relationships between classes represent a logical connection among them and can be categorized as inheritance (generalization), association, aggregation or composition. In the application, a new relationship can be added to the model by first selecting the type of relationship. Then, the user has to select super class and derived class in case of inheritance and container class, contained class and multiplicity/cardinality in the case of aggregation and composition. The XML representation of the newly created relationship is

```
<?xml version="1.0" encoding="UTF-8"?>
<Model name="" createdBy="" createdOn="" versionNo="">
    <Class name="" type="">
    </Class>
    <Relationship>
        <Inheritance>
            <Relation superclass="" derivedclass=""/>
        </Inheritance>
        <Association>
            <Relation fromclass=" " toclass=" " multiplicity=" "/>
        </Association>
        <Aggregation>
            <Relation containerclass="" containedclass="" multiplicity=""/>
        </Aggregation>
        <Composition>
            <Relation containerclass="" containedclass="" multiplicity=""/>
        </Composition>
    </Relationship>
</Model>
```

<Relationship> tag is used to declare all the different type of relationships in the model. <Relation> tag represents an individual relationship.

6.2 Comparison of Model Versions

All the versions of a model are stored in an XML file. Any type of operation, i.e., addition, deletion or edit (updation) carried out by the user is stored in the version XML file along with the timestamp and name of the user who performed the modifications. An effective differencing (delta) algorithm is used to compare the model versions and list out the changes or differences between them. Each version has a unique version number, the user name who created the particular version and the timestamp when the version was committed in the model repository. The <Version> tag contains multiple <Add> , <Edit> or <Delete> tags to define the related operations performed by the user on the model. Any particular version can be easily restored or recreated from the XML by applying the changes (delta) one after the other in a sequential order. The structure of version XML file is described as follows:

```
<?xml version="1.0" encoding="UTF-8"?>
<Model name="" createdBy="" createdOn="" versionNo="0.0">
   <Version number="1.0" createdBy="" createdOn="">
      <Add element="Class" name="" type=""/>
      <Add element="Attribute" class="" name="" type=""/>
      <Add element="Operation" class="" name="" returnType="" parameters=""/>
      <Add element="Relation" type="Inheritance" superclass="" derivedclass=""/>
      <Add element="Relation" type="Association" fromclass="" toclass=""
            multiplicity=""/>
      <Add element="Relation" type="Composition" containerclass="" containedclass=""
            multiplicity=""/>
   </Version>
   <Version number="2.0" createdBy="" createdOn="">
      <Edit element="Class" old_name="" new_name="" old_type="" new_type=""/>
      <Edit element="Attribute" class="" old_name="" new_name=""
            old_type="" new_type=""/>
      <Edit element="Operation" class="" old_name="" new_name=""
            old_returnType="" new_returnType=""
            old_parameters="" new_parameters=""/>
      <Delete element="Class" name="" />
      <Delete element="Attribute" class="" name="" />
      <Delete element="Operation" class="" name="" returnType="" parameters=""/>
      <Delete element="Relation" type="Inheritance" superclass="" derivedclass=""/>
      <Delete element="Relation" type="Association" fromclass="" toclass=""
            multiplicity=""/>
   </Version>
</Model>
```

6.3 Conflict Detection and Resolution

The model version control system supports an effective conflict resolution mechanism for resolving the contradicting and conflicting changes in the model which is developed concurrently. The process of conflict detection and resolution in ModVCS

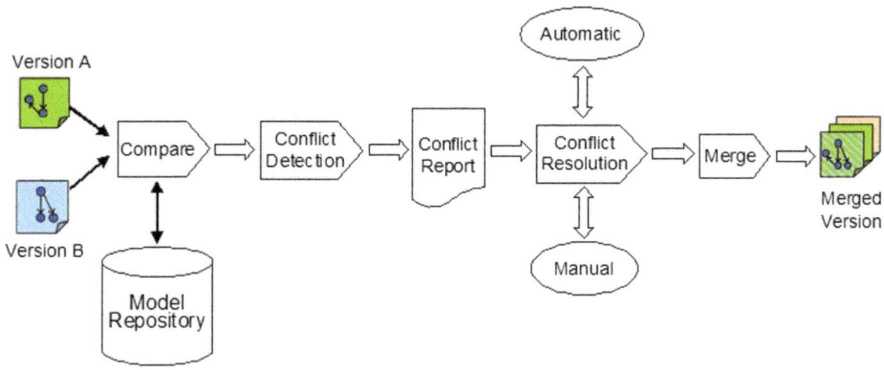

Fig. 6 Conflict detection and resolution

application is illustrated in Fig. 6. The conflict report shows details of the changes performed by the user through add, update and delete operations that have created conflicts at the time of commit. The conflicts are resolved either through automatic or manual conflict resolution process.

The equivalent changes performed by two users, i.e., when the users while working concurrently on the same model make similar changes can be resolved automatically without any manual intervention. On the other hand, certain contradicting changes require manual intervention for their resolution. A class, attribute or operation in the model can be renamed to two different names by two concurrent users. The user who is resolving the conflicts can either keep his change (this will overwrite the other users change) or discard his change and keep the existing change in the repository. Similarly, the constraint violation where an inheritance cycle has been created due to the changes done by two users can also be easily resolved manually. After all the conflicts have been successfully resolved either automatically or manually, merging technique is applied to create a consistent merged version which contains all the changes performed by the two users without any loss of data. The merged version is the latest version which is committed into the repository.

7 Results and Discussion

The model version control system implements the basic version control features for model-based artifacts such as model comparison, conflict detection, conflict resolution and model merging. We have used the model "Election Management System" as a case study to validate the functionality of the tool. The model contains an overview of the different planning activities related to district-level election management required for conducting parliamentary elections (Lok Sabha) and state-level assembly elections (Vidhan Sabha). The model is used for entering test data for the model versioning tool. A set of test cases have been designed which are executed

Table 1 Summary of model elements used as test data for the tool

Type of model elements		No. of elements
Classes		63
Attributes		216
Operations (methods)		57
Relationships	Inheritance	21
	Association	17
	Composition	11
	Aggregation	05

using the test data to check and validate the various functionalities of the tool. For every test case, we define the execution steps, the input data, the expected output, the actual output received and finally the result of test case execution (pass/fail). The following table shows the different types of model elements used in the tool.

As per Table 1, the model for election management system constituting of 63 classes was input into the versioning tool along with the related attributes, operations and relationships. The number of various model elements entered in the tool is shown in the table. The XML-based model repository was setup using LAN-driven centralized architecture. Two custom users were created in the repository to enter the test data. Different test case scenarios were designed and executed on the test model to check the behavior of the tool and its compliance with the standard version control characteristics. The different test case scenarios, number of test cases executed and the result of test case execution are mentioned in the below Table 2.

As per Table 2, total 48 test cases were executed for testing the different features of the application. 44 test cases passed and 04 test cases failed. However, the cause for their failure was detected and necessary rectifications were made in the source code to ensure 100 percent compliance of the tool with its expected and desired behavior. The login functionality was tested with multiple users, and both the version

Table 2 Results of test case execution

S. No.	Test case scenario	No. of test cases executed	Test cases passed
1.	Check the application access to authenticated users (Login)	03	03
2.	Addition of new elements in the model	16	15
3.	Updation of elements in the model	07	06
4.	Deletion of elements in the model	07	06
5.	Conflict detection and resolution	13	12
6.	Comparison of model versions and view the version history	02	02
Total		48	44

control techniques, i.e., "*Lock-Modify-Unlock*" and "*Copy-Modify-Merge*" worked as desired. The modules to add, update and delete elements were tested using total 30 test cases, and all of them reported as passed. The tool provides a comprehensive conflict detection and resolution feature that was tested using 13 different conflict scenarios which were detected successfully and completely followed their expected resolution pattern. The user has the facility to compare the model versions to view the differences/changes between any two versions and trace the complete model evolution history.

8 Related Work

Model comparison, model versioning and computation of model differences have become a key research area in the field of model-driven software development and it is essential to detect, manage and store the differences among model revisions [17]. In the survey paper [3], the approaches to accomplish model comparison and the various applications where these approaches are used have been described. UMLDiff [10] presents a type of heuristic algorithm for automatically detecting the changes in the logical structure of models as they evolve from one version to the next. A three-way merging technique for comparing the graphical structures of models is discussed in [16]. The possibility of XMI-based generic transformations of UML models is examined by Kovse and Harder [9]. A generic technique for conflict detection and intelligent conflict resolution in model versioning is presented in [7]. The challenges faced in detecting and visualizing the differences between different versions of UML diagrams using XML files have been elaborated by Ohst, Welle and Kelter [11]. The work [14] defines a mapping between the UML class diagrams and XML schema through the use of conceptual, logical and physical design levels. Gherabi and Bahaj [15] in their paper present a basic framework for converting UML class diagram into XML document using the DOM technology. A solution for effective and efficient visualization of the model differences has been described in [18]. Alanen and Porres [19] described a difference calculation routine between two models and a merging algorithm for applying the difference to the original model to produce the target model.

9 Conclusion and Future Scope

In this work, we designed a conceptual versioning framework for model-based version control systems to assist collaborative development of UML class diagrams. The tool stores the models in a centralized repository in the form of well-structured XML files. The serialization of complex UML diagrams into readable XML files eases the version control mechanism for the models. One innovative feature implemented in the tool which makes it distinct is that it allows user to choose the versioning mechanism. The "lock-modify-unlock" option allows only one user to modify the model elements at a time, whereas in the "copy-modify-merge" option, multiple users can modify the model simultaneously. An efficient conflict detection and resolution mechanism provides a conflict report to the user listing all the conflicting changes. After proper conflict resolution, the changes are merged and committed into the repository. ModVCS is a java-based tool which is platform independent and highly portable. The graphical user interface of the tool is simple and easy to use. The use of XML to store UML models improves the flexibility and simplifies data storage and data sharing. It facilitates collaborative model development where multiple users can work on the same model simultaneously. The modular approach used in the development improves the scalability of the application.

Here, we have tried to address some of the basic challenges faced in model-based version control systems. Yet, there remains a scope of improvement to further enhance the functionality developed in this work. The model version control tool developed here is used to control the version management of UML class diagrams. It can be extended to other UML diagrams such as use-case diagrams, activity diagrams, state machine diagrams, sequence diagrams and component diagrams. The models, i.e., class diagrams in the tool are represented/visualized through a hierarchical tree structure format. Component for graphical visualization of model elements and their relationships can be added in the tool to understand the model comparison and evolution in a more effective way.

References

1. Garcia JAE (2011) Software development and collaboration: version control systems and other approaches. Technical University, Berlin
2. Altmanninger K, Brosch P, Kappel G, Langer P, Seidl M, Wieland K, Wimmer M (2009) Why model versioning research is needed? An experience report. In: Joint workshop on model-driven software evolution, Model co-evolution and consistency management (MoDSE-MCCM), 2009
3. Stephan M, Cordy JR (2013) A survey of model comparison approaches and applications. MODELSWARD—international conference on model-driven engineering and software development, 2013
4. Altmanninger K, Seidl M, Wimmer M (2009) A survey on model versioning approaches. Int J Web Inf Syst 5(3):271–304
5. Rao AA, Madhavi K (2010) Framework for visualizing model-driven software evolution and its application. Int J Comput Sci Issues (IJCSI) 7(1):3
6. Scheidgen M (2016) Evaluation of model comparison for delta-compression in model persistence. In: Proceedings of the 4th workshop on scalable model driven engineering: applications and foundations (STAF), Vienna, 2016
7. Brosch P (2009) Improving conflict resolution in model versioning systems. In: Proceedings of the 31st international conference on software engineering, Vancouver, May 2009
8. Taentzer G, Ermel C, Langer P, Wimmer M (2010) Conflict detection for model versioning based on graph modifications. In: Proceedings of the 5th international conference on graph transformation, ICGT 2010
9. Kovse J, Harder T (2002) Generic XMI-based UML model transformations. International conference on object-oriented information systems. Springer, New York, pp 192–198
10. Xing Z, Stroulia E (2007) Differencing logical UML models. Springer, New York

11. Ohst D, Welle M, Kelter U (2003) Differences between versions of UML diagrams. In: Proceedings of the 9th European software engineering conference—ESEC, 2003
12. Altmanninger K (2007) Models in conflict—towards a semantically enhanced version control system for models. In: International conference on model driven engineering, Springer, 2007
13. Brosch P, Seidl M, Wimmer M, Kappel G (2012) Conflict visualization for evolving UML models. J Object Technol 11(3):1–2
14. Nicholas Routledge, Linda Bird, Andrew Goodchild, "UML and XML Schema", 13th Australasian Database Conference (ADC2002), Melbourne, 2002
15. Gherabi N, Bahaj M (2011) Robust representation for conversion UML class into XML document using DOM. Int J Comput Appl 33(9):0975–8887
16. Mehmood W, Nadir Shah MJ, Khan M, Memon MI (2016) Fine-granular model merge solution for model-based version control system. Malays J Comput Sci 29(3):225–246
17. Cicchetti A, Ciccozzi F, Leveque T (2012) A solution for concurrent versioning of meta-models and models. J Object Technol 11(3):1–32
18. van den Brand M, Protić Z, Verhoeff T (2010) Generic tool for visualization of model differences. In: Proceedings of the 1st international workshop on model comparison in practice—IWMCP, Malaga, Spain, 2010
19. Alanen M, Porres I (2003) Difference and union of models. In: 6th international conference on unified modeling language and applications, CA, USA, 2003

Jai Prakash Jyani is a M.E. Scholar in Computer Science and Engineering at M.B.M. Engineering College, Jodhpur. His field of interest is Software Engineering, Database Management Systems and Programming Languages Java, .NET, etc.

Dr. N.C. Barwar is presently working as a Professor in the Department of Computer Science and Engineering, M.B.M. Engineering College, J.N.V. University, Jodhpur. He received B.E. from MANIT Bhopal, M.E. in Digital Communication and Ph.D. from M.B.M. Engineering College, Jodhpur. He has published more than 50 papers in national and international conferences and journals and has teaching experience of 25 years at UG and PG level. His field of interest is Computer Networks, Multimedia, VOD and Information theory.

Chapter 57
Model Comparison and Individual Open Source Community Development

Dharmendra Kumar and Saroj Hiranwal

1 Designing and Comparative Procedure

Programming advancement can be testing a result of the enormous data spaces that designers must explore. Without help, engineers can move toward becoming stalled and spend an unbalanced measure of their time looking for data to the detriment of other worth delivering errands. Proposal frameworks for programming designing (RSSEs) are programming apparatuses that can help engineers with a wide scope of exercises, from reusing code to composing compelling bug reports [1].

Scores of programming makers have moved in the direction of open source licenses to improve administration for their clients. For these organizations, picking the right permit decides business achievement. At the point when the accessible open source stack and authorizing alternatives develop, so does the need to comprehend the exchange among permitting, sourcing choices, and business objectives. A model of permit decision accentuates various licenses and defends the decision of an open source programming (OSS) permit. This is urgent for littler organizations and new businesses that do not have the instruments and information to play out a careful examination of the considerable number of results of their permit decision each time they utilize OSS [2].

D. Kumar
Department of Computer Science and Engineering, Rajasthan Institute of Engineering and Technology, Jaipur, Rajasthan 302026, India

S. Hiranwal (✉)
Rajasthan Institute of Engineering and Technology, Jaipur, Rajasthan 302026, India
e-mail: dharm.1982@gmail.com

© Springer Nature Singapore Pte Ltd. 2021
S. D. Purohit et al. (eds.), *Proceedings of International Conference on Communication and Computational Technologies*, Algorithms for Intelligent Systems,
https://doi.org/10.1007/978-981-15-5077-5_57

2 Comparison Between the Models and the Single Individual Model Participation to an Open and Free Source Community

The models to be looked at are 'Advancement examples of Open Source programming frameworks and networks. Open Source qualities—normal and variable', OSS (Open Source Software) Model and 'A system investigation of the Open Source programming improvement worldview. The focal point of the initial two models was for the most part on the product improvement process, and the last two were endeavors to build up an increasingly far-reaching clarification.

3 Software Development-Based Models

The main model to be presented is 'Advancement examples of Open Source programming frameworks and networks. This model was created in the wake of summing up from contextual analyses on four distinctive open and free source ventures. The writers recommended that there was a progressive network structure beginning from inactive clients, per users, bug correspondents, bug fixers, fringe engineers, dynamic designers, center individuals, and undertaking pioneer. In addition, there are three sorts of free/open undertakings, in particular investigation arranged, utility-situated, and administration arranged, and each sort had various traits. The three sorts of open and free source undertakings could likewise advance into another kind in various improvement stages. At the point when there are new plans to be actualized, the venture may develop into investigation situated sort [3]. At the point when there were new should be fulfilled, the venture may advance into utility-situated sort. At the point when the undertaking becomes developed, it might advance into administration situated sort. From the model, there is not only one methodology in open and free source programming advancement but also three methodologies. These methodologies did influence the procedure of the improvement as well as the network structure. For instance, Cathedral-like focal control structure was found in investigation situated sort activities and Bazaar-like decentralized control in utility-arranged tasks. Contrasting and the single individual model investment to an open and free source people group, this model has a place with the co-appointment layer of the 4C model with brief notices of issues in culture and boundaries [4]. For sure, this model has an increasingly explicit depiction over the portrayal in co-appointment layer above.

The following model to be presented is 'Open source qualities—normal and variable Fig. 1. They recommended that there were basic characteristics among open and free source activities, for example, Open Source Definition, people group, inspiration and designers are clients, procedure of tolerating entries, advancement improvement cycles and measured quality of code. There were additionally factor qualities that changed from undertaking to extend, in particular decision of work region, parity of

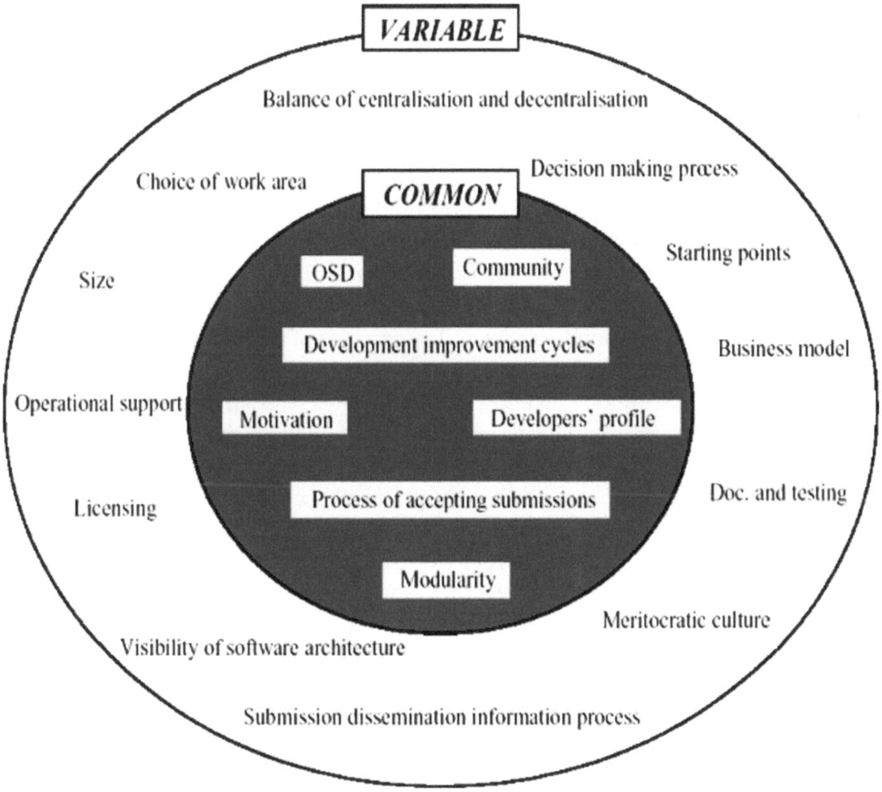

Fig. 1 Comprehensive models

centralization and decentralization, meritocratic culture, plan of action, basic leadership process, accommodation data scattering process, venture beginning stages, deceivability of programming engineering, documentation and testing, authorizing, operational help, and size [5]. Contrasting and the single individual model interest to an open and free source people group, this model for the most part has a place with the co-appointment layer of the 4C model with brief notices of issues in correspondence, culture, and inspiration. Like last model, it has an increasingly explicit portrayal over the depiction in co-appointment layer [6].

4 Comprehensive Models

The principal model in this subarea to be presented is the OSS model. They expected to devise a model of the open and free source marvel so as to propose how conventional programming improvement condition could intertwine with open and free source condition to make a mixture OSS people group and get profits by the two philosophies. The advantages that the creators wanted to acquire were decrease being developed time and time-to-showcase, improvement in quality, decrease of cost, picking up designer faithfulness, and increment engineer ability pool without extra head tally and overhead. Gotten from hierarchical hypothesis writing, the writers suggested that there were three perspectives in open and free source wonder, specifically structure, culture, and procedure. Contrasting and the single individual model cooperation to an open and free source people group, the OSS model spreads co-appointment, culture, and constructive outcomes [7].

There were additionally concise notices of rendition control framework, which has a place with the correspondence layer. Inspirations of designers and boundaries in making a half and half network were likewise talked about. By and by, the creators appeared to take less mind in dealing with the issue of adaptability. For instance, for inspirations, the creators accentuated on charitableness and philosophy. These inspirations were just one of the four kinds of significant inspirations, and countless benefactors in the review were paid to program in open and free source programming. Likewise, without referencing negative outcomes, the open and free source marvel depicted was rosier than the real world. For instance, trust and steadfastness were referenced without offsetting the image with damages in the board issues and fire wars. The last model to be introduced is an intricate model, 'A system investigation of the Open Source programming improvement worldview. This structure was determined IS (Information Systems) design system and CATWOE (Clients, Actors, Transformation, Weltanschauung (World-see), Owner, Environment) structure. Five viewpoints were proposed, in particular capability, change, partners and condition and world-see [8]. In the capability viewpoint, the creators contended that Open Source Definition was a vital and adequate definition. In the change viewpoint, which means the procedure of open and free source programming improvement, the creators recommended that seven attributes that existed in many ventures, for example, peer survey and brief input. At that point, the creators remarked on the taboos and standards of open and free source people group. Finally, the lifecycle for open and free source programming advancement was incorporated, which was for the most part embraced from a contextual investigation in FreeBSD. The phases in the lifecycle included code, audit, and pre-submit test, improvement discharge, parallel troubleshooting, and creation discharge. In the partners' angle, four bodies were considered, including engineer networks, client networks, business associations, and the non-business associations that oversaw open and free source ventures [8].

Finally, in the earth and world-see angle, three classes of inspiration were considered, in particular innovative, monetary, and socio-political. Contrasting and the single individual model interest to an open and free source people group, this model spreads correspondence, co-appointment, culture, inspiration, and constructive outcomes [9]. There was likewise a short notice on hindrances, for example, advancement from 'Designers' statue to 'Extra Contributors' required a test in CVS aptitudes; however, negative elements were not the core interest. Besides, in the clarification of the lifecycle for open and free source programming improvement, the

FreeBSD advancement lifecycle was taken as a prime model. FreeBSD was known to be one of the more sorted out activities, and it will be advantageous to have other progressively disorganized methodologies referenced [10]. Here could be as least three sorts of ventures and along these lines various methodologies being developed.

5 Comparison of the Models

Subsequent to exhibiting the four models, the consequence of the examination is organized in Table 1. The vast majority of the clarifications from the four models are more intricate than the single individual model support to an open and free source people group. For instance, the inspiration classifications proposed [10].

There are additionally regions that are excluded in the single individual model investment to an open and free source people group, for example, capability by the Open Source Definition and partners, for example, business associations. By the by, the single individual model investment to an open and free source people group is yet adaptable enough to join the vast majority of the materials in the four models. Besides, less examined zones, for example, commitments, obstructions, and positive and negative outcomes are likewise included. Additionally, however the vast majority of the substance in the four models depended on undeniable realities, a portion of the certainties may just reflect specifics of certain open and free source people group. Conversely, by being less prescriptive, the single individual model support to an open and free source people group may have the upside of enabling its clients to find choices. Reviewing the point of making the single individual model cooperation to an open and free source people group is to distinguish significant perspectives in a FOSPHost site for further examination. This point can be viewed as finished since the single individual model investment to an open and free source

Table 1 Comparison table

Authors	Communication	Co-ordination	Culture	Motivation	Barriers	Positive results	Negative results
Nakakoji et al. (2002)		√	√		√		
Gacek et al. (2001)	√	√	√	√			
Sharma et al. (2002)	√	√	√	√	√	√	
Feller and Fitzgerald (2002)	√	√	√	√	√	√	

people group incorporates a large portion of the significant perspectives that the four models talked about [11].

In addition, it likewise incorporates other critical issues that the four models have less accentuation on. Moreover, the exclusion of partners other than designers is positive as to limit the extent of examination to the most significant gathering of partners. After the looking at the focal points and detriments of the four models and the single individual model support to an open and free source people group and assessing how reasonable it is for the examination, different perceptions can be talked about. From the examination above, for far-reaching models, social speculations are utilized as a premise to determine clarifications. Notwithstanding for programming advancement-based models, talks on social issues on open and free source are incorporated [12].

Gabriel proposed that different commitments, for example, advertising and measures advancement were additionally prominent. He further remarked that progressive examination of open and free source people group dependent on expert on code could be deceiving [13, 14]. The code advancement network is only one of the numerous networks inside the open and free source wonder and the limit of a network ought to be characterized by these various types of commitments or interests so as to speak to their importance. It is then not amazing that the impacts of open and free source are less talked about in the models referenced [15–18]. In the single individual model investment to an open and free source people group, just the impacts influencing people are referenced. Likewise, negative factors, for example, hindrances and negative impacts of open and free source are less talked about. Along these lines, investigation in these regions will yield new learning. From the models introduced above, it is brought up that open and free source ventures with various co-appointment models had various qualities. Likewise it is demonstrated that there were factors between various tasks. In addition, additionally it is asserted that there were various practices in various associations and designers [19–21].

To be sure, adaptability was additionally a significant thought in the plan of the single individual model support to an open and free source people group. May be this accumulation of contrasts and factors are the place the bedlam of open and free source lies. Subsequently, further research on these factors will be gainful [22–25]. To finish up, subsequent to contrasting the single individual model support with an open and free source people group with four different models, the nature of the model is satisfactory as the reason for this examination [26–28].

6 Conclusion

The single individual model investment to an open and free source people group was in this way formulated to set up significant perspectives inside an open and free source people group, in particular correspondence, commitments, co-appointment, and culture. Basic components, for example, inspirations, hindrances, and positive and negative outcomes from the engineers' perspective were additionally included. These model were contrasted and four different models.

In the wake of contrasting the single individual model cooperation with an open and free source people group with four different models, the nature of the model is adequate as the reason for this examination and finished up to be an appropriate structure for research in FOSPHost [29]. The single individual model interest to an open and free source people group is picked as hypothetical reason for the examination of FOSPHost. Despite the fact that this model is less explicit than different models, it is extensive and adaptable enough with the end goal of this exploration.

References

1. Martin R, Robert W, Thomas Z (2009) Recommendation systems for software engineering. IEEE-2009. http://doi.ieeecomputersociety.org/10.1109/MS.2009.161
2. Juho L, Matti R, Anna P (2011) Matching open source software licenses with corresponding business models. 28(4):31–35
3. Apple Computer Inc (2018) Darwin—open source. http://developer.apple.com/darwin/
4. Arnoud E (2010) Choosing an open source license. IEEE 27:48–49
5. Bezroukov N (1999b) A second look at the cathedral and bazaar
6. Bezroukov N (2017) 4.1. Linus and linux; linus torvalds' short unauthorized biography
7. Bezroukov N (2019) California sunrise: linux hype festival and IPO gold rush
8. Choi N, Smith I, Andrew (2010) Managing first impressions of new open source soft-ware. IEEE-2010
9. Crowston K, Scozzi B (2018) Open source software projects as virtual organizations: competency rallying for software development. IEE Proc Softw
10. DiBona C, Ockman S, Stone M (1999) Appendix a: the tanenbaum-torvalds debate. In: C DiBona, S Ockman, M Stone (eds) Open sources: voices from the open source revolution
11. Eunice J (1998a) Beyond the cathedral, beyond the bazaar
12. Eunice J (1998b) Beyond the cathedral, beyond the bazaar—the key is community
13. Hauben M, Hauben R (1997) The Netizens and the wonderful world of the net: an an-thology
14. Favaro J, Pfleeger L (2011) Software as a business'. IEEE Comput Soc sp. July/August'2011:23–26
15. Johnson K (1999) Open-source software development
16. Open Source Initiative (2003a) The open source definition. http://www.opensource.org/docs/definition.php
17. Open Source Initiative (2003b) Open source initiative OSIZ—Welcome. http://www.opensource.org/
18. Pavlicek RC (2000) Embracing Insanity: open source software development, Sams, Indiana. Pennington, H Working on Free Software. http://www106.pair.com/rhp/hacking.html
19. Levy S (1984) Hackers: heroes of the computer revolution. Anchor Press/Doubleday, New York
20. Open source software engineering. In: Proceedings of 23rd Int'l conference on software engineering, Orlando, Florida. http://opensource.ucc.ie/icse2012/MockusHerbsleb.pdf
21. Open Source Initiative (2018) History of the open source initiative. http://www.opensource.org/history.html
22. Raymond ES (1999) The revenge of the hackers. In: C DiBona, S Ockman, M Stone (eds) Open sources: voices from the open source revolution, O'Reilly & Associates, Sebastopol, California. http://www.oreilly.com/catalog/opensources/book/raymond2.html
23. Raymond ES (2000a) Homesteading the Noosphere. In: Raymond ES (2000b) The cathedral and the bazaar. http://www.tuxedo.org/~esr/writings/cathedral-bazaar/cathedral-bazaar.html

24. Raymond ES (2000c) A brief history of Hackerdom. http://www.tuxedo.org/~esr/writings/hacker-history/hacker-history.html
25. Raymond ES (2001) The Jargon file. http://www.tuxedo.org/~esr/jargon/jargon.html
26. Raymond ES (2003) Problems in the culture of Unix. In: The art of UNIX programming. http://www.catb.org/~esr/writings/taoup/html/ch19s04.html
27. The UNIX vs NT Organisation (2001) The Unix vs NT Organisation. http://www.unix-vs-nt.org/
28. Scacchi W (2011) Free and open source development in the game communities. IEEE Comput Soc sp. jan./feb.'2011:59–66
29. Hamerly J, Paquin T, Walton S (1999) Freeing the source: the story of Mozilla, In: C DiBona, S Ockman, M Stone (eds) Open sources: voices from the open source revolution, O'Reilly & Associates, Sebastopol, California. http://www.oreilly.com/catalog/opensources/book/netrev.html

Chapter 58
Non-dominated Sorting Genetic Algorithm (NSGA)-III Based Fuzzy Segmentation Model

Rajani Kumari, Dinesh Kumar, and Vijay Kumar

1 Introduction

Image segmentation is a well-known image processing technique to decompose a given image to various segments [1]. Thus, it helps future vision applications to locate or monitor the objects in an image [2]. The segmentation generally results in a black-and-white image, in which white may represent foreground objects and black may represent background of a given image [3]. However, segmenting an image in an efficient manner is still a challenging research area. Because input images may contain noise and/or images with complex background which are difficult to segment. [4]. Also, color image segmentation gives unique colors to connected segments or regions. Each color represents different objects or segments [5]. Therefore, segmentation helps computer vision applications by decomposing a given image into various isolated objects [6].

To overcome the issues of image segmentation techniques, many image segmentation techniques are designed and implemented in the literature. The techniques are adaptive thresholding, region growing, Markov random field, fuzzy segmentation, watershed, etc. [7, 8]. Each technique has its own features and drawbacks [9]. However, these techniques suffer from various issues such as:

R. Kumari
Computer Science and Engineering Department, GJU&T, Hisar, India

D. Kumar
Electronics and Communication Engineering Department,
Delhi Technological University, Delhi, India

Vijay Kumar (✉)
Computer Science and Engineering Department, NIT Hamirpur, Hamirpur, India
e-mail: vijaykumarchahar@gmail.com

© Springer Nature Singapore Pte Ltd. 2021

S. D. Purohit et al. (eds.), *Proceedings of International Conference on Communication and Computational Technologies*, Algorithms for Intelligent Systems,
https://doi.org/10.1007/978-981-15-5077-5_58

1. The over-segmentation problem has been neglected, i.e., over-segmentation harms the accuracy of the segmentation outcomes. Therefore, it becomes a difficult task to decrease the consequence of over-segmentation by presenting particular preprocessing processes [10].
2. The consequence of complex regions has also been overlooked by many researchers [11].
3. The computational complexity is still found to be an issue for most of image segmentation techniques. As any improvement in the existing technique comes up with certain potential overheads, so, it is compulsory to decrease these overheads [12].

The main objectives of this paper are to segment color images and concurrently suppress over-segmentation and complex regions problems associated with the existing techniques. Therefore, a novel NSGA-III based fuzzy image segmentation model is designed. The performance of designed model is evaluated on color images with respect to some well-known performance metrics.

The rest of paper is organized as follows: Sect. 2 discusses the related work. Section 3 presents the proposed segmentation model. Section 4 presents experimental setup, result, and discussions. Finally, the concluding remarks along with future directions are given in Sect. 5.

2 Related Work

To resolve the issues associated with the existing image segmentation techniques, a novel multi-objective meta-heuristic-based image segmentation has been developed. This approach utilizes the basic concepts of fuzzy entropy and region-based active contour approach [13]. Rough-fuzzy c-means (RFCM) is implemented to segment images efficiently. In RFCM, overlapping decomposition is achieved using fuzzy logic. The issue of uncertainty is resolved by using rough set [14]. Fuzzy c-partition entropy has been utilized as a global optimization approach to evaluate optimal thresholds for unsupervised multilevel segmentation approach (UMSA) [15].

Hybrid fuzzy c-means (HFCM) been used to overcome the issue of over-segmentation with the existing image segmentation techniques [16]. Region-based convolutional neural network (RCNN) has been implemented to evaluate deep characteristics of input image to achieve segmentation in more efficient manner [17].

Although, above mentioned techniques outperform the existing image segmentation techniques, but, these techniques suffer from the hyperparameter tuning issue. Therefore, in this paper, NSGA-III is used to optimize hyperparameters.

3 The Proposed Image Segmentation Model

This section provides a mathematical formulation of the proposed image segmentation model.

3.1 Non-dominated Sorting Genetic Algorithm-III

Non-dominated sorting genetic algorithm (NSGA)-III is known as Pareto optimization method that functions well with issues that have multiple instances (more than two). It consists of six operators used in different huge data maximization problems which further signify the execution of different parts of NSGA-III [18]. It has better performance than NSGA-II for complex problems.

In NSGA-III algorithm, crossover and mutation functions with their steady corresponding amount are utilized to create the updated part for the future group. The initial population is generated. Initially, we select a definite (A) number of reference points. The selection is purely done on the basis of defined reference point's size. After selection, we initialize the points. A new reference from the existing points is then selected. Thereafter, a non-dominated sorting on the selected reference points is applied. Finally, these points are assigned to the new reference points.

Crowding gap parameter has been evaluated in order to achieve better diversity. The distance of each individual from its neighbors is calculated by it. To have significant results, the average value of crowding gap must be high. Using strength value and crowding gap, the double series of chosen value is picked by parents of community. Individuals with less strength value and more crowding gap as compared to the others will get selected. Afterward, by using the crossover and mutation functions from the chosen community, offspring are generated. With the fusion of the current population with newly generated offspring population, a new population is generated. In order to achieve the best M individuals, sorting is applied where the size of population is M. The final selection is based on crowding gap value and strength parameter on end front [19].

3.2 Fitness Function

The main goal of image segmentation model is to find maximum fitness values from two parameters such as accuracy and $F\_measure$. On the basis of these, fitness function is evaluated as:

$$\text{Maximum} \quad F(x) = \frac{1}{2} \times \left(\text{Accuracy} + F\_\text{measure} \right) \tag{1}$$

where $F\_\text{measure} \geq 8.7$

3.3 *Proposed Image Segmentation Model*

To segment an image, NSGA-III based fuzzy segmentation model is designed that includes different steps which are given below:

Step 1. Evaluate the given image (S) along with its width (b) and height (l).
Step 2. NSGA-III is used to adjust the hyperparameters of fuzzy-based segmentation model. In the early stage, random hyperparameters will be generated. Afterward, with the help of several-purpose fitness operations, the strength of each random solution is calculated. On the basis of calculated strength value, some solutions which are best are selected.
Step 3. The global best solution for NSGA-III based fuzzy segmentation model is evaluated.
Step 4. Various operators of NSGA-III are applied on the evaluated solutions. In case, any best solution is found, and then, we update the global best solution.
Step 5. Stopping criteria will be then evaluated; if stopping criteria are not met, then we move to Step 2. Step 6, otherwise.
Step 6. Based upon the obtained global optimum solution, we apply the fuzzy-based segmentation.

4 Performance Analysis

To evaluate the effectiveness of the proposed segmentation model, benchmark color images are considered. These images are resized to 256×256. The proposed model is compared with four competitive segmentation models. The designed model is simulated on Intel(R) Core(TM) $i5$-$4210U$ CPU@2.24 GHz and 8 GB RAM on MATLAB 2013a. The mask size is taken as 5×5 pixels.

4.1 *Visual Analysis*

The visual results of the proposed model are compared with four well-known segmentation models on some well-known benchmark color images.

Segmentation results in Fig. 1 demonstrate the benefits of the proposed segmentation model. It is visually founded from Fig. 1b–e that the existing segmentation techniques are not so effective to segment complex regions. Also, these techniques suffer from the over-segmentation issue. Figure 1f shows that the proposed technique is capable to effectively segment the complex regions. Also, compared to the existing techniques, it does not suffer from over-segmentation issue.

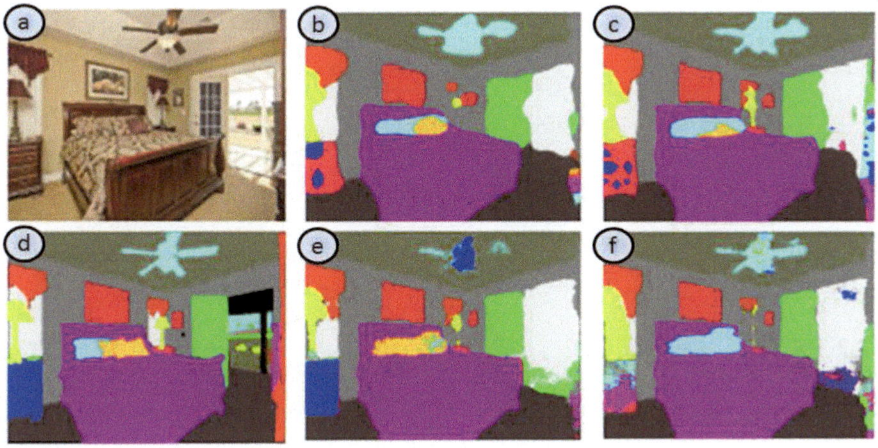

Fig. 1 Results of segmentation models. **a** Input image. **b** RFCM [14]. **c** UMSA [15]. **d** HFCM [16]. **e** RCNN [17] and **f** proposed model

4.2 Quantitative Analysis

The comparisons between the proposed and the competitive segmentation model are also considered by using various performance measures such as accuracy and F_measure.

4.2.1 Accuracy Analysis

Accuracy is defined as the detected regions same as in ground truth image of a same scene. Table 1 demonstrates accuracy analysis between the existing and the proposed technique. It has been found that the proposed model has significant accuracy values than the competitive segmentation techniques. The average improvement in accuracy using the proposed model over competitive models is approx. 0.06983.

Table 1 Accuracy analysis

Images	RFCM	UMSA	HFCM	RCNN	Proposed
1	0.9216	0.9089	0.9471	0.9643	0.9591
2	0.9132	0.9368	0.9426	0.9359	0.9652
3	0.9253	0.9350	0.9691	0.9554	0.9296
4	0.9195	0.9931	0.9211	0.9466	0.9539
5	0.9279	0.9456	0.9704	0.9750	0.9794

Table 2 F_measure analysis

Images	RFCM	UMSA	HFCM	RCNN	Proposed
1	0.9317	0.9183	0.9341	0.9216	0.8932
2	0.9648	0.9241	0.9237	0.9174	0.9264
3	0.9246	0.9219	0.9146	0.9191	0.8986
4	0.9327	0.9448	0.9215	0.9291	0.9142
5	0.9741	0.9556	0.9364	0.9447	0.9349

4.2.2 F_Measure Analysis

The accuracy is not assumed to be a good measure, as it may be biased when the given image has either maximum number of background or foreground objects. Consider, the given image has more foreground objects which means that the corresponding ground truth segmentation image has more (assume 90%) whitish pixels. Now, if your segmentation returns even a complete white image, then the corresponding accuracy will be approximately 90 %. However, the actual results are poor. Therefore, F_measure analysis has been taken on the proposed image segmentation model. Table 2 shows the F_measure analysis between the existing and the proposed technique. It reveals that the proposed model has maximum F_measure values than the competitive segmentation models. The average improvement in F_measure by using the proposed model over competitive models is approx. 0.0631.

5 Conclusion

An efficient image segmentation model has been designed in this paper. The designed model uses two concepts, namely NSGA-III based optimal parameters evaluation and optimized parameters-based fuzzy segmentation. Extensive results reveal that the designed model has an ability to segment images in more efficient manner. Experimental results reveal that the proposed segmentation model is applicable for real-time applications.

This work opens several research directions for future studies. Firstly, supervised learning-based models can be applied to estimate the parameters. Secondly, machine learning models can be integrated with segmentation models to improve their performance. It is also worth to investigate the applicability of designed model for other computer vision problems such as remote sensing imaging, outdoor video surveillance, and underwater image analysis.

References

1. Jin R, Weng G (2019) A robust active contour model driven by fuzzy c-means energy for fast image segmentation. Digit Sig Proc 90:100–109
2. Singh C, Bala A (2019) A transform-based fast fuzzy c-means approach for high brain mri segmentation accuracy. Appl Soft Comput 76:156–173
3. Hmida M, Hamrouni K, Solaiman B, Boussetta S (2018) Mammographic mass segmentation using fuzzy contours. Comput Methods Prog Biomed 164:131–142
4. Ren T, Wang H, Feng H, Xu C, Liu G, Ding P (2019) Study on the improved fuzzy clustering algorithm and its application in brain image segmentation. Appl Soft Comput 81:105503
5. Xia M, Yan W, Huang Y, Guo Y, Zhou G, Wang Y (2019) Ivus images segmentation using spatial fuzzy clustering and hierarchical level set evolution. Comput Biol Med 109:207–217
6. Hassan M, Murtza I, Hira A, Ali S, Kifayat K (2019) Robust spatial fuzzy gmm based mri segmentation and carotid artery plaque detection in ultrasound images. Comput Methods Prog Biomed 175:179–192
7. Shi J, Lei Y, Wu J, Jeon G (2019) Uncertain active contour model based on rough and fuzzy sets for auroral oval segmentation. Inf Sci 492:72–103
8. Zhao F, Liu H, Fan J, Chen CW, Lan R, Li N (2018) Intuitionistic fuzzy set approach to multi-objective evolutionary clustering with multiple spatial information for image segmentation. Neurocomputing 312:296–309
9. Choy S, Yuen K, Yu C (2019) Fuzzy bit-plane-dependence image segmentation. Sig Process 154:30–44
10. Naidu M, Kumar PR, Chiranjeevi K (2018) Shannon and fuzzy entropy based evolutionary image thresholding for image segmentation. Alexandria Eng J 57(3):1643–1655
11. Anter AM, Hassenian AE (2019) Ct liver tumor segmentation hybrid approach using neutrosophic sets, fast fuzzy c-means and adaptive watershed algorithm. Artif Intell Med 97:105–117
12. Yan Yu H, Bin Zhi X, Lun Fan J (2015) Image segmentation based on weak fuzzy partition entropy. Neurocomputing 168:994 – 1010
13. Pham TX, Siarry P, Oulhadj H (2019) A multi-objective optimization approach for brain mri segmentation using fuzzy entropy clustering and region-based active contour methods. Magn Reson Imaging 61:41–65
14. Bal A, Banerjee M, Chakrabarti A, Sharma P (2018) Mri brain tumor segmentation and analysis using rough-fuzzy c-means and shape based properties. J King Saud Univ Comput Inf Sci
15. Yin S, Qian Y, Gong M (2017) Unsupervised hierarchical image segmentation through fuzzy entropy maximization. Pattern Recogn 68:245–259
16. Alsmadi MK (2018) A hybrid fuzzy c-means and neutrosophic for jaw lesions segmentation. Ain Shams Eng J 9(4):697–706
17. Nida N, Irtaza A, Javed A, Yousaf MH, Mahmood MT (2019) Melanoma lesion detection and segmentation using deep region based convolutional neural network and fuzzy c-means clustering. Int J Med Inf 124:37–48
18. Yi J-H, Deb S, Dong J, Alavi AH, Wang G-G (2018) An improved nsga-iii algorithm with adaptive mutation operator for big data optimization problems. Future Gener Comput Syst 88:571–585
19. Ciro GC, Dugardin F, Yalaoui F, Kelly R (2016) A nsga-ii and nsga-iii comparison for solving an open shop scheduling problem with resource constraints. IFAC-Pap OnLine 49(12):1272–1277

Chapter 59
Feature Selection Algorithms for Plant Leaf Classification: A Survey

Mukesh Kumar Choudhary and Saroj Hiranwal

1 Introduction

Agriculture is the base of evolution from ancient time and now it has become a vital part of life for a major part of human beings. It is not only science but a profession also which is largely concerned with growing seeds and plants, fertilizing, irrigation and harvesting of the yield. Previously, it is concerned with cultivation only but nowadays it includes livestock, poultry, bee farming, dairy and many more for food and trade equally. Economy of some countries like Ethiopia, Nicaragua, Jamaica, Chad, Ecuador, Burundi and Costa Rica is mostly reliant on export of agricultural products. India is one of the agriculture-based country as agriculture sector contributes around 18% of its GDP. Indians are indulged in agriculture from the time of Indus valley civilization or may be before that. Presently, India is second largest country in terms of dairy and farming products.

In order to increase the productivity, it is essential to timely detect and prevent diseases in plants. For that reason, a computerized plant disease recognition system can be more advantageous for the intensive care of the plants. Usually, most diseases may be detected and classified from the symptoms seemed on the leaves. For the same, the extraction of relevant and unique features plays an important role in efficient identification of diseases in plants.

Nature-inspired approaches categorized into two major classes based on the source of inspiration: swarm intelligence-based and evolutionary algorithms. Swarm intelligence-based and evolutionary algorithms are one of the popular choices for optimization problem. Feature selection is the process of attribute selection or vari-

M. K. Choudhary (✉) · S. Hiranwal
Rajasthan Institute of Engineering and Technology,
Bhankrota Ajmer Road, Jaipur, Rajasthan 302026, India
e-mail: mukeshdesire@gmail.com

S. Hiranwal
e-mail: ersarojhiranwal@gmail.com

© Springer Nature Singapore Pte Ltd. 2021
S. D. Purohit et al. (eds.), *Proceedings of International Conference on Communication and Computational Technologies*, Algorithms for Intelligent Systems,
https://doi.org/10.1007/978-981-15-5077-5_59

able selection in order to construct an efficient classification model. This process identifies a subset of existing features and identified features play a vital role in the efficiency of the subsequent classification process. Precision, accuracy and efficiency of classification are completely dependent on the importance of selected features.

Dimensionality reduction and feature selection both are intended to reduce variables but they are not exactly same. First method organizes all the attributes in groups while second approach either picks a particular attribute or drops it if it is not important.

This paper presents state-of-the art study of nature-inspired algorithms that are used by researchers for feature selection. Next section (Sect. 2) provide brief discussion about machine learning. Section 3 discuss about nature-inspired algorithms used for the purpose of feature selection and Sect. 4 conclude the work.

2 Classification of Images

The classification of images refers to the process of collecting useful information. The classification of images may be take place by using two different approaches: supervised and unsupervised image classification. In the first type of classification, the complete process supervised by image analyst and pixel values should be provided by user or we can say that these approaches are human guided. In case of unsupervised classification process, there is no control of user creates clusters of pixels automatically. In machine learning, it is considered as supervised learning problem. These methods further may be classified based on characteristics like shape-based and motion-based and many more. During image processing, first of all we need to remove or adjust brightness, contrast, illumination and noise in order to improve overall performance.

The process of image classification includes three major steps after pre-processing: (i) feature extraction, (ii) feature selection and (iii) classification. Details of these steps mentioned in subsequent sections.

2.1 *Feature Extraction*

The feature extraction is very first step toward image classification after pre-processing. This process highlights the interesting part of images that may be helpful in efficient classification and engender high dimensional features. In this process, some new features get created to improve performance and reduce requirement of resources. An efficient feature extraction process can overcome the problem of overfitting for classification algorithms. Color, shape and texture features play an important role in feature extraction.

Andrushia and Patricia [4] extracted color, texture and shape-based features from images of grape leaves. Kour and Arora [33] used local binary pattern (LBP)approach for extraction of texture features. Sunny et al. [77] implemented gray-level co-occurrence matrix (GLCM) during feature extraction for detection of citrus canker. Features of an image are extracted using various machine learning algorithms, like convolutional neural network (CNN) [13], intra and inter-block dependencies (CHEN) [11] for Markov features, restricted Boltzmann machines [10] for spatial domain subtractive pixel adjacency model (SPAM) [59], auto-encoders [83] and bag of visual words [60], etc. Recently, Kumar et al. used SPAM method [43] and speed-edup robust features (SURF) method [42] for feature extraction during classification of plant leaves and different images of soil. Liang et al. [52] deployed CNN, LBPH and Harr-wavelet methods for feature extraction from rice blast images. Some more traditional descriptors available like: histogram of oriented gradient (HOG) [14], scale-invariant feature transform (SIFT) [53] and local binary pattern (LBP) [56].

2.2 Feature Selection

The feature extraction phase generates large number of features, some of them may be irrelevant and may be repeated. In order to get good classification accuracy, we need to identify important of them. The feature selection phase selects potentially good features that will play a critical role in classification. Swarm intelligence and evolutionary approaches are very good and efficient in optimization task. Thus, these algorithms may be used to identify optimal set of features.

Recently, Andrushia and Patricia [4] implemented ABC-based feature selection approach for an image dataset of grape leaves and get good results with 90% accuracy. Kumar et al. proposed new variant of SMO [5] algorithm for feature selection, namely exponential SMO [43] and chaotic SMO [42]. In modified SMO [42, 43], Kumar et al. introduced nonlinear perturbation rate [41]. Chuanlei et al. [12] used genetic algorithm [37] for feature selection. These stochastic algorithm play very important role in feature selection before classification. Efficiency of classification approach mainly depends on feature selection approach but most of the researchers ignore this phase or use traditional approach for feature selection.

2.3 Classification

Disease identification in plants and crops is very important task from farmer's point of view as their livelihood depends on crop yield and economy of a country heavily dependent of crop yield. Recently, researchers trying to develop some efficient and low cost approaches for farmers for disease identification using computer vision and ICT. Wallelign et al. [82] proposed a CNN-based classification method to detect disease in soyabean plants and achieved 99.32% classification accuracy. Andrushia and Patricia [4] used ABC-based feature selection approach and SVM and KNN classifiers for classification of grape leaves. They get good results with 90% accuracy.

Kour and Arora [33] used PSO for feature selection followed by SVM classifier to detect disease in apple, jamun, guava, mango, arjun, tomato and grapes leaf images. The new approach is efficient in comparison with existing techniques. The SMO-based feature selection approach is also improved the overall classification efficiency [42, 43].

Convolutional neural network (CNN) and deep CNN are also preferred by researchers during classification of handcrafted features. Deep CNN is used by Liang et al. [52] to detect disease in rice crop. Toda and Okura [80] explored the importance of CNN in plant leaf disease identification in detail. The concept of contrast limited adaptive histogram equalization (CLAHE) introduced by Sunny et al. [77] for canker detection in citrus fruits. There are number of classifiers available in literature like: LDA, ZeroR, k-nearest neighbor (kNN) [19], neural network [74], random forest (RF) [32], fisher linear discriminant (FLD) [63] and many more (Table 1).

Table 1 Comparative analysis of feature extraction/selection techniques

S. No.	Year	References	Plant/crop	Purpose	Feature extraction/selection approach(s)	Classification approach	Accuracy (%)
1	2019	[52]	Grape	Automated rice disease diagnosis systems	CNN, Harr-wavelet and LBPH feature extraction	SVM	95.83
2	2019	[16]	Tomato and maize	Disease identification in tomato and maize crop	Histogram of oriented gradient (HOG) operation	SVM and ANN	80
3	2019	[4]	Grape	Identification and classification of diseases in grape leaves	Artificial bee colony (ABC)-based feature selection	SVM, KNN	92
4	2018	[18]	Tomato	Implementing a refinement filter bank framework for tomato plant diseases and pests recognition	Region-based neural network	CNN	96
5	2018	[82]	Soybean	Soybean plant disease identification using CNN	NA	CNN	99.32
6	2018	[71]	Pine tree	Diseases identification in leaf of pine tree	Gabor filter feature extraction	KNN SVM	98.0769

(continued)

Table 1 (continued)

S. No.	Year	References	Plant/crop	Purpose	Feature extraction/selection approach(s)	Classification approach	Accuracy (%)
7	2018	[77]	Citrus plant	Citrus canker detection method	Contrast limited adaptive histogram equalization (CLAHE) in pre-processing step	KNN, SVM, NB	NA
8	2017	[86]	Cucumber	Leaf image-based cucumber disease recognition	Principal component analysis (PCA) for dimensionality	SVM, KMSNN, TF, PLI	91.25
9	2016	[81]	Grape	Identification of black rot	Local binary pattern (LBP)	SVM	96.6
10	2016	[27]	Rice	To diagnose and classify rice diseases	Color and zone wise shape features extracted	MDC and k-NN	89.23
11	2012	[26]	Cotton	Identify early scorch	K-means clustering	NA	NA
12	2010	[3]	Al-ghor	Identify late scorch	2.09E-08	Neural network classifier	93
13	2009	[7]	Sugar beet	Identify powdery mildew and leaf rust	NA	Maximum posterior classification	99

3 Nature-Inspired Algorithms for Feature Selection

Nature-inspired algorithms inspired by some natural phenomenon happening in environment. Nature has capability to solve complex problem of large size in a very easy manner. These algorithms classified in different categories based on their source of inspiration. Algorithm that based on the concept of swarm intelligence are known and swarm-based algorithm like: particle swarm optimization [17], ant colony algorithm [15], artificial bee colony algorithm [29], spider monkey optimization algorithm [5], firefly algorithm [84], bat algorithm [85] and algorithm that mimic process of natural evolution are called evolutionary algorithms like: genetic algorithm [24], differential evolution [76], etc.

3.1 Artificial Bee Colony Algorithm

Artificial bee colony (ABC) algorithm is a swarm intelligence approach and very popular among researchers working in the field of optimization. Artificial bee colony algorithm was developed by D. Karaboga [29]. Artificial bee colony algorithm is very simple and has three phases. Mathematical equations of these phases is as follow:

$$X_{ij} = X_{\min j} + \text{rand}[0, 1] \times (X_{\max j} - X_{\min j}) \tag{3.1}$$

$$V_{ij} = X_{ij} + \phi_{ij} \times (X_{ij} - X_{kj}) \tag{3.2}$$

$$\text{Prob}_i = \frac{\text{fitness}_i}{\sum_{i=1}^{n/2} \text{Fitness}_i} \tag{3.3}$$

Equation (3.1) initialize all the solutions in the given search space. The next phase (refer Eq. 3.2) update locations of current solutions. In subsequent phase (3.3), it assess the quality of solution and decides solution for the next phase. The ABC algorithm modified and implemented to solve various problems. Andrushia [4] deployed it for feature selection and get efficient results. Due to its simplicity, this algorithm used for various applications and many of its variants are very popular.

Global best guided ABC [87] is an efficient variant and modified by researchers like lbest gbest [67], improved gbest ABC [72], modified gbest ABC [8]. Memetic computing is very successful and some variant of ABC using it are memetic ABC [6], HJABC [28], etc. Some another modifications proposed for perturbation using fitness in ABC [34] is also very successful with its variants like randomized MeABC [50], weight-based ABC [79], hybrid MeABC [35, 36, 38, 44, 46, 58] and crossover-based ABC [49]. A comparative study of modifications in ABC done by Kumar et al. [45] with its drawbacks and advantages. Arrhenius ABC [40] inspired by Arrhenius equation used to solve complex problem like robot path planning [55] are some popular variant of ABC.

3.2 Particle Swarm Optimization

Particle swarm optimization (PSO) algorithm motivated by intelligent behavior of insects, fishes, birds and animals, etc. Mathematical formulation of PSO is given below [9]:

$$v_{ij} = v_{ij} + c_1 r_1 (p_{ij} - x_{ij}) + c_2 r_2 (p_{gj} - x_{ij}) \tag{3.4}$$

$$x_{ij} = x_{ij} + v_{ij} \tag{3.5}$$

Here, two D-dimensional vector X_i and V_i denote position and velocity of ith particle. PSO algorithm makes use of two position update equation (refer Eq. 3.5) and velocity update equation (3.4) for movement. Recently, Kour et al. [33] deployed PSO for disease detection in plants and get improved performance.

3.3 Genetic Algorithm

Genetic algorithm (GA) is an evolutionary algorithm and popular for solving multi-objective problems. Chuanlei et al. [12] used genetic algorithm for feature optimization and improved performance of classification process. GA mimics the biological process of gene propagation from one generation to another generation.

3.4 Differential Evolution

Differential evolution is simple evolutionary algorithm that solve complex problems easily and it has only three phases depicted in Eqs. (3.6), (3.7) and (3.8).

$$u_i(g) = x_{i_1}(g) + F \times (x_{i_2}(g) - x_{i_3}(g)) \tag{3.6}$$

$$x'_{ij}(g) = \begin{cases} u_{ij}(g), & \text{if } j \in J \\ x_{ij}(g), & \text{otherwise.} \end{cases} \tag{3.7}$$

$$x_i(g+1) = \begin{cases} x'_i(g), & \text{if } f(x'_i(g)) > f(x_i(g)). \\ x_i(g), & \text{else.} \end{cases} \tag{3.8}$$

Khushaba et al. deployed differential evolution for feature subset selection [30] and hybrid with ant colony [31]. Zorarpaci performed feature selection by a hybrid of DE and ABC [88] with improved performance.

Another popular variant of DE are: improved DE [25], memetic DE [47], PGLFDE [73] and black hole-based DE [70]. A state of art study of scaling factor used in DE are discussed in [69].

3.5 Spider Monkey Optimization

The spider monkey optimization (SMO) algorithm [5] is a swarm-based optimization algorithm that is based on extraordinary behavior of spider monkey. This class of monkey shows fission-fusion social organization. After initialization of solutions, it goes through six steps. Mathematical formulation of these phases, shown below:

$$X_{ij} = X_{\min j} + U(0, 1) \times (X_{\max j} - X_{\min j}) \tag{3.9}$$

$$X\text{new}_{ij} = X_{ij} + U(0, 1) \times (LL_{kj} - X_{ij}) + U(-1, 1) \times (X_{rj} - X_{ij}) \tag{3.10}$$

$$X\text{new}_{ij} = X_{ij} + U(0, 1) \times (GL_j - X_{ij}) + U(-1, 1) \times (X_{rj} - X_{ij}) \tag{3.11}$$

$$\text{prob}_i = \frac{\text{fitness}_i}{\sum_{i=1}^{N} \text{fitness}_i} \tag{3.12}$$

$$X\text{new}_{ij} = X_{ij} + U(0, 1) \times (GL_j - X_{ij}) + U(-1, 1) \times (X_{rj} - X_{ij}) \tag{3.13}$$

The SMO algorithm with some modifications deployed for disease detection [43] in plants and for categorization of different types of soil [42].

The SMO algorithm undergone various modification with different applications like : fitness-based SMO [39], Self-adaptive SMO [51] and modified position update in SMO [48]. Application of SMO also lies in antenna designing [2, 75], multi-level thresholding [57], PIDA controller [64], unconstrained optimization problems [21], etc. Some other variant of SMO are tournament selection-based probability in SMO [20], ageist SMO [66], limacon local search in SMO [65], hybrid with GA [1], debs technique in SMO [22], modified SMO [23] using metropolis principle and improved SMO [78]. Detailed study of SMO available in [68].

3.6 Salp Swarm Optimization

Salp swarm optimization [54] is comparatively new optimization algorithm. It mimics the swarming behavior of salp while searching for food. Salp have very close resemblance with jelly fishes. This algorithm introduced concept of leader and follower. Position of leader and follower updated using Eqs. (3.14) and (3.16).

$$L_d^1 = \begin{cases} FS_d + c_1((UB_d - LB_d))c_2 + LB_d, c_3 \geq 0 \\ FS_d - c_1((UB_d - LB_d))c_2 + LB_d, c_3 < 0 \end{cases} \tag{3.14}$$

$$c_1 = 2\mathrm{e}^{-\left(\frac{4t}{T}\right)} \tag{3.15}$$

$$F_i^d = \frac{1}{2}at^2 + v_0t^2 \tag{3.16}$$

This algorithm has potential to optimize high dimensional optimization problems. Rachapudi et al. proposed two variants of salp swarm optimizer using levy flight search as a local search strategy [61] and bag-of-words [62] for feature selection in medical images and received good results.

4 Conclusion

This work focuses on disease detection in plants using their leaf images. Role of nature-inspired algorithm in feature selection during classification is also analyzed. A detailed discussion on selected swarm and evolutionary approaches included here with various aspects. In the future, these swarm and evolutionary algorithms may be used for disease detection in plants with some parameter tuning and hybridization of these algorithms.

References

1. Agrawal A, Farswan P, Agrawal V, Tiwari DC, Bansal JC (2017) On the hybridization of spider monkey optimization and genetic algorithms. In: Proceedings of sixth international conference on soft computing for problem solving. Springer, pp 185–196
2. Al-Azza AA, Al-Jodah AA, Harackiewicz FJ (2016) Spider monkey optimization: a novel technique for antenna optimization. IEEE Antennas Wirel Propag Lett 15:1016–1019
3. Al Bashish D, Braik M, Bani-Ahmad S (2010) A framework for detection and classification of plant leaf and stem diseases. In: 2010 international conference on signal and image processing. IEEE, pp 113–118
4. Andrushia AD, Trephena Patricia A (2019) Artificial bee colony optimization (abc) for grape leaves disease detection. Evol Syst, pp 1–13
5. Bansal JC, Sharma H, Singh Jadon S, Clerc M (2014) Spider monkey optimization algorithm for numerical optimization. Memetic Comput 6(1):31–47
6. Bansal JC, Sharma H, Arya KV, Nagar A (2013) Memetic search in artificial bee colony algorithm. Soft Comput 17(10):1911–1928
7. Bauer SD, Korc F, Förstner W, Van Henten EJ, Goense D, Lokhorst C (2009) Investigation into the classification of diseases of sugar beet leaves using multispectral images. Precis Agric 9:229–238
8. Bhambu P, Sharma S, Kumar S (2018) Modified gbest artificial bee colony algorithm. In: Soft computing: theories and applications. Springer, pp 665–677
9. Bhambu P, Kumar S, Sharma K (2018) Self balanced particle swarm optimization. Int J Syst Assur Eng Manag 9(4):774–783
10. Chang H, Nayak N, Spellman PT, Parvin B (2013) Characterization of tissue histopathology via predictive sparse decomposition and spatial pyramid matching. In: International conference on medical image computing and computer-assisted intervention. Springer, pp 91–98
11. Chen C, Shi YQ (2008) Jpeg image steganalysis utilizing both intrablock and interblock correlations. In: Proceedings of IEEE international symposium on circuits and systems, pp 3029–3032
12. Chuanlei Z, Shanwen Z, Jucheng Y, Yancui S, Jia C (2017) Apple leaf disease identification using genetic algorithm and correlation based feature selection method. Int J Agric Biol Eng 10(2):74–83

13. Cruz-Roa AA, Arevalo Ovalle JE, Madabhushi A, Augusto González Osorio F (2013) A deep learning architecture for image representation, visual interpretability and automated basal-cell carcinoma cancer detection. In: International conference on medical image computing and computer-assisted intervention. Springer, pp 403–410
14. Dalal N, Triggs B (2005) Histograms of oriented gradients for human detection. In: IEEE computer society conference on computer vision and pattern recognition. CVPR 2005, Vol 1. IEEE, pp 886–893
15. Dorigo M, Di Caro G (1999) Ant colony optimization: a new meta-heuristic. In: Proceedings of the 1999 congress on evolutionary computation-CEC99 (Cat. No. 99TH8406), Vol 2. IEEE, pp 1470–1477
16. Durga NK, Anuradha G (2019) Plant disease identification using svm and ANN algorithms. Int J Recent Technol Eng 7:471–473
17. Eberhart R, Kennedy J (1995) A new optimizer using particle swarm theory. In: Proceedings of the sixth international symposium on micro machine and human science. MHS'95. IEEE, pp 39–43
18. Fuentes AF, Yoon S, Lee J, Sun Park D (2018) High-performance deep neural network-based tomato plant diseases and pests diagnosis system with refinement filter bank. Front Plant Sci 9
19. Guettari N, Capelle-Laizé AS, Carré P (2016) Blind image steganalysis based on evidential k-nearest neighbors. In: *2016 IEEE international conference on image processing (ICIP)*. IEEE, pp 2742–2746
20. Gupta K, Deep K (2016) Tournament selection based probability scheme in spider monkey optimization algorithm. In: Harmony search algorithm. Springer, pp 239–250
21. Gupta K, Deep K, Chand Bansal J (2017) Improving the local search ability of spider monkey optimization algorithm using quadratic approximation for unconstrained optimization. Comput Intell 33(2):210–240
22. Gupta K, Deep K, Chand Bansal J (2017) Spider monkey optimization algorithm for constrained optimization problems. Soft Comput 21(23):6933–6962
23. Hazrati G, Sharma H, Sharma N, Chand Bansal J (2016) Modified spider monkey optimization. In: International workshop on computational intelligence (IWCI). IEEE, pp 209–214
24. Holland JH (1975) Adaptation in natural and artificial systems, vol 53. University of Michigan Press, Michigan
25. Jain S, Kumar S, Kumar Sharma V, Sharma H (2017) Improved differential evolution algorithm. In: 2017 international conference on infocom technologies and unmanned systems (Trends and future directions) (ICTUS). IEEE, pp 627–632
26. Jaware TH, Badgujar RD, Patil PG (2012) Crop disease detection using image segmentation. World J Sci Technol 2(4):190–194
27. Joshi AA, Jadhav BD (2016) Monitoring and controlling rice diseases using image processing techniques. In: 2016 international conference on computing, analytics and security trends (CAST). IEEE, pp 471–476
28. Kang F, Junjie L, Zhenyue M, Haojin L (2011) Artificial bee colony algorithm with local search for numerical optimization. J Softw 6(3):490–497
29. Karaboga D (2005) An idea based on honey bee swarm for numerical optimization. Technical report, Technical report-tr06, Erciyes University, Engineering Faculty, Computer Engineering Department
30. Khushaba RN, Al-Ani A, Al-Jumaily A (2011) Feature subset selection using differential evolution and a statistical repair mechanism. Expert Syst Appl 38(9):11515–11526
31. Khushaba RN, Al-Ani A, AlSukker A, Al-Jumaily A (2008) A combined ant colony and differential evolution feature selection algorithm. In: International conference on ant colony optimization and swarm intelligence. Springer, pp 1–12
32. Kodovsky J, Jessica F, Vojtěch H (2012) Ensemble classifiers for steganalysis of digital media. IEEE Trans Inf Foren Secur 7(2):432–444
33. Kour VP, Arora S (2019) Particle swarm optimization based support vector machine (p-svm) for the segmentation and classification of plants. IEEE Access 7:29374–29385

34. Kumar A, Kumar S, Dhayal K, Swetank DK (2014) Fitness based position update in artificial bee colony algorithm. Int J Eng Res Technol 3(5):636–641
35. Kumar S, Sharma VK, Kumari R (2014) Improved onlooker bee phase in artificial bee colony algorithm. Int J Comput Appl 90(6):20–25
36. Kumar S, Bhambu P, Kumar Sharma V (2014) New local search strategy in artificial bee colony algorithm. Int J Comput Sci Inf Technol 5(2):2569–2565
37. Kumar S, Jain S, Sharma H (2018) Genetic algorithms. In: Advances in swarm intelligence for optimizing problems in computer science. Chapman and Hall/CRC, pp 27–52
38. Kumar S, Kumar A, Sharma VK, Sharma H (2014). A novel hybrid memetic search in artificial bee colony algorithm. In: 2014 seventh international conference on contemporary computing (IC3). IEEE, pp 68–73
39. Kumar S, Kumari R, Kumar Sharma V (2015) Fitness based position update in spider monkey optimization algorithm. Procedia Comput Sci 62:442–449
40. Kumar S, Nayyar A, Kumari R (2019) Arrhenius artificial bee colony algorithm. In: International conference on innovative computing and communications. Springer, pp 187–195
41. Kumar S, Nayyar A, Nguyen NG, Kumari R (2020) Hyperbolic spider monkey optimization algorithm. Recent Advances Comput Sci Commun (Formerly: Recent Patents on Comput Sci) 13(1):35–42
42. Kumar S, Sharma B, Kumar Sharma V, Poonia RC (2018) Automated soil prediction using bag-of-features and chaotic spider monkey optimization algorithm. Evol Intell, pp 1–12
43. Kumar S, Sharma B, Sharma VK, Sharma H, Bansal JC (2018) Plant leaf disease identification using exponential spider monkey optimization. Sustain Comput: Inform Syst
44. Kumar S, Sharma VK, Kumari R (2014) An improved memetic search in artificial bee colony algorithm. Int J Comput Sci Inform Technol 5(2):1237–1247
45. Kumar S, Sharma VK, Kumari R (2013) Comparative study of hybrids of artificial bee colony algorithm. JIMS8I-Int J Inf Commun Comput Technol 1(2):20–28
46. Kumar S, Sharma VK, Kumari R (2014) Memetic search in artificial bee colony algorithm with fitness based position update. In: International conference on recent advances and innovations in engineering (ICRAIE-2014). IEEE, pp 1–6
47. Kumar S, Sharma VK, Kumari R (2014) Memetic search in differential evolution algorithm. arXiv preprint arXiv:1408.0101
48. Kumar S, Sharma VK, Kumari R (2014) Modified position update in spider monkey optimization algorithm. Int J Emerg Technol Comput Appl Sci 2:198–204
49. Kumar S, Sharma VK, Kumari R (2014) A novel hybrid crossover based artificial bee colony algorithm for optimization problem. arXiv preprint arXiv:1407.5574
50. Kumar S, Sharma VK, Kumari R (2014) Randomized memetic artificial bee colony algorithm. arXiv preprint arXiv:1408.0102
51. Kumar S, Sharma VK, Kumari R (2014) Self-adaptive spider monkey optimization algorithm for engineering optimization problems. JIMS8I-Int J Inf Commun Comput Technol 2(2):96–107
52. Liang W, Zhang H, Zhang G, Cao H (2019) Rice blast disease recognition using a deep convolutional neural network. Sci Rep 9(1):2869
53. Lowe DG (2004) Distinctive image features from scale-invariant keypoints. Int J Comput Vis 60(2):91–110
54. Mirjalili S, Gandomi AH, Zahra Mirjalili S, Saremi S, Faris H, Mohammad Mirjalili S (2017) Salp swarm algorithm: A bio-inspired optimizer for engineering design problems. Adv Eng Softw 114:163–191
55. Nayyar A, Nguyen NG, Kumari R, Kumar S Robot path planning using modified artificial bee colony algorithm. In: Frontiers of intelligent computing: theory and application
56. Ojala T, Pietikäinen M, Harwood D (1996) A comparative study of texture measures with classification based on featured distributions. Patt Recogn 29(1):51–59
57. Pal SS, Kumar S, Kashyap M, Choudhary Y, Bhattacharya M (2016) Multi-level thresholding segmentation approach based on spider monkey optimization algorithm. In: Proceedings of the second international conference on computer and communication technologies. Springer, pp 273–287

58. Pandey S, Kumar S (2013) Enhanced artificial bee colony algorithm and its application to travelling salesman problem. HCTL Open Int J Technol Innov Res 2:137–146
59. Pevny T, Bas P, Fridrich J (2010) Steganalysis by subtractive pixel adjacency matrix. IEEE Trans Inf Foren Secur 5(2):215–224
60. Pires RDL, Gonçalves DN, Oruê JPM, Kanashiro WES, Rodrigues JF Jr, Brandoli Machado B, Nunes Gonçalves W (2016) Local descriptors for soybean disease recognition. Comput Electron Agric 125:48–55
61. Rachapudi V, Lavanya Devi G (2019) Feature selection for histopathological image classification using levy flight salp swarm optimizer. Recent Pat Comput Sci 12:329–337
62. Rachapudi V, Lavanya Devi G (2019) Optimal bag-of-features using random salp swarm algorithm for histopathological image analysis. Int J Intell Inf Database Syst
63. Ramezani M, Ghaemmaghami S (2010) Towards genetic feature selection in image steganalysis. In: 2010 7th IEEE consumer communications and networking conference (CCNC). IEEE, pp 1–4
64. Sharma A, Sharma H, Bhargava A, Sharma N, Bansal JC (2016) Optimal power flow analysis using lévy flight spider monkey optimisation algorithm. Int J Artif Intell Soft Comput 5(4):320–352
65. Sharma A, Sharma H, Bhargava A, Sharma N, Bansal JC (2017) Optimal placement and sizing of capacitor using limaçon inspired spider monkey optimization algorithm. Memetic Comput 9(4):311–331
66. Sharma A, Sharma A, Panigrahi BK, Kiran D, Kumar R, (2016) Ageist spider monkey optimization algorithm. Swarm Evol Comput 28:58–77
67. Sharma H, Sharma S, Kumar S (2016) Lbest gbest artificial bee colony algorithm. In: 2016 international conference on advances in computing, communications and informatics (ICACCI). IEEE, pp 893–898
68. Sharma H, Hazrati G, Bansal JC (2019) Spider monkey optimization algorithm. In: Evolutionary and swarm intelligence algorithms. Springer, pp 43–59
69. Sharma P, Sharma H, Kumar S, Bansal JC (2019) A review on scale factor strategies in differential evolution algorithm. In: Soft computing for problem solving. Springer, pp 925–943
70. Sharma P, Sharma H, Kumar S, Sharma K (2019) Black-hole gbest differential evolution algorithm for solving robot path planning problem. In: Harmony search and nature inspired optimization algorithms. Springer, pp 1009–1022
71. Sharma S, Kaur V, Dhillon N (2018) Plant disease classification with knn-svm classification. Int J Adv Eng Res Deve 5
72. Sharma S, Kumar S, Sharma K (2019) Improved gbest artificial bee colony algorithm for the constraints optimization problems. Evol Intell, pp 1–7
73. Sharma VP, Choudhary HR, Kumar S, Choudhary V (2015) A modified de: Population or generation based levy flight differential evolution (pglfde). In: 2015 International conference on futuristic trends on computational analysis and knowledge management (ABLAZE). IEEE, pp 704–710
74. Sheikhan M, Pezhmanpour M, Shahram Moin M (2012) Improved contourlet-based steganalysis using binary particle swarm optimization and radial basis neural networks. Neural Comput Appl 21(7):1717–1728
75. Singh U, Salgotra R, Rattan M (2016) A novel binary spider monkey optimization algorithm for thinning of concentric circular antenna arrays. IETE J Res 62(6):736–744
76. Storn R, Price K (1995) Differential evolution-a simple and efficient adaptive scheme for global optimization over continuous spaces. ICSI, Berkeley
77. Sunny S, Gandhi MPI (2018) An efficient citrus canker detection method based on contrast limited adaptive histogram equalization enhancement. Int J Appl Eng Res 13(1):809–815
78. Swami V, Kumar S, Jain S (2018) An improved spider monkey optimization algorithm. In: Soft computing: theories and applications. Springer, pp 73–81
79. Tiwari P, Kumar S (2016) Weight driven position update artificial bee colony algorithm. In: International conference on advances in computing, communication, & automation (ICACCA)(Fall). IEEE, pp 1–6

80. Toda Y, Okura F et al (2019) How convolutional neural networks diagnose plant disease. Plant Phenom 2019:9237136
81. Waghmare H, Kokare R, Dandawate Y (2016) Detection and classification of diseases of grape plant using opposite colour local binary pattern feature and machine learning for automated decision support system. In: 2016 3rd international conference on signal processing and integrated networks (SPIN). IEEE, pp 513–518
82. Wallelign S, Polceanu M, Buche C (2018) Soybean plant disease identification using convolutional neural network. In: The thirty-first international flairs conference
83. Xu J, Lei X, Liu Q, Hannah G, Wu J, Tang J, Anant M (2016) Stacked sparse autoencoder (ssae) for nuclei detection on breast cancer histopathology images. IEEE Trans Med Imaging 35(1):119–130
84. Yang X-S (2010) Firefly algorithm, stochastic test functions and design optimisation. arXiv preprint arXiv:1003.1409
85. Yang X-S (2010) A new metaheuristic bat-inspired algorithm. In: Nature inspired cooperative strategies for optimization (NICSO 2010). Springer, pp 65–74
86. Zhang S, Wu X, You Z, Zhang L (2017) Leaf image based cucumber disease recognition using sparse representation classification. Comput Electron Agric 134:135–141
87. Zhu G, Kwong S (2010) Gbest-guided artificial bee colony algorithm for numerical function optimization. Appl Math Comput 217(7):3166–3173
88. Zorarpacı E, Ayşe Özel S (2016) A hybrid approach of differential evolution and artificial bee colony for feature selection. Expert Syst Appl 62:91–103